Lecture Notes in Computer Science 14430

Founding Editors

Gerhard Goos
Juris Hartmanis

The series Lecture Notes in Computer Science (LNCS), including its subseries Lecture Notes in Artificial Intelligence (LNAI) and Lecture Notes in Bioinformatics (LNBI), has established itself as a medium for the publication of new developments in computer science and information technology research, teaching, and education.

LNCS enjoys close cooperation with the computer science R & D community, the series counts many renowned academics among its volume editors and paper authors, and collaborates with prestigious societies. Its mission is to serve this international community by providing an invaluable service, mainly focused on the publication of conference and workshop proceedings and postproceedings. LNCS commenced publication in 1973.

Qingshan Liu · Hanzi Wang · Zhanyu Ma ·
Weishi Zheng · Hongbin Zha · Xilin Chen ·
Liang Wang · Rongrong Ji
Editors

Pattern Recognition and Computer Vision

6th Chinese Conference, PRCV 2023
Xiamen, China, October 13–15, 2023
Proceedings, Part VI

Springer

Editors
Qingshan Liu (ID)
Nanjing University of Information Science
and Technology
Nanjing, China

Zhanyu Ma (ID)
Beijing University of Posts
and Telecommunications
Beijing, China

Hongbin Zha (ID)
Peking University
Beijing, China

Liang Wang
Chinese Academy of Sciences
Beijing, China

Hanzi Wang (ID)
Xiamen University
Xiamen, China

Weishi Zheng (ID)
Sun Yat-sen University
Guangzhou, China

Xilin Chen (ID)
Chinese Academy of Sciences
Beijing, China

Rongrong Ji (ID)
Xiamen University
Xiamen, China

ISSN 0302-9743   ISSN 1611-3349 (electronic)
Lecture Notes in Computer Science
ISBN 978-981-99-8536-4   ISBN 978-981-99-8537-1 (eBook)
https://doi.org/10.1007/978-981-99-8537-1

Preface

Welcome to the proceedings of the Sixth Chinese Conference on Pattern Recognition and Computer Vision (PRCV 2023), held in Xiamen, China.

PRCV is formed from the combination of two distinguished conferences: CCPR (Chinese Conference on Pattern Recognition) and CCCV (Chinese Conference on Computer Vision). Both have consistently been the top-tier conference in the fields of pattern recognition and computer vision within China's academic field. Recognizing the intertwined nature of these disciplines and their overlapping communities, the union into PRCV aims to reinforce the prominence of the Chinese academic sector in these foundational areas of artificial intelligence and enhance academic exchanges. Accordingly, PRCV is jointly sponsored by China's leading academic institutions: the Chinese Association for Artificial Intelligence (CAAI), the China Computer Federation (CCF), the Chinese Association of Automation (CAA), and the China Society of Image and Graphics (CSIG).

PRCV's mission is to serve as a comprehensive platform for dialogues among researchers from both academia and industry. While its primary focus is to encourage academic exchange, it also places emphasis on fostering ties between academia and industry. With the objective of keeping abreast of leading academic innovations and showcasing the most recent research breakthroughs, pioneering thoughts, and advanced techniques in pattern recognition and computer vision, esteemed international and domestic experts have been invited to present keynote speeches, introducing the most recent developments in these fields.

PRCV 2023 was hosted by Xiamen University. From our call for papers, we received 1420 full submissions. Each paper underwent rigorous reviews by at least three experts, either from our dedicated Program Committee or from other qualified researchers in the field. After thorough evaluations, 522 papers were selected for the conference, comprising 32 oral presentations and 490 posters, giving an acceptance rate of 37.46%. The proceedings of PRCV 2023 are proudly published by Springer.

Our heartfelt gratitude goes out to our keynote speakers: Zongben Xu from Xi'an Jiaotong University, Yanning Zhang of Northwestern Polytechnical University, Shutao Li of Hunan University, Shi-Min Hu of Tsinghua University, and Tiejun Huang from Peking University.

We give sincere appreciation to all the authors of submitted papers, the members of the Program Committee, the reviewers, and the Organizing Committee. Their combined efforts have been instrumental in the success of this conference. A special acknowledgment goes to our sponsors and the organizers of various special forums; their support made the conference a success. We also express our thanks to Springer for taking on the publication and to the staff of Springer Asia for their meticulous coordination efforts.

We hope these proceedings will be both enlightening and enjoyable for all readers.

October 2023

Qingshan Liu
Hanzi Wang
Zhanyu Ma
Weishi Zheng
Hongbin Zha
Xilin Chen
Liang Wang
Rongrong Ji

Organization

General Chairs

Hongbin Zha   Peking University, China
Xilin Chen   Institute of Computing Technology, Chinese
           Academy of Sciences, China
Liang Wang   Institute of Automation, Chinese Academy of
           Sciences, China
Rongrong Ji   Xiamen University, China

Program Chairs

Qingshan Liu   Nanjing University of Information Science and
           Technology, China
Hanzi Wang   Xiamen University, China
Zhanyu Ma   Beijing University of Posts and
           Telecommunications, China
Weishi Zheng   Sun Yat-sen University, China

Organizing Committee Chairs

Mingming Cheng   Nankai University, China
Cheng Wang   Xiamen University, China
Yue Gao   Tsinghua University, China
Mingliang Xu   Zhengzhou University, China
Liujuan Cao   Xiamen University, China

Publicity Chairs

Yanyun Qu   Xiamen University, China
Wei Jia   Hefei University of Technology, China

Local Arrangement Chairs

Xiaoshuai Sun   Xiamen University, China
Yan Yan   Xiamen University, China
Longbiao Chen   Xiamen University, China

International Liaison Chairs

Jingyi Yu   ShanghaiTech University, China
Jiwen Lu   Tsinghua University, China

Tutorial Chairs

Xi Li   Zhejiang University, China
Wangmeng Zuo   Harbin Institute of Technology, China
Jie Chen   Peking University, China

Thematic Forum Chairs

Xiaopeng Hong   Harbin Institute of Technology, China
Zhaoxiang Zhang   Institute of Automation, Chinese Academy of
                         Sciences, China
Xinghao Ding   Xiamen University, China

Doctoral Forum Chairs

Shengping Zhang   Harbin Institute of Technology, China
Zhou Zhao   Zhejiang University, China

Publication Chair

Chenglu Wen   Xiamen University, China

Sponsorship Chair

Yiyi Zhou   Xiamen University, China

Exhibition Chairs

Bineng Zhong	Guangxi Normal University, China
Rushi Lan	Guilin University of Electronic Technology, China
Zhiming Luo	Xiamen University, China

Program Committee

Baiying Lei	Shenzhen University, China
Changxin Gao	Huazhong University of Science and Technology, China
Chen Gong	Nanjing University of Science and Technology, China
Chuanxian Ren	Sun Yat-Sen University, China
Dong Liu	University of Science and Technology of China, China
Dong Wang	Dalian University of Technology, China
Haimiao Hu	Beihang University, China
Hang Su	Tsinghua University, China
Hui Yuan	School of Control Science and Engineering, Shandong University, China
Jie Qin	Nanjing University of Aeronautics and Astronautics, China
Jufeng Yang	Nankai University, China
Lifang Wu	Beijing University of Technology, China
Linlin Shen	Shenzhen University, China
Nannan Wang	Xidian University, China
Qianqian Xu	Key Laboratory of Intelligent Information Processing, Institute of Computing Technology, Chinese Academy of Sciences, China
Quan Zhou	Nanjing University of Posts and Telecommunications, China
Si Liu	Beihang University, China
Xi Li	Zhejiang University, China
Xiaojun Wu	Jiangnan University, China
Zhenyu He	Harbin Institute of Technology (Shenzhen), China
Zhonghong Ou	Beijing University of Posts and Telecommunications, China

Contents – Part VI

Vision Applications and Systems

Computational Photography, Sensing and Display Technology

RSID: A Remote Sensing Image Dehazing Network

Yuan Li(ID) and Yafeng Zhao(✉)

Northeast Forestry University, No. 26, Hexing Road, Harbin 150036, Heilongjiang,
China
liyuanwcx@163.com

Abstract. Hazy images often lead to problems such as loss of image details and dull colors, which significantly affects the information extraction of remote sensing images, so it is necessary to research image dehazing. In the field of remote sensing, remote sensing images are characterized by large-size and rich information, so the processing of remote sensing images often has the problems of GPU memory overflow and difficult removal of non-uniform haze. For remote sensing image characteristics, an efficient and lightweight end-to-end dehazing method is proposed in this paper. We use the FA attention combined with smoothed dilated convolution instead as the main structure of the encoder, which can achieve imbalanced handling of hazy images with different levels of opacity while reducing parameter count. Channel weight fusion self-attention is added in the decoder part to realize the automatic learning and pixel-level processing of different receptive field features We tested the proposed method on both public datasets RESIDE and real large-size hazy remote sensing images. The proposed method achieved satisfactory results in our experiments, which proves the effectiveness of the proposed method.

Keywords: Hazy images · FA attention · Channel weight fusion self-attention · UAV remote sensing · smoothed dilated convolution

1 Introduction

Hazy images can lead to degraded system performance due to low-quality input images. Therefore, it is essential to perform the pre-processing operation of image dehazing before feeding them to advanced computer vision systems. In recent years, image dehazing has received increasing attention from researchers as a typical underlying vision task. McCartney [1] proposed an atmospheric scattering model by analyzing the formation mechanism of haze-degraded images and introducing the scattering and absorption factors of the atmospheric medium that cause image degradation into the imaging model. After that, researchers

Supported by National Natural Science Foundation of China (61975028) and Jiao Xu.

proposed numerous dehazing algorithms [2–4] based on this model and also verified the fitting ability of the model. Image dehazing is a typical discomfort problem. For the discomfort problem, the commonly used algorithm is to convert the model into a fitness problem by extracting a prior information to impose restrictions on the model, which is defined as.

$$I(x) = J(x)t(x) + A(1 - t(x)) \tag{1}$$

where: x denotes the pixel point; $I(x)$ denotes the observed fog scene; $J(x)$ denotes the dehazing scene; A denotes the global atmospheric light intensity; $t(x)$ denotes the medium transfer function, also called transmittance, which can be further expressed as $t(x) = e^{-\beta\delta(x)}$, where β denotes the atmospheric scattering coefficient and $\delta(x)$ denotes the scene depth of pixel point x. Usually, the atmospheric light A and the transmittance $t(x)$ of each pixel point are uncertain. Therefore, image dehazing is a serious indeterminate problem.

The emergence and development of deep learning is a milestone event in the field of artificial intelligence. The powerful nonlinear mapping capability of deep learning has also been applied to image dehazing. The use of deep neural nets to eventually construct a mapping from haze images to clear images is a typical model-free image dehazing method. This type of method does not need to consider the reason for hazy image degradation and imaging process, the network inputs a hazy image, and after learning, the output is a clear image. Such as DehazeNet [5], AOD-Net [6,7], GridDehazeNet [8] and so on. These methods can achieve image dehazing by training pairs of images without complicated preprocessing and other processes, which improves the efficiency of image processing under the condition of having a large amount of arithmetic support.

All the above methods use an end-to-end dehazing model. Since the respective feature networks have different capabilities for feature extraction and utilization, the image dehazing results vary. GFNt [7]has the problem of limited ability to handle dense fog, AOD-Net[6] tends to output images with low brightness, and GridDehazeNett [8] is not effective for non-uniform fog at white backgrounds. To provide higher-quality image dehazing results, more powerful image feature extraction and utilization capabilities are required, which represents the need for more arithmetic support. In recent years some networks have produced better dehazing effects by continuously stacking feature extraction modules, but it also brings a greater arithmetic burden, among which the representative MSBDNt [9] is used, but it is suitable for general large-size remote sensing image processing.

In the field of remote sensing, remote sensing images with haze are characterized by high resolution and uneven haze distributiont [10]. Therefore when transferring the above models, there is often a problem of insufficient arithmetic power or uneven haze that is difficult to remove. In 2019, Ke et al. [11] obtained the dehazing images using deep learning methods on satellite remote sensing images of 480*480 pixels, but the brightness of the processed images was significantly reduced; in 2021, Musunuri and Kwont [12] performed experimental operations of dehazing in several sizes of images and achieve better results on small size images, but did not achieve satisfactory visual results on the large size and images with uneven haze distribution. Therefore, when processing remote sensing data, the following two main aspects are considered: (a) The model as a whole should have enough depth to balance the processing effect and the number of parameters while being as streamlined as possible. (b) For uneven distribution of haze, the multi-scale features should utilize different weights to achieve good results.

After the introduction of the Transformer, the image dehazing has been further enhanced, but Transformer requires more arithmetic supportt [13], it is often accompanied by GPU memory overflow due to insufficient arithmetic power when processing large-size images, Uformert [14] is a typical application of Transformer in image dehazing, with a much higher number of parameters than the dehazing method under CNN structure with similar results, so CNN is more suitable as the basic structure for remote sensing images. The main contributions of this paper are as follows. (a) In this paper, the proposed method is based on CNN structure with improved smoothed dilated convolution as the feature extraction module. The receptive field is changed by changing the size of the dilation rate, which saves computational resources as much as possible. (b) We use Channel weight fusion self-attention to achieve Unbalanced processing and pixel-level processing, which can better achieve the effect of dealing with non-uniform haze. (c) The residual structure is used extensively in the model proposed in the paper to deepen the model depth, the method proposed in this paper achieves satisfactory visual results on hazy images and realizes the processing of large-size 4K images in a single GPU.

2   Related Work and Model

2.1   Related Work

Improved Smoothed Dilated Convolution. Generally, downsampling is used to reduce the computation and increase the receptive field, if there is no downsampling, the receptive field has to be increased exponentially, and the information loss of downsampling often needs to be compensated by increasing the number of channels, This paper does not advocate the use of downsampling for feature extraction and learning work. In this paper, smoothed dilated convolution is invoked to balance the receptive field loss and detail feature requirements of the network. The size of the filter w of the dilated convolution is S. The output O of this dilated convolution at position i is as follows, and r is the dilated rate. The one-dimensional dilated convolution is formulated as follows [15]. In Fig. 1(a), different dilation rates correspond to different receptive field sizes.

$$O[i] = \sum_{s=1}^{s} f[i + r * s]w[i] \tag{2}$$

The process of dilated convolution is divided into three steps. Firstly the input feature maps are subsampled with r. Secondly, these intermediate feature map sets are fed into a standard convolution. This convolution has a filter

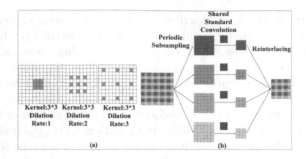

Fig. 1. Dilated convolution

with the same weights as the originally extended convolution after removing all inserted zeros. Finally, the feature maps are reinterlaced to the original resolution to produce the output [16]. The specific visualization process can be seen in Fig. 1(b). Based on the above understanding, we can perform separate shared convolution before the start of the dilated convolution, thus avoiding the appearance of grid artifacts.

The FA attention consists of two modules. The first one is Channel attention, which operates using the original feature map to reassign the features of different channels. The other module is Pixel attention, which implements pixel attention through the operation of dimensional multiplication [17]. In essence, in the detail region of the image, different thicknesses of haze correspond to different weight sizes, so different channels and pixels correspond to different weight sizes and should be treated for both channels and pixels [17]. In this paper, we use a cascade to combine the smoothed dilated convolution with the FA attention. The improved smoothed dilated convolution structure is shown below in Fig. 2.

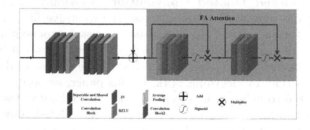

Fig. 2. Improved Smoothed Dilated Convolution

Channel Weight Fusion Self-attention. In a deep learning task, to increase the effective field of perception and thus avoid redundant information, capturing multi-scale features is a common approach adopted by current researchers [18]. In neural networks, multiscale can be concretely represented by scaling the output feature maps of different convolutional layers to a uniform size. Images of different scales are suitable for different tasks. If the image task is simple, such as determining whether the original image is a solid color or whether there is a foreground, a small-scale image is sufficient, if the image task is more difficult, such as the task requires implementing semantic segmentation or image description, etc., large-scale images are relied on to obtain good results [19]. Classical networks, such as U-Net [20] and Yolo [21], usually use jump connections to fuse different receptive fields and scale information, and each layer is fused by the same weights, thus achieving different scales and different receptive field information fusion, which makes the network have a better grasp of both details and global. For different receptive field fusions, this paper adopts the automatic learning of weights, which makes the information fused with different sizes under different receptive fields by automatic learning. The formula and the implementation are sketched as follows.

$$F = K_1F_1 + K_2F_2 + K_3F_3 + K_4F_4 \qquad (3)$$

$$\tilde{F} = F \times \sigma(Conv(\delta(Conv(F)))) \qquad (4)$$

where F_1, F_2, F_3 F_4 represent the feature information under different receptive fields, k are variable parameters, δ represents the Relu activation function, σ represents the sigmoid activation function, and *tildeF* represents the final output. Adding convolution at the end of the fused information module shown above changes the channel to $1 * H * W$, and again for the fused pixels the corresponding phase multiplication process is performed, which achieves the fusion of information under different receptive fields and the unbalanced processing of different pixels, which is beneficial to focus on the processing of information under different hazy levels.

2.2 Model

The complete structure diagram used in this paper is shown in the following Fig. 3.

In the whole structure, the main role of the forward encoder is to extract the feature maps, firstly using three normal convolutions to perform the extraction, and subsequently, to extract to different receptive fields, this paper uses a smoothed dilated convolution instead of downsampling to expand the receptive field process. A weighted automatic learning network is then used to perform the receptive field feature fusion. The feature residuals are obtained in the decoder part, and then the residuals are added to the fogged image to obtain the image dehazing.

In the paper, the first three convolution blocks can be considered as three cascaded convolution modules when the expansion rate is 0. To reduce the number of parameters, only three are used in this paper. In this paper, we do not use the method of downsampling to expand the receptive field, which can be used as appropriate in large-size images to reduce the calculation of the number of parameters and prevent the lack of memory, but in theory, it also leads to some details loss.

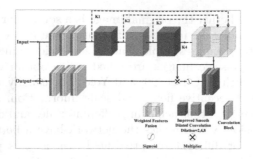

Fig. 3. Complete model structure diagram

3 Experiment

3.1 Datasets

Two different datasets are used for validation in this paper. The REalistic Single Image DEhazing (RESIDE) dataset is the small-size image used in this paper, each image is presented in pairs with corresponding clear images [22]. The real haze images taken by drone aerial photography are the large-size images used in this paper, with a minimum size of 3656*2742, all of which are real hazy images.

3.2 Evaluation and Details

In this paper, PSNR [23], and the number of parameters were adopted as evaluation criteria. the PSNR and SSIM were calculated as follows.

$$MSE = \sum_{i=1}^{D}(x_i - y_i)^2 \tag{5}$$

$$PSNR = 10 \cdot log_{10}(\frac{MAX_I^2}{MSE}) \tag{6}$$

$$SSIM(x,y) = \frac{(2\mu_x\mu_y + C_1)(2\sigma_{xy} + C_2)}{(\mu_x^2 + \mu_y^2 + C_1)(\sigma_x^2 + \sigma_y^2 + C_2)} \tag{7}$$

The $PSNR$ (Peak Signal to Noise Ratio) is the most widely used objective image evaluation index, which is based on the error between the corresponding pixel points. $SSIM$ (Structural similarity index) is a measure of the similarity between two images. The index takes a value between 0 and 1, with closer to 1 representing better image quality [24]. The number of parameters represents the size of the model and the amount of computation. A smaller number of parameters means a smaller amount of computation is required, and to a certain extent, a faster processing speed of the model.

In this paper, a total of 100 rounds are trained in the experiments, and the initial learning rate is set to 0.01. In the middle three improved smoothed dilated convolutional modules, each module contains three expanding convolutional sub-modules, and each small expanding convolutional module has the same dilation rate. The loss function in this paper uses MAE.

4 Result

4.1 Model Analysis

To compare the advantages and disadvantages of this method with other methods more objectively and fairly, we conducted tests on the public data set, and the test results are shown in Table 1. We compare the main dehazing network structures, among which the best dehazing effect is the MSBDN, the use of dense feature fusion modules leads to a large number of parameters and high computational requirements for processing large-size remote sensing data. In 2021, Uformer demonstrated the feasibility of using the Transformer structure for image dehazing, but its parameter size reaches 50.45M. In terms of model size and processing effect, it is inferior to MSBDN and the method proposed in this paper.

Table 1. Comparison of different methods

Method	PSNR	SSIM	Param
(TPAMI'10)DCP	18.75	0.8536	——
CAP	19.05	0.8428	——
(TIP'16)DehazeNet	21.14	0.8698	0.01M
(ICCV'17)AOD-NET	20.51	0.8340	0.002M
(CVPR'18)GFN	22.3	0.8845	0.499M
(ICCV'19)GridDehazeNet	32.16	0.9760	0.96M
(CVPR'20)MSBDN	33.67	0.9850	31.35M
(CVPR'21)Uformer	31.91	0.9710	50.45M
Ours	32.24	0.9841	4.96M

To show the processing effect of our method, we selected indoor and outdoor scenes and compared GridDehazeNet, MSBDN, and our method with similar processing effects. As can be seen in Fig. 4, GridDehazeNet shows significant color distortion, with less than uniform and realistic excess in the white background part, and significant black unevenness in Fig. 4(c) and Fig. 4(h). Both the MSBDN and the method proposed in this paper achieve good image dehazing effects, which are indistinguishable from the human eye. There is a subtle color difference when the details are enlarged.

Under the test of outdoor scenes, there is a clear color unevenness in the sky in Fig. 5(c) and a clear haze residue in Fig. 5(h). In Fig. 5(e) the color contrast is stronger and the outline of the tall buildings in the distance is clearer, but the trees are also relatively darker. In Fig. 5(j), the blue sky is darker and the contrast of shadows and lighting is deeper. From the overall effect, Fig. 5(i) and Fig. 5(d) are closer to the original image.

From the overall results, the images processed by the MSBDN are closer to the original images, while the method in this paper appears to have a stronger contrast in the processing of outdoor scenes. From the perspective of PSNR and SSIM, the MSBDN has a better processing effect. However, it is difficult to define whether Fig. 5 (i) or Fig. 5 (j) is better processed when subjectively evaluated. It is worth mentioning that the PSD [25] performed poorly on PNSR and SSIM at the CVPR in 2021, but received overwhelmingly favorable comments during the subjective evaluation. In terms of the number of parameters, the method in this paper compresses the parameter size to only 1/6 or smaller than MSBDN, so the method proposed in this paper is more advantageous in processing remote sensing data.

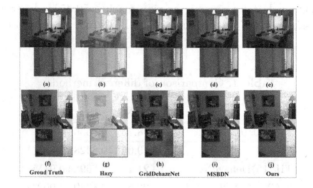

Fig. 4. Indoor scene processing effects

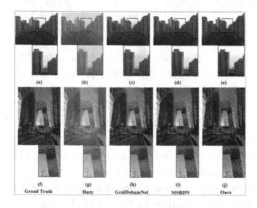

Fig. 5. Indoor scene processing effects

Table 2. Ablation experiment

FA Attention	✓	—	✓
Channel weight fusion self-attention	✓	✓	—
PSNR	32.24	31.51	31.39

In Table 2, we show the impact of the related work on the experiments.

From the table, we can see that the use of the FA attention mechanism improves PSNR by 0.73 and the use of Channel weight fusion self-attention improves PSNR by 0.89. With FA attention and Channel weight fusion self-attention, we not only give different attention to multi-channel and multi-pixel, but also we can retain the information in the lower layers and pass it to the deeper layers. Due to the weighting mechanism, it makes FA attention pay more attention to effective information such as thick fog area, high-frequency texture, and color fidelity. Therefore, it can produce a better image dehazing effect. The experiments in this subsection demonstrate the positive optimization effect of FA attention and Channel weight fusion self-attention for the network in this paper. Different weights should be used for fusion operation for channel information with different receptive fields, which is beneficial for the network.

4.2   Real Remote Sensing Image Dehazing Analysis

Finally, this paper is tested for real remote sensing images. All images are taken by UAV, which can prove the effectiveness of this paper for real remote sensing image dehazing. In this paper, a total of three different types of hazy scenarios are selected to test the effectiveness.

The image of scene one is selected with a more uniform distribution of haze.

As can be seen in Fig. 6, the removal of uniform haze is very effective. In the figure, it can be seen that the outline of the road is clearer, the car driving on the road is more obvious, the color of the farmland next to it is more vivid, and the color contrast of different tree species is more obvious, which is conducive to further work on the interpretation of the picture.

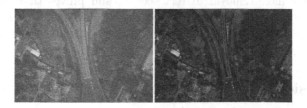

Fig. 6. Indoor scene processing effects

Scene 2 is a case of uneven haze with increased haze concentration. In Scene 2, the proposed method still demonstrates good performance. We zoom in on some details of the image, and it can be seen from Fig. 7 that the dehazing task restores the largely unseen information, which facilitates further utilization of the image.

Fig. 7. Indoor scene processing effects

Scene 3 shows the dehazing effect in the case of very high haze concentration and very uneven distribution. The haze concentration in Fig. 8 is the highest and the most uneven. The vast majority of the haze in the picture is removed, but there is still a small amount of haze that is not removed cleanly. In the blue box, there is a very small amount of residual, and in the red box, the naked eye can barely recognize the information in the cloudy scene, we recovered most of the information, but due to the excessive haze concentration, there is still some information that cannot be recovered.

Fig. 8. Indoor scene processing effects

The images with haze in all three scenes improve the image quality after processing, the contrast of colors is enhanced and details are recovered. This part of the experiments fully proves that the proposed method is still realistic and effective for haze removal in real scenes, and proves that the model has good generalizability. In addition, the remote sensing image sizes in this paper all reach the 4K standard, and there is still no GPU memory overflow in a single GPU.

5   Conclusion

In this paper, we proposed an effective end-to-end image haze removal method for the problem of large-size remote sensing image dehazing, using an improved smoothed dilated convolution and Channel weight fusion self-attention are the highlights of the model. The conducted ablation experiments successfully demonstrate the positive optimization effect of related work. Through experiments on images and visualization of the dehazing images, it is demonstrated that the proposed method in this paper can effectively perform image dehazing and shows excellent visual effects on images of multiple sizes, which provides a good foundation for further image interpretation and helps researchers to carry out related work.

References

1. Hide, R.: Optics of the atmosphere: scattering by molecules and particles. Phys. Bull. **28**(11), 521 (1977). https://doi.org/10.1088/0031-9112/28/11/025
2. Dai, C., Lin, M., Wu, X., Zhang, D.: Single hazy image restoration using robust atmospheric scattering model. Signal Process. **166**, 107257 (2020). https://doi.org/10.1016/j.sigpro.2019.107257. https://www.sciencedirect.com/science/article/pii/S0165168419303093
3. Millán, M.M.: Remote sensing of air pollutants: a study of some atmospheric scattering effects. Atmos. Environ. **14**(11), 1241–1253 (1980). https://doi.org/10.1016/0004-6981(80)90226-7. https://www.sciencedirect.com/science/article/pii/0004698180902267
4. Hong, S., Kim, M., Kang, M.G.: Single image dehazing via atmospheric scattering model-based image fusion. Signal Process. **178**, 107798 (2021). https://doi.org/10.1016/j.sigpro.2020.107798. https://www.sciencedirect.com/science/article/pii/S016516842030342X
5. Cai, B., Xu, X., Jia, K., Qing, C., Tao, D.: Dehazenet: an end-to-end system for single image haze removal. IEEE Trans. Image Process. **25**(11), 5187–5198 (2016). https://doi.org/10.1109/TIP.2016.2598681
6. Li, B., Peng, X., Wang, Z., Xu, J., Feng, D.: Aod-net: all-in-one dehazing network. In: 2017 IEEE International Conference on Computer Vision (ICCV), pp. 4780–4788 (2017). https://doi.org/10.1109/ICCV.2017.511
7. Ren, W., et al.: Gated fusion network for single image dehazing. In: 2018 IEEE/CVF Conference on Computer Vision and Pattern Recognition, pp. 3253–3261 (2018). https://doi.org/10.1109/CVPR.2018.00343
8. Liu, X., Ma, Y., Shi, Z., Chen, J.: Griddehazenet: attention-based multi-scale network for image dehazing. In: 2019 IEEE/CVF International Conference on Computer Vision (ICCV), pp. 7313–7322 (2019). https://doi.org/10.1109/ICCV.2019.00741
9. Dong, H., et al.: Multi-scale boosted dehazing network with dense feature fusion. In: 2020 IEEE/CVF Conference on Computer Vision and Pattern Recognition (CVPR), pp. 2154–2164 (2020). https://doi.org/10.1109/CVPR42600.2020.00223
10. Li, J., Hu, Q., Ai, M.: Haze and thin cloud removal via sphere model improved dark channel prior. IEEE Geosci. Remote Sens. Lett. **16**(3), 472–476 (2019). https://doi.org/10.1109/LGRS.2018.2874084
11. Ke, L., et al.: Haze removal from a single remote sensing image based on a fully convolutional neural network. J. Appl. Remote Sens. **13**(3), 036505 (2019). https://doi.org/10.1117/1.JRS.13.036505
12. Musunuri, Y.R., Kwon, O.S.: Deep residual dense network for single image super-resolution. Electronics **10**(5) (2021). https://doi.org/10.3390/electronics10050555, https://www.mdpi.com/2079-9292/10/5/555
13. Song, Y., He, Z., Qian, H., Du, X.: Vision transformers for single image dehazing. IEEE Trans. Image Process. **32**, 1927–1941 (2023). https://doi.org/10.1109/TIP.2023.3256763

14. Wang, Z., Cun, X., Bao, J., Zhou, W., Liu, J., Li, H.: Uformer: a general u-shaped transformer for image restoration. In: 2022 IEEE/CVF Conference on Computer Vision and Pattern Recognition (CVPR), pp. 17662–17672 (2022). https://doi.org/10.1109/CVPR52688.2022.01716
15. Dumoulin, V., Visin, F.: A guide to convolution arithmetic for deep learning (2018)
16. Wang, Z., Ji, S.: Smoothed dilated convolutions for improved dense prediction. Data Min. Knowl. Discov. **35**(4), 1470–1496 (2021). https://doi.org/10.1007/s10618-021-00765-5
17. Qin, X., Wang, Z., Bai, Y., Xie, X., Jia, H.: FFA-net: feature fusion attention network for single image dehazing. In: Proceedings of the AAAI Conference on Artificial Intelligence, vol. 34, no. 07, pp. 11908–11915 (2020). https://doi.org/10.1609/aaai.v34i07.6865. https://ojs.aaai.org/index.php/AAAI/article/view/6865
18. Wang, H., Kembhavi, A., Farhadi, A., Yuille, A.L., Rastegari, M.: Elastic: improving cnns with dynamic scaling policies. In: Proceedings of the IEEE/CVF Conference on Computer Vision and Pattern Recognition (CVPR) (2019)
19. Yu, X., Gong, Y., Jiang, N., Ye, Q., Han, Z.: Scale match for tiny person detection. In: 2020 IEEE Winter Conference on Applications of Computer Vision (WACV), pp. 1246–1254 (2020). https://doi.org/10.1109/WACV45572.2020.9093394
20. Kim, S.M., Shin, J., Baek, S., Ryu, J.H.: U-Net convolutional neural network model for deep red tide learning using GOCI. J. Coastal Res. **90**(sp1), 302 – 309 (2019). https://doi.org/10.2112/SI90-038.1
21. Redmon, J., Farhadi, A.: Yolov3: an incremental improvement. CoRR **abs/1804.02767** (2018). http://arxiv.org/abs/1804.02767
22. Li, B., et al.: Benchmarking single-image dehazing and beyond. IEEE Trans. Image Process. **28**(1), 492–505 (2019). https://doi.org/10.1109/TIP.2018.2867951
23. Huynh-Thu, Q., Ghanbari, M.: Scope of validity of PSNR in image/video quality assessment. Electron. Lett. **44**, 800–801 (2008). https://api.semanticscholar.org/CorpusID:62732555
24. Wang, Z., Bovik, A., Sheikh, H., Simoncelli, E.: Image quality assessment: from error visibility to structural similarity. IEEE Trans. Image Process. **13**(4), 600–612 (2004). https://doi.org/10.1109/TIP.2003.819861
25. Chen, Z., Wang, Y., Yang, Y., Liu, D.: PSD: principled synthetic-to-real dehazing guided by physical priors. In: 2021 IEEE/CVF Conference on Computer Vision and Pattern Recognition (CVPR), pp. 7176–7185 (2021). https://doi.org/10.1109/CVPR46437.2021.00710

ContextNet: Learning Context Information for Texture-Less Light Field Depth Estimation

Wentao Chao, Xuechun Wang, Yiming Kan, and Fuqing Duan[✉]

School of Artificial Intelligence, Beijing Normal University, Beijing, China
{chaowentao,wangxuechun,kanyiming}@mail.bnu.edu.cn, fqduan@bnu.edu.cn

Abstract. Depth estimation in texture-less regions of the light field is an important research direction. However, there are few existing methods dedicated to this issue. We find that context information is significantly crucial for depth estimation in texture-less regions. In this paper, we propose a simple yet effective method called ContextNet for texture-less light field depth estimation by learning context information. Specifically, we aim to enlarge the receptive field of feature extraction by using dilated convolutions and increasing the training patch size. Moreover, we design the Augment SPP (AugSPP) module to aggregate features of multiple-scale and multiple-level. Extensive experiments demonstrate the effectiveness of our method, significantly improving depth estimation results in texture-less regions. The performance of our method outperforms the current state-of-the-art methods (e.g., LFattNet, DistgDisp, OACC-Net, and SubFocal) on the UrbanLF-Syn dataset in terms of MSE ×100, Bad-Pix 0.07, BadPix 0.03, and BadPix 0.01. Our method also ranks third place of comprehensive results in the competition about LFNAT Light Field Depth Estimation Challenge at CVPR 2023 Workshop without any post-processing steps (The code and model are available at https://github.com/chaowentao/ContextNet.).

Keywords: Light field · Depth estimation · Texture-less regions

1 Introduction

Light field (LF) can simultaneously record the spatial and angular information of the scene with a single shot, which has many practical applications, for example refocusing [20, 30], super-resolution [5–7, 14, 26, 28, 36], view synthesis [15, 19, 33], semantic segmentation [22], 3D reconstruction [17], virtual reality [35], especially depth (disparity) estimation [1–4, 21, 23, 25, 27, 29, 38, 39].

By utilizing the additional angle information of LF images, researchers on LF depth estimation have made great progress. Texture-less regions are common and intractable in LF images, restricting the performance of depth estimation, especially in synthetic datasets such as UrbanLF-Syn [22]. However, there are

Q. Liu et al. (Eds.): PRCV 2023, LNCS 14430, pp. 15–27, 2024.
https://doi.org/10.1007/978-981-99-8537-1_2

few methods for texture-less LF depth estimation, specifically for large texture-less regions. We analyze the characteristics of existing methods, which can be divided into traditional and deep learning-based methods. Traditional methods [4,13,24,32,38–40] need to rely on various prior assumptions, are very vulnerable to texture-less regions, and the algorithms are time-consuming.

In recent years, with the development of deep learning, deep learning-based methods [1–3,10,11,23,25,27,29] increasingly are used for LF depth estimation tasks and have advantages over traditional methods in terms of efficiency and accuracy. Nowadays, the mainstream deep learning-based methods [1–3,25,27, 29] are mainly based on multi-view stereo matching, including four steps: feature extraction, cost volume construction, cost aggregation, and disparity regression, such as SubFocal [2], which has achieved the best overall performance on the HCI 4D benchmark [12]. Currently, existing methods [1–3,25,27,29] often adopt the LF image patch (e.g., 32×32) for efficient training, which is effective when there are few or no texture-less regions in the image. However, its side effect causes the receptive field of the model to be too limited and unable to learn abundant context information. When the texture-less region is too large, it may exceed the size of the model's receptive field, making it impossible for the model to infer reasonable results.

We find that the depth estimation results of textured regions around texture-less regions are reliable, so it is important to learn large and abundant context information effectively to alleviate the difficulty of depth estimation in large-scale texture-less regions. As for context information learning, there are two aspects to achieving it. On the one hand, we aim to enlarge the receptive field of feature extraction, including using dilated convolutions and increasing the patchsize of training images. On the other hand, we design an Augment Spatial Pyramid Pooling (AugSPP) module to aggregate features from multiple scales and levels. Combining the expanded receptive field with the AugSPP module alleviates the difficulty of depth estimation in large-scale texture-less regions. Our contributions are as follows:

- We analyze the importance of context information in texture-less regions and present a simple yet effective method called ContextNet to learn context information for texture-less LF depth estimation.
- We utilize dilated convolutions and increase the patchsize of training images to enlarge the receptive field of feature extraction. Moreover, an augment SPP (AugSPP) module is designed to effectively aggregate features from multiple scales and levels.
- Extensive experiments validate the effectiveness of our method. In comparison with state-of-the-art methods, our method achieves superior performance in terms of MSE ×100 and BadPix metrics on the UrbanLF-Syn dataset. Furthermore, our method ranks third place in the LFNAT LF Depth Estimation Challenge at the CVPR 2023 Workshop without any post-processing.

2   Related Work

The related work on LF depth estimation can generally be divided into two categories: traditional methods and deep learning-based methods. Below we review each category in detail.

2.1   Traditional Methods

Traditional methods can generally be subdivided into three categories based on the representations of LF images: multi-view stereo (MVS), epipolar plane image (EPI), and defocus-based methods. MVS-based methods [13] use multi-view information from SAIs for stereo matching to obtain depth. Jeon et al. [13] employed phase translation theory to describe the sub-pixel shift between SAIs and utilized matching processes for stereo matching. EPI-based methods [37] implicitly estimate the depth of the scene by computing the slope of the EPI. Wanner et al. [31] proposed a structure tensor that can estimate the slope of lines in horizontal and vertical EPIs, and refined these results through global optimization. Additionally, Zhang et al. [37] introduced the Spinning Parallelogram Operator (SPO), which can compute the slope of straight lines in EPI with minimal sensitivity to occlusion, noise, and spatial blending. Defocus-based methods [24] obtain depth by measuring how blurred pixels are on different focus stacks. Tao et al. [24] combined scattering and matching cues to generate a local depth map through Markov random field for global optimization. Williem et al. [32] improved the robustness of occlusion and noise for depth estimation by using information entropy between different angles and adaptive scattering. Zhang et al. [38] used the special linear structure of an EPI and locally linear embedding (LLE) for LF depth estimation. Zhang et al. [39] proposed a two-stage method for LF depth estimation that utilized graph-based structure-aware analysis. They combined an undirected graph with occluded and unoccluded SAIs in corner blocks to exploit the structural information of the LF. Han et al. [9] introduced an occlusion-aware vote cost (OAVC) to enhance the accuracy of edge preservation in the depth map. However, these methods rely on hand-designed features and subsequent optimization, which are time-consuming and have limited accuracy for text-less regions.

Deep Learning-Based Methods. Deep learning has experienced rapid development and has been widely applied in various LF processing tasks, particularly depth estimation. Heber et al.[11] were the first to use a CNN to extract features from an EPI and calculate the scene's depth. Shin et al.[23] proposed EPINet, which used four directional ($0°$, $90°$, $45°$, and $135°$) EPIs as input and a center sub-aperture image (SAI) disparity map as output. Tsai et al.[25] introduced the LFAttNet network, which employed a view selection module based on an attention mechanism to calculate the importance of each view and served as the weight for cost aggregation. Guo et al.[8] designed an occlusion region detection network (ORDNet) for explicit estimation of occlusion maps and subsequent networks focus on non-occluded and occluded regions, respectively. Chen *et al.*[3] designed the AttMLFNet, an attention-based multilevel fusion network that combines features from different perspectives hierarchically through intra-branch and inter-branch fusion strategies. Wang *et al.*[29] extended the spatial-angular interaction mechanism to the disentangling mechanism and proposed DistgDisp for LF depth estimation. They also developed the OACC-Net [27], which uses dilated convolution instead of shift-and-concat operation and itera-tive processing with occlusion masks to build an occlusion-aware cost volume. Chao *et al.*[2] proposed the SubFocal method for sub-pixel disparity distribu-tion learning by constructing a sub-pixel cost volume and leveraging disparity distribution constraints to obtain a high-precision disparity map. Chao *et al.*[1] presented a method called OccCasNet by constructing the occlusion-aware cas-cade cost volume for depth estimation and achieved a better trade-off between accuracy and efficiency. At present, existing methods often utilize LF image patches (e.g., 32×32) for training and have achieved high accuracy on textured regions of LF. However, this setting can result in a model with a too-limited receptive field, as the texture-less region may be too large and exceed the size of the model's receptive field, leading to unreasonable results and low accuracy.

We observe that the depth estimation of textured regions around texture-less regions can be accurate. Therefore, we propose a method called ContextNet to learn large and abundant context information in an efficient and effective man-ner, which will be beneficial for addressing the challenge of depth estimation in large-scale texture-less regions. To achieve this, we utilize dilated convolution and increase the patch size of training images to efficiently enlarge the recep-tive field of feature extraction. Additionally, we design an AugSPP module to effectively aggregate features at multiple-scale and multiple-level.

3   Method

3.1   Overview

An overview of our ContextNet is shown in Fig. 1(a). First, the features of each SAI are extracted and aggregated using a shared feature extraction based on dilated convolution [34] and AugSPP module [16]. Second, the sub-pixel view shift is performed to construct the sub-pixel cost volume [2]. Third, the cost aggregation module is used to aggregate the cost volume information. The pre-dicted disparity map is produced by attaching a disparity regression module. We will describe each module in detail below.

3.2   Feature Extraction and AugSPP Module

Enlarging Receptive Field. Dilated convolution expands the receptive field by adding holes in the convolution filter, and has a larger receptive field under the same parameter amount and calculation amount without using downsam-pling. Therefore, we utilize dilated convolution and increase training patchsize correspondingly to enlarge the receptive field. Figure 1(b) shows the structure

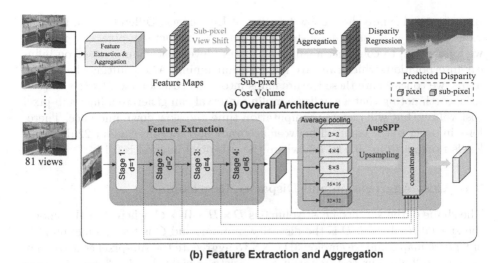

(a) Overall Architecture

(b) Feature Extraction and Aggregation

Fig. 1. The specific network design of ContextNet. (a) Overall Architecture. (b) Feature Extraction and AugSPP module.

of feature extraction. First, two 3 × 3 convolutions are employed to extract the initial feature with a channel of 4. Then, we use a feature extraction based on dilated convolution [34] module to extract multi-level features of SAI. The feature extraction module contains four stages, and the dilated ratios are set to 1, 2, 4, and 8, respectively.

Multi-scale and Multi-level Features Aggregation. Based on the original SPP module [16], we propose the AugSPP module for aggregating multi-scale and multi-level features. Specifically, we add an extra pooling size 32 × 32 in the AugSPP module different from previous methods [2,25]. So five average pooling operations at multi-scale are used to compress the features. The sizes of the average pooling blocks are 2 × 2, 4 × 4, 8 × 8, 16 × 16 and 32 × 32, respectively. Bilinear interpolation is adopted to upsample these low-dimensional feature maps to the same size. We also aggregate multi-level features of feature extraction using skip connections to further improve the discrimination of features. A 1 × 1 convolution layer is used for reducing the feature dimension. The features output by the AugSPP module contain multi-scale and multi-level discriminative context information and are concatenated to form a feature map F. The feature extraction and aggregation module incorporates additional textured features from large neighboring regions for challenging regions, such as texture-less and reflection areas.

3.3   Sub-pixel Cost Volume

Cost volume is constructed by *shift-and-concat* [1–3,25] operation within a predefined disparity range (such as from –0.5 to 1.6 in the UrbanLF-Syn dataset).

In order to alleviate the narrow baseline of LF images, Different from the previous method [13] using phase shift theorem to construct sub-pixel cost volume, we follow [1,2] to construct a sub-pixel feature level cost volume based on bilinear interpolation, which can save memory-consuming. After shifting the feature maps, we concatenate these feature maps into a 4D cost volume $D \times H \times W \times C$. It is worth noting that a smaller sampling interval can generate a finer sub-pixel cost volume but will increase computation time and slow down inference. Therefore, in order to the trade-off between accuracy and speed, we adopt 22 disparity levels ranging from –0.5 to 1.6, where the sub-pixel interval is 0.1.

3.4   Cost Aggregation and Disparity Regression

The shape of the sub-pixel cost volume is $D \times H \times W \times C$, where $H \times W$ denotes the spatial resolution, D is the disparity number, and C is the channel number of feature maps, and we employ 3D CNN to aggregate the sub-pixel cost volume. Following [2,25], our cost aggregation consists of eight $3 \times 3 \times 3$ convolutional layers and two residual blocks. After passing through these 3D convolutional layers, we obtain the final cost volume $C_f \in D \times H \times W$. We normalize C_f by using the softmax operation along dimension D. Finally, the output disparity $\hat{d}$ can be calculated as follows:

$$\hat{d} = \sum_{d_k=D_{min}}^{D_{max}} d_k \times \text{softmax}(-C_{d_k}), \tag{1}$$

where $\hat{d}$ denotes the estimated center view disparity, D_{min} and D_{max} stand for the minimum and maximum disparity values, respectively, and d_k is the sampling value between D_{min} and D_{max} according to the predefined sampling interval.

4   Experiments

In this section, we first introduce the UrbanLF-Syn datasets and implementation details. Then, we compare the performance of our method with the state-of-the-art methods. Finally, we conduct an extensive ablation study to analyze the proposed ContextNet.

4.1   Datasets and Implementation Details

UrbanLF-Syn dataset [22] contains 230 synthetic LF samples, with 170 training, 30 validation, and 30 test samples for LFNAT LF Depth Estimation Challenge at the CVPR 2023 Workshop. Each sample consists of 81 SAIs with a spatial resolution of 480×640 and an angular resolution of 9×9.

We employ the $L1$ loss as the loss function, as it is robust to outliers. We use the SubFocal as the baseline model. We follow the same data augmentation strategy as in [1,2,25] to improve the model performance, which includes random horizontal and vertical flipping, 90-degree rotation, and adding random noise. It

Table 1. Quantitative comparison results with state-of-the-art methods on the validation of UrbanLF-Syn dataset [22] in terms of BadPix 0.07, BadPix 0.03, BadPix 0.01, and MSE×100. The best results are shown in boldface.

Method	Img11	Img27	Img34	Img50	Img54	Img68	Img69	Img70	Avg. BP 0.07
LFattNet [25]	10.179	23.648	22.637	9.668	13.533	9.819	12.075	7.467	13.629
DistgDisp [29]	**6.825**	20.240	24.522	8.304	13.830	9.367	10.118	7.691	12.612
OACC-Net [27]	6.845	24.058	27.792	10.390	13.922	11.799	13.234	8.750	14.599
SubFocal [2]	9.194	22.223	22.619	6.008	11.476	8.385	9.357	6.542	11.976
Ours	8.456	**12.149**	**17.104**	**5.396**	**5.771**	**6.954**	**5.790**	**2.875**	**8.062**
Method	Img11	Img27	Img34	Img50	Img54	Img68	Img69	Img70	Avg. BP 0.03
LFattNet [25]	14.902	34.327	30.583	15.486	19.495	14.758	18.967	15.744	20.533
DistgDisp [29]	**12.240**	33.657	32.753	15.282	21.297	15.938	18.350	15.770	20.661
OACC-Net [27]	14.095	38.117	34.853	18.251	21.962	23.187	21.917	18.052	23.804
SubFocal [2]	13.342	30.742	28.696	10.661	15.413	12.176	14.103	13.739	17.359
Ours	15.032	**23.512**	**24.472**	**10.643**	**11.308**	**10.758**	**11.187**	**11.723**	**14.830**
Method	Img11	Img27	Img34	Img50	Img54	Img68	Img69	Img70	Avg. BP 0.01
LFattNet [25]	24.113	49.134	43.789	30.819	33.616	28.407	34.609	31.762	34.531
DistgDisp [29]	30.141	51.105	50.838	33.585	38.073	46.023	36.557	36.934	40.407
OACC-Net [27]	33.191	57.538	51.364	38.486	39.505	56.173	42.212	41.426	44.987
SubFocal [2]	**20.913**	41.808	37.782	19.375	22.847	19.485	23.257	25.075	26.318
Ours	23.614	**39.654**	**35.729**	**19.788**	**20.960**	**18.555**	**21.352**	**22.688**	**25.293**
Method	Img11	Img27	Img34	Img50	Img54	Img68	Img69	Img70	Avg. MSE
LFattNet [25]	0.662	1.977	2.514	0.283	1.142	1.083	0.495	0.210	1.046
DistgDisp [29]	**0.304**	1.558	2.509	0.192	0.763	0.743	0.313	0.185	0.820
OACC-Net [27]	0.342	3.025	11.739	1.799	3.527	2.790	2.095	0.273	3.199
SubFocal [2]	0.644	1.735	2.975	0.200	1.376	1.066	0.515	0.165	1.085
Ours	0.324	**0.795**	**1.038**	**0.105**	**0.252**	**0.418**	**0.172**	**0.080**	**0.398**

is important to note that the spatial and angular dimensions need to be flipped or rotated jointly to maintain the LF structures. We randomly crop LF images into 64×64 grayscale patches to provide more context information for our model. We remove texture-less regions where the mean absolute difference between the center pixel and other pixels is less than 0.02. The batchsize is set to 16, and we use the Adam optimizer [18]. The disparity range is set from –0.5 to 1.6, and the disparity interval is set to 0.1. Our ContextNet is implemented in the framework of TensorFlow on an NVIDIA A100 GPU. The learning rate is initially set to 1×10^{-3} and decreased by a factor of 0.5 after every 30 epochs. The training is stopped after 120 epochs, and we select the best model based on its performance on the validation set.

We evaluate our method using two metrics: Mean Squared Error (MSE × 100), and BadPix(ϵ). The MSE ×100 measures the mean square errors of all pixels, multiplied by 100. BadPix (ϵ) represents the percentage of pixels whose absolute disparity error exceeds a predefined threshold, commonly set to 0.01, 0.03, or 0.07.

4.2   Comparison of State-of-the-Art Methods

Qualitative Comparison. We compare our method with four state-of-the-art methods, including LFattNet [25], DistgDisp [29], OACC-Net [27], and SubFocal [2]. Figure 2 shows qualitative comparison results on *Img27*, *Img54*, *Img69*, and *Img70* scenes of UrbanLF-Syn dataset validation set. Note that these scenes contain large texture-less areas, such as road surfaces. Compared with other methods, our method has less error overall, especially in texture-less regions, which verifies the effectiveness of our method, which can help texture-less regions for depth estimation by learning context information.

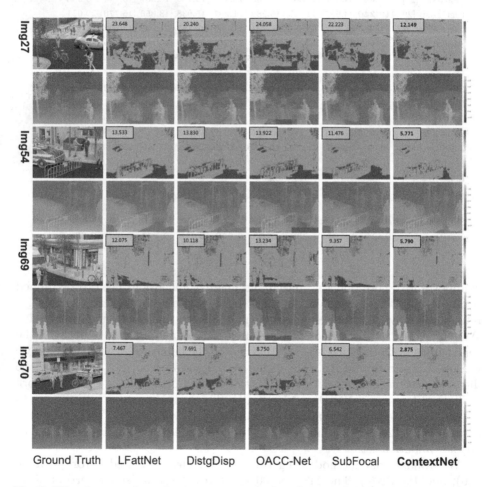

Ground Truth   LFattNet   DistgDisp   OACC-Net   SubFocal   **ContextNet**

Fig. 2. Visual comparisons between our method and state-of-the-art methods on the UrbanLF-Syn dataset [22] scenes, i.e., *Img27*, *Img54*, *Img69* and *Img70*, including LFattNet [25], DistgDisp [29], OACC-Net [27] and SubFocal [2], with the corresponding BadPix 0.07 error maps. Lower is better. The best results are shown in boldface. Please zoom in for a better comparison.

Table 2. The benchmark in the average comparison on the testing set of the UrbanLF-Syn [22] dataset in terms of BadPix 0.07, BadPix 0.03, BadPix 0.01, and MSE ×100. The best results are shown in boldface.

Method	MSE ×100	BadPix 0.07	BadPix 0.03	BadPix 0.01
MultiBranch	2.776	86.35	64.915	43.402
MTLF	1.373	41.156	21.452	13.034
UOAC	0.953	53.926	28.211	15.582
EPI-Cost	1.175	47.927	24.15	14.637
MS3D	0.559	31.066	14.664	7.917
CBPP	0.394	27.385	**12.628**	**5.907**
HRDE	**0.368**	27.802	12.825	6.205
Ours	0.416	**24.681**	12.649	6.75

Table 3. The average results of different disparity intervals and numbers on 8 scenes of the UrbanLF-Syn dataset [22] dataset validation set in terms of BadPix 0.07, and MSE×100. The best results are shown in boldface.

Disparity Interval	Disparity Number	Disparity Range	MSE ×100	BadPix 0.07
1	4	[−1, 2]	1.277	14.367
0.5	6	[−0.5, 2]	1.141	13.873
0.3	8	[−0.5, 1.6]	1.072	13.614
0.15	15	[−0.5, 1.6]	**1.063**	13.254
0.1	22	[−0.5, 1.6]	1.065	**12.946**
0.05	43	[−0.5, 1.6]	1.085	13.201

Quantitative Comparison. We have also conducted quantitative comparison experiments with four state-of-the-art methods [2, 25, 27, 29]. Table 1 shows the comparison results on the UrbanLF-Syn dataset [22] for four metrics: Bad Pix 0.07, BadPix 0.03, BadPix 0.01, and MSE ×100. Our method ranks first in most scenes and achieves the top metrics in average BadPix 0.07, BadPix 0.03, BadPix 0.01, and MSE ×100, significantly outperforming current state-of-the-art methods by a large margin. We have submitted our results to the UrbanLF-Syn dataset website. Table 2 shows that our method is competitive and also ranks third place of comprehensive results in the competition for LFNAT LF Depth Estimation Challenge at CVPR 2023 Workshop without any post-processing steps[1].

---

[1] http://www.lfchallenge.com/dp_lambertian_plane_result/. On the benchmark, the name of our method is called SF-Net.

Table 4. The average results of different variants on 8 scenes of the UrbanLF-Syn dataset [22] dataset validation set in terms of BadPix 0.07, and MSE×100. The best results are shown in boldface.

Variants	MSE ×100	BadPix 0.07
baseline	1.157	13.230
+add input training patchsize: 64×64	0.806	11.298
+change dilated ratio: [1,2,4,8]	0.416	8.758
+AugSPP: 32×32 average pooling	0.459	8.675
+AugSPP: concat multi-level feature	0.440	8.223
+finer interval: 0.1	0.460	8.184
+more feature channel: 170	**0.398**	**8.062**

4.3 Ablation Study

Disparity Interval of Sub-pixel Cost Volume. We also conduct extensive ablation experiments to validate our method. First, we experiment with different disparity intervals and disparity numbers for the sub-pixel cost volume. The training patchsize of the default settings is set to 32, and the feature channel of cost aggregation is set to 170. It can be seen from Table 3 that as the disparity interval decreases, the corresponding MSE and Badpix 0.07 are totally decreased. Considering efficiency and accuracy, we finally chose a disparity interval of 0.1.

Context Information Learning. We validate different components for context information learning. The training patchsize of the baseline model is set to 32, the feature channel of cost aggregation is set to 96, the disparity interval is 0.15, the number of disparities is 15, and the disparity range is –0.5 to 1.6. As shown in Table 4, the different components we proposed can improve the metrics step by step. Compared with the baseline model, our method achieves improvements of 65.6% and 39% on the MSE ×100 and BadPix 0.07 metrics, respectively.

5 Conclusion and Limitations

In this paper, we propose a method, namely ContextNet, to learn context information for texture-less LF depth estimation. On the one hand, we use dilated convolution and increase the patchsize of training images to enlarge the receptive field of feature extraction. On the other hand, an AugSPP module is designed to improve the overall performance of our method by effectively aggregating features from multi-scale and multi-level. Extensive experiments validate the effectiveness of our method. Our method outperforms state-of-the-art methods on the UrbanLF-Syn dataset and also ranks third place in comprehensive results in the competition about LFNAT LF Depth Estimation Challenge at the CVPR 2023 Workshop.

While our method achieves competitive results in texture-less regions, there is still room for improvement. Regarding the LF depth estimation of texture-less regions, we plan to start from the following aspects in the future. We can further expand the receptive field by fusing the results of monocular depth estimation. Additionally, we may design a post-processing step by diffusing the depth of textured regions to texture-less regions. Finally, we can utilize shape priors for texture-less depth estimation with the help of semantic segmentation maps.

Acknowledgement. This work is supported by the National Key Research and Development Project Grant, Grant/Award Number: 2018AAA0100802.

References

1. Chao, W., Duan, F., Wang, X., Wang, Y., Wang, G.: Occcasnet: occlusion-aware cascade cost volume for light field depth estimation. arXiv preprint arXiv:2305.17710 (2023)
2. Chao, W., Wang, X., Wang, Y., Wang, G., Duan, F.: Learning sub-pixel disparity distribution for light field depth estimation. TCI Early Access, 1–12 (2023)
3. Chen, J., Zhang, S., Lin, Y.: Attention-based multi-level fusion network for light field depth estimation. In: AAAI, pp. 1009–1017 (2021)
4. Chen, J., Chau, L.: Light field compressed sensing over a disparity-aware dictionary. TCSVT **27**(4), 855–865 (2017)
5. Chen, Y., Zhang, S., Chang, S., Lin, Y.: Light field reconstruction using efficient pseudo 4d epipolar-aware structure. TCI **8**, 397–410 (2022)
6. Cheng, Z., Liu, Y., Xiong, Z.: Spatial-angular versatile convolution for light field reconstruction. TCI **8**, 1131–1144 (2022)
7. Cheng, Z., Xiong, Z., Chen, C., Liu, D., Zha, Z.J.: Light field super-resolution with zero-shot learning. In: CVPR, pp. 10010–10019 (2021)
8. Guo, C., Jin, J., Hou, J., Chen, J.: Accurate light field depth estimation via an occlusion-aware network. In: ICME, pp. 1–6 (2020)
9. Han, K., Xiang, W., Wang, E., Huang, T.: A novel occlusion-aware vote cost for light field depth estimation. TPAMI **44**(11), 8022–8035 (2022)
10. He, L., Wang, G., Hu, Z.: Learning depth from single images with deep neural network embedding focal length. TIP **27**(9), 4676–4689 (2018)
11. Heber, S., Pock, T.: Convolutional networks for shape from light field. In: CVPR, pp. 3746–3754 (2016)
12. Honauer, K., Johannsen, O., Kondermann, D., Goldluecke, B.: A dataset and evaluation methodology for depth estimation on 4D light fields. In: Lai, S.-H., Lepetit, V., Nishino, K., Sato, Y. (eds.) ACCV 2016. LNCS, vol. 10113, pp. 19–34. Springer, Cham (2017). https://doi.org/10.1007/978-3-319-54187-7_2
13. Jeon, H.G., et al.: Accurate depth map estimation from a lenslet light field camera. In: CVPR, pp. 1547–1555 (2015)
14. Jin, J., Hou, J., Chen, J., Kwong, S.: Light field spatial super-resolution via deep combinatorial geometry embedding and structural consistency regularization. In: CVPR, pp. 2260–2269 (2020)

15. Jin, J., Hou, J., Chen, J., Zeng, H., Kwong, S., Yu, J.: Deep coarse-to-fine dense light field reconstruction with flexible sampling and geometry-aware fusion. TPAMI **44**, 1819–1836 (2020)
16. He, K., Zhang, X., Ren, S., Sun, J.: Spatial pyramid pooling in deep convolutional networks for visual recognition. TPAMI **37**(9), 1904–1916 (2015)
17. Kim, C., Zimmer, H., Pritch, Y., Sorkine-Hornung, A., Gross, M.H.: Scene reconstruction from high spatio-angular resolution light fields. TOG **32**(4), 73–1 (2013)
18. Kingma, D.P., Ba, J.: Adam: a method for stochastic optimization. arXiv preprint arXiv:1412.6980 (2014)
19. Meng, N., So, H.K.H., Sun, X., Lam, E.Y.: High-dimensional dense residual convolutional neural network for light field reconstruction. TPAMI **43**(3), 873–886 (2019)
20. Ng, R., Levoy, M., Brédif, M., Duval, G., Horowitz, M., Hanrahan, P.: Light field photography with a hand-held plenoptic camera. Ph.D. thesis, Stanford University (2005)
21. Peng, J., Xiong, Z., Wang, Y., Zhang, Y., Liu, D.: Zero-shot depth estimation from light field using a convolutional neural network. TCI **6**, 682–696 (2020)
22. Sheng, H., Cong, R., Yang, D., Chen, R., Wang, S., Cui, Z.: Urbanlf: a comprehensive light field dataset for semantic segmentation of urban scenes. TCSVT **32**(11), 7880–7893 (2022)
23. Shin, C., Jeon, H.G., Yoon, Y., Kweon, I.S., Kim, S.J.: Epinet: a fully-convolutional neural network using epipolar geometry for depth from light field images. In: CVPR, pp. 4748–4757 (2018)
24. Tao, M.W., Hadap, S., Malik, J., Ramamoorthi, R.: Depth from combining defocus and correspondence using light-field cameras. In: ICCV, pp. 673–680 (2013)
25. Tsai, Y.J., Liu, Y.L., Ouhyoung, M., Chuang, Y.Y.: Attention-based view selection networks for light-field disparity estimation. In: AAAI, pp. 12095–12103 (2020)
26. Van Duong, V., Huu, T.N., Yim, J., Jeon, B.: Light field image super-resolution network via joint spatial-angular and epipolar information. TCI **9**, 350–366 (2023)
27. Wang, Y., Wang, L., Liang, Z., Yang, J., An, W., Guo, Y.: Occlusion-aware cost constructor for light field depth estimation. In: CVPR, pp. 19809–19818 (2022)
28. Wang, Y., Wang, L., Liang, Z., Yang, J., Timofte, R., Guo, Y.: Ntire 2023 challenge on light field image super-resolution: dataset, methods and results. arXiv preprint arXiv:2304.10415 (2023)
29. Wang, Y., et al.: Disentangling light fields for super-resolution and disparity estimation. TPAMI **45**, 425–443 (2022)
30. Wang, Y., Yang, J., Guo, Y., Xiao, C., An, W.: Selective light field refocusing for camera arrays using bokeh rendering and superresolution. SPL **26**(1), 204–208 (2018)
31. Wanner, S., Goldluecke, B.: Variational light field analysis for disparity estimation and super-resolution. TPAMI **36**(3), 606–619 (2014)
32. Williem, W., Park, I.K.: Robust light field depth estimation for noisy scene with occlusion. In: CVPR, pp. 4396–4404 (2016)
33. Wu, G., Liu, Y., Fang, L., Dai, Q., Chai, T.: Light field reconstruction using convolutional network on epi and extended applications. TPAMI **41**(7), 1681–1694 (2018)
34. Yu, F., Koltun, V.: Multi-scale context aggregation by dilated convolutions. arXiv preprint arXiv:1511.07122 (2015)
35. Yu, J.: A light-field journey to virtual reality. TMM **24**(2), 104–112 (2017)
36. Zhang, S., Lin, Y., Sheng, H.: Residual networks for light field image super-resolution. In: CVPR, pp. 11046–11055 (2019)

37. Zhang, S., Sheng, H., Li, C., Zhang, J., Xiong, Z.: Robust depth estimation for light field via spinning parallelogram operator. CVIU **145**, 148–159 (2016)
38. Zhang, Y., et al.: Light-field depth estimation via epipolar plane image analysis and locally linear embedding. TCSVT **27**(4), 739–747 (2016)
39. Zhang, Y., Dai, W., Xu, M., Zou, J., Zhang, X., Xiong, H.: Depth estimation from light field using graph-based structure-aware analysis. TCSVT **30**(11), 4269–4283 (2019)
40. Zhu, H., Wang, Q., Yu, J.: Occlusion-model guided antiocclusion depth estimation in light field. J-STSP **11**(7), 965–978 (2017)

An Efficient Way for Active None-Line-of-Sight: End-to-End Learned Compressed NLOS Imaging

Chen Chang, Tao Yue, Siqi Ni, and Xuemei Hu$^{(\boxtimes)}$

School of Electronic Science and Engineering, Nanjing University, Nanjing, China
xuemeihu@nju.edu.cn

Abstract. Non-line-of-sight imaging (NLOS) is an emerging detection technique that uses multiple reflections of a transmitted beam, capturing scenes beyond the user's field of view. Due to its high reconstruction quality, active transient NLOS imaging has been widely investigated. However, much of the existing work has focused on optimizing reconstruction algorithms but neglected the time cost during data acquisition. Conventional imaging systems use mechanical point-by-point scanning, which requires high time cost and not utilizing the sparsity of the NLOS objects. In this paper, we propose to realize NLOS in an efficient way, based upon the theory of compressive sensing (CS). To reduce data volume and acquisition time, we introduce the end-to-end CS imaging to learn an optimal CS measurement matrix for efficient NLOS imaging. Through quantitative and qualitative experimental comparison with SOTA methods, we demonstrate an improvement of at least 1.4 dB higher PSNR for the reconstructed depth map compared to using partial Hadamard sensing matrices. This work will effectively advance the real-time and practicality of NLOS.

Keywords: Non-line-of-sight · Compressed sensing · Single pixel imaging

1 Introduction

The ability to image scenes hidden beyond the visible field of view (FOV) will significantly expand the imaging limits of existing technology, with important research implications and widespread use in machine vision, military, remote sensing, medical imaging and autonomous vehicle applications. Unlike traditional imaging techniques, NLOS imaging computationally reconstructs information of scenes outside the FOV by capturing indirect signals transmitted from the scene. In the last decade, NLOS imaging has attracted much attention, and imaging systems based on different detection principles have been proposed, including transient imaging [1–5], speckle correlations [6], thermal imaging [7], acoustic imaging [8], wave-front shaping [9, 10], and occlusion-based imaging [11, 12], with increasing image quality and reconstruction speed. Among these methods, transient imaging based on active laser scanning has become the main method to study NLOS due to its complete system construction and high spatial resolution. However, there are still many problems to be solved, such as long acquisition time,

© The Author(s), under exclusive license to Springer Nature Singapore Pte Ltd. 2024
Q. Liu et al. (Eds.): PRCV 2023, LNCS 14430, pp. 28–40, 2024.
https://doi.org/10.1007/978-981-99-8537-1_3

low SNR, limited spatial resolution, the need of accurate priors, etc. In particular, long acquisition time has become a bottleneck restricting the practical application.

To solve the problem with reasonable reconstruction quality, some research works propose to reduce the number of illumination points directly, such as random down-sampling [19] and circular scanning instead of grid scanning [20]. Besides, SPAD arrays [21, 22] are also proposed for NLOS detection, eliminating the requirement of scanning, while with a high cost and limited spatial resolution. Based upon the theory of CS, single-pixel imaging is introduced for NLOS imaging with random or Hadamard masks, e.g., single-pixel detection [18] and spatial multiplexing detection (SMD) [13]. However, these commonly utilized CS mask do not necessarily fully utilize the sparsity characteristics of transient data, which is worth exploring for efficient NLOS sampling.

In our work, we consider the advantages of single-pixel imaging in reducing acquisition time, and explore the possibility of incorporating CS into acquisition. Meanwhile, based upon the compressed measurement data, we propose a neural network called CSUNET for efficient NLOS reconstruction. The contributions of this paper can be summarized as below:

First, we construct an end-to-end neural network with a learnable module to jointly learn the optimal NLOS sensing masks and optimize reconstruction.

Second, to promote the convergence of the end-to-end NLOS imaging network, we propose a segment training strategy, and a content-aware MSE loss function for elegant convergence of the decompression module.

Third, we demonstrate the proposed NLOS CS imaging method with extensive experiments, specifically, the proposed method could realize a PSNR gain of at least 1.4 dB compared to existing methods. Besides, we demonstrate the effectiveness of the proposed method under big noise and extreme low compression ratios.

2 Sensing Methods

2.1 Forward Model for Active NLOS Imaging

Existing active transient imaging systems typically treat illumination point and detection point as the same point, called confocal mode [1]. For simplicity, we define the location of the wall as $z = 0$ plane, the system and the hidden object locates in the 3D half-space Ω where $z > 0$. The coordinates of hidden point m on the hidden object surface are defined as (x, y, z), illumination point l on the wall as $(x_l, y_l, 0)$, and the photon transient histogram corresponding to the detection point p can be expressed as:

$$\tau(x_l, y_l, t) = \int \int \int_\Omega \frac{1}{r_l^4} \rho(x,y,z) \delta(2r_l - tc) dx dy dz \tag{1}$$

where r_l is the distance between l/p and m, c is the speed of light and $\rho(x,y,z)$ is the albedo of m. The Dirac function δ relates time to space. To reconstruct the scene with a spatial resolution of $w \times w$, $w \times w$ points have to be scanned and detected, and the 1D photon histograms $\tau(x_l, y_l, t)$ obtained from each scanning point are assembled as the 3D transient data τ of size $w \times w \times t_{res}$, where t_{res} denotes the time bin of TCSPC.

2.2 Forward Model Combined with Compressed Sensing

As shown in Fig. 1(a), the transient data is sparse both spatially and temporally, making it possible to reduce mask numbers for compression. We assume that after the laser illuminates a point on the wall, DMD is used to collect photons returned from all points within the detection area. Let the set of illumination points be **L** and the detection area be **D**. For illumination point $(x_{l,i}, y_{l,j}, 0) \in \mathbf{L}$ and measurement point (x, y, z), transient histogram of the whole detection area can be expressed as:

$$\tau(x_l, y_l, t) = \sum_{(x_p, y_p, 0) \in D} \int \int \int_{\Omega} \frac{1}{r_l^2 r_p^2} \rho(x, y, z) \delta(r_l + r_p - ct) dx dy dz \tag{2}$$

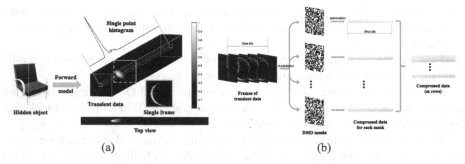

(a)   (b)

Fig. 1. Transient data and the process of compression.

Let **M** be the set of masks that DMD displays, containing m masks with a spatial resolution of $w \times w$. DMD scrolls through these masks to obtain m 1D histograms, assembled in width to form modulated photon transient data:

$$\tau'(x_{l,i}, y_{l,j}, t) = M \cdot \tau(x_{l,i}, y_{l,j}, t) \tag{3}$$

where $\cdot$ represents dot-multiplication and summation at each transient frame. The size of the modulated data is $m \times t_{res}$. When $m = w \times w$, it is theoretically possible to solve $\tau(x_{l,i}, y_{l,j}, t)$ directly from $\tau'(x_{l,i}, y_{l,j}, t)$. All scanned points $(x_{l,i}, y_{l,j}, 0) \in \mathbf{L}$ corresponds to a matrix $\tau'(x_{l,i}, y_{l,j}, t)$.

To reduce acquisition time and the number of scanned points, while maintaining a uniform distribution of returned photon signals, illumination point l is located in the center of the detection area. We define the data compression ratio (CR) as:

$$CR = \frac{m \times t_{res}}{w \times w \times t_{res}} = \frac{m}{w \times w} \times 100\% < 1 \tag{4}$$

Improvements to the original forward model have changed acquisition method from $w \times w$ mechanical scans to the current m electronic displays, significantly reducing the acquisition time.

3 Network Architecture

To reconstruct hidden objects from compressed data, we build an end-to-end neural network (CSUNET) consisting of a decompressor, an encoder and a decoder, where the decompressor serves to recover the compressed data to full size, the encoder serves to extract space-time features from the recovered transient data, and the decoder reconstruct the final depth map based on these extracted features. To optimize the CS measurement for NLOS imaging, a learnable CS module is introduced in the network, which automatically learns the optimal weights of CS masks for NLOS tasks based on the distribution of the training dataset.

3.1 Decompressor

In model (2), similar to compressed sensing for 2D images, we consider each frame of transient data to be dot-multiplied by masks of DMD. We propose a 3D CSNET for transient data with parallel processing of all time-frames. Compressed data is firstly recovered to full volume using a linear fully connected layer, and then the full-volume transient data is regained using multiple CS blocks, with each block containing three 3D convolutional layers. All convolutional layers are followed by a ReLU layer due to the non-negativity of transient data and padded zeros for different kernel sizes, ensuring that the length and width of all intermediate variables are constant (Fig. 2).

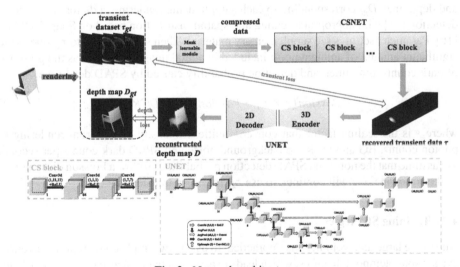

Fig. 2. Network architecture.

3.2 Encoder and Decoder

UNET has achieved good results in segmenting images using its inverted pyramidal network structure. The deeper the network layer, the larger FOV of the extracted feature map, with more attention to essential characteristics. Moreover, skip connections ensure that effective information will not be excessively lost during feature extracting level by level, increasing the robustness of gradient propagation. For NLOS imaging, which requires extraction of spatiotemporal information from 3D transient data to 2D depth maps, we use a 3D-2D coupled UNET based on [14]. On the encoder side, UNET removes temporal information by compressing 3D transient data into 2D feature vectors, and on the decoder side, the spatial information is extracted from these vectors.

3.3 Learnable Module for Weights of Masks

As for the learnable CS masks M, we use an unbiased 3D convolutional layer with a kernel size matching the spatial resolution of transient data, simulating the dot-multiplication and summation process in photon modulation. Complete transient data will be compressed by this layer and later processed by the decompressor. When fixed, unlearnable matrices are used as compression masks for comparison, we make them the initializing weight for this layer and set parameters not to update.

4 Evaluation and Results

4.1 Dataset

To test the proposed sampling model with CSUNET, we obtain the transient data τ_{gt} and depth map D_{gt} corresponding to each sample using rendering code for single-point detection [15] with appropriate scaling and spatial distribution based on ShapeNetCore [16]. Parameter settings see supplementary material. Then we add noise to τ_{gt} due to the significant amount of ambient noise included in actual imaging, as well as the presence of dark counts, time jitter, and detection probability caused by SPAD dead time.

$$\tau_{gt_{noise}} = Possion(s \times \tau_{gt} + Possion((BG+DCR) \times bin_{res})) \tag{5}$$

where s is the scaling factor that converts unitless intensity values from rendering to photon counts, BG and DCR are background noise and SPAD dark counts per second. We assume that the noise and SPAD detection probabilities follow a Poisson distribution. In the following, we will continue to use τ_{gt} for noisy transient data.

4.2 Training Strategies

Due to the large data volume, it is impractical to use a few simple convolutional layers to extract spatiotemporal features, which leads to a large network size and makes it difficult to converge if trained directly from scratch. Thus, CSNET and UNET are trained separately first. After basic convergence, the learned parameters are used as initial weights for the overall network, keeping a lower learning rate to continue training. Normalization related to the distribution of transient data needs to be done when loading the dataset to facilitate convergence.

During segment-training, for CSNET containing a compression layer, inputs and labels are both τ_{gt}. The MSE of the output τ and τ_{gt} is calculated as evaluation during testing, while the vast majority of element weights are close to 0 after normalization due to the sparsity of transient data, unreasonable to directly calculate MSE for the whole data because of small loss value with its optimization difficulty. Furthermore, only those values that are not 0 matter, so it is necessary to select a suitable threshold T and equilibrium parameter β to enhance the constraint on those non-zero values. Elements with weights less than T are noted as $\tau_{(0)}$ and those greater than T are noted as $\tau_{(1)}$. The loss function with scale β trading off the two losses can be expressed as:

$$L_{transient} = \frac{1}{N}\left(\sum_1^N \left\|\tau_{gt(0)} - \tau_{(0)}\right\|^2 + \beta\sum_1^N \left\|\tau_{gt(1)} - \tau_{(1)}\right\|^2\right) \tag{6}$$

For UNET, inputs and labels are τ_{gt} and D_{gt}. The peak signal-to-noise ratio (PSNR), structural similarity (SSIM), and mean square error (MSE) of the output D and D_{gt} of UNET are calculated as evaluation during testing. The loss function is defined as:

$$L_{depth} = \frac{1}{N}\sum_1^N ||D_{gt} - D||^2 \tag{7}$$

After segment-training, the output of CSNET is fed to UNET then the two networks are trained jointly with a loss function defined as:

$$L = L_{transient} + \lambda \cdot L_{depth} \tag{8}$$

where λ trades off transient recovery and depth reconstruction. In the selection of λ, we should choose a range, generally ranging from 0.1 to 10, to ensure that loss values of CSNET and UNET after segment-training are in the same order of magnitude, then sample from the range and compare effects.

4.3 Reconstruction of the Proposed End-to-End Network

To evaluate our CS mask-learnable end-to-end network (L-CSUNET) for reconstructing hidden scenes using compressed data, we first test it on the generated dataset by randomly selecting different types of test samples and observing the reconstructed depth maps with their PSNR, SSIM and MSE metrics against ground truth (GT) at different CRs, as shown in Fig. 3. To further validate the effectiveness of our network, we also report the mean values of reconstructed metrics for all samples in the test dataset in Table 1. There is no noticeable reduction in reconstruction quality when CR is reduced from 20% to 5%, and reconstruction is still possible at only 1%, theoretically reducing sampling time significantly.

5 Ablation and Discussion

In the following, we further discuss the proposed network. In Sect. 5.1, we perform ablation experiments to give results of depth maps reconstructed by conventional algorithms and the end-to-end network when using partial Hadamard matrices as compression masks (H-CSUNET) at different CRs, then compare them with L-CSUNET to argue for the advantages of the latter. Section 5.2 discusses the effect of different noise levels on reconstruction quality, examining the robustness of our network at low SNRs. Section 5.3 discusses the possibility of recovering hidden scenes at very low CRs.

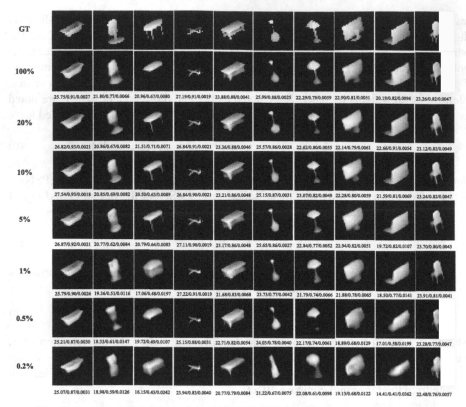

Fig. 3. Reconstructed depth maps with PSNR/SSIM/MSE metrics at different CRs.

5.1 Ablation Study

Existing reconstruction algorithms are commonly based on transient data obtained in confocal mode, whereas in our forward model, the illumination point l is always in the center of relay wall and the equivalent detection points p corresponding to DMD pixels do not coincide with l. Therefore, algorithms based on confocal mode cannot be used directly. In work [4], for the case where l does not coincide with p, the authors propose Filtered Back Projection (FBP) algorithm to solve 3D albedo of the hidden scene, which is also used in [13]. In work [5], the authors borrow the idea of Normal Moveout Correction (NMO) in seismology to approximate non-confocal data captured with a single illumination point as confocal data, then process it using f-k algorithm proposed in their paper. We index the 3D albedo obtained by the above method in depth to obtain a depth map, which is compared with H-CSUNET and L-CSUNET. With conventional algorithms, we use TVAL3 [17] to complete the process of recovering full-volume transient data from the compressed.

As shown in Fig. 4(a), the depth map reconstruction of our proposed network is significantly better than the other methods at different CRs. Conventional methods have a small reconstruction depth range, which can only locate the foremost position of the side of the hidden object towards to relay wall roughly, causing deeper information

Table 1. Mean values of reconstructed metrics for all samples in the test dataset.

CR	PSNR↑	SSIM↑	MSE↓ (×0.01)
No compression	22.43	0.7739	0.8902
20%	22.20	0.7640	0.9351
10%	22.10	0.7632	0.9524
5%	21.97	0.7567	0.9651
1%	21.49	0.7250	1.0718
0.5%	20.89	0.6965	1.2018
0.2%	20.25	0.6651	1.3749

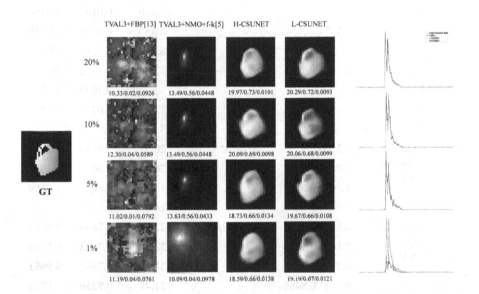

Fig. 4. (a) Depth map reconstruction based on different methods; (b) Histograms of a randomly selected point of the recovered transient data.

almost lost completely. Moreover, the methods are sensitive to noise, the resulting depth maps contain a lot of scatter noise and object contours are blurred.

On the other hand, H-CSUNET eliminates the effect of scatter noise and can recover deeper information to obtain the full shape of the hidden object, while edges and holes of the reconstructed object are not fine enough compared to L-CSUNET. The left column of Table 2 shows MSE of the recovered transient data after decompression by TVAL3 or CSNET module, performing the ability of the proposed learnable module retaining spatiotemporal information. The column on the right represents average metrics between reconstruction and depth map ground truth D_{gt}, performing the superiority of CSUNET in the reconstruction of hidden scenes. It can be seen that conventional methods have poor results just as expected, unable to reconstruct at a large depth range. On the contrary, CSUNETs perform well, and L-CSUNET results in a PSNR gain of 1.4 dB compared to H-CSUNET at least, with the gain continuing to increase as CR is reduced. In addition, MSE of the recovered transient data with and without threshold T shows that the use of T can improve decompression.

Table 2. Average metrics of four methods on testing set (from 20% to 1%), where **w/o** and **w** denote without and with respectively.

decompression		MSE of recovered transient data ($\times$ 0.01)↓	reconstruction	PSNR↑	SSIM↑	MSE↓ ($\times$0.01)
TVAL3 [17]		0.1843	FBP [13]	9.094	0.0177	12.960
				9.010	0.0183	13.166
		0.5755		8.856	0.0194	13.619
				8.272	0.0281	15.760
		1.3342	NMO + f-k [5]	12.85	0.1222	20.141
				12.77	0.1043	17.386
		5.1528		12.33	0.0728	17.848
				7.549	0.0388	35.356
Ours	Hadamard (loss **w/ w/o** T)	0.2500/0.2356	H-CSUNET	20.81	0.6999	1.2534
		0.2231/0.3787		20.40	0.6633	1.3932
		0.3694/0.5557		18.82	0.5651	1.9069
		1.4494/1.5336		14.07	0.2278	4.5532
	Learnable (loss **w/ w/o** T)	**0.1117/0.3607**	L-CSUNET	**22.20**	**0.7640**	**0.9351**
		0.1456/0.3610		**22.10**	**0.7632**	**0.9524**
		0.2315/0.4281		**21.97**	**0.7567**	**0.9651**
		0.6622/0.8468		**21.49**	**0.7250**	**1.0718**

Figure 4(b) shows histograms of a randomly selected point of the transient data recovered by TVAL3 or CSNET respectively, and the decompression effect of conventional methods in Table 2 is acceptable, indicating the poor results are caused by their reconstruction algorithms. However, L-CSUNET is satisfactory in both the intermediate decompression and the final reconstruction effect, confirming the advantages and rationality of the modules in our network.

5.2 Big Noise

A simplified noise model containing background noise and dark counts of SPAD is proposed in Sect. 4.1. In this section, we observe the robustness to noise of L-CSUNET by increasing background noise BG at different levels, considering that dark counts are related to the quality of SPAD itself, which is generally assumed to be 3000 counts/s and does not vary with the external environment. In all previous experiments, BG is set to 3×10^{11} counts/s, here we consider three noise levels with BG set to 6×10^{11}, 1.5×10^{12} and 3×10^{12} counts/s. Figure 5(a) shows reconstruction under each noise level, Fig. 5(b) shows the effect of different noise levels on a single-point transient histogram and gives the noise-free data for comparison, and Fig. 6(b–d) give the evolution of respective average metrics with CR. Since our network is trained on the dataset containing noise, metrics of no-noise data will fluctuate compared to other noisy data. We see that even under very strong noise, L-CSUNET still performs well in reconstruction. This phenomenon, combined with the different performance between single-point histograms recovered by CSNET from H-CSUNET and L-CSUNET and their corresponding reconstructions in Fig. 4, suggests that the network places more emphasis on peak locations than the amplitude of peaks since locations of peaks imply the presence of objects at corresponding depth in hidden scene.

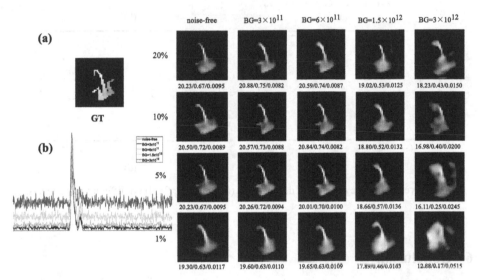

Fig. 5. (a) Reconstruction under different noise levels; (b) Effect of noise on transient data.

5.3 Extremely Low CR

In Table 1, we show that L-CSUNET is still able to perform reconstruction tasks even at a CR of 1%. To further reduce acquisition time and explore the possibility of reconstruction at an extremely low CR, we reduce CR to 0.5% and 0.2%, which means reconstructing hidden scenes will be achieved with only 5 and 2 masks at a spatial resolution of 32×32. Figure 3 and Table 1 show their reconstruction results and average metrics. The information needed for reconstruction can still be extracted by the network from transient data after extreme compression, which not only demonstrates the power of our network in extracting spatiotemporal features but also shows that the sparsity of transient data can provide ample room for maneuvers to reduce acquisition time in real experiments.

To observe the changing trend of reconstruction performance with CR, we use data from Table 1 to plot the average metric change graph of depth maps in Fig. 6(a), and find that the deterioration of reconstruction is accelerated starting from a CR of less than 5%, so for real acquisition, considering the possible errors and some uncertainty, CR should not be lower than 5% to ensure the validity of acquisition data.

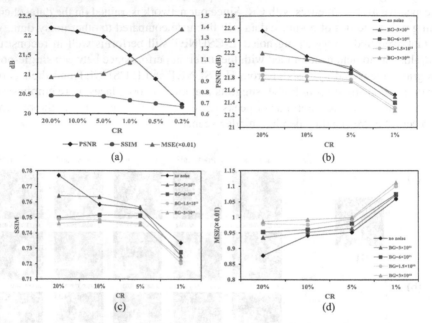

Fig. 6. Average metric changes of depth maps. (a) with CR, (b–d) on different noise levels.

6   Conclusion

In this paper, we propose an efficient NLOS imaging method through jointly optimize the CS measurement mask and the reconstruction network. Through both quantitative and qualitative ablations and experiments, we demonstrate the effectiveness of the proposed method, with at least 1.4 dB PSNR gain compared with the existing CS-based NLOS imaging methods. Experimental results under extreme low CRs and SNRs further demonstrate the potential of the proposed method for real-time NLOS.

As for the future work, based upon the end-to-end CS-based NLOS imaging framework, higher level of computer vision tasks such as target detection, classification and recognition could be realized with more efficient NLOS information acquisition and reconstruction. Besides, modeling NLOS scenes with more complex and practical geometry properties, such as BRDFs of the relay wall, occlusion in the NLOS scenes, etc., in the training dataset could enable higher generalization of the proposed NLOS imaging framework under more complex and challenging scenes.

7 Supplementary Material

For more details, you can access the following link: https://docs.foxitcloud.cn/cloudS
hare/?code=Bd7fFjQI.

Acknowledgments. This work was supported by National Key Research and Development Program of China (2022YFA-1207200), NSFC Projects 61971465, and Fundamental Research Funds for the Central Universities, China (Grant No. 0210–14380184).

References

1. O'Toole, M., Lindell, D., Wetzstein, G.: Confocal non-line-of-sight imaging based on the light-cone transform. Nature **555**, 338–341 (2018)
2. Kirmani, A., Hutchison, T., Davis, J., Raskar, R.: Looking around the corner using transient imaging. In: IEEE 12th International Conference on Computer Vision, pp. 159–166. IEEE (2012)
3. Liu, X., et al.: Non-line-of-sight imaging using phasor-field virtual wave optics. Nature **572**, 620–623 (2019)
4. Velten, A., Willwacher, T., Gupta, O., Veeraraghavan, A., Bawendi, M.G., Raskar, R.: Recovering three-dimensional shape around a corner using ultrafast time-of-flight imaging. Nat. Commun. **3**, 745 (2012)
5. Lindell, D.B., Wetzstein, G., O'Toole, M.: Wave-based non-line-of-sight imaging using fast FK migration. ACM Trans. Graph. (ToG) **38**, 1–13 (2019)
6. Bertolotti, J., Van Putten, E.G., Blum, C., Lagendijk, A., Vos, W.L., Mosk, A.P.: Non-invasive imaging through opaque scattering layers. Nature **491**, 232–234 (2012)
7. Maeda, T., Wang, Y., Raskar, R., Kadambi, A.: Thermal non-line-of-sight imaging. In: 2019 IEEE International Conference on Computational Photography (ICCP), pp. 1–11. IEEE (2019)
8. Lindell, D.B., Wetzstein, G., Koltun, V.: Acoustic non-line-of-sight imaging. In: Proceedings of the IEEE Conference on Computer Vision and Pattern Recognition, pp. 6780–6789 (2019)
9. Cao, R., de Goumoens, F., Blochet, B., Xu, J., Yang, C.: High-resolution non-line-of-sight imaging employing active focusing. Nat. Photonics **16**, 462 468 (2022)
10. Katz, O., Small, E., Silberberg, Y.: Looking around corners and through thin turbid layers in real time with scattered incoherent light. Nat. Photonics **6**, 549–553 (2012)
11. Saunders, C., Murray-Bruce, J., Goyal, V.K.: Computational periscopy with an ordinary digital camera. Nature **565**, 472–475 (2019)
12. Xu, F., et al.: Revealing hidden scenes by photon-efficient occlusion-based opportunistic active imaging. Opt. Express **26**, 9945–9962 (2018)
13. Yang, W., Zhang, C., Jiang, W., Zhang, Z., Sun, B.: None-line-of-sight imaging enhanced with spatial multiplexing. Opt. Express **30**, 5855–5867 (2022)
14. Grau Chopite, J., Hullin, M.B., Wand, M., Iseringhausen, J.: Deep non-line-of-sight reconstruction. In: Proceedings of the IEEE/CVF Conference on Computer Vision and Pattern Recognition, pp. 960–969. (2020)
15. Chen, W., Wei, F., Kutulakos, K.N., Rusinkiewicz, S., Heide, F.: Learned feature embeddings for non-line-of-sight imaging and recognition. ACM Trans. Graph. (ToG) **39**, 1–18 (2020)
16. Chang, A.X., et al.: Shapenet: An information-rich 3d model repository. arXiv preprint arXiv: 1512.03012 (2015)

17. Li, C.: An efficient algorithm for total variation regularization with applications to the single pixel camera and compressive sensing. Rice University (2010)
18. Musarra, G., et al.: Non-line-of-sight three-dimensional imaging with a single-pixel camera. Phys. Rev. Appl. **12**, 011002 (2019)
19. Ye, J.-T., Huang, X., Li, Z.-P., Xu, F.: Compressed sensing for active non-line-of-sight imaging. Opt. Express **29**, 1749–1763 (2021)
20. Isogawa, M., Chan, D., Yuan, Y., Kitani, K., O'Toole, M.: Efficient non-line-of-sight imaging from transient sinograms. In: Vedaldi, A., Bischof, H., Brox, T., Frahm, J.M., (eds.) Computer Vision–ECCV 2020: 16th European Conference, Glasgow, UK, August 23–28, 2020, Proceedings, Part VII 16, pp. 193–208. Springer, (2020). https://doi.org/10.1007/978-3-030-58571-6_12
21. Renna, M., Nam, J.H., Buttafava, M., Villa, F., Velten, A., Tosi, A.: Fast-gated 16 × 1 SPAD array for non-line-of-sight imaging applications. Instruments **4**, 14 (2020)
22. Jin, C., Tang, M., Jia, L., Tian, X., Yang, J., Qiao, K.: Scannerless non-line-of-sight three dimensional imaging with a 32 × 32 SPAD array. arXiv preprint arXiv:2011.05122 (2020)

DFAR-Net: Dual-Input Three-Branch Attention Fusion Reconstruction Network for Polarized Non-Line-of-Sight Imaging

Hao Liu, Pengfei Wang, Xin He, Ke Wang, Shaohu Jin, Pengyun Chen, Xiaoheng Jiang, and Mingliang Xu[⊠]

Zhengzhou University, Zhenzhou 450000, China
iexumingliang@zzu.edu.cn

Abstract. Polarized non-line-of-sight (NLOS) imaging is a promising visual perception technique for enhancing the visibility of occluded objects hidden behind walls. The main challenge of this task is that conventional single-angle relay wall projection polarization images provide limited effective information due to optical ill-posedness, resulting in poor imaging results. To address this problem, we designed a dual-input three-branch attention fusion reconstruction network, namely DFAR-Net, which utilizes our proposed channel-by-channel bilateral weighted fusion sub-network to fuse NLOS infrared polarization intensity and polarization degree image information and reconstruct hidden scenes. In the feature extraction part, we introduce a split kernel channel attention mechanism to emphasize or suppress features, aiming to improve the model's generalization ability and robustness. Additionally, to enhance the reconstruction quality, we employ a combination of multi-scale loss functions to optimize the model's expressive power. Experimental results on our self-collected full-optical polarization NLOS dataset, PI-ND, demonstrate the superior performance of DFAR-Net and its modules over current passive NLOS imaging methods.

Keywords: NLOS imaging · Infrared polarization · Deep learning

1 Introduction

NLOS imaging is a technique that perceives objects hidden behind obstacles by deducing the scattered photon information on relay walls [15]. Due to its capability to enhance visual perception and overcome observational limits, NLOS imaging has gained significant attention in the past decade in fields such as autonomous driving perception, public safety and disaster relief [7].

NLOS imaging can be categorized into active [5,8,21] and passive imaging [17,24] methods based on whether encoded light sources are used. In active

This work is supported by the National Natural Science Foundation of China under Grant No. 62272421, and in part supported by the NO. U21B2037 and No. 62172371.

NLOS imaging, a controllable laser is used to illuminate the surface of relay walls, and the transient response of light's multiple reflections is captured by ultra-fast single-photon detectors. Active imaging can effectively reconstruct hidden 3D scenes by utilizing information such as photon flight time and intensity. However, active techniques often involve expensive equipment, and the scanning of the light field and computation time for reconstruction algorithms can take several tens of seconds [22]. Passive NLOS imaging captures light field information from the environment or the object's own radiation using ordinary sensors and reconstructs the target scene through inverse problem optimization [16]. Due to its low cost, portability, and non-invasiveness, passive NLOS imaging is a promising visual perception technology for improving the visibility of objects hidden behind walls. In this paper, we focus on the passive imaging method, utilizing a conventional infrared camera to achieve NLOS scene reconstruction.

Due to the low signal-to-noise ratio and the lack of coupling between pixels, traditional passive NLOS imaging faces the challenge of ill-posed inverse problems. The large condition number of light transport matrices in this field makes it difficult to achieve high-quality reconstructions with limited information [2]. Previous research has proposed various methods to improve reconstruction quality, including optimizing scene prior knowledge [16,20], reducing the condition number using polarization [17], simplifying the bidirectional reflectance distribution function (BRDF) using long-wave infrared [13], and utilizing deep learning methods [1,9,24]. Polarization imaging is an effective solution that enhances image quality by encoding information about object roughness, direction, and reflection [2]. This physics-inspired approach helps mitigate the ill-posedness in passive imaging. However, existing polarization imaging methods only utilize single-angle visible light polarization information [17], which leads to high ill-conditioning and poor reconstruction quality. Deep learning methods have shown excellent performance in improving reconstruction resolution, imaging distance, and reducing reconstruction time, attracting significant attention from researchers [9,20]. However, the field of polarization-based passive NLOS imaging lacks large-scale datasets, limiting the potential advantages of deep learning. To address these challenges, the main contributions of this study are summarized as follows:

(1) We proposes a dual-input attention fusion reconstruction network architecture called DFAR-Net, which is capable of reconstructing human motion scenes in NLOS scenarios.
(2) We introduce a channel-by-channel bilateral weighted fusion sub-network called CBCF Sub-Net, which effectively combines the polarized intensity and degree features by applying the split kernel attention mechanism (SKAM) for feature weighting. To enhance reconstruction quality, a combination of multi-scale gradient and content loss functions is defined.
(3) We conducted comparative and ablation experiments using a self-constructed full-optical NLOS polarimetric dataset (PI-ND). The results validate the effectiveness of our method.

2   Related Work

2.1   Passive NLOS Imaging

Due to the lack of programmable illumination sources in passive NLOS imaging, the projected scene is easily affected by external environmental interference, leading to a sharp decrease in the signal-to-noise ratio of the reconstructed image. Passive NLOS imaging captures only a single reflected light signal, lacking time and phase information, so most existing works only perform 2D reconstruction or localization [16,24]. Many attempts have been made to address the high complexity and ill-posedness of light propagation in passive NLOS imaging, including the use of scene prior knowledge, where methods using occluders [16] and polarization [17] reduce the condition number of the light transmission matrix by creating high-frequency shadows to enhance reconstruction. Physically-inspired models include coherence [3,4], which accomplishes sensing and localization of hidden targets by introducing spatial and temporal coherence information missing in passive NLOS imaging. For long-wave infrared (LWIR) with stronger reflective characteristics, Kaga et al. [11] use the object's own thermal radiation to recover its temperature and position from the angular distribution of specular and diffuse reflections. Maeda et al. [13] demonstrate real-time 2D and 3D localization of hidden objects by using the hidden object's heat source as the light source and redesigning the LWIR optical transmission framework. In recent years, deep learning-based methods have faster reconstruction times and better performance, including Wang et al. [20] who use ResNet to achieve rapid recognition and classification of hidden handwritten digits and letters. Geng et al. [9]use the OT network to transform the reconstruction task into low-dimensional manifold mapping by optimal transportation, thereby reconstructing hidden scenes. However, physics-based methods can only accomplish simple scene and short-range reconstruction, while the reconstruction performance of deep learning methods depends more on the dataset and network architecture, but currently there is a lack of relevant NLOS imaging datasets.

2.2   Polarimetric NLOS Imaging

The polarimetric NLOS imaging technique can reduce the condition number of the light transport matrix by filtering out scattered and background interference light, thus enhancing the effectiveness of NLOS imaging. In recent years, many attempts have been made to use polarization as a means of NLOS imaging, including Hassan et al. [10] research on the sparsity of the light transport matrix under polarization state to simplify the solution of inverse problems, achieving the reconstruction of hidden scenes without modifying the scene. Tanaka et al. [17] introduced the effective angle theory of polarization into the field of computer vision, created high-frequency half-shadows by adjusting the effective angle of polarization to capture the linear components of polarization, and thus improved the condition number of the light transport matrix to enhance NLOS imaging.

3 Method

3.1 Network Architecture

To effectively utilize the information of the polarization full dimension and enhance the information richness, we propose DFAR-Net, a dual-input attention bridging fusion reconstruction network, whose architecture is shown in Fig. 1. Under the condition of polarization passive NLOS imaging, the network takes as inputs the NLOS polarization intensity (S0) image and degree of polarization (DoLP) image in the LWIR band. Inspired by the U-shaped encoder-decoder structure of U-Net [6], DFAR-Net aims to maintain a similar structure. The network has a downsampling encoding branch that can fully extract features, an upsampling decoding branch that can restore image clarity, and a channel-by-channel bilateral weighted fusion subnet (CBCF Sub-Net) branch that can fully fuse S0 and DoLP features.

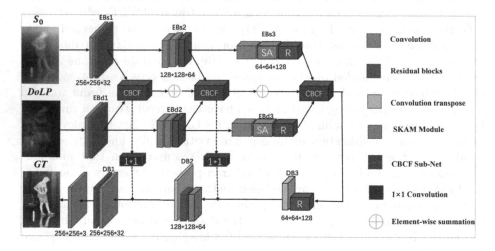

Fig. 1. The architecture of the DFAR-Net.

3.2 Encoder/Decoder

During the encoding stage, our encoder adopts a three-stage structure, using different branches to perform the same processing on two polarized NLOS images. Taking the polarization intensity branch as an example in the feature extraction part, we call the output of the k-th stage of the encoder EB_{sk}^{out}. First, we use a 3×3 convolution layer and a Residual block [6] consisting of 8 modified Resblocks to extract shallow features and obtain the output of the first stage of the encoder EB_{s1}^{out}. Then, in the second and third stages, we still use a 3×3 convolution layer to extract shallow features and change the size and number of channels of the previous stage's output EB_{sk-1}^{out}. We then use our improved SKAM module to emphasize or suppress the output features and highlight the target region of interest. We will detail the implementation of the SKAM module in Sect. 3.3. Finally, we deepen the feature extraction through a Residual block, obtaining EB_{s2}^{out} and EB_{s3}^{out}. It is worth noting that in each encoding stage, our output features have half the size and double the number of channels. In the decoder section, we decode and reconstruct fused features from the upper layer input. To prevent feature degradation and gradient disappearance during training, we apply feature weighting using a 1×1 convolution layer on the outputs generated in each fusion stage.

3.3   Split Kernel Channel Attention Mechanism (SKAM)

In this section, we will provide a detailed introduction to the construction process of the SKAM module, which is inspired by the group attention mechanism [23]. The structure of SKAM is shown in Fig. 2. First, the input feature $X \in \mathbb{R}^{C \times H \times W}$ is divided into g groups along the channel direction, where C is the number of feature channels, and H and W are the height and width, respectively. Thus, X = $[X_1, \dots, X_g]$, where $X_i \in \mathbb{R}^{C/g \times H \times W}$, and each group in X gradually captures specific feature responses during the model training process. Next, unlike the work in the group attention mechanism [23], we use 3×3 and 5×5 convolutional layers, batch normalization, and ReLU functions to process the two branches within each group separately. The group features with different receptive fields are adjusted by the channel attention mechanism, and the output results are aggregated using the concat operation to obtain the adjusted features with the same size and number of channels.

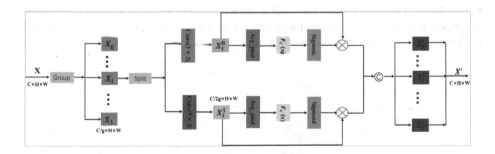

Fig. 2. The architecture of the split kernel channel attention mechanism.

In the channel attention module, we use the module mentioned in the grouped attention mechanism [23] to process it. We use global average pooling to provide global average information and generate channel-wise statistics $M \in \mathbb{R}^{C/2g \times 1 \times 1}$. The calculation method of M is shown in Eq. 1.

$$M = \mathcal{F}_{gp}\left(X_i^0\right) = \frac{1}{H \times W} \sum_{m=1}^{H} \sum_{n=1}^{W} X_i^0\left(m, n\right) \tag{1}$$

where $\mathcal{F}_{gp}(.)$ is the global average pooling function, X_i^0 is one of the channel features of the i-th group's feature split, and H and W are the height and width, respectively. The final output of the channel attention mechanism can be defined using Eq. 2.

$$\widehat{X_i^0} = \sigma\left(\Gamma_c\left(M\right)\right) \; . \; X_i^0 = \sigma\left(wM + b\right) \; . \; X_i^0 \tag{2}$$

Among them, $\widehat{X_i^0}$ represents the feature adjusted by the channel attention mechanism, $\sigma(.)$ is the sigmoid activation function, and $\Gamma_c(.)$ is the mapping function trained by the model in the form of $\Gamma_c(.) = (wx+b)$, where b is the trained bias and w is the trained weight. By weighting the global features using the trained weights and biases, then activating them with the sigmoid function, and finally adjusting the original features using the weighted global features.

We performed the same operation on another group, obtaining channel attention weights under different receptive field conditions, effectively highlighting the detailed and overall structural features in the target area and improving the reconstruction effect and image structural integrity. The effectiveness of SKAM was validated through ablation experiments.

3.4   CBCF Sub-Net

To increase the richness of image information and fully integrate image features, we propose a channel-by-channel bilateral weighted fusion sub-network, for fusing polarimetric information in the full optical dimension to improve reconstruction results. The structure of the CBCF Sub-Net is shown in Fig. 3, and we divide its processing into two phases, including a feature bilateral weighting phase and a channel-by-channel feature fusion phase.

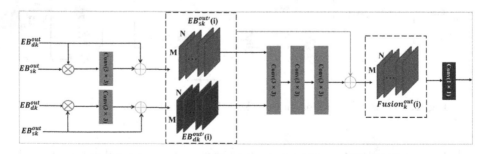

Fig. 3. The architecture of the channel-by-channel bilateral weighted fusion sub-net.

The first stage is the feature weighting stage, where we set the encoder output of the k-th stage in the S0 branch as EB_{sk}^{out}, and the encoder output of the k-th stage in the DoLP branch as EB_{dk}^{out}. We multiply EB_{sk}^{out} and EB_{dk}^{out} element-wise, and then pass them through a 3×3 convolutional layer with ReLU activation and batch normalization, followed by weighted fitting with the encoder output of the other branch. This ensures that the adjusted feature sets $EB_{sk}^{out\prime}$ and $EB_{dk}^{out\prime}$ have their unique features while also incorporating the feature weighting of the other branch, thus compensating for the lack of intensity information in DoLP images and the unclear contour information in S0 images, and improving information richness. The specific implementation method is shown in Eq. 3.

$$\begin{cases} EB_{sk}^{out\prime} = EB_{dk}^{out} + o\left(EB_{dk}^{out} \times EB_{sk}^{out}\right) \\ EB_{dk}^{out\prime} = EB_{sk}^{out} + f\left(EB_{sk}^{out} \times EB_{dk}^{out}\right) \end{cases} \tag{3}$$

where o(.) and f(.) are mapping functions in two separate branch training processes, $EB_{sk}^{out\prime}$ and $EB_{dk}^{out\prime}$ are the weighted feature representations of the S0 and DoLP images respectively, with k = [1, 2, 3].

In the channel-wise feature fusion stage, as shown in Fig. 3, for each channel obtained, $EB_{sk}^{out\prime}(i)$ and $EB_{dk}^{out\prime}(i)$ are fed into the fusion part of the CBCF Sub-Net to generate the desired fusion feature $Fusion_k^{out}(i)$, where i = [1, 2 ... , C], and C is the number of channels of the input feature. The fusion part is implemented by three 3×3 convolutional layers followed by ReLU activation. The output channels of the three convolutional layers are set to 2, 2, and 1, respectively, and a residual structure is used to obtain the final output. The fusion result and input have the same number of channels and size, so they can be merged, and all channels are concatenated in the end. The specific implementation is shown in Eq. 4.

$$Fusion_k^{out}(i) = EB_{sk}^{out\prime}(i) + EB_{dk}^{out\prime}(i) + Fuse\left(EB_{sk}^{out\prime}(i); EB_{dk}^{out\prime}(i)\right) \tag{4}$$

Among them, i represents the i-th channel, Fuse(.) represents the mapping function learned in the fusion stage. After two stages of processing, we obtain the final fusion result that contains more information, and then feed the fusion result into the encoder part for image clarity restoration. We verified the effectiveness of the CBCF Sub-Net through ablation experiments.

3.5 Loss Function

The first part of the loss function is set as the multi-scale pixel content loss. Since the output of our network is a list of outputs at different scales for each decoder stage, we first downsample our ground truth image to match the output size at each decoder stage. Then, we calculate the pixel intensity difference between the reconstructed image at different scales and the corresponding ground truth image using the L1 norm [14]. The pixel content loss, denoted as $MLoss_{pixel}$, is shown in Eq. 5.

$$MLoss_{pixel} = \sum_{i=1}^{3} \frac{1}{H \times W} \sum_{j=1}^{H \times W} \left| y_{label}^i - y_{pred}^i \right| \tag{5}$$

where y_{pred}^i and y_{lable}^i represent the output and corresponding ground truth image at scale i of the i-th decoder layer, respectively. H and W are the height and width of the image.

The second part of the loss function is the multi-scale image gradient loss calculated using the Sobel operator [18]. As polarization images have relatively prominent image gradient features, the designed multi-scale gradient loss function can make the reconstructed image have higher image gradients. In other words, we optimize the disparity constraint between the edge features of the encoder output list and the multi-scale real images, so that the image has a more visually effective structure. The $MLoss_grad$ is shown in Eq. 6.

$$MLoss_{grad} = \sum_{i=1}^{3} \frac{1}{H \times W} \sum_{j=1}^{H \times W} \left| \nabla y_{label}^i - \nabla y_{pred}^i \right| \tag{6}$$

Here, ∇ represents the gradient operation function obtained using the Sobel operator, and H and W are the height and width of the image, respectively.

Finally, a joint multi-scale loss function is defined to guide the network training, as shown in Eq. 7.

$$Mloss = \alpha MLoss_{pixel} + \beta MLoss_{grad} \tag{7}$$

where α and β are hyperparameters defined by ourselves to balance the multi-scale content loss and multi-scale gradient loss, and in our experiment, we set them to 0.8 and 0.2, respectively.

4　Experiment

4.1　Experimental Setup

We trained our model using a self-collected PI-ND full optical dimension polarization NLOS dataset. The dataset consists of 2000 pairs of NLOS polarization intensity images, NLOS polarization degree images, and ground truth images computed by the Stokes vector [19]. The training, testing, and validation sets contain 1500, 400, and 100 pairs of polarization images and ground truth images, respectively. For each iteration, we cropped the images to 256*256 to reduce the number of network training parameters and increase the training speed. To ensure the network fully converges, we set the training epochs to 2000 with an initial learning rate of 10^{-4}, decreasing by half every 200 epochs. Our experiments were conducted on an i7 Intel(R) Xeno W-2275 CPU and an NVIDIA RTX A4000.

4.2　Experimental Results

We compared our model with several existing deep learning-based NLOS imaging methods [9,12,24] to validate the superiority of our model. Then, since all current imaging methods are single-input methods, we changed the input of our network to produce two variants: (1) DFAR-Single, where both feature extraction branches have the same input, including S0 or DoLP images, and (2) DFAR-Double, where both feature extraction branches have different inputs. We used peak signal-to-noise ratio (PSNR) and structural similarity index (SSIM) to evaluate the experimental results and the runtime of each model code tested.

S0 Quantitative Results. In this section, we change the inputs of both branches of our model to S0 images, and the quantitative results are shown in Table 1. The quantitative data from the table shows that our method outperforms several existing imaging methods in both PSNR and SSIM. This is because our dual-input network architecture and fusion structure complement the deficiencies in NLOS image information. In terms of reconstruction time, although our method did not achieve the best reconstruction efficiency, the reconstruction time is only slightly worse than Phong [24] method.

Table 1. Quantitative results of several NLOS imaging methods with S0 as input and our method.

Model	PSNR	SSIM	RUNTIME
CA-GAN [12]	18.97	0.8220	0.1476
Phong [24]	12.24	0.6814	**0.0075**
NLOS-OT [9]	23.34	0.8748	0.0384
DFAR-Single	23.51	0.8934	0.0232
DFAR-Double	**24.66**	**0.9157**	0.0238

S0 Qualitative Results. We used S0 as the input and qualitatively evaluated the imaging performance of several reconstruction methods. The evaluation results are shown in Fig. 4. From a qualitative perspective, both DFAR-Single with the same input and DFAR-Double with different inputs achieved visually superior results. Our method has better reconstruction effects on the background and human body outline, and even achieves good reconstruction for the mobile phone placed on the chair, the letters in the model's hand, and the chair's support.

DoLP Quantitative Results. In this section, we change the input of the network to DoLP images for experimentation. According to the data in Table 2, our method still has higher PSNR and SSIM. However, the values of PSNR and SSIM have slightly decreased. This is because the individual DoLP images are insensitive to the intensity information of light and have a larger amount of noise, which makes the reconstruction more difficult due to the lack of information on pixel intensity in the image.

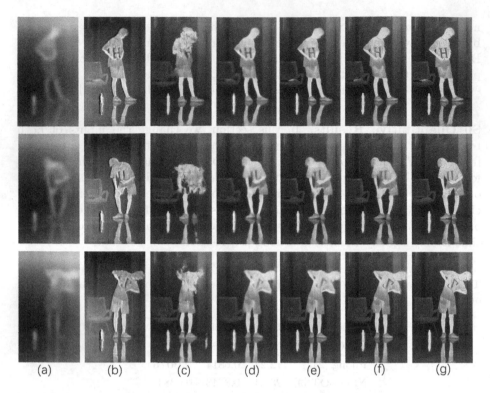

Fig. 4. Our method and several existing imaging methods qualitative results. (a) S0 NLOS images, (b) CA-GAN, (c) Phong, (d) NLOS-OT, (e) DFAR-Single, (f) DFAR-Double, (g) Ground truth images.

Table 2. Quantitative results of several NLOS imaging methods with DoLP as input and our method.

Model	PSNR	SSIM
CA-GAN [12]	18.13	0.8127
Phong [24]	10.34	0.5707
NLOS-OT [9]	21.85	0.8593
DFAR-Single	22.31	0.8736
DFAR-Double	**24.66**	**0.9157**

4.3 Ablation Study

We conducted ablation experiments to validate the effectiveness of each component of DFAR-Net on the PI-ND dataset. First, we trained an optimal model using the CBCF Sub-Net, SKAM, and our multi-scale composite loss function, achieving the highest performance in Table 3. Next, we tested DFAR-Net without CBCF, SKAM, and/or MLoss. In Table 3, we compare our model with a baseline

Table 3. Results of the ablation experiments.

CBCF	SKAM	MLoss	PSNR
			19.35
		√	20.57
√		√	22.64
	√	√	22.18
√	√		22.72
√	√	√	24.66

model that lacks the three components. The baseline model achieved an average PSNR of 19.35. However, our model outperformed the baseline with significant improvements. MLoss improved the average PSNR by 1.22 dB, CBCF increased it by 3.29 dB, and SKAM further enhanced it by 2.83 dB. When utilizing the first two components, the PSNR increased by 3.37 dB. Finally, when all components were employed, our model achieved the optimal PSNR of 24.66 dB. These results demonstrate the effectiveness and superiority of our proposed model structure and modules.

5 Conclusion

In this paper, we introduce a deep learning-based method for full optical dimension infrared polarization NLOS imaging. Our proposed DFAR-Net network improves image quality and reduces reconstruction time. We also propose a channel-by-channel bilateral weighted fusion subnet and a split kernel channel attention mechanism to increase information richness and improve feature information. We use a multi-scale loss function to improve the model's representation ability. Our method is validated through experiments on the PI-ND dataset.

References

1. Aittala, M., et al.: Computational mirrors: blind inverse light transport by deep matrix factorization. Adv. Neural Inf. Process. Syst. **32** (2019)
2. Baek, S.H., Heide, F.: Polarimetric spatio-temporal light transport probing. ACM Trans. Graph. (TOG) **40**(6), 1–18 (2021)
3. Beckus, A., Tamasan, A., Atia, G.K.: Multi-modal non-line-of-sight passive imaging. IEEE Trans. Image Process. **28**(7), 3372–3382 (2019)
4. Boger-Lombard, J., Katz, O.: Passive optical time-of-flight for non line-of-sight localization. Nat. Commun. **10**(1), 3343 (2019)
5. Chen, W., Wei, F., Kutulakos, K.N., Rusinkiewicz, S., Heide, F.: Learned feature embeddings for non-line-of-sight imaging and recognition. ACM Trans. Graph. (ToG) **39**(6), 1–18 (2020)
6. Cho, S.J., Ji, S.W., Hong, J.P., Jung, S.W., Ko, S.J.: Rethinking coarse-to-fine approach in single image deblurring. In: Proceedings of the IEEE/CVF International Conference on Computer Vision, pp. 4641–4650 (2021)

7. Faccio, D., Velten, A., Wetzstein, G.: Non-line-of-sight imaging. Nat. Rev. Phys. **2**(6), 318–327 (2020)
8. Feng, X., Gao, L.: Ultrafast light field tomography for snapshot transient and non-line-of-sight imaging. Nat. Commun. **12**(1), 2179 (2021)
9. Geng, R., et al.: Passive non-line-of-sight imaging using optimal transport. IEEE Trans. Image Process. **31**, 110–124 (2021)
10. Hassan, B.: Polarization-informed non-line-of-sight imaging on diffuse surfaces. University of California, Los Angeles (2019)
11. Kaga, M., Kushida, T., Takatani, T., Tanaka, K., Funatomi, T., Mukaigawa, Y.: Thermal non-line-of-sight imaging from specular and diffuse reflections. IPSJ Trans. Comput. Vision Appl. **11**(1), 1–6 (2019)
12. Kupyn, O., Budzan, V., Mykhailych, M., Mishkin, D., Matas, J.: Deblurgan: blind motion deblurring using conditional adversarial networks. In: Proceedings of the IEEE Conference on Computer Vision and Pattern Recognition, pp. 8183–8192 (2018)
13. Maeda, T., Wang, Y., Raskar, R., Kadambi, A.: Thermal non-line-of-sight imaging. In: 2019 IEEE International Conference on Computational Photography (ICCP), pp. 1–11. IEEE (2019)
14. Nah, S., Hyun Kim, T., Mu Lee, K.: Deep multi-scale convolutional neural network for dynamic scene deblurring. In: Proceedings of the IEEE Conference on Computer Vision and Pattern Recognition, pp. 3883–3891 (2017)
15. Ramesh, R., Davis, J.: 5D time-light transport matrix: what can we reason about scene properties? Technical report (2008)
16. Saunders, C., Murray-Bruce, J., Goyal, V.K.: Computational periscopy with an ordinary digital camera. Nature **565**(7740), 472–475 (2019)
17. Tanaka, K., Mukaigawa, Y., Kadambi, A.: Polarized non-line-of-sight imaging. In: Proceedings of the IEEE/CVF Conference on Computer Vision and Pattern Recognition, pp. 2136–2145 (2020)
18. Tang, L., Yuan, J., Ma, J.: Image fusion in the loop of high-level vision tasks: a semantic-aware real-time infrared and visible image fusion network. Inf. Fusion **82**, 28–42 (2022)
19. Tyo, J.S., Goldstein, D.L., Chenault, D.B., Shaw, J.A.: Review of passive imaging polarimetry for remote sensing applications. Appl. Opt. **45**(22), 5453–5469 (2006)
20. Wang, Y., et al.: Accurate but fragile passive non-line-of-sight recognition. Commun. Phys. **4**(1), 88 (2021)
21. Wu, C., et al.: Non–line-of-sight imaging over 1.43 km. Proc. Natl. Acad. Sci. **118**(10), e2024468118 (2021)
22. Xin, S., Nousias, S., Kutulakos, K.N., Sankaranarayanan, A.C., Narasimhan, S.G., Gkioulekas, I.: A theory of fermat paths for non-line-of-sight shape reconstruction. In: Proceedings of the IEEE/CVF Conference on Computer Vision and Pattern Recognition, pp. 6800–6809 (2019)
23. Zhang, Q.L., Yang, Y.B.: Sa-net: shuffle attention for deep convolutional neural networks. In: ICASSP 2021-2021 IEEE International Conference on Acoustics, Speech and Signal Processing (ICASSP), pp. 2235–2239. IEEE (2021)
24. Zhou, C., Wang, C.Y., Liu, Z.: Non-line-of-sight imaging off a phong surface through deep learning. arXiv preprint arXiv:2005.00007 (2020)

EVCPP:Example-Driven Virtual Camera Pose Prediction for Cloud Performing Arts Scenes

Jucheng Qiu[1], Xiaoyu Wu[2(✉)], and Boshu Jia[2]

[1] State Key Laboratory of Media Convergence and Communication, Communication University of China, Beijing, China
[2] Communication University of China, Beijing, China
wuxiaoyu@cuc.edu.cn

Abstract. Studying intelligent virtual shooting in cloud performing arts scenes is of great significance to the sustainable development of the cloud performing arts industry. Difficulties in summarizing the language of shots and the need to consider the attributes of cameras and actor movements are key concerns in stage filming. We propose EVCPP: Example-driven Virtual Camera Pose Prediction for cloud performing arts that uses existing shooting videos to guide the shooting of virtual scenes. By using the camera behavior information from reference videos as external guidance weights, along with the camera intrinsic parameters and actor state information as covariates, we combine them with the historical pose of the camera in the virtual scene to predict the future camera pose. Meanwhile, we propose a combined loss function for our task. Our method has achieved promising results in virtual 3D cloud performing arts scenes.

Keywords: Camera Pose Prediction · Virtual Cloud Performing Arts · Deep learning

1 Introduction

Cloud performing arts, as a new kind of digital cultural format, rises with the development of modern technology. It is a transformation and upgrading of traditional performing arts and serves as a new artistic expression form that follows the laws of performing arts. With the arrival of the post-pandemic era and the iterative upgrade of cutting-edge technologies, there is an increasing demand from the traditional performing arts industry for cloud-based performances. Consequently, the key technologies of cloud performance have also faced higher requirements and expectations.

In traditional video production for performing arts shows, various factors such as scene design, performance arrangement, plot design, and camera positioning influence each shot. Capturing the ideal shots can be challenging due to limitations and uncertainties. Pre-rehearsal filming is often chosen, but it involves high costs and limited control over the outcome. Recently, virtual engines like Unity and Unreal have emerged, allowing realistic simulations of the real world. Leveraging AI in a virtual environment

Q. Liu et al. (Eds.): PRCV 2023, LNCS 14430, pp. 53–64, 2024.
https://doi.org/10.1007/978-981-99-8537-1_5

to generate optimal camera shot options before actual shooting could help directors quickly determine a better shooting plan, enhance efficiency, and reduce costs.

Proper filming techniques can enhance the emotional and content aspects of the performance for viewers. High-quality professional program production often requires expertise in the language of cinematography. Even for the same scene, different shooting techniques may be chosen by directors for different narratives. Designing a model that can summarize shooting techniques across different scenes remains a challenging task.

In this paper, we propose EVCPP, which focuses on camera pose prediction in cloud performing arts scenes, where different shooting styles are controlled by selecting reference videos. Specifically, we start by performing human pose estimation on a given reference video, from which we estimate camera and human features. These two features are then combined to obtain implicit camera behavior features. However, directly applying the behavior features only results in a simple replication of camera angles or actions, leading to poor performance in virtual environments [1]. To address this, we use the implicit behavior features as expert weights to guide the shooting process. We further incorporate features from the virtual scene's characters, the camera's intrinsic properties, and the camera's historical trajectories to predict the future camera pose.

The contributions of our work are:

1. A modified time series prediction network is employed to process the relevant input sequences separately, followed by camera pose prediction. The model considered both the intrinsic properties of the camera and the actions of the captured actors for subsequent camera motion prediction.
2. A combined loss function of Mean Squared Error (MSE) and Jensen-Shannon Divergence (JS divergence) is utilized to enable the model to simultaneously consider the shape and position of the predicted trajectories.

2 Related Work

Traditional approaches have mostly used machine learning methods and scripted techniques [2] extracting shot knowledge from videos, a shot library is constructed, and when using a shot, the relevant parameters are directly applied to the camera through a script. Wang et al. [3] extract keywords from user input scripts and select matching candidate clips from a material library. Finally, these shots are combined to complete video editing. Xiong et al. [4] propose a weakly supervised framework that automatically creates video sequences from a diverse set of shots using scripts as input. Chen et al. [5, 6] train a 3DOF camera pose predictor using a recurrent decision tree network to automatically shoot sports matches. It predicts the best shooting angle for the next camera based on the motion of objects and camera poses, but it simplifies the camera parameters and can only be applied to fixed indoor camera positions.

With the rise of deep learning, Huang et al. [7] use a seq2seq structure for automatic shooting of single-person outdoor sports videos. By combining temporal and spatial information, they predict the camera's optical flow in the next frame based on the current position and motion status, and then calculate the camera coordinates based on the optical flow and camera parameter matrix. However, this approach is more complex and has larger errors. In recent research on virtual cameras, Jiang et al. [8, 9] use toric space coordinates [10] instead of 6DOF coordinates to reduce errors caused by coordinate systems. They extract the shooting styles from real movie videos and apply them to drive the camera's shooting in virtual scenes. They also incorporate keyframe techniques to allow users to participate in camera control. But due to coordinate system limitations, it can only be applied to scenes with two actors, limiting its applicability.

In recent research, techniques such as reinforcement learning and neural radiation field have started to be applied in automatic camera shooting. RT2A [11] introduces a reward function to guide the algorithm in finding the best shooting strategies and imitates the director's decision-making process for camera selection in each scene. [12] proposes a reinforcement learning-based automatic filming program for unmanned aerial vehicles (drones) that can track moving actors in real-time and make real-time decisions based on shot design. These decisions are based on the experience obtained through reinforcement learning. Dang et al. [13] propose an end-to-end imitation learning framework for drone cinematography systems. They introduce a Path Analysis-based Reinforcement Learning (PABRL) algorithm to derive aesthetic shooting strategies related to human motion from motion information, image composition features, and camera motion matrices. [14] presents an integrated aerial photography system for automatically capturing cinematic shots of action scenes. It learns to predict the best viewpoint for the next camera based on imitating the observed subject's motion. Xi Wang et al. [1] innovatively utilizes Neural Radiance Fields (NeRF) to transfer the cinematography features of a reference in-the-wild video clip to new clips or virtual scenes, achieving highly successful results. However, a drawback of this approach is that the speed and generation time of the clips are still limited.

Inspired by Jiang's work [8, 9], we designed a camera pose prediction framework for cloud performing arts. Different from [8, 9], we address the limitation on the number of individuals in the input video, and we utilize a hybrid loss combining JS (Jensen-Shannon) divergence and MSE (Mean Squared Error) to simultaneously constrain the position and scale of the predicted camera poses. Additionally, we separately consider the camera's intrinsic properties and the actors' states in the virtual scene.

3 Methods

We propose our method, EVCPP, as shown in Fig. 1. The framework consists of four main components: human pose estimation, cinematographic feature extraction, camera behavior extraction, and camera pose prediction.

We only need to provide an effective reference video and perform human body pose estimation on the individuals within it, and the network will estimate the camera and human features based on the results of the pose estimation. Subsequently, the network analyzes the camera behavior, and these behavior features are used to guide and influence the camera's shooting pose in the virtual scene. Moreover, the features of the actors in the virtual stage scene serve as past and future dynamic covariates, while the camera's intrinsic properties (focal length) act as static covariates. Together, they are used to predict the camera's future motion pose. The network combines the guidance video and existing virtual scene information to make reasonable predictions about the camera's future motion. Once the camera's motion pose is predicted, it is applied and rendered in the virtual scene to obtain the final camera shooting video.

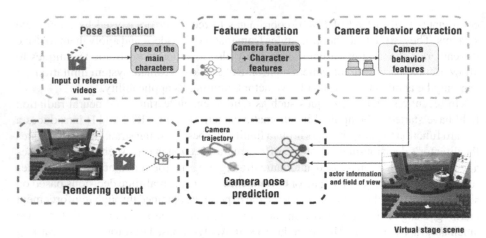

Virtual stage scene

Fig. 1. Overview of EVCPP pipeline

3.1 Human Pose Estimation Module

The human pose estimation module first estimates the keypoint information of the main actors based on the selected reference video. The reference video can be directly uploaded as specified, and we also refer to the approach in [2] to create a small video library with annotations of camera shooting styles. This allows users to conveniently specify the reference shooting style based on a few annotated textual items when it is difficult to find suitable reference videos. Once the video is successfully obtained, we use DEKR [15] to extract the keypoint information of the actors and determine the number of people in the video. Due to the specific nature of the camera coordinate system we use (see Sect. 3.2), in this module, we only expect to output the keypoint information of the two main actors. Therefore, we need to process and refine the keypoint detection data. For reference videos with multiple actors, we use the method in [16] to determine the importance scores of each person in each frame. The importance scores are primarily based on the size of the person's face, the distance between the person's face and the center of the frame, and the clarity of the person's face obtained using Canny edge detection (used to determine whether the camera is focused on that area). Ultimately, we output the keypoint

(a)   (b)

Fig. 2. (a) Results of human pose estimation (shown as blue dots) (b) Results of primary person detection (shown as green bounding boxes) (color figure online)

information of the top two individuals with the highest importance scores in each frame. For single-person videos, we calculate the mirror symmetry plane of the human body based on the keypoints of the individual. We then introduce a slight displacement along the direction perpendicular to this plane and add noise to all keypoints. This effectively creates a virtual "dummy person" and allows the framework to handle single-person data in the same manner. The results of person keypoints detection and primary person detection can be seen in Fig. 2.

3.2 Cinematic Feature Extraction Module

The purpose of the cinematic feature extraction module is to map the raw six degrees of freedom coordinate system representing actors and camera positions to a specific coordinate space, providing usable data for subsequent networks to learn the camera behavior style or rules inherent in the video. In the field of performing arts camera shooting, the story script of the filmed characters and their relationship status, as well as the position of the characters in the camera frame, should be the primary factors constraining the camera behavior. Taking these factors into account, the output of this module includes actor state features and camera state features.

For the extracted dual-person body keypoints, we utilize the feature estimation network from Jiang's work [8], which is a simple convolutional neural network, to map them into a low-dimensional camera feature space. This mapping successfully captures the relationships between the characters and between the characters and the camera, calibrating the actor and camera states in three-dimensional space and normalizing them in both actor and camera spaces.

In this camera feature space, the actor state feature represents the positional and orientational relationship between the two actors to capture the changes in their states during the cloud performance. The actor state features are modeled based on the distance and angles between the actors, capturing the positional and state relationship between the two actors. This modeling also implicitly reflects the storyline of the performance process. In the ith frame, the actor's state feature is represented by a 9-dimensional vector x_i^v:

$$x_i^v = \{d_{AB}, h_{AB}, S_A, S_B, h_A, h_B, M\} \in \mathbb{R}^9 \tag{1}$$

A and B represent the two actors in the performance scene, while AB is the line passing through the midpoints of their bodies. h_A and h_B respectively represent the orientations of the heads of the two actors and the angle between the head orientations and line AB, h_{AB} represents the angle between the head orientations. d_{AB} represents the distance between the two actors. s_A and s_B represent the normal vectors of the line connecting their shoulders and the angle between the line AB. s_{AB} represents the angle between the normal vectors of the shoulder lines. M is a two-dimensional vector specifying the main character in the current scene.

The camera features are based on the positions of the actors captured by the camera, represented in the form of a Toric hyperboloid coordinate system [10] to characterize the camera pose and state. It describes the relative relationship between the camera and two actors. In the ith frame, the camera state is represented by a 5-dimensional hyperboloid coordinate x_i^c:

$$x_c^i = \{x_A, x_B, mean(y_A, y_B), \theta, \varphi\} \in \mathbb{R}^5 \tag{2}$$

where (x_A, y_A), (x_B, y_B) represent the coordinates of the two actors' head positions in the frame, taking the average of the y-coordinate ensures that the camera remains level in the coordinate system. θ, φ represent the yaw angle and pitch angle of the camera in space, respectively.

The final output of this module is a 14-dimensional vector x_i:

$$x_i = Concat(x_i^v, x_i^c) \in \mathbb{R}^{14} \tag{3}$$

3.3 Camera Behavior Extraction Module

The camera motion extraction module takes a sequence of frames from the reference video as input, where each frame contains the extracted camera features and actor features. The reference video sequence is then fed into a standard encoder of Informer [17]. After the input embeddings and positional encodings are applied, the data is processed using a self-attention encoding layer with a distillation mechanism. The self-attention mechanism evaluates the correlation between multiple features in each frame of the input sequence, enhancing the smoothness of the predicted data. We take the output of the last time step in the Informer encoder as the feature representation of the entire input sequence. Then the output features from the encoder are further processed through a fully connected layer and a Softmax layer to obtain an expert vector $\mathbf{e}_i$ with a sum of 1. This expert vector consists of a total of m dimensions:

$$\mathbf{e}^i = \mathbf{F}(x_i^h) \in \mathbb{R}^{\mathbf{m}} \tag{4}$$

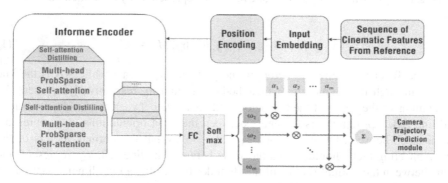

Fig. 3. The Camera Behavior Extraction Module

where x_i^h represents the feature of the input sequence from the reference videos, and $\mathbf{F}(\cdot)$ represents the network processing.

The expert vector successfully integrates the camera motion features into a compressed representation of the user-selected 2D video camera behavior. The compressed camera behavior vector is then fed into a Mixture of Experts (MOE) algorithm [18]. It attempts to classify the example video into one or several combinations of camera behaviors from the training dataset, assigning different weights to each combination. The weighted inputs are then passed to the subsequent camera pose prediction network for further linear blending. The specific module structure is depicted in Fig. 3.

3.4 Camera Pose Prediction Module

In the camera pose prediction module, we utilize and enhance Google's time-series prediction model TiDE (Time-series Dense Encoder) [19] to predict future camera poses, making it more suitable for our camera pose prediction task. Specifically, we introduce the camera behavior expert vectors extracted from the camera behavior extraction module as guided weights, which are incorporated into certain linear layers in residual blocks. This allows the network to consider the camera motion behavior of the reference videos during prediction. The modified residual block is depicted in the right half of Fig. 4. We also remove the feature projection module as our actor features are only of dimension 9, eliminating the need for additional dimensionality reduction. The specific network architecture is illustrated in the left half of Fig. 4.

For our task, TiDE considers both the corresponding static covariates and the future dynamic covariates, both of which are potential predictors that may influence the dependent variable (camera motion pose). Static covariates do not change over time in time series forecasting, while dynamic covariates vary as the time series progresses and can be further classified as known future and unknown future dynamic covariates. In the stage setting, the camera's shooting pose is closely related to the state of the actors being filmed and the camera's focal length. Therefore, we set the static covariate as a property of the camera itself. Since the focal length of our virtual camera remains unchanged in virtual scenes, we use the camera's field of view $a^{(i)}$ as a substitute for the focal length as the static covariate. In virtual scenes, the actions of the actors are pre-determined, and we know the state of the actors at any given time point, making the dynamic covariates correspond to the state features of the filmed actors $x_{1:L+H}^{(i)}$, from past L time steps to future H time steps. So, combining the camera pose $c_{1:L}^{(i)}$ from the past L time steps, the entire model input is $(c_{1:L}^{(i)}, x_{1:L+H}^{(i)}, a^{(i)})$, and the output of the model is $\hat{c}_{L+1:L+H}^{(i)}$.

Specifically, the input vector first passes through an encoder and decoder consisting of two layers of weighted residual blocks to obtain the output vector $g^{(i)}$:

$$g^{(i)} = \text{Decoder}(\text{Encoder}(c_{1:L}^{(i)}; x_{1:L+H}^{(i)}; a^{(i)})) \in \mathbb{R}^{p \cdot H} \tag{5}$$

where p is the output dim of the decoder. It is then reshaped into a matrix $D^{(i)} \in \mathbb{R}^{p \times H}$. $d_t^{(i)}$ is the tth column of the $D^{(i)}$. We concatenate the decoding vector at time t $d_t^{(i)}$ and dynamic covariate $x_{1:L+t}^{(i)}$, and then pass them through a Temporal Decoder to obtain the final prediction result $\hat{c}_{L+t}^{(i)}$.

In the stage setting, there are instances where a specific action by the actor prompts the camera to adopt a particular shooting style. Therefore, we combine the prediction

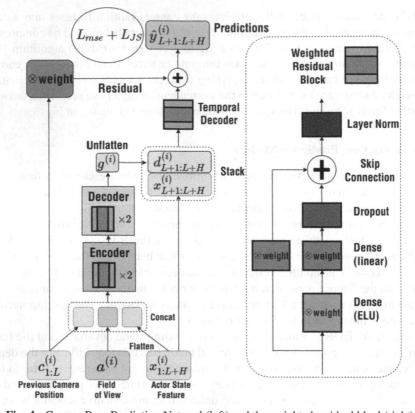

Fig. 4. Camera Pose Prediction Network(left) and the weighted residual block(right)

results at time-step L + t with the actor state features at time-step L + t, aiming for the model to pay special attention to certain abrupt changes in the actor's actions.

Finally, we also applied a global residual connection, where we added the predicted results to the linearly transformed historical camera pose, weighted with guidance weights. This was done to incorporate long-term trends and further refine the predictions using the historical camera pose and reference video's camera behavior.

$$\hat{c}_{L+t}^{(i)} = \text{TemporalDecoder}(d_t^{(i)}; x_{L+t}^i) \forall t \in [\mathbf{H}] \tag{6}$$

3.5 Object Function

The loss function we use is shown in formula (7):

$$L = \lambda \cdot L_{MSE} + L_{JS} \tag{7}$$

L_{MSE} is the mean square error loss (MSE), which measures the difference between the predicted pose and the ground truth. L_{JS} is Jensen Shannon (JS) divergence, which measures the difference between our predicted trajectory and the real trajectory. λ is a learnable parameter that assigns the ratio of weight between two losses. Since we only use the first frame when rendering, we give more weight to the first frame when calculating L_{MSE}. The calculation of the two losses is shown in formulas (8) and (9):

$$L_{MSE} = \frac{\sigma \|\hat{c}_i - \tilde{c}_i\| + \sum\limits_{j=i+1}^{i+N-1} \|\hat{c}_j - \tilde{c}_j\|}{N} \tag{8}$$

$$L_{JS} = \frac{1}{2} KL\left(\text{softmax}(\hat{c}_i) \| \text{softmax}\left(\frac{\hat{c}_i + \tilde{c}_i}{2}\right) \right) + \frac{1}{2} KL\left(\text{softmax}(\tilde{c}_i) \| \text{softmax}\left(\frac{\hat{c}_i + \tilde{c}_i}{2}\right) \right) \tag{9}$$

where $\hat{c}_i$ is the predicted camera coordinates, $\tilde{c}_i$ is the ground truth of the camera coordinates, N is the length of the predicted number of frames, σ is the weight of the first frame. KL($\cdot$) means KL divergence.

4 Experiment and Results

4.1 Experiment Details

We use the hybrid dataset provided by Jiang in his work [8]. The dataset includes 30 virtual scene data with pre-defined camera trajectories, each of which has a length of 1500 frames. Additionally, it comprises 62 preprocessed real film data. However, the test data of the dataset only provided 30 frames of the predicted label, so we selected the first 30 frames (N = 30) from the 60 frames predicted for the calculation of loss.

Our experiments and scene rendering were conducted on Ubuntu 20.04. The CPU used was the Intel Xeon Silver 4116, and the GPU was Nvidia GeForce RTX 2080Ti. The rendering engine employed was Unity, specifically version 2020.3.31f1.

We set the number of epochs to 300, the batch size to 256, and the loss weight of the first frame o to 10. The weight decay coefficient was set to 1e–5, and the initial learning rate was set to 0.001. We used an exponential learning rate decay strategy, where the learning rate was decayed by a factor of 0.97 after each epoch. The Adam optimizer was utilized. In the camera behavior extraction module, the length of the input feature sequence from the reference video was set to 400. We employed a single-layer Informer encoder with standard parameters, producing a 512-dimensional output. The dimensionality of the expert vector was set to m = 9.In the camera pose prediction module, the length of the input camera history sequence was set to L = 60, and the length of the output sequence was set to H = 60. The feature dimensions of the historical camera pose, static covariate, and dynamic covariate are 5, 1, and 9, respectively. The hidden layer dimensionality of the encoder-decoder was set to 512. The decoder output dimension p = 32, and the temporal decoder part had a hidden layer dimensionality of 128.

In the actual application process, a similar autoregressive method is used, using the forecast result of the first frame of 60 frames to render on Untiy (too few forecast frames will cause the camera to be unable to fully use the existing information for prediction).

Table 1. Comparative experiment

Methods	RMSE ($\times 10^{-3}$)↓
Jiang [9]	32.115
Jiang [8]	20.309
EVCPP(ours)	**13.705**

4.2 Experimental Results

We use the Root Mean Squared Error (RMSE) of the predicted results using 30 frames as the evaluation metric, and also assign a weight of 10 to the prediction results of the first frame and a weight of 1 to the rest 29 frames. Since there are few studies on the example-driven virtual camera shooting methods, we only reproduce two papers [8, 9] with open code for comparison, and the experimental results are shown in Table 1, the camera pose predicted by our method has the smallest RMSE.

To verify the effectiveness of each module, we conducted ablation experiments. We first replaced the hybrid loss with a single loss respectively, and the experimental results are shown in the first three rows of Table 2. It can be seen from the experimental results that the loss used in this paper achieves the best performance. This is mainly because the MSE loss function primarily focuses on the positional information of predicted trajectories, which is the difference between coordinates point by point, but ignores the similarity of the overall trajectory. By using the hybrid loss, the optimization of trajectory shape and position information can be balanced, allowing the model to learn and predict the pose more comprehensively.

Table 2. Ablation experiments

Parts	Methods	RMSE ($\times 10^{-3}$)↓
Loss	L_{JS}	25.462
	L_{MSE}	14.066
	$L_{MSE} + L_{JS}$(EVCPP)	**13.705**
prediction module	Transformer Decoder	25.992
	Fully connected network	14.278
	modified TiDE(EVCPP)	**13.705**

We replaced the modified TiDE in the camera pose prediction module with a three-layer fully connected neural network with weight guidance or a transformer [20] decoder. Specific experimental results are shown in the last three rows of Table 2. Modified TiDE achieved the best results. This is primarily due to the transformer's multi-head self-attention mechanism, which is not affected by changes in sequence order. However, for camera pose prediction, the order itself often plays a crucial role. And simply using a fully connected neural network does not separately consider the effects of individual covariates on the camera pose.

4.3 Visualization Results

By capturing corresponding reference guide videos, we achieved common camera operations in cloud performing arts scenes. In the upper part of Fig. 5, we show that the camera takes a close shot to achieve the shooting effect of "pan-tilt" from the left side of the actor to the right side of the actor. In the bottom part, we show the effect of "push-pull" from a camera's wide shot.

Fig. 5. Camera trajectory graph and actual rendered video sequence clips. Video sequence order: from left to right, from top to bottom.

5 Conclusions

This paper proposes EVCPP, focuses on the cloud performing scenes, and emphasizes the state of actors and the attributes of the camera itself, combined with the weighted guidance from reference videos, to predict the future pose of the camera. Additionally, a hybrid loss is used to reduce the pose error and balance the shape of the trajectory. The experimental results demonstrate the favorable performance of our method.

References

1. Wang, X., Courant, R., Shi, J., et al.: JAWS: just a wild shot for cinematic transfer in neural radiance fields. In: Proceedings of the IEEE/CVF Conference on Computer Vision and Pattern Recognition, pp. 16933–16942 (2023)
2. Lou, Z.: The Design and Application of Knowledge-based Intelligent Shot Planning System. Zhejiang University (2011)

3. Wang, M., Yang, G.W., Hu, S.M., et al.: Write-a-video: computational video montage from themed text. ACM Trans. Graph. **38**(6), 1–13 (2019)
4. Xiong, Y., Heilbron, F.C., Lin, D.: Transcript to video: efficient clip sequencing from texts (2021)
5. Chen, J., Carr, P.: mimicking human camera operators In: 2015 IEEE Winter Conference on Applications of Computer Vision. IEEE (2015)
6. Chen, J., Le, H.M., Carr, P., et al.: Learning online smooth predictors for realtime camera planning using recurrent decision trees. In: Computer Vision & Pattern Recognition. IEEE (2016)
7. Huang, C., Lin, C.E., Yang, Z., et al.: Learning to film from professional human motion videos. In: 2019 IEEE/CVF Conference on Computer Vision and Pattern Recognition (CVPR). IEEE (2019)
8. Jiang, H., Wang, B., Wang, X., et al.: Example-driven virtual cinematography by learning camera behaviors. ACM Trans. Graph. (TOG), **39**(4), 45:1–45:14 (2020)
9. Jiang, H., Christie, M., Wang, X., et al.: Camera keyframing with style and control. ACM Trans. Graph. (TOG) **40**(6), 1–13 (2021)
10. Lino, C., Christie, M.: Intuitive and efficient camera control with the toric space. ACM Trans. Graph. **34**(4CD):82.1–82.12 (2015)
11. Yu, Z., Yu, C., Wang, H., et al.: Enabling automatic cinematography with reinforcement learning. In: 2022 IEEE 5th International Conference on Multimedia Information Processing and Retrieval (MIPR). IEEE, pp. 103–108 2022
12. Gschwindt, M., Camci, E., Bonatti, R., et al.: Can a robot become a movie director? learning artistic principles for aerial cinematography. In: 2019 IEEE/RSJ International Conference on Intelligent Robots and Systems (IROS), pp. 1107–1114 IEEE (2019)
13. Dang, Y., Huang, C., Chen, P., Liang, R., Yang, X., Cheng, K.T.: Path-analysis-based reinforcement learning algorithm for imitation filming. IEEE Trans. Multimedia (2022).https://doi.org/10.1109/TMM.2022.3151463
14. Dang, Y., Huang, C., Chen, P., Liang, R., Yang, X., Cheng, K.-T.: Imitation learning-based algorithm for drone cinematography system. IEEE Trans. Cogn. Dev. Syst. **14**(2), 403–413 (2022). https://doi.org/10.1109/TCDS.2020.3043441
15. Geng, Z., Sun, K., Xiao, B., et al.: Bottom-up human pose estimation via disentangled keypoint regression (2021)
16. Seker, M., Mnnist, A., Losifidis, A., et al.: automatic main character recognition for photographic studies (2021). https://doi.org/10.48550/arXiv.2106.09064[P]
17. Zhou, H., Zhang, S., Peng, J., et al.: Informer: Beyond efficient transformer for long sequence time-series forecasting. In: Proceedings of the AAAI Conference on Artificial Intelligence, vol. 35. no. 12, pp. 11106–11115 (2021)
18. Jacobs, R.A., Jordan, M.I., Nowlan, S.J., Hinton, G.E.: Adaptive mixtures of local experts. Neural Comput. **3**(1), 79–87 (1991). https://doi.org/10.1162/neco.1991.3.1.79
19. Das, A., Kong, W., Leach, A., et al.: Long-term Forecasting with TiDE: Time-series Dense Encoder arXiv preprint arXiv:2304.08424 (2023)
20. Vaswani, A., Shazeer, N., Parmar, N., et al.: Attention is all you need.In: Advances in Neural Information Processing Systems, vol. 30 (2017)

RBSR: Efficient and Flexible Recurrent Network for Burst Super-Resolution

Renlong Wu, Zhilu Zhang, Shuohao Zhang, Hongzhi Zhang[⊠],
and Wangmeng Zuo

Harbin Institute of Technology, Harbin, China
zhanghz0451@gmail.com, wmzuo@hit.edu.cn

Abstract. Burst super-resolution (BurstSR) aims at reconstructing a
high-resolution (HR) image from a sequence of low-resolution (LR) and
noisy images, which is conducive to enhancing the imaging effects of
smartphones with limited sensors. The main challenge of BurstSR is to
effectively combine the complementary information from input frames,
while existing methods still struggle with it. In this paper, we suggest
fusing cues frame-by-frame with an efficient and flexible recurrent network. In particular, we emphasize the role of the base-frame and utilize
it as a key prompt to guide the knowledge acquisition from other frames
in every recurrence. Moreover, we introduce an implicit weighting loss to
improve the model's flexibility in facing input frames with variable numbers. Extensive experiments on both synthetic and real-world datasets
demonstrate that our method achieves better results than state-of-the-
art ones. Codes and pre-trained models are available at https://github.
com/ZcsrenlongZ/RBSR.

Keywords: Burst Super-Resolution · Recurrent Network ·
Super-Resolution

1 Introduction

The rising prevalence of smartphones has led to an increasing demand for capturing high-quality images. However, inherent disadvantages of the smartphone
camera are inevitable in order to integrate it into the thin-profile device, including tiny sensor size, fixed aperture, and restricted zoom [11]. Such physical
demerits not only result in limited image spatial resolution, but also easily
include noise [14,43], especially in the low-light environment. With the development of deep learning [17,27,40,44], many single-image denoising [8,26,49,53,54]
and super-resolution [13,30,48,57] methods have attempted to address the issue,
achieving great progress. Nevertheless, they are severely ill-posed and difficult
to restore high-quality images with realistic and rich details when dealing with

Supplementary Information The online version contains supplementary material
available at https://doi.org/10.1007/978-981-99-8537-1_6.

Q. Liu et al. (Eds.): PRCV 2023, LNCS 14430, pp. 65–78, 2024.
https://doi.org/10.1007/978-981-99-8537-1_6

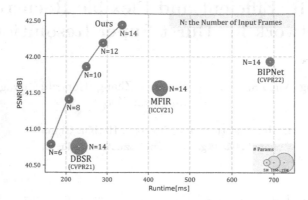

Fig. 1. Performance and runtime comparison on SyntheticBurst dataset [3].

badly degraded images. Instead, burst super-resolution (BurstSR) [1–3] can relax the ill-posedness and give a chance to bridge the imaging gap with DSLRs, by reconstructing a high-resolution (HR) sRGB image from continuously captured low-resolution (LR) and noisy RAW frames.

Two critical processes for BurstSR are inter-frame alignment and feature fusion [3,4,14,15,35–37]. The former has been more fully explored in various ways, involving the optical flow [42] alignment [3,4,6], deformable convolution [10] alignment [35,37,47], and cross-attention alignment [28]. But for the latter, existing approaches still have certain limitations in merging complementary information from multiple frames.

Among those, the weighted-based fusion [3,4] combines aligned features by predicting element-wise weights, which only pays attention to the communication between base-frame and non-base frames, while ignoring information exchange among non-base frames. The pseudo-burst fusion [14] concatenates burst features channel by channel, being more sufficient yet only processing input frames with a fixed number. The attention-based fusion [15,36,37] exploits inter-frame cross-attention mechanism to enhance feature interaction on the channel or spatial dimension, but it is computationally extensive. Thus, it is still worth exploring a multi-frame fusion method that can be efficient and adaptive to variable numbers of frames in BurstSR.

In this work, we suggest fusing cues frame-by-frame, which can specifically merge the beneficial information of each frame and has no limit on the number of frames. A recurrent manner is a natural and suitable choice, and we propose an efficient and flexible recurrent network for burst super-resolution, dubbed RBSR. RBSR processes the base-frame first and aggregates other aligned temporal features sequentially to reconstruct the HR result. In particular, we find that the base-frame plays an important role in such a recurrent approach. Thus, we emphatically utilize the base-frame as a key prompt to guide the acquisition of knowledge from other frames in every recurrence. Moreover, although the recurrent network itself has the ability to deal with variable lengths of inputs, simply constraining the output during the last recurrence will lead to a flexibility decrease in handling fewer frames. Thus, we further propose an implicit weighting loss to enhance the model's ability in facing fewer frames while maintaining the performance in processing the longest frames.

Experiments are conducted on both synthetic and real-world datasets [3]. Benefiting from the compact and efficient method design, the results show that our RBSR not only achieves better fidelity as well as perceptual performance than state-of-the-art methods, but also has favorable efficiency. In comparison with BIPNet [14], our RBSR obtains 0.51 dB PSNR gain while only taking less than half of its inference time, as shown in Fig. 1.

The main contributions can be summarized as follows:

1. We focus on the efficient and flexible fusion manner in BurstSR, and propose a recurrent network named RBSR, where the base-frame is emphatically utilized to guide the knowledge acquisition from other frames in every recurrence by the suggested KFGR module.
2. An implicit weighting loss is introduced to further enhance the model's ability in facing input frames with variable numbers.
3. Experiments on both synthetic and real-world datasets demonstrate that our method not only outperforms state-of-the-art methods quantitatively and qualitatively, but also has favorable inference efficiency.

2   Related Work

Single Image Super-Resolution. With the development of deep learning [17, 27,40,44], single-image super-resolution (SISR) has achieved great success in terms of both performance [13,29,55,56] and efficiency [22,30,34,45,52]. SISR-oriented network designs [21,23,29,50] and optimization objectives [20,21,33] are widely explored. However, these methods still struggle with heavily degraded images and sometimes generate artifacts due to the severe ill-posedness of SISR.

Multi-Frame Super-resolution. Compared to SISR methods, multi-frame super-resolution (MFSR) approaches [3,4,12,14,15,36,37,43,51] aggregate multiple aliased images for better reconstruction. Tsai [43] first proposes a frequency domain method with the assumption that the input image translations are known. HighResNet [12] focuses on satellite imagery, aligning each frame to a reference frame implicitly and performing recursive fusion. DBSR [3] introduces a weighted-based fusion mechanism, where element-wise weights between the base-frame and other frames are predicted. MFIR [4] further extends this fusion mechanism by learning the image formation model in deep feature space. However, only paying attention to communication between the base-frame and non-base frames limits complementary information exchange among non-base ones. Recently, BIPNet [14] proposes a pseudo-burst fusion strategy by fusing temporal features channel-by-channel, being more sufficient but requiring fixing the input frame number. A few works [15,36,37] exploits inter-frame attention-based

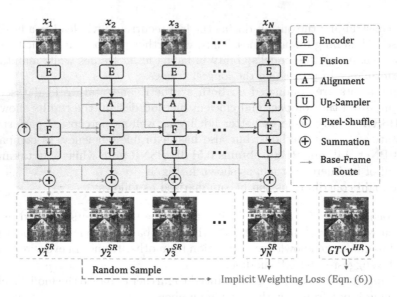

Fig. 2. The architecture of the proposed RBSR. Each LR frame is first fed into the encoder to extract features which are then warped to the base ones by the alignment module. Next, the previous merged features and current aligned ones are passed into our proposed recurrent fusion module, which is shown in Fig. 3. Finally, the up-sampler is employed to generate the SR result. Moreover, an implicit weighting loss is present to enhance the model's flexibility in processing input frames with variable lengths.

mechanism to enhance feature interaction via cross-attention on the channel or spatial dimension, yet are computationally extensive. In this work, we suggest fusing cues frame-by-frame, which enables sufficient integration of advantageous information from each frame and is not constrained by the number of frames.

Recurrent Neural Network. Recurrent Neural Network(RNN) [41] provides an elegant way for temporal modeling and can be naturally applied to multi-frame restoration [6,7,9,16,19,25,39,46]. For instance, BasicVSR [6] employs bidirectional propagation, achieving high performance. BasicVSR++ [7] further enhances BasicVSR [6] with second-order bidirectional propagation. Rong *et al..* [39] proposes a temporally shifted wavelet transform to combine spatial and frequency information for burst denoising. In our work, we explore a specific recurrent manner suitable for BurstSR, and propose a compact, efficient, and flexible recurrent neural network RBSR.

3 Method

3.1 Problem Formation and Pipeline

Problem Formation. BurstSR aims at generating a clean and high-resolution sRGB image $y_N^{SR} \in \mathbb{R}^{sH \times sW \times 3}$ from multiple low-resolution and noisy RAW images $\{x_i\}_{i=1}^{N}$, where $x_i \in \mathbb{R}^{H \times W}$. N and s denote the input frame number and super-resolution factor, respectively. Inter-frame alignment and feature fusion are two important aspects of BurstSR. Existing methods have more fully explored the former [3,4,6,28,35,37,47], but are limited in the effectiveness [3,4], efficiency [15,36,37], and flexibility [14] of the latter. In this work, we expect to explore a fusion method that is efficient and adaptive to variable frame numbers. Specifically, we suggest merging features frame-by-frame to utilize the beneficial information of each frame as much as possible, and propose an efficient and flexible recurrent network named RBSR.

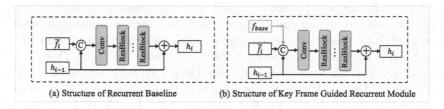

(a) Structure of Recurrent Baseline   (b) Structure of Key Frame Guided Recurrent Module

Fig. 3. Structure of the proposed recurrent baseline and key frame guided recurrent (KFGR) module.

Pipeline. RBSR includes the encoder, alignment, recurrent fusion, and up-sampler module, as show in Fig. 2. First, LR images are fed into the encoder $\mathcal{E}$ to obtain deep feature representations $\{f_i\}_{i=1}^{N}$. To merge the features, the inevitable spatial misalignment issue between frames should be pre-addressed. Thus, we deploy the alignment model $\mathcal{A}$ based on flow-guided deformable strategy [7] to warp the non-base features to the base ones f_1, which can be written as,

$$\bar{f}_i = \mathcal{A}(f_{base}, f_i), \tag{1}$$

where $f_{base} = f_1$, $\bar{f}_i$ denotes the aligned f_i. Then, the recurrent fusion module $\mathcal{F}$ combines the aligned information frame-by-frame, which can be written as,

$$h_N = \mathcal{F}(f_{base}, \bar{f}_2, ..., \bar{f}_N). \tag{2}$$

Finally, the merged feature h_N is taken into up-sampler to obtain HR result y_N^{SR}.

In this paper, we focus on the fusion manner and propose a key frame guided recurrent (KFGR) module in Sect. 3.2. In comparison with common recurrent networks [6,7], KFGR utilizes the base-frame as a key prompt to guide the acquisition of knowledge from other frames. Moreover, although recurrent networks can naturally generalize to variable input frame numbers, simply constraining the final output y_N^{SR} limits the model flexibility, dropping the performance in handling shorter input sequences. To address this issue, in Sect. 3.3, we propose an implicit weighting loss, which enhances the ability in processing shorter sequences while maintaining performance on the longest one.

3.2 Recurrent Fusion Module

In this sub-section, we first design our basic recurrent fusion scheme. Then we further enhance the scheme by utilizing the base-frame more sufficiently, proposing a key frame guided recurrent (KFGR) module.

Recurrent Scheme. The base-frame mainly serves as spatial position guidance in BurstSR. However, the non-base features processed by the alignment model generally are not perfectly aligned with the base-frame, since the misalignment cases are varied and complex. Such an inherent problem drives the fusion module to consider the base-frame as guidance, which can ensure that the merged features will not have position shifts, alleviating artifacts. Thus, in this work, the recurrent fusion scheme is designed to propagate information from the base-frame to others, rather than from others to the base-frame. The scheme can make the base-frame provide implicit guidance for others in every recurrence, and it can be also seen as a gradual restoration of the base-frame. Specifically, we consider a recurrent network incorporating residual learning strategy with two inputs: the current aligned feature $\bar{f}_i$ and the previous merged feature h_{i-1}, *i.e.*,

$$h_i = \mathcal{F}(h_{i-1}, \bar{f}_i) + h_{i-1}. \tag{3}$$

where $\mathcal{F}$ is recurrent baseline module. h_N will merge all frame information, and then be passed to the up-sampler for generating the final result. For simplicity and efficiency, after concatenating inputs along channel dimension, we only use a 3×3 convolutional layer for channel reduction and several residual blocks [17] for feature enhancement, as illustrated in Fig. 3(a).

Key Frame Guided Recurrent Module. In the above recurrent fusion scheme, the base-frame is only utilized directly at the beginning. As the propagation progresses, the valuable information from base-frame may be lost, resulting in a weakened guiding effect. Thus, we additionally provide the feature of baseframe as a key condition to guide the acquisition of knowledge from other frames in every recurrence. Eq. (3) can be modified as,

$$h_i = \mathcal{F}(h_{i-1}, \bar{f}_i | f_{base}) + h_{i-1}, \tag{4}$$

where $\mathcal{F}$ is KFGR module. Without bells and whistles, KFGR simply concatenates the base-frame feature f_{base}, the current aligned feature $\bar{f}_i$, and the previous merged feature h_{i-1} along channel dimension, as shown in Fig. 3(b). And the backbone of KFGR remains the same as that of the recurrent baseline. Such a more explicit scheme can enable more sufficient utilization of the base-frame, thereby improving the performance.

Discussion. There are several advantages of our proposed recurrent fusion module KFGR. First, merging cues frame-by-frame makes it effective, achieving favorable performance. Moreover, our experiments in Sect. 4.3 show that such a unidirectional propagation is more suitable than a bidirectional one for BurstSR. Second, the compact network design makes it efficient, costing low inference time. Third, the recurrent manner makes it flexible to deal with input frames of varying lengths, which is detailed in the next section.

3.3   Implicit Weighting Loss

Existing BurstSR methods [3,4,14,15,36,37] are generally optimized by minimizing the ℓ_1 distance between the output y_N^{SR} and the ground truth y^{HR}, i.e.,

$$\mathcal{L} = \left\| y_N^{SR} - y^{HR} \right\|_1 . \tag{5}$$

However, Eq. (5) only constrains the result with the input of all frames, and it leads to a significant performance decrease in facing fewer ones. One straightforward solution is to apply loss in every recurrence, but it increases training time and weakens the ability to handle the longest frames. The related experiments will be shown in Sect. 4.3.

To address the above issues, we propose an implicit weighting loss, which keeps the constraint in the last recurrence and randomly adds one in the previous recurrences. The loss can be written as,

$$\mathcal{L} = \left\| y_i^{SR} - y^{HR} \right\|_1 + \left\| y_N^{SR} - y^{HR} \right\|_1 , \tag{6}$$

where i is randomly and uniformly sampled from 1 to $N - 1$. On the one hand, the proposed loss implicitly imposes constraints in every recurrence, thus bringing performance improvement in facing fewer frames. On the other hand, it also implicitly applies the strongest weight to the last result, thus making performance maintenance in processing the longest frames. In addition, the implicit weighting loss only slightly increases the training cost compared with Eq. (5).

4   Experiments

4.1   Experimental Settings

Datasets. The experiments are conducted on synthetic SyntheticBurst and real-world BurstSR datasets [3]. SyntheticBurst dataset consists of 46839 bursts for training and 300 for testing. Each burst contains 14 LR RAW images that are synthesized from a single sRGB image captured by a Canon camera. Specifically, the sRGB image is first converted to the linear RGB values using an inverse camera pipeline [5]. Next, random translations and rotations are applied to generate a shifted burst. Then the transformed images are downsampled by the bilinear kernel and added noise, obtaining the low-resolution and noisy burst. Finally, Bayer mosaicking is deployed to get the input RAW burst. BurstSR dataset involves 5405 patches for training, and 882 patches for testing. The LR RAW images and the corresponding HR sRGB image are captured with a Samsung smartphone camera and a Canon DSLR camera, respectively.

Implement Details. The scale factor s for super-resolution is set to 4. The RAW burst images are packed into four channels (*i.e.*, RGGB) as inputs according to the Bayer filter pattern. The batch size is set to 16. The input patch size is set to 48×48 for synthetic experiments and 56×56 for real-world experiments. For the SyntheticBurst dataset [3], the model is trained with AdamW optimizer [32] with $\beta_1 = 0.9$ and $\beta_2 = 0.999$ for 400k iterations. Cosine annealing strategy [31] is employed to steadily decrease the learning rate from 10^{-4} to 10^{-6}. For the BurstSR dataset [3], following [3,4,14], we fine-tune the pre-trained model on SyntheticBurst dataset for additional 50k iterations. All experiments are conducted with PyTorch [38] on an Nvidia GeForce RTX 2080Ti GPU.

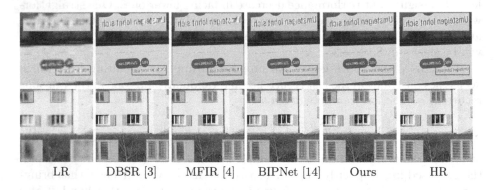

| LR | DBSR [3] | MFIR [4] | BIPNet [14] | Ours | HR |

Fig. 4. Qualitative comparison on synthetic SyntheticBurst dataset [3].

4.2   Comparison with State-of-the-Art Methods

We compare the proposed RBSR with five BurstSR methods, *i.e.*, HighResNet [12], DBSR [3], MFIR [4], LKR [24], and BIPNet [14]. PSNR and SSIM [18] are adopted as evaluation metrics. Additionally, the number of model parameters, #FLOPs and inference time are also reported when generating a 1568×1024 HR image.

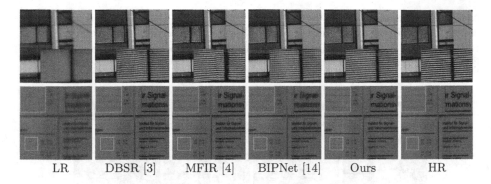

LR   DBSR [3]   MFIR [4]   BIPNet [14]   Ours   HR

Fig. 5. Qualitative comparison on real-world BurstSR dataset [3].

Table 1. Quantitative comparison on SyntheticBurst and BurstSR datasets [3].

Method	SyntheticBurst PSNR/SSIM	BurstSR PSNR/SSIM	#Params (M)	#FLOPs (G)	Time (ms)
HighResNet [39]	37.45/0.924	46.64/0.980	34.78	96	–
DBSR [3]	40.76/0.959	48.05/0.984	13.01	103	231
LKR [24]	41.45/0.953	–	–	–	–
MFIR [4]	41.56/0.964	48.33/0.985	12.13	121	428
BIPNet [14]	41.93/0.967	48.49/0.985	6.66	326	691
RBSR (Ours)	42.44/0.970	48.80/0.987	6.42	158	336

Results on Synthetic SyntheticBurst Dataset. Table 1 shows the quantitative results. It shows that our RBSR performs satisfactorily both in effectiveness and efficiency. In comparison with the recent state-of-the-art method BIPNet [14], RBSR yields a PSNR gain of 0.51 dB without increasing the number of parameters. Besides, RBSR runs much faster, benefiting from concise and efficient design. The qualitative results in Fig. 4 show that our RBSR restores more rich and realistic textures and fewer artifacts than others.

Results on Real-World BurstSR Dataset. The LRs and HR in the real-world dataset are slightly misaligned, since they are captured with different cameras in BurstSR dataset [3]. To overcome this issue, following previous works [3,4,14], we train RBSR with the aligned ℓ_1 loss and evaluate with the aligned PSNR as well as SSIM metrics in Table 1. In comparison with BIPNet, we achieve 0.64% PSNR gains. The qualitative results are shown in Fig. 5. And the results demonstrate that our model still performs better, and recovers more details in the real world.

Handling Inputs with Variable Frame Lengths. We compare with DBSR [3] and MFIR [4] to verify our method's flexibility in facing inputs with variable frame numbers. We did not compare with BIPNet [14] as it is only used in the case with a fixed input frame number. Table 2 shows that our method significantly outperforms others across all numbers of input frames. Especially, for our RBSR, taking 6 frames as inputs can obtain a similar performance with DBSR [3] that is fed 14 frames.

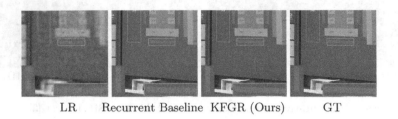

LR   Recurrent Baseline KFGR (Ours)   GT

Fig. 6. The result comparison when taking recurrent baseline and KFGR module (see Fig. 3) as fusion manner.

Table 2. Qualitative comparison with other approaches when facing variable input frame numbers. We show 'PSNR/SSIM' metrics in the table.

Method	N = 2	N = 4	N = 6	N = 8	N = 10	N = 12	N = 14
DBSR [3]	34.78/0.892	37.42/0.930	38.90/0.945	39.73/0.952	40.22/0.956	40.55/0.958	40.78/0.959
MFIR [4]	36.37/0.915	38.70/0.942	39.81/0.952	40.50/0.958	40.98/0.961	41.32/0.963	41.55/0.964
Ours	38.08/0.935	39.85/0.952	40.79/0.959	41.41/0.963	41.86/0.966	42.19/0.968	42.44/0.970

4.3   Ablation Study

Comparison of Unidirectional and Bidirectional Propagation. We compare our unidirectional recurrent fusion model with two typical propagation schemes, *i.e.*, bidirectional [6] and enhanced bidirectional propagation [7]. Although they have been widely used in video-restoration tasks, they have not shown any superiority for BurstSR, only receiving 42.34 dB and 42.33 dB PSNR respectively, while our RBSR achieves 42.44 dB on SyntheticBurst dataset [3].

Effect of Base-Frame Utilization. Our sufficient utilization of the base-frame is reflected in two aspects: propagating information starting from the base-frame to others and providing the base-frame feature as an explicit prompt in KFGR module. Here we conduct an experiment by propagating information in a reversed order, which leads 0.09 dB PSNR drop. Moreover, when we replace the recurrent baseline with KFGR module (see Fig. 3), it enables 0.12 dB PSNR gain. Figure 6 also shows that the utilization of the base-frame can significantly reduce artifacts.

Table 3. Qualitative comparison of different loss calculation approaches. We show 'PSNR/SSIM' metrics in the table.

Method	N = 6	N = 10	N = 14
Constraining Last Result	39.79/0.953	41.53/0.965	42.45/0.970
Constraining Every Result	40.81/0.959	41.80/0.966	42.35/0.969
Implicit Weighting Loss (Ours)	40.79/0.959	41.86/0.966	42.44/0.970

Effect of Implicit Weighting Loss. We compare our implicit weighting loss with two loss calculation approaches. One only constrains the last result as shown in Eq. (5), and the other constrains the results in every recurrence. As shown in Table 3, our implicit weighting loss can improve performance for shorter sequences while maintaining performance for the longest ones.

5   Conclusion

BurstSR aims at reconstructing a high-resolution sRGB image from a sequence of low-resolution and noisy RAW images. In this work, we focus on the multi-frame fusion manner, proposing an efficient and flexible recurrent model for BurstSR. On the one hand, the model merges cues frame-by-frame and emphatically utilizes the base-frame to guide the acquisition of knowledge from other frames in every recurrence. On the other hand, implicit weighting loss is present to enhance the model's flexibility in processing inputs with variable lengths. Experiments on both synthetic and real-world datasets demonstrate our method achieves better results than state-of-the-art ones.

References

1. Bhat, G., Danelljan, M., Timofte, R.: Ntire 2021 challenge on burst super-resolution: methods and results. In: CVPRW (2021)
2. Bhat, G., et al.: Ntire 2022 burst super-resolution challenge. In: CVPRW (2022)
3. Bhat, G., Danelljan, M., Van Gool, L., Timofte, R.: Deep burst super-resolution. In: CVPR (2021)
4. Bhat, G., Danelljan, M., Yu, F., Van Gool, L., Timofte, R.: Deep reparametrization of multi-frame super-resolution and denoising. In: CVPR (2021)
5. Brooks, T., Mildenhall, B., Xue, T., Chen, J., Sharlet, D., Barron, J.T.: Unprocessing images for learned raw denoising. In: CVPR (2019)
6. Chan, K.C., Wang, X., Yu, K., Dong, C., Loy, C.C.: Basicvsr: the search for essential components in video super-resolution and beyond. In: CVPR (2021)
7. Chan, K.C., Zhou, S., Xu, X., Loy, C.C.: Basicvsr++: improving video super-resolution with enhanced propagation and alignment. In: CVPR (2022)
8. Chen, H., et al.: Pre-trained image processing transformer. In: CVPR (2021)
9. Chen, X., Song, L., Yang, X.: Deep rnns for video denoising. In: Applications of Digital Image Processing. SPIE (2016)
10. Dai, J., et al.: Deformable convolutional networks. In: ICCV (2017)

11. Delbracio, M., Kelly, D., Brown, M.S., Milanfar, P.: Mobile computational photography: a tour. Ann. Rev. Vision Sci. **7**, 571–604 (2021)
12. Deudon, M., et al.: Highres-net: recursive fusion for multi-frame super-resolution of satellite imagery. arXiv:2002.06460 (2020)
13. Dong, C., Loy, C.C., He, K., Tang, X.: Image super-resolution using deep convolutional networks. TPAMI **38**, 395–307 (2015)
14. Dudhane, A., Zamir, S.W., Khan, S., Khan, F.S., Yang, M.H.: Burst image restoration and enhancement. In: CVPR (2022)
15. Dudhane, A., Zamir, S.W., Khan, S., Khan, F.S., Yang, M.H.: Burstormer: burst image restoration and enhancement transformer. CVPR (2023)
16. Fuoli, D., Gu, S., Timofte, R.: Efficient video super-resolution through recurrent latent space propagation. In: ICCVW. IEEE (2019)
17. He, K., Zhang, X., Ren, S., Sun, J.: Deep residual learning for image recognition. In: CVPR (2016)
18. Hore, A., Ziou, D.: Image quality metrics: PSNR vs. SSIM. In: ICPR. IEEE (2010)
19. Huang, Y., Wang, W., Wang, L.: Video super-resolution via bidirectional recurrent convolutional networks. TPAMI **40**, 1015–1028 (2017)
20. Johnson, J., Alahi, A., Fei-Fei, L.: Perceptual losses for real-time style transfer and super-resolution. In: Leibe, B., Matas, J., Sebe, N., Welling, M. (eds.) ECCV 2016. LNCS, vol. 9906, pp. 694–711. Springer, Cham (2016). https://doi.org/10. 1007/978-3-319-46475-6_43
21. Kim, J., Lee, J.K., Lee, K.M.: Accurate image super-resolution using very deep convolutional networks. In: CVPR (2016)
22. Kong, X., Zhao, H., Qiao, Y., Dong, C.: ClassSR: a general framework to accelerate super-resolution networks by data characteristic. In: ECCV (2021)
23. Lai, W.S., Huang, J.B., Ahuja, N., Yang, M.H.: Deep laplacian pyramid networks for fast and accurate super-resolution. In: CVPR (2017)
24. Lecouat, B., Ponce, J., Mairal, J.: Lucas-kanade reloaded: end-to-end super-resolution from raw image bursts. In: ICCV (2021)
25. Li, D., Zhang, Y., Law, K.L., Wang, X., Qin, H., Li, H.: Efficient burst raw denoising with variance stabilization and multi-frequency denoising network. IJCV **130**, 2060–2080 (2022)
26. Li, J., et al.: Spatially adaptive self-supervised learning for real-world image denoising. In: CVPR (2023)
27. Liang, J., Cao, J., Sun, G., Zhang, K., Van Gool, L., Timofte, R.: Swinir: image restoration using swin transformer. In: ICCV (2021)
28. Liang, J., et al.: Recurrent video restoration transformer with guided deformable attention. In: NeurIPS (2022)
29. Lim, B., Son, S., Kim, H., Nah, S., Mu Lee, K.: Enhanced deep residual networks for single image super-resolution. In: CVPRW (2017)
30. Liu, M., Zhang, Z., Hou, L., Zuo, W., Zhang, L.: Deep adaptive inference networks for single image super-resolution. In: Bartoli, A., Fusiello, A. (eds.) ECCV 2020. LNCS, vol. 12538, pp. 131–148. Springer, Cham (2020). https://doi.org/10.1007/978-3-030-66823-5_8
31. Loshchilov, I., Hutter, F.: SGDR: stochastic gradient descent with warm restarts. arXiv:1608.03983 (2016)
32. Loshchilov, I., Hutter, F.: Decoupled weight decay regularization. arXiv:1711.05101 (2017)
33. Lugmayr, A., Danelljan, M., Van Gool, L., Timofte, R.: SRFlow: learning the super-resolution space with normalizing flow. In: Vedaldi, A., Bischof, H., Brox,

T., Frahm, J.-M. (eds.) ECCV 2020. LNCS, vol. 12350, pp. 715–732. Springer, Cham (2020). https://doi.org/10.1007/978-3-030-58558-7_42

34. Luo, X., et al.: Adjustable memory-efficient image super-resolution via individual kernel sparsity. In: ACMMM, pp. 2173–2181 (2022)

35. Luo, Z., et al.: BSRT: improving burst super-resolution with swin transformer and flow-guided deformable alignment. In: CVPR (2022)

36. Luo, Z., et al.: EBSR: feature enhanced burst super-resolution with deformable alignment. In: CVPR (2021)

37. Mehta, N., Dudhane, A., Murala, S., Zamir, S.W., Khan, S.: Gated multi-resolution transfer network for burst restoration and enhancement. In: CVPR (2023)

38. Paszke, A., Gross, S., Massa, F., Lerer, A., Bradbury, J., et al.: Pytorch: an imperative style, high-performance deep learning library. In: NeurIPS (2019)

39. Rong, X., Demandolx, D., Matzen, K., Chatterjee, P., Tian, Y.: Burst denoising via temporally shifted wavelet transforms. In: Vedaldi, A., Bischof, H., Brox, T., Frahm, J.-M. (eds.) ECCV 2020. LNCS, vol. 12358, pp. 240–256. Springer, Cham (2020). https://doi.org/10.1007/978-3-030-58601-0_15

40. Ronneberger, O., Fischer, P., Brox, T.: U-net: convolutional networks for biomedical image segmentation. In: Navab, N., Hornegger, J., Wells, W.M., Frangi, A.F. (eds.) MICCAI 2015. LNCS, vol. 9351, pp. 234–241. Springer, Cham (2015). https://doi.org/10.1007/978-3-319-24574-4_28

41. Schuster, M., Paliwal, K.K.: Bidirectional recurrent neural networks. IEEE TSP **45**, 2673–2681 (1997)

42. Sun, D., Yang, X., Liu, M.Y., Kautz, J.: PWC-NET: CNNs for optical flow using pyramid, warping, and cost volume. In: CVPR (2018)

43. Tsai, R.Y., Huang, T.S.: Multiframe image restoration and registration (1984)

44. Vaswani, A., et al.: Attention is all you need. In: NeurIPS (2017)

45. Wang, L., et al.: Exploring sparsity in image super-resolution for efficient inference. In: CVPR (2021)

46. Wang, R., et al.: Benchmark dataset and effective inter-frame alignment for real-world video super-resolution. In: CVPRW (2023)

47. Wang, X., Chan, K.C., Yu, K., Dong, C., Change Loy, C.: EDVR: video restoration with enhanced deformable convolutional networks. In: CVPRW (2019)

48. Wang, X., Xie, L., Dong, C., Shan, Y.: REAL-ESRGAN: training real-world blind super-resolution with pure synthetic data. In: ICCV (2021)

49. Wang, Z., Cun, X., Bao, J., Zhou, W., Liu, J., Li, H.: Uformer: a general u-shaped transformer for image restoration. In: CVPR (2022)

50. Wei, P., et al.: Component divide-and-conquer for real-world image super-resolution. In: Vedaldi, A., Bischof, H., Brox, T., Frahm, J.-M. (eds.) ECCV 2020. LNCS, vol. 12353, pp. 101–117. Springer, Cham (2020). https://doi.org/10.1007/978-3-030-58598-3_7

51. Wronski, B., et al.: Handheld multi-frame super-resolution. TOG **38**, 1–18 (2019)

52. Xie, W., Song, D., Xu, C., Xu, C., Zhang, H., Wang, Y.: Learning frequency-aware dynamic network for efficient super-resolution. In: ICCV (2021)

53. Zamir, S.W., Arora, A., Khan, S., Hayat, M., Khan, F.S., Yang, M.H.: Restormer: efficient transformer for high-resolution image restoration. In: CVPR (2022)

54. Zhang, K., Zuo, W., Chen, Y., Meng, D., Zhang, L.: Beyond a gaussian denoiser: residual learning of deep CNN for image denoising. TIP **36**, 3142–3155 (2017)

55. Zhang, K., Zuo, W., Zhang, L.: Learning a single convolutional super-resolution network for multiple degradations. In: CVPR (2018)

56. Zhang, Y., Zhang, Z., DiVerdi, S., Wang, Z., Echevarria, J., Fu, Y.: Texture hallucination for large-factor painting super-resolution. In: Vedaldi, A., Bischof, H., Brox, T., Frahm, J.-M. (eds.) ECCV 2020. LNCS, vol. 12352, pp. 209–225. Springer, Cham (2020). https://doi.org/10.1007/978-3-030-58571-6_13
57. Zhang, Z., Wang, H., Liu, M., Wang, R., Zhang, J., Zuo, W.: Learning RAW-to-SRGB mappings with inaccurately aligned supervision. In: ICCV (2021)

WDU-Net: Wavelet-Guided Deep Unfolding Network for Image Compressed Sensing Reconstruction

Xinlu Wang[1] , Lijun Zhao[1]([⊠]) , Jinjing Zhang[2], Yufeng Zhang[1],
and Anhong Wang[1]

[1] Institute of Digital Media and Communication, Taiyuan University of Science and
Technology, Taiyuan 030024, China
leejun@tyust.edu.cn
[2] Data Science and Technology, North University of China, Jiancaoping District,
Taiyuan 030051, China

Abstract. More and more deep unfolding networks have been studied to obtain good interpretability for high-quality image Compressed Sensing (CS) reconstruction. However, most of these networks simply focus on transmitting information across adjacent stages in the image-domain and ignore that frequency-domain information also can greatly restrain the solutions of CS optimization model. Motivated by this observation, frequency-domain consistency constraint is proposed to be inserted into image CS reconstruction. Concretely, we build a new CS optimization model based on discrete wavelet transform, in which frequency-domain information is used to guide image CS reconstruction. This model is divided into two sub-problems, which are optimized in an iterative manner. The iterative optimization procedure is expanded as a Wavelet-guided Deep Unfolding Network (WDU-Net) for image CS reconstruction. In view of the fact that denoising is the key step of the CS reconstruction problem, a dual-domain guided filtering block and a self-guided filtering enhancement block are proposed to remove noises for image reconstruction. Experimental results have shown that the reconstruction performance of our method is beyond many explainable CS reconstruction methods.

Keywords: Compressed sensing · Discrete wavelet transform · Deep unfolding network · Image reconstruction

1 Introduction

Compressed Sensing (CS) [1] has caused a research upsurge in academia because it can acquire and reconstruct signals at a very low sampling ratio, if signals can be sparsely represented. In mathematics, given original signal $x \in R^N$ and sampling matrix $\Phi \in R^{M \times N}$, the goal of CS is to infer the original signal x from the random CS measurement vector $y = \Phi x (y \in R^M)$. In general, M is

© The Author(s), under exclusive license to Springer Nature Singapore Pte Ltd. 2024
Q. Liu et al. (Eds.): PRCV 2023, LNCS 14430, pp. 79–91, 2024.
https://doi.org/10.1007/978-981-99-8537-1_7

far smaller than N. This is an ill-posed inverse problem, which can be converted into the following optimization problem:

$$\arg\min_x \frac{1}{2}\|\Phi x - y\|_2^2 + \lambda\|F(x)\|_1, \tag{1}$$

where $F(x)$ is a regularization function, which can sparsely represent natural images, and λ is a hyper-parameter that can make trade-off between data fidelity term and regularization term.

Traditional CS reconstruction methods [2,3] are built upon sparse prior knowledge, and iteratively reconstruct the original signal by solving the prior-regularized optimization problem. Although these methods are interpretable, natural image does not meet the sparsity in the transform domain all the time. Therefore, the reconstruction algorithm based on sparsity prior may reduce the accuracy of reconstruction when applied to some real-world signals. Additionally, its calculation cost is very high due to many iterative steps, which greatly limits the practical applications of CS. In recent years, Deep Neural Network (DNN) has become extremely popular since it has a strong representation ability. Therefore, a series of image CS methods based on DNN [4–6] are designed to greatly promote the quality of image reconstruction. Compared with traditional image CS methods, DNN-based methods solve the problem of image CS reconstruction by directly learning the inverse mapping from the compressed measurement values to the original image, which can achieve high-quality reconstruction. However, those methods regard the DNN as a black box model, which lacks theoretic basis. In recent years, with the development of Deep Learning (DL) technology, a whole lot of researchers have tried to combine traditional algorithms with DL. For instance, a large number of unfolding networks [7–14] are designed to boost image reconstruction performance. Among them, ISTA-Net [7] is the most popular network to combine traditional algorithms with DL, which utilizes the iterative shrinkage-thresholding algorithm to solve CS reconstruction problem and expands it into a network. On this basis, OPINE-Net [8] applies orthogonal and binary constraints to the sampling matrix. MADUN [9], DGUNet [12] and FSOINet [11] aim to promote the reconstruction quality by designing new network architectures. On the other hand, COAST [10] makes its framework more generalized by designing a novel kind of sampling matrix. LR-CSNet [13] and LG-Net [14] improve the CS model by adding some constraint terms on the basis of the original CS mathematical model, so as to improve generalization ability of CS model. All in all, these methods are the combination of traditional algorithms and DL, which makes deep unfolding networks have good interpretability and reconstruct high-quality images.

Most of the existing deep unfolding networks focus on transmitting information in the image-domain and they ignore that frequency-domain information can greatly limit the solution of the image CS reconstruction. This leads to the decline of high-frequency detail reconstruction performance. As we all know, wavelet transform can separate low-frequency and high-frequency information from signals, which is reversible. It is widely used in signal processing tasks [15,16] and shows excellent performance. In this paper, Discrete Wavelet

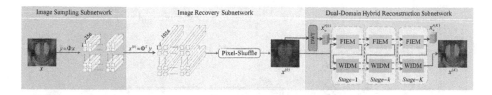

Fig. 1. The diagram of the proposed WDU-Net.

Transform (DWT) is introduced into image CS optimization model, in which frequency-domain information is used to guide CS reconstruction.

The main contributions of this paper are summarized as follows: 1) We build a DWT-based CS optimization model, in which frequency-domain information is used to guide image CS reconstruction. In the proposed model, frequency-domain consistency constraint is introduced to regularize image CS reconstruction. 2) The proposed optimization model can be decomposed into two sub-problems, which are solved in an iterative manner. The iterative optimization process is unfolded as a Wavelet-guided Deep Unfolding Network (WDU-Net) for image CS reconstruction. 3) A dual-domain guided filtering block and a self-guided filtering enhancement block are proposed to achieve feature denoising, given that the core issue of image CS reconstruction inherently involves image denoising, so as to obtain better reconstruction quality. 4) A large number of experimental results have shown that the reconstruction performance of WDU-Net is superior to that of the latest interpretable image CS reconstruction methods, especially in restoring textural details of images. We organize the rest as follows. First, we will present the proposed method in Sect. 2. Secondly, experimental results are provided and analyzed in Sect. 3. Finally, conclusions are given in Sect. 4.

2 The Proposed Method

2.1 Problem Formulation

The wavelet coefficients of natural image x can be obtained using the equation $x_w = DWT(x)$, where $DWT(\cdot)$ represents discrete wavelet transform. Because this process is reversible, the original image x can be restored from the wavelet coefficients, namely, $x = IDWT(x_w)$, where $IDWT(\cdot)$ represents the inverse discrete wavelet transform. In this paper, we introduce DWT into the CS optimization model, in which frequency-domain consistency constraint is inserted to regularize image CS reconstruction. Concretely, the proposed optimization model can be written as follows:

$$\arg \min_{x, x_w'} \frac{1}{2}\|\Phi x - y\|_2^2 + \frac{\alpha}{2}\|IDWT(x_w') - x\|_2^2 + \frac{\beta}{2}\|x_w' - x_w\|_2^2 + \lambda\|F(x)\|_1 + \mu\|\Omega(x_w')\|_1. \tag{2}$$

In Eq. (2), the first three terms are data fidelity terms, and the last two terms are regularization terms. $F(\cdot)$ and $\Omega(\cdot)$ are two regularization functions, and

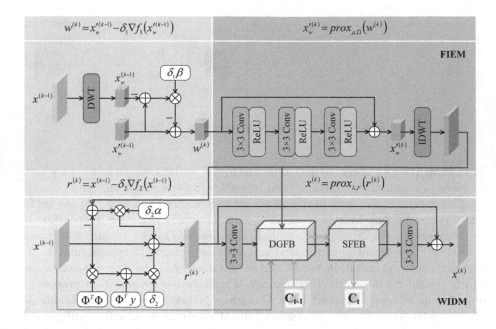

Fig. 2. The stage-k architecture of WDU-Net.

α, β, λ and μ are four hyper-parameters. x_w is the wavelet coefficient of the input image, and x'_w is the updated wavelet coefficient. It is worth mentioning that $IDWT(x'_w)$ is obtained under the condition that x'_w is given. Since solving Eq. (2) directly is a thorny problem, Eq. (2) is decomposed into the following two optimization sub-problems:

$$\arg\min_{x'_w} \frac{\beta}{2}\|x'_w - x_w\|_2^2 + \mu\|\Omega(x'_w)\|_1, \tag{3}$$

$$\arg\min_x \frac{1}{2}\|\Phi x - y\|_2^2 + \frac{\alpha}{2}\|IDWT(x'_w) - x\|_2^2 + \lambda\|F(x)\|_1. \tag{4}$$

Update x'_w: The Proximal Gradient Descent (PGD) algorithm is used to solve the optimization sub-problem given in Eq. (3). The iterative optimization process can be described as the following two steps:

$$\begin{cases} \nabla f_1(x'^{(k-1)}_w) = \beta(x'^{(k-1)}_w - x^{(k-1)}_w), \\ x'^{(k)}_w = prox_{\mu,\Omega}(x'^{(k-1)}_w - \delta_1\nabla f_1(x'^{(k-1)}_w)), \end{cases} \tag{5}$$

where $\nabla f_1(x'^{(k-1)}_w)$ denotes the gradient operation, δ_1 is the update step-size, k is the index value of the iteration, and $prox_{\mu,\Omega}(\cdot)$ is the proximal operator about the implicit prior $\mu\|\Omega(x'_w)\|_1$.

Update x: The PGD algorithm is also used to solve the optimization sub-problem given in Eq. (4). The iterative optimization process can be described as the following two steps:

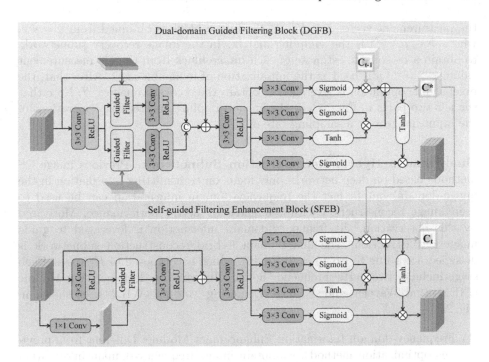

Fig. 3. The structures of Dual-domain Guided Filtering Block (DGFB) and Self-guided Filtering Enhancement Block (SFEB).

$$\begin{cases} \nabla f_2(x^{(k-1)}) = \Phi^T(\Phi x^{(k-1)} - y) - \alpha(IDWT(x'^{(k)}_w) - x^{(k-1)}), \\ x^{(k)} = prox_{\lambda,F}(x^{(k-1)} - \delta_2 \nabla f_2(x^{(k-1)})), \end{cases} \tag{6}$$

where $\nabla f_2(x^{(k-1)})$ represents the gradient operation, δ_2 is the step-size of the update, and $prox_{\lambda,F}(\cdot)$ is the proximal operator about the implicit prior $\lambda\|F(x)\|_1$. In summary, Eq. (5) and Eq. (6) can be alternatively updated to obtain the solution of CS reconstruction problem given in Eq. (2).

2.2 Wavelet-Guided Deep Unfolding Network (WDU-Net)

According to joint alternate updating of Eq. (5) and Eq. (6), iterative optimization procedures are expanded into an interpretable WDU-Net. The overall framework of WDU-Net is shown in Fig. 1. The proposed WDU-Net consists of image sampling subnetwork, image recovery subnetwork and dual-domain hybrid reconstruction subnetwork. Next, we will introduce these three subnetworks.

Image Sampling Subnetwork and Image Recovery Subnetwork. The image sampling subnetwork first divides the input image $x \in R^{H \times W}$ into $\frac{H}{\sqrt{B}} \times \frac{W}{\sqrt{B}}$ non-overlapping image blocks $\bar{x}$, where B is the size of the image block. Then, the measurement vector $\bar{y} \in R^M$ of the image block is obtained from $\bar{y} = \Phi\bar{x}$, where $\Phi \in R^{M \times N}$ is the sampling matrix. In the image recovery subnetwork, to obtain a reasonable estimate of each image block from the CS measurement value, $\bar{x}^{(0)} = \Phi^T \bar{y}$ is used as the initialization of image blocks $\bar{x}$. After that, the pixel-shuffle operation is leveraged to obtain the recovery image $x^{(0)}$. Note that the left two parts in Fig. 1 show the image sampling and recovery operations of the input image x at a sampling rate of 25%.

Dual-Domain Hybrid Reconstruction Subnetwork. Previous image CS methods based on deep networks only focus on transmitting information in the image-domain, but in fact the frequency-domain information can be used to keep image reconstruction within bounds for better performance. Motivated by this observation, the frequency-domain information is leveraged to guide image reconstruction. The dual-domain hybrid reconstruction subnetwork is designed according to Eq. (5) and Eq. (6), which is composed of K stages. Each stage includes two parts: Frequency-domain Information Enhancement Module (FIEM) and Wavelet-guided Image Denoising Module (WIDM), as shown in Fig. 2.

(1) Frequency-domain Information Enhancement Module: Different from previous optimization methods, we obtain image frequency-domain information through DWT and enhance it to improve network representation ability, as shown in Fig. 2. Equation (5) is expanded into FIEM. In this module, we first obtain the frequency feature $w^{(k)}$ of the image through the Gradient Descent Algorithm (GDA). Since the regularization term $\Omega(\cdot)$ is intractable in the proximal operator $prox_{\mu,\Omega}(\cdot)$, we use residual block to approximate the proximal operator. Residual block consists of three groups of convolution and activation operations, which are used to extract shallow features and enhance frequency-domain information.

(2) Wavelet-guided Image Denoising Module: The frequency-domain information updated by FIEM is embedded into image reconstruction process by using IDWT. The unfolding network about Eq. (6) is named WIDM. The network architecture of WIDM is shown in Fig. 2. In this module, we first obtain image feature $r^{(k)}$ through the GDA. Since the regularization function $F(\cdot)$ exists in $prox_{\lambda,F}(\cdot)$, which cannot be explicitly solved, we use convolutional neural network to replace it. To better remove noises in $r^{(k)}$ and improve reconstruction performance, we propose Dual-domain Guided Filtering Block (DGFB) and Self-guided Filtering Enhancement Block (SFEB), as shown in Fig. 3. Among them, DGFB uses the frequency-domain reconstruction map and image-domain reconstruction map from the previous stage as guide maps to guide [17] convolutional features processed by a set of convolution and activation operations. Then, the filtered features are processed by a set of convolution and activation operations to obtain enhanced features, and the two sets of enhanced features are cascaded along the channel dimension. After that, residual connections are used to supplement the

enhanced features. Considering the problem of information loss in the process of feature transmission, we use a simplified Long Short Term Memory (LSTM) mechanism to effectively control the flow of information and capture the long-term dependence during iterations. In addition, the cell state in simplified LSTM can retain important information and can be modified and updated through the gating mechanism, effectively reducing information loss during transmission. SFEB is mainly used to further enhance and restore the features output by DGFB. Specifically, in SFEB, one-channel feature map from the DGFB output is firstly obtained by 1×1 convolution operations, which is used as a guidance map. Then, guided filtering [17] is performed on the features processed by convolution and activation operations, and the filtered features are then processed by convolution and activation operations again. Subsequently, residual connections are used. Finally, the simplified LSTM mechanism is used to selectively reserve valid features. At the end of WIDU, the 3×3 convolution is used to transform multi-channel feature maps into one-channel image, and then $r^{(k)}$ is integrated with the reconstructed image by using skip-connection to obtain the final reconstructed image.

2.3 Loss Function

In this section, we add the orthogonal constraint of sampling matrix by following the literature of OPINE-Net [8] and insert consistency constraint of wavelet coefficients in the WDU-Net. For orthogonal constraints $\Phi\Phi^T = I$, we add an orthogonal loss term, expressed as $\mathscr{L}_{orth}$. Given the training dataset $\{(x_i)\}_{i=1}^{N_b}$, in order to reduce the difference between x_i and the reconstructed image $x_i^{(K)}$, as well as between original image's wavelet coefficient x_{wi} and the reconstructed image's wavelet coefficient $x_{wi}'^{(K)}$, we use mean absolute error function as the loss function of CS reconstruction, namely,

$$\mathscr{L}_{rec} = \frac{1}{N_a N_b} \sum_{i=1}^{N_b} \|x_i - x_i^{(K)}\|_1, \tag{7}$$

$$\mathscr{L}_{wave} = \frac{1}{N_a N_b} \sum_{i=1}^{N_b} \|x_{wi} - x_{wi}'^{(K)}\|_1, \tag{8}$$

$$\mathscr{L}_{orth} = \frac{1}{M^2} \|\Phi\Phi^T - I\|_1^2, \tag{9}$$

where N_a is the size of each image, N_b is the number of training images, and K is the number of stages of the network. I represents the identity matrix. The total loss function can be expressed as:

$$\mathscr{L}(\Theta) = \mathscr{L}_{rec} + \gamma\mathscr{L}_{wave} + \eta\mathscr{L}_{orth}, \tag{10}$$

where $\Theta = \{\Phi, \alpha, \beta, \delta_1, \delta_2\}_{k=1}^K$ is the learnable parameter in WDU-Net. Empirically, both γ and η are set to be 0.01.

Table 1. The average PSNR/SSIM comparison of interpretable CS methods on Set11, CBSD68 and Urban100 datasets. (The best results are shown in bold, and the second best results are displayed in underline.)

Datasets	Methods	CS Ratios				
		10%	25%	30%	40%	50%
Set11	ISTA-Net+[7]	26.53/0.8066	32.44/0.9239	33.77/0.9388	35.99/0.9580	38.03/0.9704
	OPINE-Net+[8]	29.81/0.8902	34.83/0.9514	36.04/0.9601	38.24/0.9721	40.20/0.9799
	MADUN[9]	27.87/0.8470	33.28/0.9373	34.66/0.9494	36.85/0.9644	38.60/0.9739
	COAST[10]	28.67/0.8608	-	35.07/0.9507	37.22/0.9655	39.03/0.9749
	DGUNet+[12]	28.86/0.8791	31.53/0.9266	34.26/0.9508	36.44/0.9662	39.13/0.9782
	FSOINet[11]	29.80/0.8911	34.84/0.9526	36.18/0.9618	38.29/0.9727	40.17/0.9806
	WDU-Net	**30.12/0.8954**	**35.38/0.9559**	**36.56/0.9635**	**38.66/0.9740**	**40.65/0.9816**
CBSD68	ISTA-Net+[7]	25.33/0.7011	29.29/0.8505	30.34/0.8777	32.18/0.9161	34.02/0.9423
	OPINE-Net+[8]	27.81/0.8047	31.49/0.9064	32.51/0.9238	34.41/0.9487	36.32/0.9657
	MADUN[9]	25.52/0.7191	29.35/0.8575	30.46/0.8837	32.22/0.9183	33.94/0.9424
	COAST[10]	26.42/0.7434	-	31.15/0.8939	33.02/0.9273	34.82/0.9502
	DGUNet+[12]	26.89/0.7824	28.68/0.8639	30.97/0.9063	32.82/0.9351	35.09/0.9592
	FSOINet[11]	27.64/0.8021	31.22/0.9025	32.32/0.9217	34.24/0.9474	36.03/0.9640
	WDU-Net	**28.01/0.8082**	**31.76/0.9096**	**32.78/0.9271**	**34.70/0.9513**	**36.61/0.9674**
Urban100	ISTA-Net+[7]	23.52/0.7218	28.84/0.8830	30.17/0.9073	32.31/0.9371	34.41/0.9574
	OPINE-Net+[8]	26.64/0.8375	31.44/0.9280	32.59/0.9417	34.69/0.9602	36.67/0.9728
	MADUN[9]	25.08/0.7870	30.23/0.9079	31.53/0.9265	33.48/0.9483	35.28/0.9634
	COAST[10]	25.80/0.8012	-	31.93/0.9305	34.11/0.9528	35.97/0.9670
	DGUNet+[12]	25.61/0.8152	28.99/0.9020	30.86/0.9278	32.96/0.9501	35.24/0.9674
	FSOINet[11]	26.56/0.8412	31.49/0.9303	32.66/0.9438	34.77/0.9618	36.55/0.9728
	WDU-Net	**27.10/0.8506**	**32.17/0.9371**	**33.25/0.9486**	**35.39/0.9653**	**37.37/0.9762**

3   Experimental Results

3.1   Implementation Details

During training, we use the 91 images from [8] to obtain the training dataset. The brightness components of 88912 image blocks with a size of 32×32 are obtained by random cropping, and then the training dataset $\{(x_i)\}_{i=1}^{N_b}$ is obtained, where $N_b = 88912$. In WDU-Net, the number of feature channels is 32, the batch size is set to 64, the number of stages K is 9, and the convolution kernel size is set to 3×3. The Adam optimizer [18] is chosen to optimize the proposed network. The initial learning rate is 5e–4, and the cosine annealing strategy is used to reduce it to 5e–5, and the warm-up time is 3. The network is trained by a total of 100 epochs. Our experimental simulations are implemented by PyTorch [21], and the workbench is a computer configured with the NVIDIA RTX A6000 GPU and the Intel(R) Core(TM) i9-10980XE CPU @ 3.00 GHz. For testing, we use three publicly available test datasets, including Set11 [4], CBSD68 [20] and Urban100 [19]. In addition, we use Set14 [23] as the validation dataset. We use two commonly used image evaluation criteria, namely Peak Signal-to-Noise Ratio (PSNR) and Structural Similarity (SSIM) [22], to compare the performance of CS reconstruction.

Groundtruth PSNR/SSIM   ISTA-Net$^+$ 35.32/0.9510   OPINE-Net$^+$ 37.65/0.9705   MADUN 36.58/0.9651

COAST 36.80/0.9650   FSOINet 38.00/0.9731   DGUNet$^+$ 36.73/0.9679   WDU-Net 38.55/0.9756

Fig. 4. The visual comparison between the proposed WDU-Net and several recent interpretable CS methods on the Set11 dataset.

3.2 Comparison with the Latest Interpretable CS Methods

We compare the proposed WDU-Net with the latest six interpretable CS reconstruction methods, including ISTA-Net$^+$ [7], OPINE-Net$^+$ [8], MADUN [9], COAST [10], DGUNet$^+$ [12] and FSOINet [11]. For a fair comparison, we retrain these six CS reconstruction methods on our workbench by using the source code provided by the author, and use the same training dataset as done in WDU-Net. Table 1 shows the average PSNR/SSIM comparison of these methods on the Set11, CBSD68 and Urban100 datasets. It can be observed from Table 1, ISTA-Net$^+$ has the worst performance, because its network is too simple and there is no constraint on the sampling matrix. On the basis of ISTA-Net$^+$, OPINE-Net$^+$ applies binary and orthogonal constraints to the sampling matrix, which improves the network reconstruction performance. By designing more complex network architecture, the reconstruction performance of MADUN, COAST, DGUNet$^+$ and FSOINet are improved in different degrees compared to ISTA-Net$^+$. Different from these methods, our proposed WDU-Net guides image reconstruction by introducing frequency-domain information. Table 1 shows that under five sampling ratios, the reconstruction performance of our WDU-Net is superior to that of several latest interpretable CS reconstruction methods, which clearly verifies the advantages of WDU-Net.

Figure 4 shows the visual comparison results of several CS methods for restoring "boats" on the Set11 dataset with a sampling rate of 30%. From this figure, it can be seen that the image restored by DGUNet$^+$ has serious artifacts, and the image details restored by ISTA-Net$^+$, OPINE-Net$^+$, MADUN, COAST and FSOINet are poor. The image restored by WDU-Net outperforms competitive methods in terms of texture and detail preservation. Figure 5 shows the visual

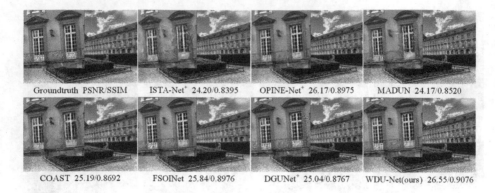

Fig. 5. The visual comparison between the proposed WDU-Net and several recent interpretable CS methods on the Urban100 dataset.

Table 2. The comparison of ablation study between different modules of WDU-Net when CS sampling ratio=25%. (The best results are shown in bold.)

FIEM	DGFB	SFEB	Set11	Set14	CBSD68
			PSNR/SSIM	PSNR/SSIM	PSNR/SSIM
✗	✔	✔	31.51/0.9208	30.68/0.8900	29.87/0.8795
✔	✗	✔	35.10/0.9540	33.34/0.9233	31.64/0.9086
✔	✔	✗	34.94/0.9530	33.29/0.9231	31.64/0.9088
✔	✔	✔	**35.38/0.9559**	**33.54/0.9249**	**31.76/0.9096**

comparison of several CS reconstruction methods to restore the image named "img_077" from the Urban100 dataset. It can be found from this figure that when the CS sampling ratio is 30%, the images reconstructed by ISTA-Net[+], COAST and DGUNet[+] suffer from different degrees of artifact distortion. The details of images reconstructed by OPINE-Net[+], MADUN and FSOINet are blurred. Obviously, compared with other interpretable CS methods, our method retains more detail information and has clearer edges, which clearly shows that frequency information can improve representation ability of the proposed WDU-Net for image reconstruction. It can be seen from the above experiments that compared with other CS methods, the proposed WDU-Net has achieved higher improvement in both objective quality and subjective quality.

3.3   Ablation Study

To evaluate the contribution of each module of the proposed WDU-Net, we perform ablation studies on FIEM, DGFB and SFEB. Table 2 shows the ablation results of these three modules on Set11, Set14 and CBSD68 datasets with CS sampling rate of 25%. Among them, "✗" means that the module is removed from the original baseline, and "✔" means that the module is retained from the original baseline. It is worth mentioning that when FIEM is discarded, the loss term of wavelet coefficient "$\mathscr{L}_{wave}$" is removed at the same time, and the wavelet guide map in DGFB is replaced by the output map of the previous stage as the guide map. As you can see from Table 2, network performance deteriorates dramatically when FIEM is removed. It can be seen that inserting DWT into the reconstruction model of CS can promote the reconstruction performance of the network. In addition, both DGFB and SFEB improve the reconstruction capability of the network to varying degrees. In general, the FIEM, DGFB and SFEB can improve the performance of CS reconstruction for the proposed WDU-Net.

4 Conclusion

In this paper, we propose a new CS optimization model based on DWT, and unfold the optimization process of this model as WDU-Net. The proposed WDU-Net guides the CS reconstruction process by introducing frequency-domain information. In addition, in order to remove the noise more effectively in the process of image reconstruction, we use the guided filter to carry out dual-domain guided denoising and self-guided denoising. Experimental results show that compared with other advanced image CS methods, frequency-domain information can effectively improve image reconstruction performance. Considering the high complexity of existing deep models, in future work, we plan to use knowledge distillation to build lightweight, efficient and accurate models for image CS reconstruction.

Acknowledgements. This work was supported by National Natural Science Foundation of China Youth Science Foundation Project (No. 62202323), Fundamental Research Program of Shanxi Province (No. 202103021223284), Taiyuan University of Science and Technology Scientific Research Initial Funding (No. 20192023, No. 20192055), Graduate Education Innovation Project of Taiyuan University of Science and Technology in 2022 (SY2022027), National Natural Science Foundation of China (No. 62072325).

References

1. Donoho, D.L.: Compressed sensing. IEEE Trans. Inf. Theory **52**(4), 1289–1306 (2006)
2. Liu, X.J., Xia, S.T., Fu, F.W.: Reconstruction guarantee analysis of basis pursuit for binary measurement matrices in compressed sensing. IEEE Trans. Inf. Theory **63**(5), 2922–2932 (2017)
3. Ji, S., Xue, Y., Carin, L.: Bayesian compressive sensing. IEEE Trans. Signal Process. **56**(6), 2346–2356 (2008)

4. Kulkarni, K., Lohit, S., Turaga, P., Kerviche, R., Ashok, A.: ReconNet: non-iterative reconstruction of images from compressively sensed measurements, pp. 449–458 (2016)
5. Sun, Y., Chen, J., Liu, Q., Liu, B., Guo, G.: Dual-path attention network for compressed sensing image reconstruction. IEEE Trans. Image Process. **29**, 9482–9495 (2020)
6. Shi, W., Jiang, F., Liu, S., Zhao, D.: Scalable convolutional neural network for image compressed sensing. In: Proceedings of the IEEE/CVF Conference on Computer Vision and Pattern Recognition, pp. 12290–12299 (2019)
7. Zhang, J., Ghanem, B.: ISTA-Net: interpretable optimization-inspired deep network for image compressive sensing. In: Proceedings of the IEEE Conference on Computer Vision and Pattern Recognition, pp. 1828–1837 (2018)
8. Zhang, J., Zhao, C., Gao, W.: Optimization-inspired compact deep compressive sensing. IEEE J. Sel. Topics Signal Process. **14**(4), 765–774 (2020)
9. Song, J., Chen, B., Zhang, J.: Memory-augmented deep unfolding network for compressive sensing. In: Proceedings of the 29th ACM International Conference on Multimedia, pp. 4249–4258 (2021)
10. You, D., Zhang, J., Xie, J., Chen, B., Ma, S.: COAST: controllable arbitrary-sampling network for compressive sensing. IEEE Trans. Image Process. **30**(1), 6066–6080 (2021)
11. Chen, W., Yang, C., Yang, X.: FSOINET: feature-space optimization-inspired network for image compressive sensing. In: ICASSP 2022-2022 IEEE International Conference on Acoustics, Speech and Signal Processing (ICASSP), pp. 2460–2464. IEEE (2022)
12. Mou, C., Wang, Q., Zhang, J.: Deep generalized unfolding networks for image restoration. In: Proceedings of the IEEE/CVF Conference on Computer Vision and Pattern Recognition, pp. 17399–17410 (2022)
13. Zhang, T., Li, L., Igel, C., Oehmcke, S., Gieseke, F., Peng, Z.: LR-CSNet: low-rank deep unfolding network for image compressive sensing. 2022 IEEE 8th International Conference on Computer and Communications (ICCC), pp. 1951–1957 (2022)
14. Lian, Q., Su, Y., Shi, B., Zhang, D.: LG-Net: local and global complementary priors induced multi-stage progressive network for compressed sensing. Signal Process. **202**, 108737 (2023)
15. Zhang, H., Jin, Z., Tan, X., Li, X.: Towards lighter and faster: learning wavelets progressively for image super-resolution. In: Proceedings of the 28th ACM International Conference on Multimedia, pp. 2113–2121 (2020)
16. Zou, W., Chen, L., Wu, Y., Zhang, Y., Xu, Y., Shao, J.: Joint wavelet sub-bands guided network for single image super-resolution. IEEE Trans. Multimedia **25**, 4623–4637 (2022)
17. He, K., Sun, J., Tang, X.: Guided image filtering. IEEE Trans. Pattern Anal. Mach. Intell. **35**(6), 1397–1409 (2013). https://doi.org/10.1109/TPAMI.2012.213
18. Kingma, D.P., Ba, J.: Adam: a method for stochastic optimization, vol. 1, no. 1, pp. 1–15 (2014). arXiv preprint arXiv:1412.6980
19. Huang, J.B., Singh, A., Ahuja, N.: Single image super-resolution from transformed self-exemplars. In: Proceedings of the IEEE Conference on Computer Vision and Pattern Recognition, pp. 5197–5206 (2015)
20. Martin, D., Fowlkes, C., Tal, D., Malik, J.: A database of human segmented natural images and its application to evaluating segmentation algorithms and measuring ecological statistics. In: Proceedings Eighth IEEE International Conference on Computer Vision, ICCV 2001, vol. 2, pp. 416–423. IEEE (2001)

21. Paszke, A., et al.: PyTorch: an imperative style, high-performance deep learning library. ArXiv arxiv:1912.01703 (2019)
22. Wang, Z., Bovik, A., Sheikh, H., Simoncelli, E.: Image quality assessment: from error visibility to structural similarity. IEEE Trans. Image Process. **13**(4), 600–612 (2004). https://doi.org/10.1109/TIP.2003.819861
23. Cui, W., Xu, H., Gao, X., Zhang, S., Jiang, F., Zhao, D.: An efficient deep convolutional laplacian pyramid architecture for CS reconstruction at low sampling ratios. In: 2018 IEEE International Conference on Acoustics, Speech and Signal Processing (ICASSP), pp. 1748–1752. IEEE (2018)

Video Analysis and Understanding

Memory-Augmented Spatial-Temporal Consistency Network for Video Anomaly Detection

Zhangxun Li[1], Mengyang Zhao[1], Xinhua Zeng[1(✉)], Tian Wang[2,3], and Chengxin Pang[4]

[1] Academy for Engineering and Technology, Fudan University, Shanghai, China
zengxh@fudan.edu.cn
[2] Institute of Artificial Intelligence, SKLSDE, Beihang University, Beijing, China
[3] Zhongguancun Laboratory, Beijing, China
[4] School of Electronics and Information Engineering, Shanghai University of Electric Power, Shanghai, China

Abstract. Video anomaly detection (VAD) in intelligent surveillance systems is a crucial yet highly challenging task. Since appearance and motion information is vital for identifying anomalies, existing unsupervised VAD methods usually learn normality from them. However, these approaches tend to consider appearance and motion separately or simply integrate them while ignoring the consistency between them, resulting in sub-optimal performance. To address this problem, we propose a Memory-Augmented Spatial-Temporal Consistency Network, aiming to model the latent consistency between spatial appearance and temporal motion by learning the unified spatiotemporal representation. Additionally, we introduce a spatial-temporal memory fusion module to record spatial and temporal prototypes of regular patterns from the unified spatiotemporal representation, increasing the gap between normal and abnormal events in the feature space. Experimental results on three benchmarks demonstrate the effectiveness of the spatial-temporal consistency for VAD tasks. Our method performs comparably to the state-of-the-art methods with AUCs of 97.6%, 89.3%, and 73.3% on the UCSD Ped2, CUHK Avenue, and ShanghaiTech datasets, respectively.

Keywords: Video anomaly detection · Spatial-temporal consistency · Unified spatiotemporal representation · Memory network

1 Introduction

Video anomaly detection (VAD) aims to identify the events that diverge from expected behavior, which has been extensively researched in recent years [1].

This work is partially supported by the National Natural Science Foundation of China (Grant No. 61972016) and the Science and Technology Commission of Shanghai Municipality Research Fund (Grant No. 21JC1405300).

Q. Liu et al. (Eds.): PRCV 2023, LNCS 14430, pp. 95–107, 2024.
https://doi.org/10.1007/978-981-99-8537-1_8

Compared to regular events, anomalies are usually characterized by rarity and uncertainty, making it almost unfeasible to gather all kinds of abnormal samples for training. Therefore, many VAD methods [2–6] perform to train the model in an unsupervised mode that learns regular patterns using only normal data. Then the instances deemed as outliers by this trained model are anomalies.

In recent years, various methods [2–4,7,8] have been proposed to tackle VAD tasks. Nevertheless, most of them either focus on learning only spatial appearance [3,7] or simply integrate the spatial and temporal information [2,4] to detect anomalies while ignoring the underlying correlation between them (i.e., the spatial-temporal consistency), leading to sub-optimal performance. Specifically, the spatial-temporal consistency takes into account the inner correlation between spatial appearance and temporal motion, e.g., a person and a bag are both common objects, but it could be abnormal when the person is robbing the bag. Without considering the consistent correlation between spatial appearance and temporal motion, the anomaly detector may fail to detect these anomalous events [8]. Recently, some approaches [8,9] attempt to explore the relationship between them. However, they still have difficulty modeling spatial-temporal consistency since they learn from spatial and temporal information separately, resulting in limited performance gains.

In this paper, we propose a united framework named Memory-Augmented Spatial-Temporal Consistency Network (MASTC-Net), which learns the spatial appearance and temporal motion simultaneously to consider the spatial-temporal consistency. Specifically, we first capture the unified spatiotemporal representation via an integration module, which utilizes the spatial-temporal network inspired by PredRNN [10] to extract spatial and temporal features within the same memory cell. Furthermore, we introduce a spatial-temporal memory fusion module that contains a memory pool [11] to memorize normal patterns from unified spatiotemporal representation, aiming to restrict the model's ability to represent anomalies. Finally, the obtained spatiotemporal fusion feature is decoupled as appearance and motion, performing to predict the future frame and estimate RGB difference, respectively. Our proposed MASTC-Net accounts for spatial-temporal consistency, which enables it to understand normal patterns more comprehensively and thus effectively distinguish anomalies from regular events. In summary, our main contributions can be summarized as follows:

– We propose a simple yet effective method to make full use of spatial-temporal consistency by learning unified spatiotemporal representation using an integration module.
– We design a spatial-temporal memory fusion module for recording spatiotemporal patterns of normal events, increasing the gap between normal and abnormal patterns in the feature space.
– Extensive experiments and ablation studies on benchmark datasets demonstrate that our proposed MASTC-Net is competitive compared with the state-of-the-art methods.

2   Related Work

Reconstruction-Based Methods. The reconstruction-based methods tend to measure the deviation between the reconstructed frame and the input frame. They take the assumption that the model learned on only regular events cannot represent anomalies well, leading to large reconstruction errors. Some researchers adopted Generative Adversarial Networks (GANs) [12] or deep auto-encoders [3,13,14] to learn the representation of normality. For example, Ravanbakhsh et al. [12] proposed a VAD model based on GANs to learn the distribution of normal data, then the abnormal instances are regarded as outliers in respect of the distribution. Hasan et al. [13] devised a deep CNN-based network by using an autoencoder to learn normal patterns, then the instances with large reconstruction errors are deemed as anomalies at test time. However, sometimes anomalies can be reconstructed well due to the powerful representation of convolution neural networks. To deal with this problem, Gong et al. [3] introduced a memory module to memorize the prototypes of normal regularities. They retrieve memorized items and use them to boost the model's ability to reconstruct normal data while constraining the abnormal ones. Nevertheless, most of them almost ignore the temporal characteristics, leading to poor performance.

Prediction-Based Methods. Compared with reconstruction-based methods, the prediction-based methods [2,8,9,15,16] typically exhibit superior performance since they consider both spatial appearance and temporal motion. For example, Liu et al. [2] devised a prediction framework that performs to predict a future frame using consecutive video frames. Then the anomalies could be detected according to the prediction errors during the testing phase. Lee et al. [15] proposed a hybrid network that combined ConvLSTM with GAN to learn spatiotemporal features of normal events. However, these methods only utilize the temporal features as the motion constraints while not considering the inner correlation between the spatial appearance and temporal motion, which makes it hard to identify anomalies effectively. Recently, these state-of-the art approaches [8,9] are proposed to explore the relationship between them. For example, Cai et al. [8] devised a two-stream framework termed AMMC-Net to model spatial-temporal consistency. Notice that they adopt RGB appearance and optical flow as appearance and motion signals, respectively, and perform to learn appearance patterns and motion patterns separately. On the one hand, adopting optical flow leads to additional computational costs. On the other hand, the fused spatiotemporal features may be inconsistent to some extent since there are specific differences between the optical flow and the RGB appearance. Furthermore, Chang et al. [9] proposed to utilize spatial and temporal autoencoders to decouple the spatial-temporal representation. They focus on separately capturing the spatial and temporal information while paying little attention to the consistency between them. Although these methods have achieved some good performance for VAD tasks, they still have some shortcomings in modeling spatial-temporal consistency, thus resulting in far less than expected performance.

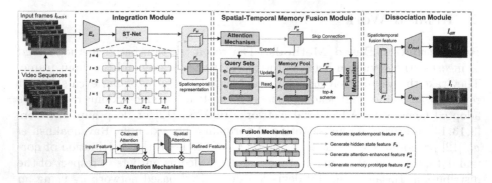

Fig. 1. Overview of the proposed Memory-**A**ugmented **S**patial-**T**emporal **C**onsistency Network (**MASTC-Net**).

3 Method

3.1 Overall Framework

As shown in Fig. 1, the proposed MASTC-Net is composed of three modules: an integration module, a spatial-temporal memory fusion module, and a dissociation module. Firstly, we input consecutive video frames into the integration module to simultaneously consider the spatial appearance and temporal motion and learn the unified spatiotemporal features. After that, the unified spatiotemporal features are put into the spatial-temporal memory fusion module to record spatiotemporal prototypes of normal patterns. Finally, we obtain the fused spatiotemporal feature and decouple it by the dissociation module, performing to predict the future frame and estimate RGB difference.

3.2 Integration Module

To model the inner consistency between the spatial appearance and temporal motion, we introduce an integration module that mainly contains two components: the spatial encoder E_s and the Spatial-Temporal Network (ST-Net). Initially, the E_s is utilized to compress the input frame sequence $\{I_{t-n}, \ldots, I_{t-1}\}$ into latent space features $\{z_{t-n}, \ldots, z_{t-1}\}$, where n denotes the number of frame in the sequence. Then every z is fed into the ST-Net in time order. The ST-Net (Fig. 2) is a convolutional recurrent neural network that consists of four ST-LSTM [10] units, which utilizes a zigzag memory stream and captures spatial and temporal representation within a unified memory cell M_{t-1}^l. The key equations of ST-Net are shown as Eq. (1), where $*$ and $\odot$ denote the convolution operator and the Hadamard product, respectively.

$$C^l_{t-1} = f_{t-1} \odot C^l_{t-2} + i_{t-1} \odot \tanh\left(W^x_g * x_{t-1} + W^h_g * H^l_{t-2}\right)$$

$$M^l_{t-1} = f'_{t-1} \odot M^{l-1}_{t-1} + i'_{t-1} \odot \tanh\left(W^{x'}_g * x_{t-1} + W^m_g * M^{l-1}_{t-1}\right) \quad (1)$$

$$H^l_{t-1} = o_{t-1} \odot \tanh\left(W_{1\times1} * \left[C^l_{t-1}, M^l_{t-1}\right]\right)$$

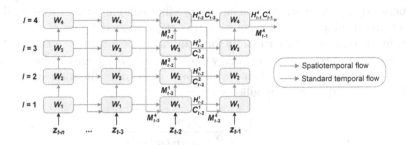

Fig. 2. Details of the ST-Net. The W_l is defined as the parameters of ST-LSTM units. The orange and green arrows denote the transmission path of spatiotemporal memory M^l_{t-1} and standard temporal memory C^l_{t-1} respectively. (Color figure online)

After processing the input features $\{z_{t-n}, \ldots, z_{t-1}\}$, we obtain the final spatiotemporal memory M^4_{t-1}, which contains the spatial and temporal information from the past time step to the current time step. Furthermore, we also adopt H^4_{t-1} as the final output of the ST-Net, which is the fusion of standard temporal memory C^4_{t-1} and spatiotemporal memory M^4_{t-1} at the current time step. Note that we denote M^4_{t-1} and H^4_{t-1} as F_{st} and F_h respectively in Fig. 1, and both of them are considered as the unified spatiotemporal representation.

3.3 Spatial-Temporal Memory Fusion Module

The spatial-temporal memory fusion module is designed to record spatial and temporal prototypes of normal patterns from the unified spatiotemporal representation. Firstly, we perform an attention mechanism to focus on the crucial information of the spatiotemporal feature F_{st} and obtain the attention-enhanced feature F^a_{st}. The attention mechanism is conducted by the convolutional block attention module [17] and the attention process can be summarized as Eq. (2), where $c(\cdot)$ and $s(\cdot)$ refer to channel attention and spatial attention respectively.

$$\begin{aligned} F' &= c\left(F_{st}\right) \otimes F_{st} \\ F^a_{st} &= s\left(F'\right) \otimes F' \end{aligned} \quad (2)$$

After that, we expand $F^a_{st} \in \mathbb{R}^{H \times W \times C}$ into query maps $q_k \in \mathbb{R}^C (k = 1, \ldots, K,$ where $K = H \times W$) along the channel dimension and send them to the memory pool [11]. The memory pool is a two-dimensional matrix $M \in \mathbb{R}^{N \times C}$, where N denotes the number of memory items and C is the dimension of each memory item $p_m \in \mathbb{R}^C (m = 1, ..., N)$ as same as the dimension of q_k. By performing the read and update operation, the memory pool can record prototypical patterns of normal events and restrict the model to represent anomalies. The details are described in the following paragraphs.

Read. The read operation aims to use the memorized items p_m to represent the queries q_k approximately. Specifically, we first calculate the cosine similarity between query maps q_k and memory items p_m, denoted by $w^{k,m}$. To further constrain the representation capacity for anomalies, we introduce a top-k scheme to retrieve several relevant items, i.e., we retain the top k largest values in $w^{k,m}$ and set other values to 0. Then we perform a softmax function on $w^{k,m}$ to calculate the weights $\bar{w}^{k,m}$ as follows:

$$\bar{w}^{k,m} = \frac{\exp\left(w^{k,m}\right)}{\sum_{m'=1}^{N} \exp\left(w^{k,m'}\right)} \tag{3}$$

As shown in Eq. 4, by using the combination of the prototype items p_m according to $\bar{w}^{k,m}$, we obtain the weighted item $\hat{p}_r^k \in \mathbb{R}^C$ that represents q_k approximately. Eventually, we perform the read operation to each query q_k and aggregate all weighted items $\hat{p}_r^k$ to obtain the memory prototype feature $F_{st}^m \in \mathbb{R}^{H \times W \times C}$, which represents the attention-enhanced feature F_{st}^a approximately.

$$\hat{p}_r^k = \sum_{m=1}^{N} \bar{w}^{k,m} p_m \tag{4}$$

Update. The update operation is to update each memory item p_m using the queries q_k. Specifically, when we update an item, we search for all the queries declared that the item is the most relevant one based on the similarity weights $w^{k,m}$, and the indexes of all these selected queries are recorded in the $\mathbb{U}$. Then we perform a softmax function horizontally to compute matching probabilities and apply a max normalization to renormalize it using the queries indexed by the set $\mathbb{U}$ to obtain the $v_u^{k,m}$. After that, we update each item with corresponding queries as follows:

$$\hat{q}_w^k = \sum_{k \in \mathbb{U}} v_u^{k,m} q_k \tag{5}$$

$$p_m = \left\| p_m + \hat{q}_w^k \right\|_2$$

Since the memory pool stores the compressed representation of the input features, some detailed spatiotemporal information may be lost. To this end, we perform a fusion mechanism between the attention-enhanced feature F_{st}^a, the memory prototype feature F_{st}^m and the hidden state feature F_{st}^h to obtain the final spatiotemporal fusion feature $F_{st}^f \in \mathbb{R}^{H \times W \times 3C}$. Specifically, the fusion mechanism is conducted by the channel shuffle operation [18] and pointwise group convolution operation sequentially, which is computed as follows:

$$F_{st}^{f} = f\left(F_{st}^{a} \oplus F_{st}^{h} \oplus F_{st}^{m}\right) \tag{6}$$

where $f(\cdot)$ refers to the feature fusion operation, and $\oplus$ denotes the channel concatenation.

3.4 Dissociation Module

Due to the complex nature of anomalies, it is difficult to detect abnormal events from normality only by appearance or motion [9]. Therefore, we introduce a dissociation module to decouple the unified spatiotemporal fusion feature, aiming to detect anomalies from appearance and motion. The dissociation module consists of two decoders with the same structure: appearance decoder D_{app} and motion decoder D_{mot}. Specifically, the appearance decoder D_{app} is utilized to generate predicted frame $\hat{I}_t$ while the motion decoder D_{mot} is adopted to predict RGB difference between I_t and I_{t+1}. Note that we adopt the RGB difference between two adjacent frames as our motion information rather than utilizing optical flow, which proved to be effective [9] and also reduce computational cost.

3.5 Training Loss

To better optimize our model, we construct loss functions that include appearance loss $\mathcal{L}_{app}$, motion loss $\mathcal{L}_{mot}$, feature compactness loss $\mathcal{L}_c$ and feature separateness loss $\mathcal{L}_s$, balanced by parameters λ_m, λ_c and λ_s as follows:

$$\mathcal{L} = \mathcal{L}_{app} + \lambda_m \mathcal{L}_{mot} + \lambda_c \mathcal{L}_c + \lambda_s \mathcal{L}_s \tag{7}$$

The appearance loss $\mathcal{L}_{app}$ is defined as the L_2 distance between the predicted frame $\hat{I}_t$ and ground truth I_t, as follows:

$$\mathcal{L}_{app} = \|\hat{I}_t - I_t\|_2 \tag{8}$$

We denote the smoothed L_1 loss between the predicted RGB difference $\hat{I}_{diff}$ and ground truth I_{diff} as the motion loss $\mathcal{L}_{mot}$, as follows:

$$\mathcal{L}_{mot} = \mathcal{S}_{L1}(\hat{I}_{diff} - I_{diff})$$
$$\mathcal{S}_{L1}(x) = \begin{cases} \frac{1}{2}x^2, & \text{if } |x| < 1 \\ |x| - \frac{1}{2}, & \text{otherwise} \end{cases} \tag{9}$$

Additionally, we adopt a feature compactness loss $\mathcal{L}_c$ proposed in [11] to ensure the representation ability of the memory items as follows, where c is the index of the closest memory item for the query q_k:

$$\mathcal{L}_c = \sum_{k=1}^{K} \|q_k - p_c\|_2 \tag{10}$$

To prevent memory items from tending to be similar during the update operation, we introduce a separateness loss $\mathcal{L}_s$ inspired by [11] as follows:

$$\mathcal{L}_s = \sum_{k}^{K}[(\|\boldsymbol{q}_k - \boldsymbol{p}_c\|_2 - \|\boldsymbol{q}_k - \boldsymbol{p}_s\|_2 + \alpha)_+$$
$$+ (\|\boldsymbol{q}_k - \boldsymbol{p}_c\|_2 - \|\boldsymbol{p}_s - \boldsymbol{p}_t\|_2 + \beta)]_+ \tag{11}$$

where $\boldsymbol{p}_c$ is the closest item to $\boldsymbol{q}_k$, $\boldsymbol{p}_s$ and $\boldsymbol{p}_t$ are the second and third nearest item, respectively. α and β are the margin and we set $\alpha = 1$, $\beta = 0.5$ respectively.

3.6 Anomaly Detection

Since the model trained only on regular events cannot represent anomalies well, we argue that the query maps obtained from abnormal frames could cause a large distance from the memory items. Therefore, we calculate the L_2 distance between each query map $\boldsymbol{q}_k$ and its closest item as follows:

$$R(\boldsymbol{q}, \boldsymbol{p}) = \frac{1}{K} \sum_{k=1}^{K} \|\boldsymbol{q}_k - \boldsymbol{p}_c\|_2 \tag{12}$$

Besides, we adopt the PSNR followed by [2] to measure the quality of the predicted frame as follows, where N is the number of pixels in the video frame:

$$P(\hat{\boldsymbol{I}}_t, \boldsymbol{I}_t) = 10 \log_{10} \frac{\max(\hat{\boldsymbol{I}}_t)}{\frac{1}{N}\|\hat{\boldsymbol{I}}_t - \boldsymbol{I}_t\|_2^2} \tag{13}$$

We also consider the prediction error of the motion decoder and calculate the $L2$ distance between $\hat{\boldsymbol{I}}_{diff}$ and $\boldsymbol{I}_{diff}$ as follows:

$$D(\hat{\boldsymbol{I}}_{diff}, \boldsymbol{I}_{diff}) = \|\hat{\boldsymbol{I}}_{diff} - \boldsymbol{I}_{diff}\|_2 \tag{14}$$

Finally, we combine the R, P, and D together to detect anomalies, and the normality score S_t can be derived as follows:

$$S_t = \lambda_p \mathcal{N}(P(\hat{\boldsymbol{I}}_t, \boldsymbol{I}_t)) + \lambda_d(1 - \mathcal{N}(D(\hat{\boldsymbol{I}}_{diff}, \boldsymbol{I}_{diff})))$$
$$+ (1 - \lambda_p - \lambda_d)(1 - \mathcal{N}(R(\boldsymbol{q}, \boldsymbol{p}))) \tag{15}$$

where λ_p and λ_d denote the weight of P and D, respectively, and $\mathcal{N}(\cdot)$ is the maximum-minimum normalization to scale the values to $[0, 1]$ as follows:

$$\mathcal{N}(z) = \frac{z - \min(z)}{\max(z) - \min(z)} \tag{16}$$

Therefore, the final normality score S_t indicates the normal level of predicted frames, i.e., a frame with higher S_t means more normal, and lower means more abnormal. We can select a threshold to determine regular or irregular frames.

4   Experiments

4.1   Datasets and Evaluation Metrics

To qualitatively evaluate the performance of our method, we conduct extensive experiments on three benchmark datasets, which are described below:

- UCSD Ped2 [19]. The Ped2 dataset contains 16 training videos and 12 testing videos with a resolution of 240 × 360. In the dataset, walking pedestrians are deemed as regularities, while riding and driving are defined as anomalies.
- CUHK Avenue [20]. The Avenue dataset consists of 16 training videos and 21 testing videos with a resolution of 640 × 360, and it includes a total of 47 anomalous events such as throwing and skipping.
- ShanghaiTech [2]. The ShanghaiTech dataset is one of the most complicated and largest datasets for VAD tasks. It contains 130 abnormal events from 13 different scenes, such as running, fighting, and robbing.

Following previous work [2], we adopt the frame-level area under the ROC curve (AUC) as the evaluation metric, and a higher AUC indicates better performance.

4.2   Implementation Details

The experiments are performed on an Nvidia Geforce RTX 2080Ti GPU. Each video frame is resized to 256 × 256 and normalized to the range $[-1, 1]$. n is empirically set to 4. The dimension of the latent feature z_{t-1} is set to $64 \times 64 \times 128$, and we set the H, W and C to 64, 64 and 128, respectively. The number of memory items N is set to 50, 100, and 200 for Ped2, Avenue, and ShanghaiTech, respectively. For the parameters of loss functions, λ_m, λ_c and λ_s is empirically set to 0.5, 1 and 1, respectively. For Ped2, Avenue, and ShanghaiTech, λ_p is set to 0.2, 0.4, and 0.6, λ_d is set to 0.6, 0.1, and 0.2, respectively. The number of retrieved memory items k in the read operation is set to 30, 60, and 120 for Ped2, Avenue, and ShanghaiTech, respectively.

4.3   Ablation Study

We implement corresponding experiments to verify the effectiveness of devised components on AUC performance, and the results are presented in Table 1. The model $\mathcal{A}$ is our baseline model, which contains the ST-Net and appearance decoder D_{app}. When combining the appearance decoder and motion decoder, the AUC performance significantly improves by 1.4% ($\mathcal{B}$ vs. $\mathcal{A}$), which indicates that considering both the spatial appearance and temporal motion information is crucial for the VAD task. Notice that the adoption of the memory pool enhances the AUC performance by 1.2% ($\mathcal{C}$ vs. $\mathcal{B}$), demonstrating that the memory pool can effectively improve performance using the recorded normal prototypes. Besides, model $\mathcal{E}$ is equipped with the attention and fusion mechanism based on model $\mathcal{C}$, which brings an AUC gain of 1.1% ($\mathcal{E}$ vs. $\mathcal{C}$), demonstrating that the proposed attention and fusion strategies are effective.

Table 1. Results of ablation studies. We report frame-level AUC (%) performance on the CUHK Avenue dataset. The bold entries show the best results.

Model	ST-Net	Memory Pool	Attention	Fusion	D_{app}	D_{mot}	AUC(%)
$\mathcal{A}$	✓				✓		85.6
$\mathcal{B}$	✓				✓	✓	87.0
$\mathcal{C}$	✓	✓			✓	✓	88.2
$\mathcal{D}$	✓	✓	✓		✓	✓	88.7
$\mathcal{E}$	✓	✓	✓	✓	✓	✓	**89.3**

4.4 Comparison with State-of-the-Art Methods

As shown in Table 2, the methods in our comparison experiments can be categorized as reconstruction-based methods (Recon-based) [3,7,14] and prediction-based methods (Pred-based) [2,6,8,9,11,16,21]. Compared with reconstruction-based methods, our method outperforms them due to making full use of both spatial and temporal information. Although the prediction-based methods have made a lot of effort to learn the representation of normal events from both spatial and temporal aspects and achieve superior performance, they still lack the exploration of the consistent correlation between the two modalities. In comparison, we attempt to model the inherent consistency between spatial appearance and temporal motion, which achieves competitive performance on all benchmarks.

Table 2. Results of The frame-level AUC (%) comparison. The bold entries indicate the best performance.

Types	Methods	Ped2	Avenue	ShanghaiTech
	Stacked-RNN (2017 ICCV) [7]	92.2	81.7	68.0
Recon-based	MemAE (2019 ICCV) [3]	94.1	83.3	71.2
	MESDnet (2021 TMM) [14]	95.6	86.3	73.2
	Frame-Pred (2018 CVPR) [2]	95.4	84.9	72.8
	MNAD (2020 CVPR) [11]	97.0	88.5	70.5
	AMMC-Net (2021 AAAI) [8]	96.6	86.6	73.7
Pred-based	STC (2022 ICPR) [16]	96.7	87.8	73.1
	AMAE (2022 TCAS-II) [6]	97.4	88.2	73.6
	STD (2022 PR) [9]	96.7	87.1	73.7
	STCEN (2022 PR) [21]	96.9	86.6	**73.8**
	MASTC-Net (Ours)	**97.6**	**89.3**	73.3

4.5   Visualization Analysis

As shown in Fig. 3, we visualize the normality score curves over time for frames in some testing videos on three benchmark datasets. We can see that the normality score drops sharply when abnormal events occur, such as bicycling and running, and remains high and stable during normal events, indicating that our proposed method can effectively identify anomalies.

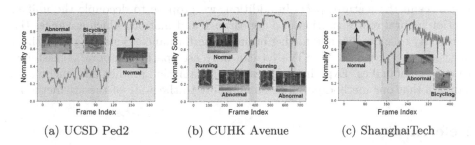

(a) UCSD Ped2   (b) CUHK Avenue   (c) ShanghaiTech

Fig. 3. Normality score curves of some testing video clips. The blue regions represent ground truth regular frames, while the red regions indicate anomalous frames. (Color figure online)

To display the gap in the prediction results between normal and abnormal instances, we draw the prediction error map as shown in Fig. 4. For simplicity, we denote the prediction results of the appearance decoder and motion decoder as appearance and motion, respectively. Notice that the abnormal objects in anomalies are highlighted in the error map, which indicates the larger prediction error. In contrast, the error map of normal frames is not prominent, further demonstrating the effectiveness of our proposed MASTC-Net.

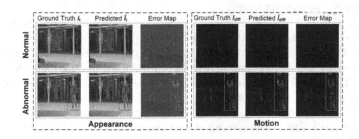

Fig. 4. Visualization of prediction results on the CUHK Avenue dataset. The first row is for regular streets, while the second includes anomalies that a man throwing a bag.

5 Conclusion

In this paper, we propose a framework named MASTC-Net to tackle unsupervised VAD tasks. By learning spatial appearance and temporal motion simultaneously, we model the intrinsic consistent correlation between them and capture the unified spatiotemporal representation. Further, we design a spatial-temporal memory fusion module for recording normal prototypes from the unified spatiotemporal representation, increasing the gap between normal and abnormal instances in the feature space. In addition, we propose to decouple the unified spatiotemporal representation into appearance and motion, which enhances the sensitivity to anomalies and further improves performance. Benefiting from these methods, the proposed MASTC-Net can fully use spatial-temporal consistency to detect anomalies. Experimental results and ablation studies on benchmarks validate that our method performs comparably to the state-of-the-art methods.

References

1. Liu, Y., Yang, D., Wang, Y., Liu, J., Song, L.: Generalized video anomaly event detection: systematic taxonomy and comparison of deep models. arXiv preprint arXiv:2302.05087 (2023)
2. Liu, W., Luo, W., Lian, D., Gao, S.: Future frame prediction for anomaly detection - a new baseline. In: CVPR, pp. 6536–6545 (2018)
3. Gong, D., Liu, L., Le, V., Saha, B., Mansour, M.R., Venkatesh, S., Van Den Hengel, A.: Memorizing normality to detect anomaly: memory-augmented deep autoencoder for unsupervised anomaly detection. In: ICCV, pp. 1705–1714 (2019)
4. Nguyen, T.N., Meunier, J.: Anomaly detection in video sequence with appearance-motion correspondence. In: ICCV, pp. 1273–1283 (2019)
5. Liu, Y., Liu, J., Zhao, M., Yang, D., Zhu, X., Song, L.: Learning appearance-motion normality for video anomaly detection. In: ICME, pp. 1–6 (2022)
6. Liu, Y., Liu, J., Lin, J., Zhao, M., Song, L.: Appearance-motion united autoencoder framework for video anomaly detection. IEEE Trans. Circ. Syst. II Express Briefs **69**(5), 2498–2502 (2022)
7. Luo, W., Liu, W., Gao, S.: A revisit of sparse coding based anomaly detection in stacked rnn framework. In: ICCV, pp. 341–349 (2017)
8. Cai, R., Zhang, H., Liu, W., Gao, S., Hao, Z.: Appearance-motion memory consistency network for video anomaly detection. In: AAAI, pp. 938–946 (2021)
9. Chang, Y., et al.: Video anomaly detection with spatio-temporal dissociation. Pattern Recogn. **122**, 108213 (2022)
10. Wang, Y., Long, M., Wang, J., Gao, Z., Yu, P.S.: Predrnn: recurrent neural networks for predictive learning using spatiotemporal LSTMs. In: NeurIPS, pp. 879–888 (2017)
11. Park, H., Noh, J., Ham, B.: Learning memory-guided normality for anomaly detection. In: CVPR, pp. 14360–14369 (2020)
12. Ravanbakhsh, M., Sangineto, E., Nabi, M., Sebe, N.: Training adversarial discriminators for cross-channel abnormal event detection in crowds. In: WACV, pp. 1896–1904 (2019)
13. Hasan, M., Choi, J., Neumann, J., Roy-Chowdhury, A.K., Davis, L.S.: Learning temporal regularity in video sequences. In: CVPR, pp. 733–742 (2016)

14. Fang, Z., Zhou, J.T., Xiao, Y., Li, Y., Yang, F.: Multi-encoder towards effective anomaly detection in videos. IEEE Trans. Multimedia **23**, 4106–4116 (2021)
15. Lee, S., Kim, H.G., Ro, Y.M.: Stan: spatio-temporal adversarial networks for abnormal event detection. In: ICASSP, pp. 1323–1327 (2018)
16. Zhao, M., Liu, Y., Liu, J., Zeng, X.: Exploiting spatial-temporal correlations for video anomaly detection. In: ICPR, pp. 1727–1733 (2022)
17. Woo, S., Park, J., Lee, J.-Y., Kweon, I.S.: CBAM: convolutional block attention module. In: Ferrari, V., Hebert, M., Sminchisescu, C., Weiss, Y. (eds.) ECCV 2018. LNCS, vol. 11211, pp. 3–19. Springer, Cham (2018). https://doi.org/10.1007/978-3-030-01234-2_1
18. Zhang, X., Zhou, X., Lin, M., Sun, J.: Shufflenet: an extremely efficient convolutional neural network for mobile devices. In: CVPR, pp. 6848–6856 (2018)
19. Li, W., Mahadevan, V., Vasconcelos, N.: Anomaly detection and localization in crowded scenes. IEEE Trans. Pattern Anal. Mach. Intell. **36**(1), 18–32 (2014)
20. Lu, C., Shi, J., Jia, J.: Abnormal event detection at 150 fps in matlab. In: ICCV, pp. 2720–2727 (2013)
21. Hao, Y., Li, J., Wang, N., Wang, X., Gao, X.: Spatiotemporal consistency-enhanced network for video anomaly detection. Pattern Recogn. **121**, 108232 (2022)

Frequency and Spatial Domain Filter Network for Visual Object Tracking

Manqi Zhao[1,2,3], Shenyang Li[1,2,3]([✉]), and Han Wang[1,2,3]

[1] Key Laboratory of Space Utilization, CAS, Beijing 100094, China
[2] Technology and Engineering Center for Space Utilization, CAS,
Beijing 100094, China
{zhaomanqi19,shyli}@csu.ac.cn
[3] University of Chinese Academy of Sciences, Beijing 100049, China

Abstract. Cross-correlation serves as the core similarity calculation operation in Siamese-based trackers, and generally produces response maps with high values at the target center. During this process, global context, including boundary and surrounding background of the target, which is conducive to target localization and bounding box regression, has been overlooked. In this work, we propose a Frequency and Spatial domain Filter Network (FSFNet) for visual object tracking, which exploits abundant global context in the frequency domain and enhances target representation in the spatial domain. First, frequency filters generated from template and search patches are applied to the target, capturing and enhancing valuable frequency components. These enhanced frequency components describe the global regions of interest in the spatial domain. Second, spatial domain convolutions are adopted to highlight local details of the target. Compared with mechanisms including depthwise correlation, pixel-wise correlation, and transformer, our method provides more accurate tracking results. Experiments on five benchmarks show that our tracker obtains competitive results. For example, our tracker achieves an AUC score of 81.2% on TrackingNet, outperforming the state-of-the-art two-stream tracker TrDiMP by 2.8% while running at 50 FPS.

Keywords: Visual tracking · Frequency filter · Global context

1 Introduction

Visual object tracking is one of the fundamental tasks in computer vision with broad applications, such as autonomous driving, video surveillance, and human-computer interaction. Given the initial state of the target in the first frame, visual object tracking aims to predict the target's state in consecutive frames. Although many well-performed trackers [2,13,23,28] have been proposed in recent decades, accurate tracking is still challenging due to factors such as severe occlusion, deformation, and background interference.

Q. Liu et al. (Eds.): PRCV 2023, LNCS 14430, pp. 108–120, 2024.
https://doi.org/10.1007/978-981-99-8537-1_9

Recently, cross-correlation has drawn much attention in the tracking community, especially for Siamese-based trackers [13,28] that calculate cross-correlation similarities between a template and a search patch. Most of the Siamese-based trackers crop the template patch into small regions, serving as the correlation kernel. The small kernel size provides a limited receptive field and makes it challenging to model the global context, including the target's boundary and surrounding background information in the search region. This context information is helpful for target localization and bounding box regression. As a result, cross-correlation may misestimate the target state, which can not meet the demand for accurate tracking. Some previous trackers introduce additional modules, including pixel-wise correlation [6,14], self-attention and cross-attention [2,23], to capture the global information and improve the semantic representation ability. However, the pixel-wise correlation would be distracted when pixels in the background are similar to the target region.

In this paper, we aim to design a time-saving tracker with the capabilities of integrating abundant global context information in the search region. We analyze the advantages of frequency domain processing and find that the problems can be adequately tackled. First, the Fourier transform maps images and features in the spatial domain to the frequency domain, with each frequency component naturally describing the global information of images. For example, the target center and boundary generally correspond to low-frequency and high-frequency components, respectively. Each component is treated equally in the frequency domain without local biases. Second, element product in the frequency domain is equivalent to convolution operation in the spatial domain, but it eliminates the limitation of small receptive field and has a significant speed advantage.

Following the analysis above, we propose a Frequency and Spatial domain Filter Network (FSFNet) for visual object tracking. Basically, we formulate the tracking problem as a two-stage filtering framework. In the first stage, we leverage the template feature to construct a frequency filter in the frequency domain. This filter is then applied to the search feature, allowing us to capture and model global information within the search patch. In the second stage, we adopt convolutions in the spatial domain to focus on the local details in the search region. Note that our proposed framework includes both a search branch and a symmetrical template branch, ensuring comprehensive and balanced tracking performance. The main contributions of our work are three-fold.

- We propose a novel Frequency and Spatial domain tracking framework based on the observation that frequency domain provide a convenience for global context utilization. The proposed algorithm formulates the tracking task as a filtering process and further explores the capability of frequency filtering in visual object tracking.
- We design a global and local context enhancement module, which first performs frequency filtering to capture global information and then utilizes spatial convolution to focus on local details. Compared with depth-wise correlation, pixel-wise correlation, and transformer, our designed module extracts

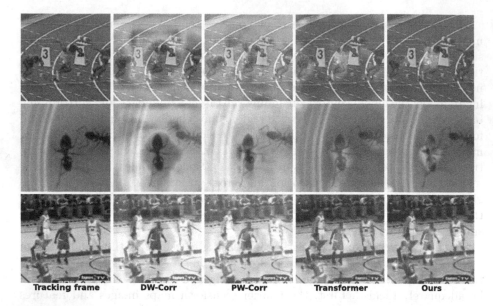

Fig. 1. Visualization of attention feature maps for different feature enhancement modules, which include depth-wise correlation (DW-Corr), pixel-wise correlation (PW-Corr), Transformer, and our proposed GLCE module.

the information of the search region more effectively and provides better tracking results, as shown in Fig. 1.

– Our FSFNet achieves comparable performance with the state-of-the-art trackers on five benchmark datasets, including GOT-10k [11], LaSOT [7], TrackingNet [18], UAV123 [17], and NFS [12]. For example, our tracker achieves an AUC score of 81.2% on TrackingNet, outperforming the state-of-the-art real-time tracker TrDiMP [23] by 2.8% while running at 50 FPS.

2   Related Work

Visual Tracking with Correlation. In recent years, correlation has played an important role in the tracking community [6,13,28]. Using the given first frame as a template, trackers adopt cross-correlation to integrate the features of the template and target to obtain high-response regions in consecutive frames. The Siamese-based trackers employed the Siamese network with depth-wise correlation operation in the spatial domain. Further improvements such as more sophisticated backbone networks [13], attention mechanisms [2,6,23] and anchor-free frameworks [25,28] have been applied to Siamese trackers to obtain accurate tracking results. However, the small receptive field in spatial convolution operation makes the Siamese-based trackers easy to fall into the local optimum. In this work, we design a filter in the frequency domain and perform frequency filtering to capture global information in the target region.

Attention Mechanisms in Tracking. More recently, attention mechanisms, including channel-wise attention [10], spatial attention [24], and notably, the transformer [22], became popular in the tracking field. Building on the success of the transformer structure, some trackers [2,3,23] adopted and applied it to tracking frameworks. To enhance feature interaction, certain algorithms concatenate the template and search patches [15,26], establishing unified tracking pipelines [4,27]. Moreover, the utilization of advanced pretrained models such as MAE [9] has further contributed to performance improvement. Although transformer-based trackers achieve state-of-the-art on current benchmarks, spatial domain matrix multiplication is more time-consuming with quadratic complexity, while element-wise filtering in the frequency domain has log-linear complexity. Therefore, our frequency filtering exhibits well-balanced accuracy and complexity.

Frequency Domain Learning. Fourier transform is a linear transformation between the spatial and frequency domain. Many works, including the DCF trackers, utilize the convolution theorem to accelerate the equivalent convolution operation in deep neural networks [20]. Besides, some works explore the attention mechanisms and operators in the frequency domain for effective feature representation. GFNet [21] proposed a global filter that learns long-term spatial dependencies in the frequency domain. AFNO [8] designed an adaptive Fourier neural operator for efficient feature representation. Motivated by the GFNet and AFNO, we adopt the core idea of capturing global context in the frequency domain and design a robust frequency filter for the tracking task.

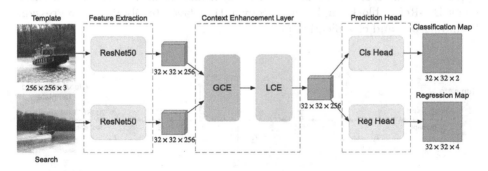

Fig. 2. Architecture of the proposed FSFNet. It consists of three components: a backbone network to extract features, a global and local context enhancement (GLCE) layer to capture global context, focus on local details and enhance feature representation, and a classification and regression prediction head to estimate the state of the target.

3 Proposed Method

This section presents the proposed Frequency and Spatial domain Filter Network (FSFNet) for visual object tracking in detail. As shown in Fig. 2, our proposed FSFNet consists of three components: a weight-sharing backbone network, context enhancement layer, and a classification and regression head. First, the backbone network extracts features from template and search regions. Then, the extracted features are fed into the context enhancement layer to capture global context and highlight local details of the target. Finally, a classification and regression head is applied to the enhanced features to generate final tracking results. We describe the details of the overall framework and introduce the global and local context enhancement layer in the following subsections.

3.1 Overall Architecture

Similar to Siamese-based trackers [13], FSFNet takes image pairs as inputs of the weight-sharing backbone network. The input template and search patches share the same shape $z, x \in \mathbb{R}^{3 \times H \times W}$, which differs from most Siamese-based trackers with a minor template patch. The backbone processes the input template and search patches to corresponding features $f_z, f_x \in \mathbb{R}^{C \times H' \times W'}$. Here, we employ layers 1–3 of the modified ResNet50 for feature extraction, $H', W' = \frac{H}{8}, \frac{W}{8}$ and $C = 256$. Then, the extracted features f_z and f_x are fed into GLCE. The GLCE layer enhances and fuses f_z and f_x to $f_e \in \mathbb{R}^{C \times H' \times W'}$. As shown in Fig. 2, global context enhancement (GCE) is first applied to f_z and f_x to filter and capture global context in the frequency domain. After the global enhancement, local context enhancement (LCE) module focuses on the local region of interest and enhances feature representation in the spatial domain. Section 3.2 provides a detailed implementation of GCE and LCE. Finally, the enhanced feature f_e is passed to the regression and classification heads. Each head consists of three Conv-BN-ReLU blocks and a final convolution layer to obtain the regression bounding box and classification map.

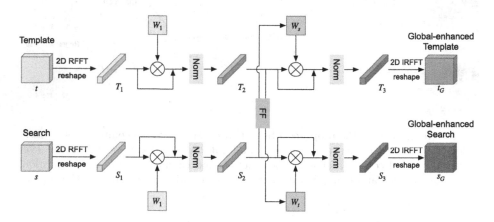

Fig. 3. The structure of our proposed global context enhancement (GCE) module. It takes a pair of template and search features as input, $\otimes$ denotes matrix multiplication. FF stands for Feature-Filter block, which uses the template and search feature to generate the frequency filters. After a self and a cross enhancement process, features are transformed back to the spatial domain to obtain globally enhanced features.

3.2 Global and Local Context Enhancement

With the initial state of the target, most existing correlation-based trackers treat the tracking task as a template matching process [13,28]. As for accurate object tracking, this template matching process needs to be robust for challenges, including object appearance changes and occlusion. However, in depth-wise correlation [13,25,28], templates are always cropped into small regions to facilitate spatial domain cross-correlation, with response map concentrating on the target's center while neglecting other positions. Although pixel-wise correlation [6,14] applies spatial attention mechanisms and provides more precise target locations and boundaries, pixels similar to target regions in the background may obtain high response values. To overcome these issues, we reconsider the tracking task as a two-stage filtering process: in the first stage, regions of interest are captured in the frequency domain; in the second stage, the extracted global regions are refined to highlight local part of the target in the spatial domain.

Feature-Filter Block (FF). From the perspective of the frequency domain, images are composed of different frequencies. These frequency components are natural representations of global context. Therefore, we aim to design a frequency filter to enhance the frequency of interest and filter out irrelevant frequencies in the search patch. GFNet [21] and AFNO [8] proposed randomly initialized frequency filters for the classification task. Taking inspiration from them and tending to integrate the template and search information into the frequency filter, we design a Feature-Filter block (FF). The input of FF is a frequency domain feature, taking template $\mathbf{T} \in \mathbb{C}^{N \times d}$ as example. Inspired by the attention matrix [22], we apply a learnable matrix $\mathbf{A} \in \mathbb{C}^{N \times N}$ to $\mathbf{T}$ to highlight the significant frequency component, and multiply with the transpose of the original features to get $\mathbf{T}^T \mathbf{A} \mathbf{T} \in \mathbb{C}^{d \times d}$. Further, we introduce Complex ReLU ($\mathbb{C}$ReLU) [19], a simple yet effective complex non-linear activation that applies separate ReLU functions on the real and imaginary part:

$$\mathbb{C}\text{ReLU}(z) = \text{ReLU}(\mathscr{R}(z)) + i\text{ReLU}(\mathscr{I}(z)). \tag{1}$$

Then, two weight matrix $\mathbf{W}_{t1}, \mathbf{W}_{t2} \in \mathbb{C}^{d \times d}$ are absorbed to enhance the filtering ability. The basic Feature-Filter block can be summarized as:

$$\mathbf{F}_T = \mathbb{C}\text{ReLU}(\mathbb{C}\text{ReLU}(\mathbf{T}^T \mathbf{A} \mathbf{T})\mathbf{W}_{t1})\mathbf{W}_{t2}, \tag{2}$$

where $\mathbf{F}_T \in \mathbb{C}^{d \times d}$ represents the generated frequency filter using template feature, and $\mathbf{F_S}$ is similar using search feature.

Global Context Enhancement (GCE). The basic structure of GCE is shown in Fig. 3. Based on the designed FF blocks, GCE adaptively enhances the global context through frequency filtering. Specifically, GCE takes both template and search region features $t, s \in \mathbb{R}^{H \times W \times d}$ as input. The spatial domain features are first transformed to the frequency domain using 2D Fast Fourier Transform

(FFT) along the spatial dimensions. Since input features are real tensors, real-valued Fast Fourier Transform (RFFT) is adopted for saving computation. This process can be expressed as:

$$\mathbf{T}_1 = \mathrm{RFFT}[t], \mathbf{S}_1 = \mathrm{RFFT}[s] \in \mathbb{C}^{H \times \hat{W} \times d}, \tag{3}$$

where $\hat{W} = \lfloor \frac{W}{2} \rfloor + 1$. Then, we flatten $\mathbf{T}_1, \mathbf{S}_1$ in spatial dimensions and obtain $\mathbf{T}_1, \mathbf{S}_1 \in \mathbb{C}^{N \times d}$, where $N = H \times \hat{W}$. The flattened features, $\mathbf{T}_1$ and $\mathbf{S}_1$, first apply shared filter weights $\mathbf{W}_1 \in \mathbb{C}^{d \times d}$ for self enhancement, and obtains $\mathbf{T}_2, \mathbf{S}_2$ through the residual structure. $\mathbf{T}_2$ and $\mathbf{S}_2$ are fed into FF blocks to generate global context frequency filters $\mathbf{F}_T$ and $\mathbf{F}_S$, and multiplied with $\mathbf{S}_2$ and $\mathbf{T}_2$ respectively for cross enhancement:

$$\mathbf{T}_3 = \mathrm{Norm}(\mathbf{T}_2 \mathbf{F}_S + \mathbf{T}_2), \mathbf{S}_3 = \mathrm{Norm}(\mathbf{S}_2 \mathbf{F}_T + \mathbf{S}_2), \tag{4}$$

Finally, 2D inverse RFFT (IRFFT) is applied to transform $\mathbf{T}_3$ and $\mathbf{S}_3$ back to the global enhanced template and search features in the spatial domain:

$$t_G = \mathrm{IRFFT}[\mathbf{T}_3], s_G = \mathrm{IRFFT}[\mathbf{S}_3]. \tag{5}$$

Without the limitation of small receptive fields, the filtering process can adaptively focus on useful global context without local biases.

Local Context Enhancement (LCE). With the global enhanced feature t_G obtained in GCE, model needs to further focus on the local regions of interest. The spatial domain convolution operation has priority on assembling the local regions. Besides, it can also bring in the interactions between channels, which further enhances the semantic representation capability. Here, we adopt two 3×3 convolution layers with a Batch Normalization layer and a ReLU layer in between to perform local context enhancement. The local enhanced feature t_L is added to t_G as a residual term, that is,

$$t' = \mathrm{BN}(t_L + t_G), \tag{6}$$

where t' is the final enhanced feature and $\mathrm{BN}(\cdot)$ denotes the batch normalization.

4   Experiments

Our tracking method is implemented in Python using PyTorch. FSFNet achieves a tracking speed of 50 FPS on a single Nvidia RTX 2080ti GPU.

4.1   Implementation Details

Offline Training. To optimize the proposed network, we employ the binary cross-entropy (BCE) loss for the classification branch, and the generalized IoU

(GIoU) loss and $L1$ loss for the regression branch. The joint loss function is the linear combination of these three loss functions:

$$\mathcal{L} = \lambda_1 \mathcal{L}_{BCE} + \lambda_2 \mathcal{L}_{GIoU} + \lambda_3 \mathcal{L}_{L1}, \tag{7}$$

where $\lambda_1, \lambda_2, \lambda_3$ are hyper-parameters that are respectively set to 4, 1, and 2 to banlance each component in our experiments.

Similar to [5], we use the training splits of LaSOT [7], GOT-10k [11], TrackingNet [18] and COCO [16] datasets. The sizes of input template and search patches are both 256×256 pixels. The backbone network is initialized with ImageNet-pretrained ResNet-50 parameters. We train the network on two Nvidia Titan RTX GPUs for 200 epochs in total with 1000 iterations per epoch and 64 image pairs per batch. The ADAM optimizer is employed with learning rate decay of 0.5 every 50 epochs. We set backbone's learning rate to $2e-5$, other parameter's learning rate to $2e-4$, and train the whole network end-to-end.

Online Tracking. In the online tracking stage, the window penalty with a hanning window is applied to suppress the large displacement in the classification score map as in [2]:

$$s' = \alpha \cdot s + (1 - \alpha) \cdot s_h, \tag{8}$$

where s is the original classification score, s_h denotes the corresponding value of hanning window, and s' is the final classification score. The weight of the penalty window α is set to 0.6 in our experiments.

4.2 Analysis of the Proposed Method

Comparison with Other Feature Enhancement Module. To evaluate the effectiveness of GLCE in our designed filter network structure, we make a comparison with popular feature enhancement modules, including depth-wise correlation (DW-Corr) [13,28], pixel-wise correlation (PW-Corr) [6] and transformer mechanism [3,22,23]. Specifically, we absorb both self-attention and cross-attention with a feed-forward network [22] to formulate the transformer module, and we use the PW-Corr module proposed in [6]. In the comparison experiments, GLCE is replaced with the modules above, while other network components are kept the same. The experiments are performed on the GOT-10k test set [11]. Average overlap (AO) is the primary evaluation metric. Besides, all experiments are conducted under the same settings for a fair comparison.

As shown in Table 1, our GLCE outperforms DW-Corr, PW-Corr and Transformer on AO score with a notable performance gain of 12.5%, 4.8%, and 3.2%, respectively. These outstanding advantages benefit from the designed pipeline for global and local context enhancement.

Ablation Analysis. We analyze the impact of each component in our proposed GLCE through an ablation study. The results of the comparison are shown in Table 1. We set DW-Corr as the baseline structure, with an AO score of 55.0%.

Table 1. Comparison results of different feature enhancement modules on the GOT-10k test set. Average overlap (AO) is the evaluation metric.

	DW-Corr	PW-Corr	Transformer	GCE	GLCE	GLCE×3
AO(%)	55.0	62.7	64.3	65.0	**67.5**	**69.5**
Speed (FPS)	79.7	66.8	49.5	58.1	56.3	50.0

A single GCE module can improve AO by 10.0%. With an LCE module, our method further provides an improvement of 2.5% on AO, showing the significance of focusing on local details from the global-enhanced feature. The experiments so far reveal the effectiveness of our designed GLCE, which first captures long-term feature associations in the frequency domain and concentrates on local details in the spatial domain.

4.3 State-of-the-Art Comparison

In this subsection, we compare our proposed FSFNet with recent state-of-the-art trackers on five tracking benchmarks including GOT-10k [11], TrackingNet [18], LaSOT [7], UAV123 [17], and NFS [12].

Table 2. State-of-the-art comparison on the TrackingNet test set in terms of success (AUC), precision (Prec.), and normalized precision (N-Prec.). The best two results are highlighted in red and blue fonts.

	CGACD [6]	SiamRPN++ [13]	KYS [1]	PrDiMP-50 [5]	TREG [3]	TrDiMP [23]	**FSFNet**
AUC	71.1	73.3	74.0	75.8	**78.5**	78.4	81.2
Prec.	69.3	69.4	68.8	70.4	**75.0**	73.1	78.6
N-Prec.	80.0	80.0	80.0	81.6	**83.8**	83.3	85.9
Speed (FPS)	70	49	20	30	30	26	**50**

TrackingNet [18]. TrackingNet is a large-scale dataset that consists of 30,643 video sequences with upright bounding boxes and 15 attributes, in which the test set contains 511 videos. As shown in Table 2, our FSFNet obtains a success score of 81.2%, a precision score of 78.6%, and a normalized precision score of 85.9% while running at 50 FPS, which surpasses the previous state-of-the-art real-time trackers such as TrDiMP [23] and TREG [3] both in terms of accuracy and speed.

GOT-10k [11]. GOT-10k is a large-scale dataset with more than 10,000 video segments. The test subset consists of 180 sequences and has different classes with the training sets. Table 3 provides the comparison results of our method with state-of-the-art trackers in terms of average overlap (AO) and success rates (SR) at overlap thresholds 0.5 and 0.75. Our tracker achieves the highest AO and

Table 3. State-of-the-art comparison on the GOT-10k test set in terms of average overlap (AO), and success rates (SR) at overlap thresholds 0.5 and 0.75. The best two results are highlighted in red and blue fonts.

	SiamRPN++ [13]	Ocean [28]	KYS [1]	PrDiMP-50 [5]	TREG [3]	TrDiMP [23]	**FSFNet**
AO(%)	51.7	61.1	63.6	63.6	67.2	**68.8**	69.5
$SR_{0.5}$(%)	61.6	72.1	75.1	73.9	78.6	80.5	**78.9**
$SR_{0.75}$(%)	32.5	47.3	51.5	52.9	56.9	**59.7**	64.4
Speed (FPS)	**49.8**	–	20.0	30.0	24.3	13.6	50.0
Hardware	2080ti	–	2080	–	2080ti	1080ti	2080ti

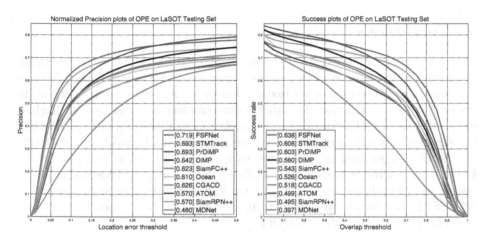

Fig. 4. Comparison with the state-of-the-art trackers on the test set of LaSOT. Trackers are evaluated using the success and normalized precision metrics.

Table 4. Comparison with the state-of-the-art trackers on the NFS and UAV123 datasets in terms of AUC score. The best two results are highlighted in red and blue fonts.

	SiamRPN++ [13]	SiamFC++ [25]	KYS [1]	PrDiMP-50 [5]	TrDiMP [23]	**FSFNet**
NFS	50.2	–	63.5	63.5	**66.5**	66.6
UAV123	61.3	63.2	–	68.0	**67.5**	67.5
Speed (FPS)	49	64	20	30	26	**50**

$SR_{0.75}$ score of 69.5% and 64.4%, outperforming the state-of-the-art trackers, including TrDiMP [23] and TREG [3] with gains of 0.7% and 2.3% on AO score.

LaSOT [7]. LaSOT is a recent large-scale dataset gathering 1,400 videos with an average of 2,512 frames per sequence, which demands higher robustness of the trackers. We evaluate our tracker on the test set with 280 long videos. Results in Fig. 4 shows that the proposed tracker achieves highest success score of 63.8% and normalized precision score of 71.9%. The remarkable results further demonstrate the ability of our method when handling long sequences.

NFS [12]. NFS provides a dataset of 100 video sequences with fast-moving objects. We evaluate our FSFNet on the 30 FPS version of the dataset. As shown in Table 4, our approach achieves an AUC score of 66.6%, outperforming the state-of-the-art tracker TrDiMP [23], while running at high speed of 50 FPS.

UAV123 [17]. UAV123 consists of 123 aerial sequences captured by UAVs or generated by a simulator. The proposed method provides a competitive performance compared to the recent remarkable trackers in Table 4. Specifically, our FSFNet achieves an area-under-the-curve (AUC) at 67.5%, which is the same score as the state-of-the-art tracker TrDiMP [23].

5   Conclusions

In this work, we propose a novel Frequency and Spatial domain Filter Network (FSFNet) for visual object tracking. Our model constructs a frequency filter to capture the global context in the frequency domain and focuses on local details in the spatial domain. Compared to other feature enhancement mechanisms, including depth-wise correlation, pixel-wise correlation, and transformer, our method can strongly enhance target representation capability and provide robust results against challenges such as appearance variations, fast motion, and distractors. Experiment results on five benchmarks show that our proposed method performs competitively with the state-of-the-art trackers while running at high speed of 50 FPS. In the future, we intend to further explore the connections between video frames for more robust tracking.

References

1. Bhat, G., Danelljan, M., Van Gool, L., Timofte, R.: Know your surroundings: exploiting scene information for object tracking. In: Vedaldi, A., Bischof, H., Brox, T., Frahm, J.-M. (eds.) ECCV 2020. LNCS, vol. 12368, pp. 205–221. Springer, Cham (2020). https://doi.org/10.1007/978-3-030-58592-1_13
2. Chen, X., Yan, B., Zhu, J., Wang, D., Yang, X., Lu, H.: Transformer tracking. In: Proceedings of the IEEE/CVF Conference on Computer Vision and Pattern Recognition, pp. 8126–8135 (2021)
3. Cui, Y., Jiang, C., Wang, L., Wu, G.: Target transformed regression for accurate tracking. arXiv preprint arXiv:2104.00403 (2021)
4. Cui, Y., Jiang, C., Wang, L., Wu, G.: Mixformer: end-to-end tracking with iterative mixed attention. In: Proceedings of the IEEE/CVF Conference on Computer Vision and Pattern Recognition, pp. 13608–13618 (2022)
5. Danelljan, M., Gool, L.V., Timofte, R.: Probabilistic regression for visual tracking. In: Proceedings of the IEEE/CVF Conference on Computer Vision and Pattern Recognition, pp. 7183–7192 (2020)
6. Du, F., Liu, P., Zhao, W., Tang, X.: Correlation-guided attention for corner detection based visual tracking. In: Proceedings of the IEEE/CVF Conference on Computer Vision and Pattern Recognition, pp. 6836–6845 (2020)
7. Fan, H., et al.: Lasot: a high-quality benchmark for large-scale single object tracking. In: Proceedings of the IEEE/CVF Conference on Computer Vision and Pattern Recognition, pp. 5374–5383 (2019)

8. Guibas, J., Mardani, M., Li, Z., Tao, A., Anandkumar, A., Catanzaro, B.: Efficient token mixing for transformers via adaptive fourier neural operators. In: The Tenth International Conference on Learning Representations, ICLR 2022, Virtual Event, 25–29 April 2022. OpenReview.net (2022)
9. He, K., Chen, X., Xie, S., Li, Y., Dollár, P., Girshick, R.: Masked autoencoders are scalable vision learners. In: Proceedings of the IEEE/CVF Conference on Computer Vision and Pattern Recognition, pp. 16000–16009 (2022)
10. Hu, J., Shen, L., Sun, G.: Squeeze-and-excitation networks. In: Proceedings of the IEEE Conference on Computer Vision and Pattern Recognition, pp. 7132–7141 (2018)
11. Huang, L., Zhao, X., Huang, K.: Got-10k: a large high-diversity benchmark for generic object tracking in the wild. IEEE Trans. Pattern Anal. Mach. Intell. **43**, 1562–1577 (2019)
12. Kiani Galoogahi, H., Fagg, A., Huang, C., Ramanan, D., Lucey, S.: Need for speed: a benchmark for higher frame rate object tracking. In: Proceedings of the IEEE International Conference on Computer Vision, pp. 1125–1134 (2017)
13. Li, B., Wu, W., Wang, Q., Zhang, F., Xing, J., Yan, J.: Siamrpn++: evolution of siamese visual tracking with very deep networks. In: Proceedings of the IEEE/CVF Conference on Computer Vision and Pattern Recognition, pp. 4282–4291 (2019)
14. Liao, B., Wang, C., Wang, Y., Wang, Y., Yin, J.: PG-Net: pixel to global matching network for visual tracking. In: Vedaldi, A., Bischof, H., Brox, T., Frahm, J.-M. (eds.) ECCV 2020. LNCS, vol. 12367, pp. 429–444. Springer, Cham (2020). https://doi.org/10.1007/978-3-030-58542-6_26
15. Lin, L., Fan, H., Zhang, Z., Xu, Y., Ling, H.: Swintrack: a simple and strong baseline for transformer tracking. Adv. Neural. Inf. Process. Syst. **35**, 16743–16754 (2022)
16. Lin, T.-Y., et al.: Microsoft COCO: common objects in context. In: Fleet, D., Pajdla, T., Schiele, B., Tuytelaars, T. (eds.) ECCV 2014. LNCS, vol. 8693, pp. 740–755. Springer, Cham (2014). https://doi.org/10.1007/978-3-319-10602-1_48
17. Mueller, M., Smith, N., Ghanem, B.: A benchmark and simulator for UAV tracking. In: Leibe, B., Matas, J., Sebe, N., Welling, M. (eds.) ECCV 2016. LNCS, vol. 9905, pp. 445–461. Springer, Cham (2016). https://doi.org/10.1007/978-3-319-46448 0_27
18. Muller, M., Bibi, A., Giancola, S., Alsubaihl, S., Ghanem, B.: Trackingnet: a large-scale dataset and benchmark for object tracking in the wild. In: Proceedings of the European Conference on Computer Vision (ECCV), pp. 300–317 (2018)
19. Nitta, T.: An extension of the back-propagation algorithm to complex numbers. Neural Netw. **10**(8), 1391–1415 (1997)
20. Pratt, H., Williams, B., Coenen, F., Zheng, Y.: FCNN: fourier convolutional neural networks. In: Ceci, M., Hollmén, J., Todorovski, L., Vens, C., Džeroski, S. (eds.) ECML PKDD 2017. LNCS (LNAI), vol. 10534, pp. 786–798. Springer, Cham (2017). https://doi.org/10.1007/978-3-319-71249-9_47
21. Rao, Y., Zhao, W., Zhu, Z., Lu, J., Zhou, J.: Global filter networks for image classification. In: Advances in Neural Information Processing Systems (NeurIPS) (2021)
22. Vaswani, A., et al.: Attention is all you need. Adv. Neural Inf. Process. Syst. **30**, 5998–6008 (2017)
23. Wang, N., Zhou, W., Wang, J., Li, H.: Transformer meets tracker: exploiting temporal context for robust visual tracking. In: Proceedings of the IEEE/CVF Conference on Computer Vision and Pattern Recognition, pp. 1571–1580 (2021)

24. Woo, S., Park, J., Lee, J.Y., Kweon, I.S.: CBAM: convolutional block attention module. In: Proceedings of the European Conference on Computer Vision (ECCV), pp. 3–19 (2018)
25. Xu, Y., Wang, Z., Li, Z., Yuan, Y., Yu, G.: Siamfc++: towards robust and accurate visual tracking with target estimation guidelines. In: AAAI, pp. 12549–12556 (2020)
26. Yan, B., Peng, H., Fu, J., Wang, D., Lu, H.: Learning spatio-temporal transformer for visual tracking. arXiv preprint arXiv:2103.17154 (2021)
27. Ye, B., Chang, H., Ma, B., Shan, S., Chen, X.: Joint feature learning and relation modeling for tracking: a one-stream framework. In: European Conference on Computer Vision, pp. 341–357. Springer, Heidelberg (2022). https://doi.org/10.1007/978-3-031-20047-2_20
28. Zhang, Z., Peng, H., Fu, J., Li, B., Hu, W.: Ocean: object-aware anchor-free tracking. In: Vedaldi, A., Bischof, H., Brox, T., Frahm, J.-M. (eds.) ECCV 2020. LNCS, vol. 12366, pp. 771–787. Springer, Cham (2020). https://doi.org/10.1007/978-3-030-58589-1_46

Enhancing Feature Representation for Anomaly Detection via Local-and-Global Temporal Relations and a Multi-stage Memory

Xuan Li, Ding Ma, and Xiangqian Wu[✉]

Faculty of Computing, Harbin Institute of Technology, Harbin, China
xuanli@stu.hit.edu.cn, martin3436@yeah.net, xqwu@hit.edu.cn

Abstract. Weakly supervised video anomaly detection is a challenging task because frame-level labels are not accessible at the training time. Effectively tackling this task necessitates models to learn discriminative feature representation. To address this challenge, we propose a multi-stage memory-augmented feature discrimination learning (MMFDL) method. The first stage obtains the preliminary abnormal probabilities of clip features. In the second stage, an easy normal pattern memory (ENPM) are proposed to store normal patterns with low abnormal probabilities. In the last stage, we bring clip features with high abnormal probabilities in normal videos close to ENPM and away from the clip features with high probabilities of being abnormal in abnormal videos to make models learn more discriminative features for anomaly detection. Furthermore, we propose a local-and-global temporal relations modeling (LGTRM) module to enhance clip features by aggregating local and global contexts. Our LGTRM module can be divided into two subnetworks: DW-Net and TF-Net. DW-Net integrates the current clip feature with its adjacent clip features to capture local-range temporal dependencies. TF-Net utilizes the multi-head self-attention mechanism of the transformer to capture global-range temporal dependencies. Experiments on two datasets demonstrate that our method outperforms state-of-the-art approaches. The code is available at https://github.com/xuanli01/PRCV347.

Keywords: Video anomaly detection · Weak supervision · Feature representation enhancing · Temporal relations · Multi-stage memory

This work was supported in part by the National Key Research and Development Program of China under Grant 2020AAA0106502, in part by the Natural Science Foundation of China under Grant 62073105, in part by the Natural Science Foundation of Heilongjiang Province of China under Grant ZD2022F002, and in part by the Heilongjiang Touyan Innovation Team Program.

Q. Liu et al. (Eds.): PRCV 2023, LNCS 14430, pp. 121–133, 2024.
https://doi.org/10.1007/978-981-99-8537-1_10

1 Introduction

Video anomaly detection, which aims to detect whether there is an anomaly in the video frame, is widely used in intelligent surveillance [26]. Due to the rarity and diversity of anomalous events, anomalies are generally defined as behaviors that deviate from the normal distribution of training samples in previous works [16,23]. Some works [4,16,23] employ methods based on semi-supervised learning to address this problem. They only use normal data to learn a model describing normality in training. Then, video frames that cannot be described by the model are treated as anomalies in testing. Although these methods have made some progress, they have a disadvantage that it is difficult for the training data to contain all kinds of normal events, which makes it difficult for the model to accurately judge unseen normal events. To solve this problem, weakly supervised video anomaly detection (WS-VAD) is introduced by Sultani *et al.* [15]. The training data for WS-VAD contains both abnormal and normal videos with only video-level labels. A video is marked as abnormal if it contains anomalous events and vice versa. In WS-VAD, a video can be viewed as a bag and a video clip which contains several consecutive frames in this video can be treated as an instance of this bag. Recently, some researchers [7,11,12,17,22] focus on enhancing the pre-extracted clip features (e.g., I3D [3] RGB features) to make the classifier more easily distinguish between abnormal and normal features to solve WS-VAD. They augment the clip features via two ways, *i.e.*, temporal relations modeling and feature discrimination learning.

For temporal relations modeling, local and global contexts of video clips need to be perceived together due to the uncertain length of an event. Local context refers to the neighboring clips of the current clip. Global context refers to the entire video. To capture local and global temporal cues, we introduce a local-and-global temporal relations modeling (LGTRM) module which contains two subnetworks: DW-Net and TF-Net. DW-Net aims to capture local temporal relations by integrating the current clip feature with its adjacent clip features. TF-Net aims to enrich global context. We implement TF-Net based on the transformer [19], which is inspired by transformer's powerful ability to capture long-range temporal dependencies. Compared with [11,15,17,22,25], our TF-Net has two characteristics in WS-VAD. The first one is that with the help of transformer's [19] powerful sequence processing capability, our model can take any number of video clips less than or equal to 512 as input during the training phase. Previous works [7,15,17,25] process videos into fixed-length video clips, which might not be a good choice. Because anomalies rarely occur and the normal frames dominate the video, abnormal patterns are easy to drown in a video with short-term anomalies if the video is compressed down to a short length, which leads to the difficulty of learning abnormal patterns. The second characteristic is that our TF-Net adds a learnable positional encoding. A video is not just a stack of images but a sequence with order information, which is ignored by previous works [7,11,17]. Although Wu *et al.* [22] have considered the order information and propose a CTR module with a position prior to model local temporal relations, their position prior is handcrafted so that it is hard to generalize

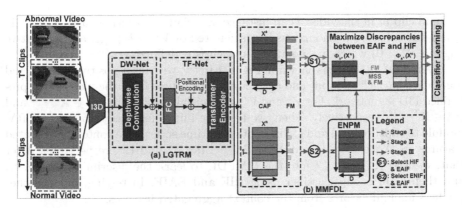

Fig. 1. Overall architecture of our method. CAF, FM and MSS are short for the context aggregated features, feature magnitudes and memory similarity scores, respectively.

to complex scenarios. Moreover, the CTR module is based on non-local network [21] which is less powerful in obtaining long-range temporal dependencies than the multi-head self-attention mechanism [19] of the transformer.

For feature discrimination learning, researchers [6,15,17] promote the separability of videos and clips by enlarging the inter-class distance between hard instance and easy abnormal instance features. Hard instance features refer to the instance features which may generate false alarm in normal videos. Easy abnormal instance features refer to the instance features with high probabilities of being abnormal in the abnormal videos. For the sake of discussion, HIF and EAIF are short for hard instance features and easy abnormal instance features, respectively. These methods have a problem that instance features in EAIF may not be all from abnormal clips due to normal clips in the abnormal videos. This problem may mislead the models to produce more hard instance features for normal clips, causing small margins between abnormal and normal clips. To alleviate this issue, we propose a multi-stage memory-augmented feature discrimination learning (MMFDL) method. In the first stage, we produce elementary abnormal probabilities of each input video clips. The second stage aims to obtain an easy normal pattern memory (ENPM) which records easy normal patterns with low abnormal probabilities. To achieve this, we update our ENPM by bringing it close to easy normal instance features (*i.e.* the instance features that produce low abnormal probabilities in normal videos) and away from the most likely anomalous features in abnormal videos. ENIF is the abbreviation for easy normal instance features. In the last stage, we bring HIF close to the ENPM and away from EAIF. Our MMFDL method addresses the issue above, as follows: when EAIF contain normal instance features, our ENPM can help enforce large margins between abnormal and normal clips.

2   Proposed Method

Given a video $V = \{v_t\}_{t=1}^{T}$ with T clips, the video-level label $Y \in \{0, 1\}$ denotes whether an anomalous event exists in this video. We consider a video V as a bag and clip v_t in the video as an instance. Specially, a positive bag (*i.e.* $Y = 1$) which has at least one anomalous clip is denoted by $V^a = \{v_t^a\}_{t=1}^{T^a}$ and a negative bag (*i.e.* $Y = 0$) which has no anomalous clips is defined as $V^n = \{v_t^n\}_{t=1}^{T^n}$.

Overall architecture is shown in Fig. 1. Firstly, we take a positive bag V^a and a negative bag V^n as input and extract initial features from the videos by a pretrained I3D [3] network. For V^a and V^n, the initial features are $F^a \in \mathbb{R}^{T^a \times D}$ and $F^n \in \mathbb{R}^{T^n \times D}$, respectively. Then, we feed the initial features into the LGTRM module to capture local and global context in time series. The context aggregated features are $X^a \in \mathbb{R}^{T^a \times D}$ and $X^n \in \mathbb{R}^{T^n \times D}$ for V^a and V^n, respectively. After that, we feed X^a and X^n into the MMFDL to learn the discriminative power, and then train a clip classifier using HIF and EAIF. In testing, our classifier predicts anomaly scores from the context aggregated features.

2.1   Local-and-Global Temporal Relations Modeling

The LGTRM module which is composed of DW-Net and TF-Net aims to model the local and global temporal relations from initial features. In practice, we first pass initial features through the DW-Net and obtain local aggregated features which are $\tilde{F}^a = \{\tilde{f}_t^a\}_{t=1}^{T^a}$ and $\tilde{F}^n = \{\tilde{f}_t^n\}_{t=1}^{T^n}$ for V^a and V^n, respectively. Then, we feed the local aggregated features into the TF-Net and get the context aggregated features which are $X^a = \{x_t^a\}_{t=1}^{T^a}$ and $X^n = \{x_t^n\}_{t=1}^{T^n}$ for V^a and V^n.

DW-Net aims to capture local context from the initial features. We take the dynamic changes of clip features as local context, which is motivated by the fact that video clip features change slowly within an event while change obviously at the boundary between an abnormal event and a normal event. Specifically, we use difference between the current and its adjacent clip initial features to represent the feature dynamics, integrating the difference into the current clip feature via element-wise addition. For the t-th clip feature $f_t \in \mathbb{R}^D$ in the initial features $F = \{f_t\}_{t=1}^{T} \in \mathbb{R}^{T \times D}$, the t-th local aggregated feature $\tilde{f}_t \in \mathbb{R}^D$ is

$$\tilde{f}_t = \alpha_t \cdot f_t + \beta_t \cdot (f_t - f_{t-1}) + \gamma_t \cdot (f_t - f_{t+1}) \tag{1}$$

where α_t, β_t and $\gamma_t \in \mathbb{R}^D$ are learnable parameters to adjust the contribution of initial feature of the current clip, difference between the current and its previous clip and difference between the current and its next clip. $\cdot$ is a Hadamard product. To implement the DW-Net with a network, we rewrite the Eq. 1 as Eq. 2.

$$\tilde{f}_t = f_t - \beta_t \cdot f_{t-1} + (\alpha_t + \beta_t + \gamma_t - 1) \cdot f_t - \gamma_t \cdot f_{t+1} \tag{2}$$

In this way, we can implement the DW-Net with a one-dimensional depth-wise convolution [5] followed by a skip connection which aims to preserve the initial features of the current clip. $(-\beta_t, \alpha_t + \beta_t + \gamma_t - 1, -\gamma_t)$ can be treated as the kernel weights of the depth-wise convolution.

TF-Net aims to capture the global-range temporal dependencies from the local aggregated features $\tilde{F} = \{\tilde{f}_t\}_{t=1}^{T} \in \mathbb{R}^{T \times D}$ in a video sequence. We first pass $\tilde{F}$ through a linear projection layer FC to map the second dimension of $\tilde{F}$

into a D-dimension feature space to fuse its cross-channel information. Then, we embed positional encoding $E \in \mathbb{R}^{T \times D}$ via element-wise addition to add order information to the local aggregated features. Positional encoding comes in two forms: handcrafted input (e.g., sinusoidal encoding) and learnable embedding. In our model, we apply the learnable positional encoding because of its flexibility. The input $F^{in} \in \mathbb{R}^{T \times D}$ of the transformer encoder is $F^{in} = FC(\tilde{F}) + E$. Then, the input F^{in} is fed into a one-layer vanilla transformer encoder [19] to capture long-range contextual information in the temporal dimension of F^{in}. We denote the output context aggregated features as $X = \{x_t\}_{t=1}^{T} \in \mathbb{R}^{T \times D}$. X is

$$H = MSA(LN(F^{in})) + F^{in} \tag{3}$$
$$X = FFN(LN(H)) + H \tag{4}$$

where MSA, LN and FFN represent multi-head self-attention [19], layernorm [1] and a two-layered feed-forward network.

2.2 Multi-stage Memory-Augmented Feature Discrimination Learning and Classifier Learning

Our MMFDL method aims to learn discriminative features by enlarging the discrepancies between EAIF and HIF so that the classifier can accurately predict anomaly scores of the input context aggregated features. As shown in Fig. 1(b), our designed MMFDL method is a multi-stage algorithm.

The first stage aims to acquire the preliminary abnormal probabilities of context aggregated features and use the abnormal probabilities to identify HIF, EAIF and ENIF. Inspired by RTFM [17] where feature magnitudes and abnormal probabilities are positively correlated, we implement the first stage based on the feature magnitude learning and classifier learning parts of RTFM [17] so that feature magnitudes can represent abnormal probabilities to help us find HIF, EAIF and ENIF. Similar to [17], we first select instance features $\Phi_k(X)$ with top-k feature magnitudes from the context aggregated features $\{x_t\}_{t=1}^{T}$, where k is equal to $\lfloor T/16 + 1 \rfloor$. k depends on the length of the video, rather than a fixed value [17], which can help our model adapt to videos of different lengths. Feature magnitude is ℓ_2-norm of the feature. Therefore, the mean feature magnitude $g_k(X)$ of the selected features is

$$g_k(X) = \max_{\Phi_k(X) \subseteq \{x_t\}_{t=1}^{T}} \frac{1}{k} \sum_{x_t \in \Phi_k(X)} \|x_t\|_2 \tag{5}$$

For V^a and V^n, the selected features are denoted by $\Phi_{k^a}(X^a) \in \mathbb{R}^{k^a \times D}$ and $\Phi_{k^n}(X^n) \in \mathbb{R}^{k^n \times D}$, respectively. According to RTFM [17], normal clip features have smaller feature magnitudes than abnormal ones in this learning method. Therefore, we can regard $\Phi_{k^n}(X^n)$ as HIF, $\Phi_{k^a}(X^a)$ as EAIF and instance features with bottom-$(T^n - k^n)$ magnitudes in X^n as ENIF (*i.e.* $\Phi_{k^n}^c(X^n) = X^n \backslash \Phi_{k^n}(X^n) \in \mathbb{R}^{(T^n - k^n) \times D}$). The above operation of selecting the features with top-k^a and top-k^n feature magnitudes from X^a and X^n can be seen

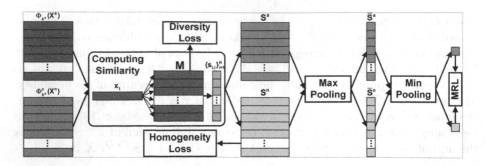

Fig. 2. The second stage of our MMFDL method. MRL denotes margin ranking loss.

as the process of selecting EAIF and HIF. The discrepancy between EAIF and HIF is defined based on feature magnitudes by $d(X^a, X^n) = g_{k^a}(X^a) - g_{k^n}(X^n)$. Following [17], we maximize the discrepancy to preliminarily guarantee the inter-class dispersion between abnormal and normal clips, which becomes

$$\mathcal{L}_\theta(X^a, X^n) = \max(0, \theta - d(X^a, X^n)) \tag{6}$$

where θ is a predefined margin. The classifier learning is the same as RTFM [17], so the classification loss function is denoted by

$$\mathcal{L}_c(X^a, X^n) = \underbrace{\sum_{x^a \in \Phi_{k^a}(X^a)} -\log(f_c(x^a))}_{\textcircled{1}}^{T^a} + \underbrace{\sum_{x^n \in \Phi_{k^n}(X^n)} -\log(1 - f_c(x^n))}_{\textcircled{2}}$$
$$+ \lambda_1 \sum_{t=1}^{T^a} |f_c(x_t^a)| + \lambda_2 \sum_{t=1}^{T^a-1} (f_c(x_t^a) - f_c(x_{t+1}^a))^2 \tag{7}$$

where $\textcircled{1}$ represents the sparsity regularization and $\textcircled{2}$ indicates the temporal smoothness regularization. The loss function $\mathcal{L}_f$ of the first stage is expressed as

$$\mathcal{L}_f = \frac{1}{B} \sum_{i=1}^{B} \mathcal{L}_\theta(X_i^a, X_i^n) + \mathcal{L}_c(X_i^a, X_i^n) \tag{8}$$

where B is the mini-batch size (Fig. 2).

The second stage aims to construct a memory $M = \{m_i\}_{i=1}^N \in \mathbb{R}^{N \times D}$ to store easy normal patterns. To achieve this, we first calculate the similarity score $s_{t,i}$ between each easy instance feature x_t (*i.e.* each feature in $\Phi_{k^a}(X^a)$ and $\Phi_{k^n}^c(X^n)$) and each slot m_i in the memory by cosine similarity $s_{t,i} = \frac{x_t m_i^\mathsf{T}}{\|x_t\|_2 \|m_i\|_2}$ For the EAIF and ENIF, the similarity scores are denoted by $S^a = \{s_{t,i}^a\}_{t=1,i=1}^{k^a,N} \in \mathbb{R}^{k^a \times N}$ and $S^n = \{s_{t,i}^n\}_{t=1,i=1}^{T^n-k^n,N} \in \mathbb{R}^{(T^n-k^n) \times N}$, respectively.

To measure the distance between an instance feature and the memory, we calculate the maximum similarity score between the instance feature and each slot in the memory. The maximum similarity score which ranges from -1 to 1 is also called the memory similarity score of the instance feature in this paper. In practice, we apply a maxpooling operation along the second dimension of S^a and S^n to get the memory similarity scores $\bar{S}^a \in \mathbb{R}^{k_a}$ and $\bar{S}^n \in \mathbb{R}^{T^n - k^n}$, respectively. A large memory similarity score means a small distance.

During the memory updating, the memory is expected to remember easy normal patterns rather than abnormal patterns. An intuitive idea is that we use ranking loss to force the distance between the memory and ENIF to be smaller than the one between the memory and abnormal feature. Since the instance feature with the lowest memory similarity score in $\Phi_{k^n}^c(X^n)$ is most likely not to be remembered and the instance feature with the lowest memory similarity score in $\Phi_{k^a}(X^a)$ has the potential to be the most abnormal, we propose the following ranking objective function:

$$\min \bar{S}^a < \min \bar{S}^n \tag{9}$$

Then, we combine Eq. 9 with a margin ranking loss [2] and set the margin to 2 according to the range of memory similarity scores, which is

$$\mathcal{L}_r = \frac{1}{B} \sum_{i=1}^{B} \max(0, 2 - \min \bar{S}_i^n + \min \bar{S}_i^a) \tag{10}$$

Since the slots of our memory is supposed to be unique, we apply a diversity loss function [10] $\mathcal{L}_d = \left\| M^\mathsf{T} M - I \right\|_F$ to update the memory, where $I \in \mathbb{R}^{D \times D}$ is the identity matrix and $\|\cdot\|_F$ is the Frobenius norm of a matrix. Meanwhile, each memory slot should really store an easy normal pattern. Therefore, we apply a homogeneity loss function [10] $\mathcal{L}_h = \left\| \frac{1}{B} \sum_{i=1}^{B} \mathrm{softmax}(\sum_{t=1}^{T^n - k^n} S_i^n(t)) \right\|_2$ to guarantee that no slot is useless, where $S_i^n(\cdot)$ denotes the first dimension of S^n extracted from the i-th sample in the batch training data with the batch size B. At last, the loss function $\mathcal{L}_s$ of the second stage becomes $\mathcal{L}_s = \mathcal{L}_r + \lambda_3 \mathcal{L}_d + \lambda_4 \mathcal{L}_h$, where λ_3 and λ_4 are hyper-parameters to balance each loss function.

The third stage. Given that only video-levels are available, EAIF of the first stage may contain normal features, which will reduce the difference between normal and abnormal clips. The third stage aims to alleviate this problem and further enlarge the inter-class separability on the basis of the first stage via the ENPM to learn discriminative features. To increase the inter-class dispersion, we enlarge the discrepancies between EAIF and HIF based on feature magnitudes and memory similarity scores (see Fig. 1(b)). Formally, the discrepancy based on the memory similarity scores is defined as

$$\mathcal{D}(\tilde{S}^a, \tilde{S}^n) = \frac{1}{k^n} \sum_{\tilde{s}^n \in \tilde{S}^n} \tilde{s}^n - \frac{1}{k^a} \sum_{\tilde{s}^a \in \tilde{S}^a} \tilde{s}^a \tag{11}$$

where $\tilde{S}^a \in \mathbb{R}^{k_a}$ and $\tilde{S}^n \in \mathbb{R}^{k_n}$ represent the memory similarity scores of EAIF and HIF, respectively. To maximize the discrepancy, we apply the Eq. 12 similar to the Eq. 6 with the margin set to 2 so that we can make the HIF close to our ENPM and away from the EAIF. In this way, our ENPM can assist to guarantee large inter-class dispersion when EAIF contain normal features.

$$\mathcal{L}_m = \frac{1}{B} \sum_{i=1}^{B} \max(0, 2 - \mathcal{D}(\tilde{S}_i^a, \tilde{S}_i^n)) \tag{12}$$

To ensure that we can find HIF and EAIF, the feature magnitude learning and classifier learning are applied the same as the first stage. Thus, loss function of the third stage $\mathcal{L}_t$ is the combination of $\mathcal{L}_m$ and $\mathcal{L}_f$, i.e., $\mathcal{L}_t = \mathcal{L}_m + \mathcal{L}_f$.

3   Experiments

3.1   Experimental Setup

Datasets and Evaluation Metrics. We conduct experiments on UCF-Crime [15] and ShanghaiTech [8] datasets. Following [7,17], we calculate the frame-level area under the curve (AUC) of the ROC curve to evaluate our model.

Implementation Details. Following [17], the video clip features of size $D = 2048$ are obtained from the $Mixed\_5c$ layer of pre-trained I3D [3] network. The maximum number (MN) of input video clips is set to 112 for ShanghaiTech and 405 for UCF-Crime. If the number of input video clips is greater than the MN during training, we divide the video into clips of the MN following [15], otherwise we pad the input clips to the MN with zero matrices. The depth-wise convolution in DW-Net uses a 1×3 filter with a stride of 1. We set the head number as 16 and query dimension as 2048 for the transformer encoder of TF-Net. The classifier (Sect. 2.2) has the same setting as [17]. Following [17], we set the margin $\theta = 100$, weights $\lambda_1 = 8 \times 10^{-3}$, $\lambda_2 = 8 \times 10^{-4}$. We set $N = 2000$ and the weights $\lambda_3 = 0.2$, $\lambda_4 = 0.5$. Our method is trained in a multi-stage approach with a mini-batch size of $B = 32$ using the Adam optimizer. In the second stage, we freeze all parameters except the ENPM. In the third stage, we only freeze the parameters of the ENPM. In each stage, we train the model for 50 epochs with a learning rate of 0.001.

3.2   Results on ShanghaiTech and UCF-Crime

We compare our model with existing semi-supervised [4,16,23], and weakly supervised [6,7,9,11–15,17,20,22,24–26] methods on UCF-Crime and Shang-haiTech. [6,7,17,20,26] use different types of pre-trained features, for example, [6] reports the results of I3D [3] and C3D [18] RGB features. In this paper, we only take their best results and list them in the table. It can be seen from Table 1 that our method achieves the best performance on both datasets.

3.3 Ablation Study

Impact of each module is studied by using different combinations of components. The LGTRM module is decomposed into DW-Net and TF-Net. The baseline model replaces DW-Net and TF-Net with a 1×1 convolutional layer and is only trained in the first stage of the MMFDL method. It can be seen from Table 2 that our model can be boosted by each module and obtain the best result with the combination of all components, which proves that all components are necessary for the proposed method.

Impact of the positional encoding is studied by using different types of the positional encoding (PE) in Table 3. The model with learnable positional encoding achieves the best AUC value. Compared to the model without positional encoding, the model with sinusoidal one performs better on ShanghaiTech while worse on UCF-Crime. This indicates that handcrafted positional encoding is hard to adapt to different scenes and even harms the model.

Impact of the training manner is investigated by using different training methods, *i.e.*, end-to-end and multi-stage training methods in Table 4. For the end-to-end training approach, we add together the loss functions of the second and third stages as the loss function for training, *i.e.*, $\mathcal{L}_{e2e} = \mathcal{L}_s + \mathcal{L}_t$. This means that we optimize the memory and classifier together. The end-to-end training approach performs worse on both datasets. The reason is that end-to-end training can't guarantee that the patterns memorized by the ENPM are all easy and normal, leading to wrongly guiding the direction of model convergence.

Table 1. Frame-level AUC (%) value on two datasets. **Bold** and *italic* indicate the best and second-best performances, respectively.

Supervision	Method	Feature	Year	ShanghaiTech	UCF-Crime
Semi-supervised	Sun *et al.* [16]	TCN RGB	2020	74.7	72.7
	Yu *et al.* [23]	3DCNN RGB	2021	–	81.84
	Chen *et al.* [4]	–	2022	78.1	–
Weakly supervised	Sultani *et al.* [15]	C3D RGB	2018	–	75.41
	Zhang *et al.* [25]	TCN RGB	2019	82.50	78.66
	Zhong *et al.* [26]	TSN RGB	2019	84.44	82.12
	CLAWS [24]	C3D RGB	2020	89.67	83.03
	AR-Net [20]	I3D RGB & Flow	2020	91.24	–
	MIST [6]	I3D RGB	2021	94.38	82.30
	RTFM [17]	I3D RGB	2021	97.21	84.30
	Wu *et al.* [22]	I3D RGB	2021	*97.48*	84.89
	Lv *et al.* [11]	TSN RGB	2021	–	85.38
	Purwanto *et al.* [13]	Relation-aware	2021	96.85	85.00
	BN-SVP [14]	I3D RGB	2022	96.00	83.39
	Liu *et al.* [9]	I3D RGB	2022	90.2	83.0
	DDL [12]	I3D RGB	2022	–	85.12
	Li *et al.* [7]	VideoSwin RGB	2022	97.32	*85.62*
	Ours	I3D RGB	2023	**97.94**	**87.16**

Table 2. Ablation studies of our method on two datasets using AUC (%) value.

Baseline	DW-Net	TF-Net	MMFDL	ShanghaiTech	UCF-Crime
✓				94.38	83.38
✓	✓			94.42	83.71
✓		✓		97.68	85.22
✓	✓	✓		97.81	85.90
✓			✓	94.80	83.49
✓	✓		✓	95.06	83.86
✓		✓	✓	97.89	85.44
✓	✓	✓	✓	97.94	87.16

Table 3. Ablation studies of positional encoding using AUC (%) value.

PE Type	ShanghaiTech	UCF-Crime
Without PE	96.32	84.53
Sinusoidal PE	96.94	82.46
Learnable PE	97.94	87.16

Table 4. Ablation studies of training manner on two datasets using AUC (%) value.

Training Manner	ShanghaiTech	UCF-Crime
End-to-end	96.78	81.46
Multi-stage	97.94	87.16

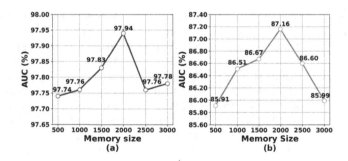

Fig. 3. Ablation studies of memory size N on (a) ShanghaiTech and (b) UCF-Crime.

Impact of the memory size is studied by using different memory size settings. In Fig. 3, the performance will gradually rise first and then slowly fall with the increase of N. When the memory size is small, the ENPM cannot remember all easy normal patterns, causing the model to misjudge normal as

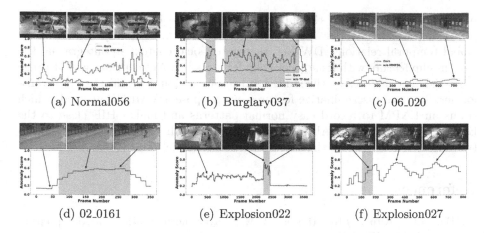

(a) Normal056   (b) Burglary037   (c) 06_020

(d) 02_0161   (e) Explosion022   (f) Explosion027

Fig. 4. Visual results of our model and ablation studies on (a) (c) (d) ShanghaiTech and (b) (e) (f) UCF-Crime. Orange blocks indicate frame-level labels of anomalous events. (a), (b) and (c) compares our model with our model without DW-Net, TF-Net and MMFDL method, respectively. (d), (e) show good cases. (f) shows a failure case. (Color figure online)

abnormal. However, the effect of increasing the memory size without limit is not good, because the ENPM can easily memorize abnormal patterns in this setting.

3.4   Visual Analysis

Visual results are shown in Fig. 4. In Fig. 4(a), the model with DW-Net can output a smooth curve with low anomaly scores when detecting video frames in a normal video. In Fig. 4(b), the model without TF-Net fails to detect anomalies in the video. Figure 4(b) also shows that our model can well detect a video with multiple anomalous events. In Fig. 4(c), the model with MMFDL method can produce lower anomaly scores of the video frames in a normal video. In Fig. 4(d) and (e), our model has the ability to detect the anomalous events which take up a large and small temporal part of abnormal videos respectively. We provide a failure case in Fig. 4(f). Although our model has detected the explosion event, it predicts high anomaly scores of the fire fighting and smoke marked with red boxes. The reason is that fire fighting and smoke caused by burning do not happen in normal videos of the training set, but they always appear in explosion and arson events of the training set. It is hard to classify these two events without frame-level supervision, which indicates the great challenge of WS-VAD.

4   Conclusions

In this work, we propose a LGTRM module and a MMFDL method to augment the clip feature representation for WS-VAD so that the classifier can accurately predict anomaly scores of the input video clips. The LGTRM module is composed of two subnetworks: DW-Net and TF-Net, aiming to capture local and global temporal relations. DW-Net aggregates local context by integrating the current clip feature with its adjacent clip features. TF-Net captures global-range temporal dependencies by the multi-head self-attention mechanism of the transformer. To learn discriminative features, we propose a MMFDL method, which trains an ENPM to record easy normal patterns and makes HIF close to the ENPM and away from EAIF to ensure inter-class dispersion. Experiments on two datasets demonstrate the superiority of our model over SOTA methods.

References

1. Ba, J.L., Kiros, J.R., Hinton, G.E.: Layer normalization. arXiv preprint arXiv:1607.06450 (2016)
2. Bordes, A., Usunier, N., Garcia-Duran, A., Weston, J., Yakhnenko, O.: Translating embeddings for modeling multi-relational data. In: NIPS, vol. 26 (2013)
3. Carreira, J., Zisserman, A.: Quo vadis, action recognition? A new model and the kinetics dataset. In: CVPR, pp. 4724–4733 (2017)
4. Chen, C., et al.: Comprehensive regularization in a bi-directional predictive network for video anomaly detection. In: AAAI, vol. 36, pp. 230–238 (2022)
5. Chollet, F.: Xception: deep learning with depthwise separable convolutions. In: CVPR, pp. 1800–1807 (2017)
6. Feng, J., Hong, F., Zheng, W.: MIST: multiple instance self-training framework for video anomaly detection. In: CVPR, pp. 14009–14018 (2021)
7. Li, S., Liu, F., Jiao, L.: Self-training multi-sequence learning with transformer for weakly supervised video anomaly detection. In: AAAI, pp. 1395–1403 (2022)
8. Liu, W., Luo, W., Lian, D., Gao, S.: Future frame prediction for anomaly detection - a new baseline. In: CVPR, pp. 6536–6545 (2018)
9. Liu, Y., Liu, J., Zhu, X., Wei, D., Huang, X., Song, L.: Learning task-specific representation for video anomaly detection with spatial-temporal attention. In: ICASSP, pp. 2190–2194 (2022)
10. Luo, W., et al.: Action unit memory network for weakly supervised temporal action localization. In: CVPR, pp. 9969–9979 (2021)
11. Lv, H., Zhou, C., Cui, Z., Xu, C., Li, Y., Yang, J.: Localizing anomalies from weakly-labeled videos. IEEE TIP **30**, 4505–4515 (2021)
12. Pu, Y., Wu, X.: Locality-aware attention network with discriminative dynamics learning for weakly supervised anomaly detection. In: IEEE ICME, pp. 1–6 (2022)
13. Purwanto, D., Chen, Y., Fang, W.: Dance with self-attention: a new look of conditional random fields on anomaly detection in videos. In: ICCV, pp. 173–183 (2021)
14. Sapkota, H., Yu, Q.: Bayesian nonparametric submodular video partition for robust anomaly detection. In: CVPR, pp. 3212–3221 (2022)
15. Sultani, W., Chen, C., Shah, M.: Real-world anomaly detection in surveillance videos. In: CVPR, pp. 6479–6488 (2018)
16. Sun, C., Jia, Y., Hu, Y., Wu, Y.: Scene-aware context reasoning for unsupervised abnormal event detection in videos. In: ACMMM, pp. 184–192 (2020)
17. Tian, Y., Pang, G., Chen, Y., Singh, R., Verjans, J.W., Carneiro, G.: Weakly-supervised video anomaly detection with robust temporal feature magnitude learning. In: ICCV, pp. 4955–4966 (2021)

18. Tran, D., Bourdev, L.D., Fergus, R., Torresani, L., Paluri, M.: Learning spatiotemporal features with 3d convolutional networks. In: ICCV, pp. 4489–4497 (2015)
19. Vaswani, A., et al.: Attention is all you need. In: NIPS, pp. 5998–6008 (2017)
20. Wan, B., Fang, Y., Xia, X., Mei, J.: Weakly supervised video anomaly detection via center-guided discriminative learning. In: IEEE ICME, pp. 1–6 (2020)
21. Wang, X., Girshick, R., Gupta, A., He, K.: Non-local neural networks. In: CVPR, pp. 7794–7803 (2018)
22. Wu, P., Liu, J.: Learning causal temporal relation and feature discrimination for anomaly detection. IEEE TIP **30**, 3513–3527 (2021)
23. Yu, J., Lee, Y., Yow, K.C., Jeon, M., Pedrycz, W.: Abnormal event detection and localization via adversarial event prediction. In: IEEE TNNLS, pp. 1–15 (2021)
24. Zaheer, M.Z., Mahmood, A., Astrid, M., Lee, S.-I.: CLAWS: clustering assisted weakly supervised learning with normalcy suppression for anomalous event detection. In: Vedaldi, A., Bischof, H., Brox, T., Frahm, J.-M. (eds.) ECCV 2020. LNCS, vol. 12367, pp. 358–376. Springer, Cham (2020). https://doi.org/10.1007/978-3-030-58542-6_22
25. Zhang, J., Qing, L., Miao, J.: Temporal convolutional network with complementary inner bag loss for weakly supervised anomaly detection. In: ICIP, pp. 4030–4034 (2019)
26. Zhong, J.X., Li, N., Kong, W., Liu, S., Li, T.H., Li, G.: Graph convolutional label noise cleaner: train a plug-and-play action classifier for anomaly detection. In: CVPR, pp. 1237–1246 (2019)

DFAformer: A Dual Filtering Auxiliary Transformer for Efficient Online Action Detection in Streaming Videos

Shicheng Jing[1,2] and Liping Xie[1,2(✉)]

[1] Key Laboratory of Measurement and Control of Complex Systems of Engineering, Ministry of Education, School of Automation, Southeast University, Nanjing 210096, China
scjing10@gmail.com
[2] Southeast University Shenzhen Research Institute, Shenzhen 518063, China
lpxie@seu.edu.cn

Abstract. Online action detection (OAD) aims to identify the specific type of ongoing action frame by frame without future information. The full exploration of historical memory with limited yet redundant information constraints for potential patterns thus becomes an important yet challenging problem. We propose a novel transformer-based framework called **Dual Filtering Auxiliary Transformer** (DFAformer) to achieve this goal. In DFAformer, a two-stage filtering mechanism filters impurities related to background and uninterested actions in the historical memory at the frame and element levels. To make the model concentrate on the ongoing action, we elaborate an auxiliary task, Jaccard Summary Unit, explicitly correlates the past with the future. This auxiliary task guide the learning of model weights without extra computational costs during inference. Experiments on three real-world benchmark datasets demonstrate the superiority of the proposed method.

Keywords: Online action detection · Two-stage filtering mechanism · Auxiliary task

1 Introduction

The task of online action detection (OAD), which requires the identification of the category of an ongoing action from a streaming video in the absence of future information, presents significant challenges [3]. These challenges are exacerbated by the subtle inter-class distinctions observable during the early stages of action

This work was supported in part by the National Natural Science Foundation of China under Grant 62372104 and in part by the Guangdong Basic and Applied Basic Research Foundation under Grant 2022A1515110518. Besides, we thank the Big Data Computing Center of Southeast University for providing the facility support on the numerical calculations.
Supplementary material is available online at https://github.com/Prot-debug/Dual-Filtering-Auxiliary-Transformer.

Q. Liu et al. (Eds.): PRCV 2023, LNCS 14430, pp. 134–145, 2024.
https://doi.org/10.1007/978-981-99-8537-1_11

and the severe class imbalance between the background and actions of interest. Consequently, OAD research has recently garnered considerable attention within the computer vision community [1, 4, 17–22].

Traditional OAD approaches use Recurrent Neural Network (RNN) to progressively distill the window-based input sequence into a hidden state for temporal modeling [4, 10, 12, 20, 23]. However, these methods often fall short in adequately capturing global dependencies and suffer from catastrophic forgetting in long sequences. Recent advancements in transformer-based models offer a robust alternative, utilizing effective attention mechanisms for one-step global information aggregation [16]. Several models have attained state-of-the-art performance in the OAD domain [1, 6, 17, 21, 22].

One specific challenge in OAD is the presence of a majority of frames constituting background and irrelevant actions in long streaming videos. To tackle this, GateHUB [1] proposed an asymmetric module, the Gated History Unit, that enhances or suppresses each frame based on its relative importance to the current frame. However, we contend that this asymmetric gated mechanism induces a model bias towards enhancing frames over suppression, which may not be rational. Furthermore, this frame-level filter fails to completely filter out all redundant frames, and its application becomes inappropriate once self-attention layers disrupt and reorganize tokens.

Another approach, implemented by TRN [20] and OadTR [17], involves predicting future ℓ-step action categories during training and using corresponding ground-truth labels as supervision signals. This implicitly guides the model to learn future distributions, adjusting model weights favorably for OAD. Nonetheless, we regard this approach as a sub-optimal solution due to the underutilization of rich semantic future information. We argue for a more explicit correlation between past and future events.

In this work, inspired by the aforementioned limitations, we introduce a novel encoder-decoder transformer-based framework, termed DFAformer, **D**ual **F**iltering **A**uxiliary **Trans**former, as depicted in Fig. 1. DFAformer incorporates a two-stage filtering mechanism to filter out frame-level and element-level impurities. The first stage, the Equalized Gated Unit, filters uninformative frames by an ingenious symmetric gated mechanism. The second stage, the Recurrent Filtering Unit, selects applicable parts of tokens disrupted and reorganized by self-attention layers. To continually enhance model accuracy without incurring additional inference costs, we elaborate an auxiliary task, Jaccard Summary Unit (JSU), which is solely utilized during training. JSU leverages future rich semantic information during training and explicitly correlates the past with the future through Jaccard similarity losses, which provides additional evidence for model weight learning during training.

Our experimental results on three publicly available datasets, THUMOS14 [9], TVSeries [3], and HDD [12], reveal that DFAformer outperforms all existing methods. An extensive ablation study further demonstrates the effectiveness and efficiency of DFAformer.

To encapsulate, our work contributes the following:

- We introduce a two-stage filtering mechanism as an encoder, composed of an Equalized Gated Unit and a Recurrent Filtering Unit, which effectively filter frame-level and element-level impurities in tokens.
- Designed as an auxiliary task, the Jaccard Summary Unit (JSU) explicitly correlates past and future actions, enhancing model accuracy without incurring additional computational cost during inference.
- Our DFAformer model outperforms all existing methods and achieves new state-of-the-art in terms of accuracy and efficiency on three benchmarks. An extensive ablation study demonstrates its effectiveness and efficiency.

2   Related Work

Temporal Action Detection. The temporal action detection(TAD) algorithm aims to locate all action instances from untrimmed videos, where videos here are complete. Due to the great success of object detection in images, the mainstream approaches in TAD are divided into one-stage [11] and two-stage [7] anchor-based approaches. Most recent methods [7] tend to be two-stage proposal generation methods. The two-stage approaches generate multiple candidate action proposals and perform category classification and boundary regression. However, online action detection is not able to access future frames during inference, so these procedures above for TAD are unsuitable for OAD.

Online Action Detection. Geest *et al.* [3] initially introduced the concept of online action detection to the mainstream, sparking numerous subsequent works that utilized CNNs [3], recurrent networks [4,10,12,20,23], and the more recent transformers [1,6,17,21,22]. For instance, TRN [20] employs an RNN-based architecture for sequential processing, supplemented with an additional prediction task during training, coercing the model to implicitly learn future distributions. OadTR [17], transformer-based framework, also predicts future outputs, combining them with a class token for OAD execution. Unlike TRN and OadTR, our proposed DFAformer harnesses a novel transformer-based architecture that captures extensive time-range contextual dependencies and explicitly establishes a correlation between past and future distributions through the JSU module, without additional computational cost during inference. Furthermore, while GateHUB [1] develops a frame-level filter to sift out impurities among frames, DFAformer advances a two-stage filtering mechanism, capable of purging impurities at both frame and element levels.

Sequential Data Modeling. Traditional methods [13] like RNNs and LSTMs, adept at local dependency modeling, suffer from gradient issues and long-term learning challenges. Attention-based models [14,15] like Transformers offer global modeling solutions by addressing long-term dependencies effectively. However, Perceiver and PerceiverIO [1,21], despite utilizing fixed-sized latent encoding to cut computational complexity, miss local focus on current frames and global attention to historical frames. Our method thus merges Transformers' strengths with LSTM-like gating to enhance current frame handling and elevate long-sequence learning for online action detection.

3 Methodology

In this section, we first introduce the notational definition for OAD, then describe in detail each component of our proposed DFAformer as shown in Fig. 1, and finally illustrate the process for training and inference.

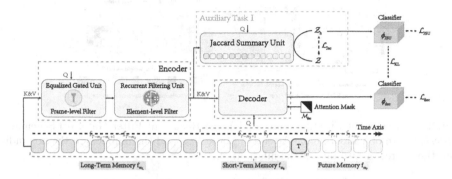

Fig. 1. Overview of the proposed Dual Filtering Auxiliary Transformer.

3.1 Notational Definition

OAD aims to classify ongoing action using only past and current observations. Following [1,21], given a sampled streaming video sequence $\mathbf{F} = [F_t]_{t=-m_L-m_S+1}^{m_F}$, where m_L, m_S, and m_F represent the duration of the long-term, short-term, and future memory, respectively, and F_0 denotes the current frame, we extract the features $\mathbf{f} = [f_t]_{t=-m_L-m_S+1}^{m_F}$ for each frame in $\mathbf{F}$ using an out-of-the-box pre-trained model. We use $\mathbf{f_{m_L}}$, $\mathbf{f_{m_S}}$, and $\mathbf{f_{m_F}}$ to denote long-term, short-term, and future memory features, respectively, and perform OAD based on this setting.

3.2 Encoder

It is widely acknowledged that the dense nature of video data leads to significant redundant information such as background and irrelevant actions. In response, we propose a two-stage filtering encoder comprising an Equalized Gated Unit and a Recurrent Filtering Unit, designed for frame-level and element-level filtering respectively, aiming to mitigate the aforementioned challenge.

Equalized Gated Unit (EGU), as a frame-level filter, automatically select frames beneficial to OAD. Compared with GHU in GateHUB [1], which uses a learnable scalar factor g mapped into [-inf, 1] for attention weight calibration, our proposed EGU with minimal changes effectively balance the relationship between enhancement and suppression more effectively, thus achieving improved performance. The formula for EGU can be expressed as,

$$\mathbf{f_g} = \sigma\left((\mathbf{f_{m_L}}\mathbf{W}_a + \mathbf{E_{pos}})\mathbf{W}_b\right) \tag{1}$$

$$g = \mathbf{EqualizedGate}(\log(\mathbf{f_g}) + \mathbf{f_g}) \tag{2}$$

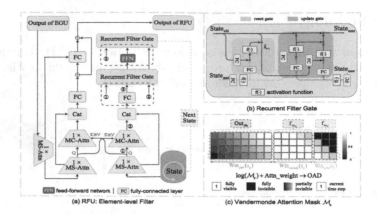

Fig. 2. (a) Recurrent Filtering Unit (RFU). (b) Recurrent Filter Gate. (c) Schematic representation of Vandermonde Attention Mask $\mathcal{M}_z$, and the numbers in the box indicate the degree of visibility of the information.

where $\mathbf{f_{m_L}} \in \mathbb{R}^{m_L \times T}$, $\mathbf{W}^a \in \mathbb{R}^{T \times D}$, $\mathbf{W}^b \in \mathbb{R}^{D \times 1}$, $\mathbf{f^g} \in \mathbb{R}^{m_L \times 1}$. T and D denote the dimensions of extracted features and projecting features, respectively. $\mathbf{E_{pos}}$ represent relative position encoding [2], relative to the current time. $\sigma(\cdot)$ is sigmoid activation function. $\mathbf{f_g}$ measures the importance of each frame in $\mathbf{f_{m_L}}$. **EqualizedGate**$(\cdot)$ is a range limiting function to limit Eq. (2) to [-inf, ln2]. Thus, it is natural for g to make the attention weight scaled by a factor in [0, 2], i.e. [0, 1] suppression and [1, 2] enhancement in EGU. This symmetric design is proved to be effective in Sect. 4.3. Due to n_0 learnable tokens as Query, the output of EGU is $\mathrm{Out_{egu}} \in \mathbb{R}^{n_0 \times D}$. More details refer to § Supp.1.

Recurrent Filtering Unit (RFU), as an element-level filter, targets redundant elements within tokens. Our belief is that EGU performs coarse filtration, primarily filtering out background and irrelevant actions as much as possible. However, after the self-attention layers within EGU, tokens become disrupted and reorganized. To address this, we design RFU, a more granular element-level filter factoring in both global and local considerations. Transformers excel at global history-based time step calculations, yet their efficiency lags concerning local temporal dependencies. We adopt and extend the principal concept from a robust model in [8], which combines LSTM and transformer for time series modeling, capturing both global and local temporal dependencies. Specifically, RFU employs $\mathrm{Out_{egu}}$ to cross-attend to a recurrent state and vice versa in one forward process. Generally, the internal structure of RFU can be clearly shown in Fig. 2a, where $\mathrm{Out_{rfu}} \in \mathbb{R}^{n_0 \times D}$ is the output of RFU. MS-Attn and MC-Attn are Multi-Head Self-Attention and Multi-Head Cross-Attention, respectively. In addition, inspired by [4], the branch for updating state uses neither a direct residual connection used in input branch nor a simple gating mechanism used in [8], but a relatively complex yet effective filtering gate,

$$\mathrm{State_{next}} = \mathbf{FilterGate}(\mathrm{State_{old}}, \mathrm{State_{new}}, \mathrm{State_{mid}}) \tag{3}$$

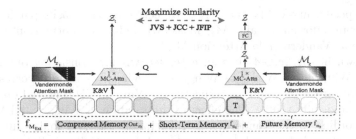

Fig. 3. Illustration of Jaccard Summary Unit (JSU) module. JSU adopts two separate decoders to decode ongoing action from extended memory $\mathbf{f}_{M_{Ext}}$, one of which could see future information and the other not. And the output distributions of the two decoders will be narrowed by Jaccard similarity measures.

where **FilterGate** represents the recurrent filter gate, as shown in Fig. 2b. The reset gate can effectively drop or take past information based on the correlation between State_{old} and State_{mid}. The update gate updates out State_{next} according to State_{new} and $\tilde{h}_{t-1}$.

3.3 Decoder

The decoder decodes ongoing action from the compressed memory Out_{rfu}, then classifies it through the linear classifier ϕ_{dec}. Our decoder here continues to follow the previous design of [1,21]. Specifically, we use $\mathbf{f}_{ms}$ as Query to cross-attend to Out_{rfu}. Note that the attention operation here introduces a Square Attention Mask $\mathcal{M}_{dec}$ to realize causal setting. With this causal setting, it is possible to make each frame in $\mathbf{f}_{ms}$ seem like the latest frame for the model, thus increasing the supervision signals.

3.4 Auxiliary Task — Jaccard Summary Unit

The aforementioned encoder-decoder setup competently addresses OAD tasks and yields competitive outcomes. However, we are convinced that the information in $\mathbf{f}_{m_L}$ remains underutilized. To enhance data exploitation, we propose an auxiliary task aimed at uncovering the latent information within $\mathbf{f}_{m_L}$.

To our knowledge, existing OAD methods do not explicitly leverage future abundant features. Instead, DFAformer leverages them during training and elaborates Jaccard Summary Unit (JSU, Fig. 3) to correlate the past with the future. The core idea of JSU is to narrow the gap between past and future distributions, forcing the model during inference as if it could 'see into the future'. To this end, we employ two separate decoders to decode ongoing action from an extended memory, one with access to future information and the other without. More specifically, we first concatenate Out_{rfu} with projecting features $\mathbf{f}'_{ms}$ and $\mathbf{f}'_{mF}$, thus forming extended memory $\mathbf{f}_{M_{Ext}} = \text{Cat}(\text{Out}_{rfu}, \mathbf{f}'_{ms}, \mathbf{f}'_{mF}) \in \mathbb{R}^{(n_0+m_S+m_F) \times D}$. Then we cross-attend n_1 learnable tokens with $\mathbf{f}_{M_{Ext}}$ using two task-specific mask $\mathcal{M}_z$ and $\mathcal{M}_{z_t}$, respectively. $\mathcal{M}_z$ masks partial historical and future information and $\mathcal{M}_{z_t}$ masks whole future information, which are referred to as Vandermonde Attention Mask.

As shown in Fig. 2c, $\mathcal{M}_z$ has an extended window consisting of three sub-windows ($\text{Win}_{\text{Ext}_z} = \text{Win}_{\text{past}} + \text{Win}_{\text{current}} + \text{Win}_{\text{future}}$), whose corresponding window sizes are s_p, s_c, and s_f, respectively. $\text{Win}_{\text{current}}$ is centered on the current frame and fully visible. The effect of information in Win_{past} and $\text{Win}_{\text{future}}$ decays according to the Vandermonde rule. In general, $\mathcal{M}_z$ makes information of $s_p + \frac{1}{2}s_c$ steps before and $\frac{1}{2}s_c + s_f$ steps later partially available according to the Vandermonde rule for the decoder at the current time step. On the contrary, $\mathcal{M}_{z_t}$ has no access to future information. In other words, its extended window is $\text{Win}_{\text{Ext}_{zt}} = \text{Win}_{\text{past}} + \text{Win}_{\text{current}}$. Building upon previous work [5], we adopt the Jaccard similarity measures proposed therein to assess the similarity between two distributions, integrating it into our loss function. Compared with [5], our proposed JSU is different in two keys aspects. Firstly, [5] is tailored for early action recognition (one video inputs, one label outputs), which is not compatible with frame-level labeling tasks such as OAD. Thus, we implement a Vandermonde matrix to minimize the influence of extraneous frames, thereby focusing attention on the current frame. Secondly, the architectural backbone of our JSU is grounded entirely in transformer models.

The output distributions of the two decoders will be narrowed by Jaccard similarity measures. We send the output of the decoder with $\mathcal{M}_{z_t}$ to the linear classifier ϕ_{JSU}. Note that JSU used for training does not introduce extra computation costs during inference, no matter how complex it is.

3.5   Training and Inference

In DFAformer, the output of decoder is the classification score $\mathbf{s} \in \mathbb{R}^{m_S \times (C+1)}$, where C denotes the number of categories. Then we employ the standard cross-entropy loss over $\mathbf{s}$, as follows,

$$\mathcal{L}_{\text{OAD}} = - \sum_{i=t-m_S+1}^{0} \sum_{j=0}^{C} \boldsymbol{y}_{i,j} \log \hat{\mathbf{s}}_{i,j} \tag{4}$$

where $\hat{\mathbf{s}}_{i,j}$ indicates scores after softmax operation. $\boldsymbol{y}_{i,j}$ is the ground-truth of the ith frame, which is one-hot encoding. In JSU, to force the model to see into the future, we use $\mathcal{L}_{\text{Jac}}$ to correlate the past with the future,

$$\begin{aligned}
\mathcal{L}_{\text{Jac}} &= \text{JVS}(Z_t, Z) + \text{JCC}(Z_t, Z) + \text{JFIP}(Z_t, Z) \\
&= \exp\left(-2\frac{\mathbf{z}_t \cdot \mathbf{z}}{\mathbf{z}_t \cdot \mathbf{z}_t + \mathbf{z} \cdot \mathbf{z}}\right) \\
&\quad + \left\|\exp\left(-2\frac{E\left[Z^T \times Z_t\right]}{E\left[Z^T \times Z\right] + E\left[Z_t^T \times Z_t\right]}\right)\right\|_{\text{mean}} \\
&\quad + \exp\left(-\frac{2\langle C_t, C\rangle_F}{\langle C_t, C_t\rangle_F + \langle C, C\rangle_F}\right)
\end{aligned} \tag{5}$$

where Z_t and Z denote the outputs of decoder with $\mathcal{M}_{z_t}$ and $\mathcal{M}_z$, respectively. $\mathbf{z}_t$ and $\mathbf{z}$ are the one-to-one corresponding vectors in Z_t and Z, respectively. The cross-correlation matrix of Z_t and Z is obtained by $E\left[Z^T \times Z_t\right]$ where $E[]$ is the expectation. $\|A\|_{\mathrm{mean}}$ is the mean norm of the matrix. Let us define the covariance matrix of Z_h by C_h and the covariance matrix of Z is C. $\langle C_a, C_b \rangle_F$ denotes the Frobenius inner product between matrices C_a and C_b. Like $\mathbf{s}$ in decoder, JSU calculates $\mathbf{s}^{\mathrm{JSU}}$ to assist in classification,

$$\mathcal{L}_{\mathrm{JSU}} = - \sum_{i=t-m_S+1}^{0} \sum_{j=0}^{C} \boldsymbol{y}_{i,j} \log \hat{\boldsymbol{s}}_{i,j}^{\mathrm{JSU}} \qquad (6)$$

Two classification scores should be consistent, as decoder and JSU tackle the same frames. Hence, we introduce the KL-divergence loss to enable mutual guidance among two modules,

$$\mathcal{L}_{KL} = \mathcal{L}_{KL}\left(\hat{\mathbf{s}}\|\hat{\mathbf{s}}^{\mathrm{JSU}}\right) + \mathcal{L}_{KL}\left(\hat{\mathbf{s}}^{\mathrm{JSU}}\|\hat{\mathbf{s}}\right) \qquad (7)$$

Given the above, the final joint training loss is,

$$\mathcal{L} = \alpha \mathcal{L}_{\mathrm{OAD}} + \beta \mathcal{L}_{\mathrm{Jac}} + \gamma \mathcal{L}_{\mathrm{JSU}} + \delta \mathcal{L}_{KL} \qquad (8)$$

where $\alpha, \beta, \gamma, \delta$ are trade-off parameters.

4   Experiments

4.1   Dataset and Setup

THUMOS14 contains more than 20 h of sports videos with 20 actions. Following the prior work [1,17,20–22], we train on the validation set (200 untrimmed videos) and inference on the test set (213 untrimmed videos).

TVSeries contains 30 daily actions from 6 TV series, for a total of 16 h. TVSeries exhibits some challenging characteristics, e.g., a wide variety of backgrounds and unconstrained perspectives.

HDD is a driving dataset recorded from a front-mounted camera. About 104 h of 137 driving sessions are captured, containing 11 driving behaviors and some readings from non-visual sensors on the vehicle. As in the previous works [1,17, 20,22], we use 100 driving sessions for training and the rest for inference.

Metric. To evaluate the performance of DFAformer, following the previous methods [1,17,22], we report mean Average Precision (mAP) on THUMOS14 and HDD, and calibrated mean Average Precision [3] (cmAP) on TVSeries.

Feature Encoding. For THUMOS14 and TVSeries, following [1,21], we sample the videos at 24 FPS (frames per second). For HDD, following [17], we extract the sensor data at 3 FPS. As for feature extractor, following [1,21], we use the TSN models. We experiment with feature extractors pretrained on two datasets, ActivityNet and Kinetics.

Implementation Details. We implement DFAformer in PyTorch and conduct experiments with NVIDIA GeForce RTX 3090 graphics cards. Following [17,21], we use Adam optimizer with weight decay 5×10^{-5}, and DFAformer is optimized for 25 epochs using batch size of 16. For transformer units, we set number of heads as 8 and d_model as 768. For THUMOS14 and TVSeries, we set m_L, m_S, m_F=1024, 3, 2, representing 1024, 3 and 2 s, respectively. For HDD, we set m_L, m_S, m_F=24, 3, 6, representing 24, 3 and 6 s, respectively. The main reason for this choice is that duration of the actions in HDD is relatively short. For joint training loss, we set α=65, β=10, γ=1, δ=1.

4.2   Compared with State-of-the-Art Methods

Table 1. Comparison between DFAformer and other state-of-the-art online action detection methods on the THUMOS14 [9], HDD [12], TVSeries [3] dataset in terms of mAP and cmAP(%), using Kinetics pretrained features.

Method	Reference	Architecture	mAP(%)		cmAP(%)
			THUMOS14	HDD	TVSeries
CNN [3]	ICLR'15	RNN	-	22.7	-
LSTM [12]	CVPR'18		-	23.8	-
FATS [10]	PR'21		59.0	-	84.6
IDN [4]	CVPR'20		60.3	-	86.1
TRN [20]	ICCV'19		62.1	29.2	86.2
PKD [23]	PR'22		64.5	-	86.4
OadTR [17]	ICCV'21	Transformer	65.2	29.8	87.2
Colar [22]	CVPR'22		66.9	30.6	88.1
WOAD [6]	CVPR'21		67.1	-	-
LSTR [21]	NeurIPS'21		69.5	-	89.1
GateHUB [1]	CVPR'22		70.7	32.1	89.6
DFAformer	-	Transformer	**72.1**	**33.7**	**89.8**

We compare our proposed DFAformer with existing state-of-the-art (SOTA) methods on the THUMOS14, HDD and TVSeries datasets. As demonstrated in Table 1, DFAformer, based on Kinetics features, achieves SOTA performance on THUMOS14, surpassing all prior SOTA methods with a 1.4% improvement over the previously best-performing method, GateHUB [1]. This underscores the efficacy and practicality of the two-stage filtering mechanism and the auxiliary task in DFAformer.

Table 1 also presents results on the HDD dataset using sensor data. Here, DFAformer outstrips other methods by a significant margin, showing an improvement of 1.6% (32.1% *vs.* 33.7%) with sensor data input. These findings suggest the applicability of our approach extends beyond the realm of visual data, potentially applicable across other modalities.

From Table 1, we observe that DFAformer outperforms all existing methods on TVSeries dataset by at least 0.2% when using Kinetics features. Considering the results from the other two datasets, we have reason to believe that the performance on TVSeries dataset is nearly saturated, mainly due to the ambiguity of the action boundaries.

These empirical results, spread across three diverse datasets, underscore the scalability and wide applicability of DFAformer. We refer the reader to § Supp.2 for a more in-depth analysis and discussion on additional results.

4.3   Ablation Studies

Table 2. Ablation study compares different variants of DFAformer, measures by mAP(%) on THUMOS14. The ablation in Encoder and Auxiliary Task is carried out at an inter-module level, whereas the ablation in JSU and RFU is executed at an intra-module level.

Encoder	mAP(%)
LSTR [21]	69.5
GateHUB [1]	70.7
w/ GHU [1]	71.0
w/ EGU	71.2
w/ RFU	71.1
w/ EGU+RFU(ours)	**72.1**

JSU	mAP(%)
w/ JVS	71.7
w/ JVS+JCC	71.9
w/ JVS+JCC+JFIP(ours)	**72.1**
Mask in JSU	mAP(%)
w/ vanilla Mask	71.9
w/ Vandermonde Mask(ours)	**72.1**

Auxiliary Task	mAP(%)
GateHUB [1]	70.7
w/o JSU	71.5
w/ JSU(ours)	**72.1**

RFU	mAP(%)
w/ residual connection	71.4
w/ simple gating mechanism [8]	71.8
w/ recurrent filtering gate(ours)	**72.1**

(a) Ablation Study among Modules   (b) Ablation Study within Modules

We carry out a comprehensive ablation study to probe the effects of each DFAformer component. Unless otherwise stated, all tests are conducted on THUMOS14 using Kinetics features. Additional ablations are discussed in § Supp.3.

Effect of Encoder. We experimentally validate the effectiveness of the two-stage filtering mechanism in encoder. We employ LSTR [21], a transformer-based encoder-decoder paradigm without filtering modules, as our baseline. 'w/ EGU' implies exclusion of the RFU, retaining only the EGU within the encoder. Conversely, w/ RFU' denotes the removal of EGU while preserving RFU. Besides, like in GateHUB [1], 'w/ GHU' further removes EqualizedGate on top of 'w/ EGU'. As depicted in Table 2a, DFAformer outperforms all variants, elevating mAP from 69.5% to 72.1%, thereby affirming the indispensability of the two-stage filtering mechanism. Remarkably, even with a singular filter, the model surpasses the foremost SOTA solution [1] by at least 0.4%. This suggests that EGU and RFU are both crucial for OAD, attributed to the ability of DFAformer to filter frame-level and element-level impurities. Another comparison of EGU and GHU reveals that our proposed EqualizedGate for symmetric design is valid.

Effect of Auxiliary Task. To explore how the auxiliary task affect performance, we conduct experiments. In Table 2a, 'w/o JSU' means that the model is unable to see into the future during training and do not correlate past with future. As shown in Table 2a, when JSU is unavailable, the training costs reduce, but 'w/o JSU' observes performance drop. Specifically, 'w/o JSU' observes 0.6% performance drops. However, 'w/ JSU' achieves higher accuracy than 'w/o JSU'. It indicates that JSU can lead to better results. It could be attributed to the ability of our DFAformer to dig into the memory for latent information.

Ablations about Jaccard Summary Unit. To further investigate the efficacy of Jaccard similarity losses and JSU internal structure, we add the Jaccard similarity losses step by step and replace Vandermonde Mask with vanilla Mask. As illustrated in Table 2b, with the addition of JVS, JCC and JFIP, the performance of the model is gradually improved, indicating that these individual losses provide different properties into the learned representation and it is interesting to see that they are complementary. Moreover, Vandermonde Mask is more effective than the vanilla Mask, since Vandermonde Mask allows the attention to automatically focus on the information of the current moment, which makes it more discriminative for the learned representation.

Ablations About Recurrent Filtering Unit. We further explore ways to update the state. We can use a direct residual connection as in the input branch, a simple gating mechanism as in literature [8], or a relatively complex but effective filtering gate like in DFAformer. In Table 2b, having a direct residual connection and a simple gating mechanism leads to 0.7% and 0.3% lower performance, respectively, further validating our design of recurrent filter gate. This may be since our recurrent filter gate filters elemental-level noise more thoroughly.

5   Conclusion

This paper presents DFAformer, a novel encoder-decoder framework founded upon transformers for online action detection. DFAformer elaborates a two-stage filtering mechanism to filter out impurities at both frame and element levels within the historical memory. To further boost accuracy without introducing extra inference costs, we propose an auxiliary task, i.e., Jaccard Summary Unit, to force DFAformer to dig into the memory for latent information. Remarkably, DFAformer outperforms all existing methods and achieves SOTA performance.

References

1. Chen, J., Mittal, G., Yu, Y., Kong, Y., Chen, M.: Gatehub: gated history unit with background suppression for online action detection. In: CVPR, pp. 19925–19934 (2022)

2. Dai, Z., Yang, Z., Yang, Y., Carbonell, J.G., Le, Q.V., Salakhutdinov, R.: Transformer-xl: attentive language models beyond a fixed-length context. In: ACL, pp. 2978–2988 (2019)
3. De Geest, R., Gavves, E., Ghodrati, A., Li, Z., Snoek, C., Tuytelaars, T.: Online action detection. In: Leibe, B., Matas, J., Sebe, N., Welling, M. (eds.) ECCV 2016. LNCS, vol. 9909, pp. 269–284. Springer, Cham (2016). https://doi.org/10.1007/978-3-319-46454-1_17
4. Eun, H., Moon, J., Park, J., Jung, C., Kim, C.: Learning to discriminate information for online action detection. In: CVPR, pp. 809–818 (2020)
5. Fernando, B., Herath, S.: Anticipating human actions by correlating past with the future with Jaccard similarity measures. In: CVPR, pp. 13224–13233 (2021)
6. Gao, M., Zhou, Y., Xu, R., Socher, R., Xiong, C.: WOAD: weakly supervised online action detection in untrimmed videos. In: CVPR, pp. 1915–1923 (2021)
7. He, B., Yang, X., Kang, L., Cheng, Z., Zhou, X., Shrivastava, A.: ASM-Loc: action-aware segment modeling for weakly-supervised temporal action localization. In: CVPR, pp. 13925–13935 (2022)
8. Hutchins, D., Schlag, I., Wu, Y., Dyer, E., Neyshabur, B.: Block-recurrent transformers. In: NeurIPS, pp. 33248–33261 (2022)
9. Jiang, Y.G., et al.: THUMOS challenge: action recognition with a large number of classes (2014). http://crcv.ucf.edu/THUMOS14/
10. Kim, Y.H., Nam, S., Kim, S.J.: Temporally smooth online action detection using cycle-consistent future anticipation. PR **116**, 107954 (2021)
11. Liu, X., et al.: End-to-end temporal action detection with transformer. IEEE TIP **31**, 5427–5441 (2022)
12. Ramanishka, V., Chen, Y.T., Misu, T., Saenko, K.: Toward driving scene understanding: a dataset for learning driver behavior and causal reasoning. In: CVPR, pp. 7699–7707 (2018)
13. Tong, J., Xie, L., Fang, S., Yang, W., Zhang, K.: Hourly solar irradiance forecasting based on encoder-decoder model using series decomposition and dynamic error compensation. Energy Convers. Manage. **270**, 116049 (2022)
14. Tong, J., Xie, L., Yang, W., Zhang, K., Zhao, J.: Enhancing time series forecasting: a hierarchical transformer with probabilistic decomposition representation. Inf. Sci. **647**, 119410 (2023)
15. Tong, J., Xie, L., Zhang, K.: Probabilistic decomposition transformer for time series forecasting. In: SDM, pp. 478–486 (2023)
16. Vaswani, A., et al.: Attention is all you need. In: NeurIPS (2017)
17. Wang, X., et al.: OadTR: online action detection with transformers. In: ICCV, pp. 7565–7575 (2021)
18. Xie, L., Luo, Y., Su, S.F., Wei, H.: Graph regularized structured output SVM for early expression detection with online extension. IEEE Trans. Cybern. **53**(3), 1419–1431 (2023)
19. Xie, L., Tao, D., Wei, H.: Early expression detection via online multi-instance learning with nonlinear extension. IEEE TNNLS **30**(5), 1486–1496 (2019)
20. Xu, M., Gao, M., Chen, Y.T., Davis, L.S., Crandall, D.J.: Temporal recurrent networks for online action detection. In: ICCV, pp. 5532–5541 (2019)
21. Xu, M., et al.: Long short-term transformer for online action detection. In: NeurIPS, pp. 1086–1099 (2021)
22. Yang, L., Han, J., Zhang, D.: Colar: effective and efficient online action detection by consulting exemplars. In: CVPR, pp. 3160–3169 (2022)
23. Zhao, P., Xie, L., Wang, J., Zhang, Y., Tian, Q.: Progressive privileged knowledge distillation for online action detection. PR **129**, 108741 (2022)

Relation-Guided Multi-stage Feature Aggregation Network for Video Object Detection

Tingting Yao[✉], Fuxiao Cao, Fuheng Mi, and Danmeng Li

College of Information Science and Technology, Dalian Maritime University,
Dalian 116026, China
ytt1030@dlmu.edu.cn

Abstract. Video object detection task has received extensive research attention and various methods have been proposed. The quality of single frame in the original video is usually deteriorated by motion blur and object occlusion, which leads to the failure of detection. Although some methods have attempted to enhance the feature representation of each frame by aggregating temporal context information from other frames, the existing methods are usually sensitive to the change of object appearance and scale, which lead to false or missing detection. Therefore, in this paper, we propose a Relation-guided Multi-stage Feature Aggregation (RMFA) network for video object detection. First, a Multi-Stage Feature Aggregation (MSFA) framework is devised to aggregate the feature representation of global and local support frames in each stage. In this way, both global semantic information and local motion information could be better captured. Furthermore, a Multi-sources Feature Aggregation (MFA) module is proposed to enhance the quality of support frames, hence the feature representation of current frame could be improved. Finally, a Temporal Relation-Guided (TRG) module is proposed to improve the feature aggregation perception by supervising the semantic similarity relationships between different object proposals. Therefore, the model adaptability to selectively store valuable features could be enhanced. Qualitative and quantitative experimental results on the ImageNet VID dataset demonstrate that our model could achieve superior video object detection results against a number of the state-of-the-art ones. Especially, when object is occluded or under fast motion, our model shows outstanding performances.

Keywords: Video object detection · Temporal context information · Feature aggregation · Temporal relation-guided

1 Introduction

Video object detection task is aims to determine the location, size and category of objects in each frame from a continuous video. With the development of computer vision and deep learning technology, this task has received extensive

Q. Liu et al. (Eds.): PRCV 2023, LNCS 14430, pp. 146–157, 2024.
https://doi.org/10.1007/978-981-99-8537-1_12

research attention [1]. It has been applied to a variety of fields such as military reconnaissance, intelligent surveillance, and driver assistance [2]. In the early research, the still-image object detectors have been directly applied to the field of video object detection [3,4]. However, the quality of single frame in the original video is usually deteriorated by motion blur and out of view, which leads to false detection. In order to solve this problem, some video object detection methods have been proposed [5–7]. These methods extract the temporal context information from the adjacent frames (support frames) to enhance the feature representation of each detected frame (current frame) with feature aggregation, thus improving the detection performance of each frame in the video.

Based on the different aggregation strategies, the current video object detection methods could be broadly classified into two categories: pixel-based feature aggregation method and region-based feature aggregation method. The former ones usually utilize optical flow to extract inter-frame motion information for feature aggregation [8]. These methods enhance the feature representation power of each frame at pixel lever by making full use of local motion information, which could obtain favourable detection performance with a fast operating speed. However, these methods are usually sensitive to complex background clutter and scale variation. In order to pay more attention to object areas and fully explore their relationship in the feature aggregation process, region-based feature aggregation methods have been proposed. This kind of method focuses on the region of interest in each frame and utilizes region proposal network (RPN) to reinforce the feature representation of object proposals. Although they improve the detection result of single frame in the case of deterioration, the performance of such methods depends heavily on the quality of proposals, which usually results in drifting due to object occlusion. To solve the above-mentioned problems of two kinds of feature aggregation strategies, in this paper, we propose a relation-guided multi-stage feature aggregation network for video object detection task. The main contributions of this paper could be summarized as follows:

(1) A multi-stage feature aggregation framework is devised. Both global semantic information and local motion information are aggregated through a cascaded network structure, which enhances the network adaptability to the different objects with various scales.
(2) A multi-sources feature aggregation module is proposed to enhance the feature representation of support frames, thus the object proposals quality of support frames could be guaranteed, which will further improve the detection robustness in continuous frames.
(3) A temporal relation-guided module is designed to increase the attention weights of the similar object in the support frames, hence significant characterization of objects could be captured in the feature aggregation process.

2   Related Work

The object detectors for still-image could be grouped into two categories: two-stage detectors and one-stage detectors. The former ones firstly generate a set of candidate boxes for regions of interest, and then determine the object location and category for each candidate region [3]. The later ones, on the other hand, predict the regions of interest directly from the extracted feature representation without an additional candidate boxes generation stage [9]. Compared with still-image, the video contains richer temporal information, hence the video object detection methods mainly focus on how to effectively aggregate temporal contextual information from support frames to the current frame to improve the detection accuracy of basic detector selected from still-image object detection tasks.

2.1 Pixel-Based Aggregation

The first class of methods which boost the detection result of single frame by utilizing pixel-based feature aggregation. Such as FGFA [10] extracted the motion information by optical flow to align the inter-frame features and then solved the false detection problem of current frame with adaptive weight. To further improve the video object detection speed, DFF [11] proposed the concept of key-frames. It only needs to extract the features of key-frames. For non key-frames, the features are obtained based on key-frames through the feature propagation network, which effectively improved the computational efficiency. Furthermore, THP [12] proposed a sparse recursive feature aggregation method based on FGFA [10] and DFF [11]. Since the feature aggregation was only performed on key frames, this method took into account the speed while improving the detection performance. MANet [13] subdivided inter-frame motion into rigid motion and non-rigid motion based on FGFA [10], and utilized different feature aggregation strategy for these two kinds of motion respectively, which effectively improved the detection performance in the case of object occlusion.

2.2 Region-Based Aggregation

Another kind of methods attempt to aggregate the region-based features across frames. For example, SELSA [14] treated the video as a series of unordered frames, and proposed a sequence-level semantic aggregation module to improve the feature representation of each frame by capturing semantic information between the current frame and the support frames throughout the video. Based on this method, LLTR [15] captured the similarity information between objects over a longer time range, and further introduced a supervision loss function to adjust the attention weights of different objects, thus improving the inter-frame feature aggregation capability of original method. To further improve the detection accuracy of current frame, MEGA [16] reused the features extracted from the RPN and integrated the global and local information to enhance the feature representation of each frame. TF Blender [17] further proposed three sub-modules: temporal relation module, feature adjustment module and feature blending module, which could be effortlessly plugged into any detection network. By modeling lower-level temporal relationships between current frame and support frames, the feature representation of each support frame could be enhanced.

The detection performances of the above methods are significantly improved. However, most of these methods ignore the ability of the model to perceive the salient features in the process of local-global feature aggregation, which prone to occur false detection under fast motion. Hence, we propose a relation-guided multi-stage feature aggregation framework to better improve the detection performances.

3 Proposed Method

3.1 Overview

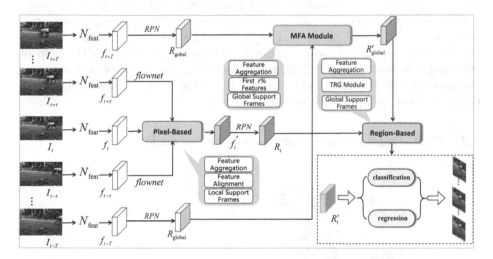

Fig. 1. The architecture of our proposed relation-guided multi-stage feature aggregation network.

The architecture of our proposed network is shown in Fig. 1. The sequence length of input video is defined as N. For each detected frame I_t in video, we select τ frames before and after it as the local support frames $I_{local} = \{I\}_{t-\tau}^{t+\tau}$, and sample T frames at equal intervals throughout the whole video sequence as global support frames $I_{global} = \{I\}_{t-T}^{t+T}$. In this paper, the ResNet-101 is chosen as the backbone network to obtain the original feature representation of each current frame f_t, local support frames $f_{local} = \{f\}_{t-\tau}^{t+\tau}$, and global support frame feature $f_{global} = \{f\}_{t-T}^{t+T}$.

We select the Faster-RCNN [3] as the basic detector, hence a set of proposals of object in each frames will be generated by RPN. Inspired by LGFN [18], we combine pixel-based and region-based feature aggregation, and construct the multi-stage feature aggregation framework with a local-global support frames sampling strategy. Firstly, in order to improve the quality of object proposals generated by RPN, we obtain the pixel-level enhanced features f'_t of each frame with pixel-based aggregation. This operation aggregates the feature extracted from local support frames to the current frame. Then, based on the pixel-level enhanced features, the object proposals are obtained through the RPN for the following region-based feature aggregation. Since the global support frames may also suffer from quality deterioration, which will seriously affect aggregation quality of global features R_{global} to the current frame, we further extract semantic dependencies among support frames with devised multi-sources aggregation module. Hence, the feature representation of global support frames are enhanced with generated R'_{global}. Finally, We aggregate the feature extracted from global support frames to the current frame via region-based feature aggregation. Since the object proposals with high objectness scores are likely to contain the main area or salient parts of the objects, we propose a temporal relation-guided aggregation module to adjust the attention weights of the different object proposals and generate R'_t based on R_t and R'_{global}. In this way, the obtained features include both global and local significant information. Then, with the help of still-image detector Faster-RCNN, the object detection result of each frame is obtained with classification and regression. In the following, we present the detection process of each proposed module in detail.

3.2   Multi-stage Feature Aggregation Framework

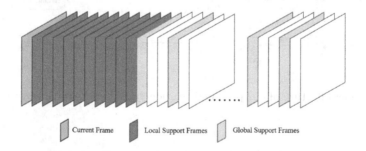

Current Frame   Local Support Frames   Global Support Frames

Fig. 2. Sampling local and global support frames.

For pixel-based feature aggregation, we use optical flow to extract motion information between neighboring frames and achieve feature alignment. As shown in blue part of Fig. 2, the adjacent τ frames before and after the current frame are set as local support frames (Only the part after the current frame is drawn in the figure, and the same for the part before). Given the current frame f_t and one of the support frame $f_{t-\tau}$, the feature of this support frame that could be mapped to the current frame with motion offset $M_{(t-\tau)\to t}$ calculated by optical flow. This mapping process can be formulated as follows:

$$f_{(t-\tau)\to t} = \xi(f_{t-\tau}, M_{(t-\tau)\to t}) \tag{1}$$

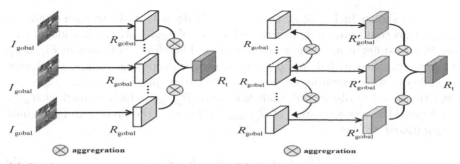

(a) Single source aggregation framework (b) Multi-sources aggregation framework

Fig. 3. Comparison of different aggregation frameworks

where ξ is the bilinear warping function applied on all the locations for each channel in the feature maps.

Then, all of the mapped features are aggregated to the current frame to obtain f'_t, which can be formulated as follows:

$$f'_t = \sum_{j=t-\tau}^{t+\tau} w_{j\to t} f_{j\to t} \tag{2}$$

where $w_{j\to t}$ denotes the attentive weight and is calculated by the cosine similarity:

$$w_{j\to t} = exp(\frac{f_{j\to t} \cdot f_t}{\mid f_{j\to t} \mid \cdot \mid f_t \mid}) \tag{3}$$

After the above pixel-based feature aggregation, f'_t contains the spatiotemporal information of the local support frames in the pixel-level. For region-based feature aggregation, the global support frames are sampled as shown in yellow part of Fig. 2 (Only the part after the current frame is drawn in the figure, and the same for the part before). We sample T global support frames at equal intervals throughout the video sequence, thus ensuring that diverse semantic feature information is obtained. In addition, to enhance the effectiveness of region-level feature aggregation, we introduce a temporal relation-guided module in the aggregation process (described in details in Sect. 3.4).

3.3 Multi-sources Feature Aggregation

With the global support frames, the existing methods usually directly use the object proposals feature of global support frames R_{global} to aggregate to the current frame, which could be formulated as follows:

$$\Upsilon(R_t, R_{global}) = R_{global\to t} \tag{4}$$

where Υ stands for aggregation operation, which employs different weights at different spatial locations and let all feature channels share the same spatial weight. As shown in Fig. 3(a), these methods do not apply any constraint to R_{global}, thus could not guarantee the reliability of aggregation when global support frames suffer from quality deterioration. Therefore, as shown in Fig. 3(b), we aggregate each global support frames object proposals feature to improve its quality before region-based feature aggregation process. In this paper, we select the first $1/r$ object proposals feature of R_{global}, which defined as R_{other} to enhanced the representation of R_{global}. In this way, the process in Eq. 5 could be formulated as follows:

$$\Upsilon(R_t, R_{other \to global}) = R_{global' \to t} \tag{5}$$

3.4   Temporal Relation-Guided Aggregation Module

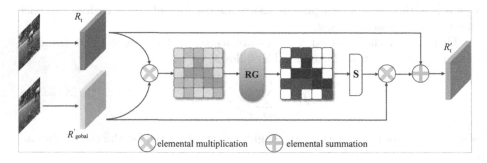

Fig. 4. The architecture of our proposed temporal relation-guided aggregation module.

The detailed structure of this module is shown in Fig. 4, where RG presents the relation-guided process and S denotes the softmax operation. First, with the object proposals feature for the object of the current frame R_t and the global support frames R'_{global}, the initial attention weight matrix A is calculated by the cosine similarity calculation after applying the matrix element multiplication:

$$A = R_t \otimes R'_{global} \tag{6}$$

where $\otimes$ denotes matrix multiplication. Then, we use the relationship guidance coefficient to amplify the positive weight values with high correlation and decrease the negative weight value with low correlation in A, and then obtain the augmented weight matrix A_{RE}, which could be formulated as follows:

$$A_{RE} = softmax(\alpha \cdot A), \alpha \geq 1 \tag{7}$$

where α denotes the relation-guided factor. A_{RE} amplifies the attention weights between the same objects and suppresses the weights between different objects, so that it could better represent the semantic similarity between object proposals while retaining all the semantic information of the supporting frames.

With the above operation, the feature of all pixels in R_t' has been enhanced independently. Finally, with optimized weights A_{RE}, R_{global}' is aggregated into the current frame, the aggregated features of the current frame could be formulated as follows:

$$R_t' = R_t \oplus A_{RE} \cdot R_{global}' \tag{8}$$

4   Experiments and Discussions

4.1   Dataset and Experimental Setting

In order to verify the effectiveness of our proposed model, we evaluate it on the ImageNet VID dataset [19]. This dataset contains 30 basic object classes, including 3862 videos in training set and 555 videos in validation set. For the training process, similar to the setting in FGFA [10], we introduce the data from ImageNet DET [19] to jointly train our model, and the global sampling frame number T is set to 14, the local sampling frame number τ is set to 30. The $1/r$ is set to 1/3 in Sect. 3.3. The mean Average Precision (mAP) (IOU = 0.5) is chosen as metrics for quantitative evaluation. Following by FGFA [10] and SELSA [14], all comparison methods are evaluated on the same testing dataset.

4.2   Quantitative Analysis

As shown in Table 1, we conduct the comparison experiment on ImageNet VID dataset [19] with 12 state-of-the-art methods. Benefiting from our proposed feature aggregation framework, we achieve the best precision 83.6%. Compared with the classical video object detection framework FGFA [10], our method has 5.6% improvement. Compared with MANet [13], we combine the advantages of different aggregation stages through a local-global support frames sampling strategy and improve the precision by 3.3%. Compared with SELSA [14], we append the pixel-based feature aggregation into local support frames and propose a temporal relation-guided enhancement module which outperforms the precision by 2.1%. The experimental results demonstrate the effectiveness of our model.

4.3   Qualitative Analysis

Figure 5 shows the qualitative results of our model against other state-of-the art ones on the representative video sequences selected from the ImageNet VID dataset [19]. In the first and second rows, since the appearance and scale of object are changed drastically under fast motion, the DFF [11] and FGFA [10] are prone to cause failure of detection. They do not fully consider the change of object in detection process, which leads to a large scale estimation deviation. In the third and fourth rows, the detection object is really small in each frame, SELSA [14] and TROI [23] occur drifting problems. With our proposed model, the different objects with various scales could be better detected. In the fifth row, the object has a severely rare posture, all other methods have completely failed.

Table 1. The comparison results of our model with other state-of-the-art ones on ImageNet VID dataset

Methods	Backbone	Detector	mAP(%)
FGFA [10]	ResNet-R101	Faster R-CNN	78.4
D&T [20]	ResNet-R101	Faster R-CNN	79.8
MANet [13]	ResNet-R101	Faster R-CNN	80.3
STMN [21]	ResNet-R101	Faster R-CNN	80.5
SELSA [14]	ResNet-R101	Faster R-CNN	81.5
RDN [22]	ResNet-R101	Faster R-CNN	81.8
TF-Blende [17]	ResNet-R101	Faster R-CNN	82.5
TROI [23]	ResNet-R101	Faster R-CNN	82.6
MEGA [16]	ResNet-R101	Faster R-CNN	82.9
HVR-Net [24]	ResNet-R101	Faster R-CNN	83.2
MST [25]	ResNet-R101	Faster R-CNN	83.3
TAFA [26]	ResNet-R101	Faster R-CNN	83.4
Ours	ResNet-R101	Faster R-CNN	**83.6**

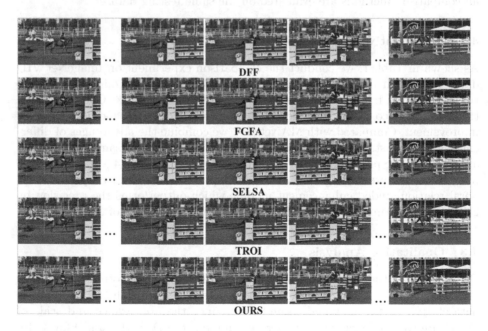

Fig. 5. The qualitative results of our method against other state-of-the-art ones.

However, with the help of RMFA, our model can effectively solve the problems of motion blur and object occlusion, which successfully detects the objects in each frame without any post-processing techniques.

4.4 Parameter Setting Analysis and Ablation Experiment

In Sect. 3.2, we introduce a relation-guided factor α to adjust the attention weights of the same and different objects in the support frames. As shown in Table 2, the best experimental results were obtained by our method when α was set to 6.5. Thus, we chose the value of $\alpha = 6.5$ in all of our experiments.

Table 2. The detection results on ImageNet VID dataset with different α

α	mAP(%)
3.5	81.9
4.5	81.9
6.5	**82.3**
7.5	82.1
8.5	81.5

To investigate the effectiveness of each module in our method, we further conducted ablation experiments on the ImageNet VID dataset. As shown in Table 3, when all modules are removed at the same time, the mean average precision is 81.5. When only introduce the MSFA, MFA, and TRG, the model detection precision is improved by 1%, 0.4%, and 0.8% respectively. Furthermore, When two modules are introduced at the same time, the detection accuracy of model is improved by 0.4% ,1.6% and 1.7%. It demonstrates that each module could effectively enhance the detection results.

Table 3. Ablation study on ImageNet VID dataset

MSFA	MFA	TRG	mAP(%)
			81.5
✓			82.5
	✓		81.9
		✓	82.3
	✓	✓	81.9
✓	✓		83.1
✓		✓	83.2
✓	✓	✓	**83.6**

5   Conclusion

In this paper, we propose a relation-guided multi-stage feature aggregation network for video object detection task. The MSFA framework is conducted to effectively combine pixel-based and region-based feature aggregation in a local-global sampling strategy. Furthermore, the feature quality of supported frame is improved by a MFA module. Finally, the TRG module is proposed to further improve the feature perception of model by adjusting the attention weights of different object proposals in the support frames, which in turn improves the detection precision of current frame. The experimental results on ImageNet VID dataset demonstrate that our proposed model is able to achieve superior detection results compared with other state-of-the-art ones.

Acknowledgment. This work was partially supported by National Natural Science Foundation of China (No. 62001078) and Fundamental Research Funds for the Central Universities (No. 3132023249).

References

1. Kang, K., Li, H., Yan, J., et al.: T-CNN: tubelets with convolutional neural networks for object detection from videos. IEEE Trans. Circ. Syst. Video Technol. **28**(10), 2896–2907 (2017)
2. Xiong, B., Kalantidis, Y., Ghadiyaram, D., et al.: Less is More: learning highlight detection from video duration. In: CVPR, pp. 1–13 (2019)
3. Ren, S., He, K., Girshick, R., et al.: Faster R-CNN: towards real-time object detection with region proposal networks. In: IEEE Transactions on Pattern Analysis and Machine Intelligence, pp. 1137–1149 (2017)
4. Dai, J., Li, Y., He, K., et al.: R-FCN: object detection via region-based fully convolutional networks. arXiv preprint arXiv:1605.06409 (2016)
5. Zhou, J.T., Du, J., Zhu, H., et al.: Anomalynet: an anomaly detection network for video surveillance. IEEE Trans. Inf. Forensics Secur. **14**(10), 2537–2550 (2019)
6. Afchar, D., Nozick, V., Ghadiyaram, D., et al.: MesoNet: a compact facial video forgery detection network. In: IEEE International Workshop on Information Forensics and Security, pp. 1–7(2018)
7. Sengar, S.S., Mukhopadhyay, S.: Moving object detection based on frame difference and Background Cumulant. IEEE Trans. Image Video Process. **11**(7), 1357–1364 (2017)
8. Brox, T., Bruhn, A., Papenberg, N., Weickert, J.: High accuracy optical flow estimation based on a theory for warping. In: Pajdla, T., Matas, J. (eds.) ECCV 2004. LNCS, vol. 3024, pp. 25–36. Springer, Heidelberg (2004). https://doi.org/10.1007/978-3-540-24673-2_3
9. Tian, Z., Shen, C., Chen, H., et al.: FCOS: fully convolutional one-stage object detection. In: ICCV, pp. 9627–9636 (2019)
10. Zhu, X., Wang, Y., Dai, J., et al.: Flow-guided feature aggregation for video object detection. In: ICCV, pp. 408–417 (2017)
11. Zhu, X., Xiong, Y., Dai, J., et al.: Deep feature flow for video recognition. In: CVPR, pp. 2349–2358 (2017)

12. Zhu, X., Dai, J., Yuan, L., et al.: Towards high performance video object detection. In: CVPR, pp. 7210–7218 (2018)
13. Wang, S., Zhou, Y., Yan, J., et al.: Fully motion-aware network for video object detection. In: ECCV, pp. 542–557 (2018)
14. Wu, H., Chen, Y., Wang, N., et al.: Sequence level semantics aggregation for video object detection. In: ICCV, pp. 9217–9225 (2019)
15. Shvets, M., Liu, W., Berg, A.: Leveraging long-range temporal relationships between proposals for video object detection. In: ICCV, pp. 9756–9764 (2019)
16. Chen, Y., Cao, Y., Hu, H., et al.: Memory enhanced global-local aggregation for video object detection. In: CVPR, pp. 10337–10346 (2020)
17. Cui, Y., Yan, L., Cao, Z., et al.: TF-blender: temporal feature blender for video object detection. In: ICCV, pp. 8138–8147 (2021)
18. Su, D., Wang, H., Jin, L., et al.: Local-global fusion network for video super-resolution. IEEE Access **8**, 172443–172456 (2020)
19. Russakovsky, O., Deng, J., Su, H., et al.: ImageNet large scale visual recognition challenge. Int. J. Comput. Vis. **115**(3), 211–252 (2015)
20. Feichtenhofer, C., Pinz, A., Zisserman, A.: Detect to track and track to detect. In: ICCV, pp. 3038–3046 (2017)
21. Bertasius, G., Torresani, L., Shi, J.: Object detection in video with spatio temporal sampling networks. In: ECCV, pp. 331–346 (2018)
22. Deng, J., Pan, Y., Yao, T., et al.: Relation distillation networks for video object detection. In: ICCV, pp. 7023–7032 (2019)
23. Gong, T., Chen, K., Wang, X., et al.: Temporal ROI align for video object recognition. In: American Association for Artificial Intelligence (AAAI), vol. 35, no. 2, pp. 1442–1450 (2021)
24. Han, M., Wang, Y., Chang, X., Qiao, Yu.: Mining inter-video proposal relations for video object detection. In: Vedaldi, A., Bischof, H., Brox, T., Frahm, J.-M. (eds.) ECCV 2020. LNCS, vol. 12366, pp. 431–446. Springer, Cham (2020). https://doi.org/10.1007/978-3-030-58589-1_26
25. Xu, C., Zhang, J., Wang, M., et al.: Multi-level spatial-temporal feature aggregation for video object detection. IEEE Trans. Circ. Syst. Video Technol. **32**(11), 7809–7820 (2022)
26. He, F., Li, Q., Zhao, X.: Temporal-adaptive sparse feature aggregation for video object detection. Pattern Recogn. **127**, 108587 (2022)

Multimodal Local Feature Enhancement Network for Video Summarization

Zhaoyun Li, Xiwei Ren$^{(\boxtimes)}$, and Fengyi Du

School of Electronic Information and Artificial Intelligence,
Shaanxi University of Science and Technology, Xi'an 710021, China
renxiwei@126.com

Abstract. Multimodal information processing has garnered considerable attention in recent years. Due to the inherent multimodal information in videos, multimodal learning has been introduced in the domain of video summarization, leading to a significant improvement in summary accuracy. However, previous multimodal video summarization methods have overlooked the local fine-grained features of the fused multimodal information, constraining the representational capacity of the models. Therefore, we propose a Multimodal Local Feature Enhancement Network (MLFEN) to address this issue. Firstly, we utilize multi-head attention to encode single-modal information, capturing their mutual dependencies, and then fuse multiple modalities. Secondly, we further process the multimodal information by employing local attention to explore local fine-grained interdependencies and global attention to capture global dependencies. Finally, a regression network is employed to predict the frame-level importance scores used in generating video summaries. In addition, we explored different fusion methods for integrating different modalities of information. Our method demonstrates superior performance on both evaluation metrics, achieving F-measure of 0.528, Kendall's τ of 0.076, and Spearman's ρ of 0.099 on the SumMe dataset, as well as F-measure of 0.612, Kendall's τ of 0.155, and Spearman's ρ of 0.204 on the TVSum dataset. These results surpass the performance of most existing video summarization methods.

Keywords: Multimodal learning · Video summarization · Multi-head attention · Local attention

1 Introduction

The primary goal of video summarization techniques is to automatically generate a short and informative video briefing that encompasses the majority of the original video content. This technology enables efficient browsing of key information within the video, facilitating tasks such as video storage, retrieval, and comprehension [5].

In recent years, deep learning approaches have demonstrated astonishing performance, significantly propelling the development of video summarization

Q. Liu et al. (Eds.): PRCV 2023, LNCS 14430, pp. 158–169, 2024.
https://doi.org/10.1007/978-981-99-8537-1_13

technology [3, 19]. Most of these methods involve modeling and analyzing visual features extracted from video frames, predicting the importance of frames, and generating key shots based on these predictions [22]. However, video data consists of multimodal information [20]. Most current video summarization algorithms primarily focus on visual information processing while neglecting the role of other modalities such as audio, which differs from how humans summarize videos in real-world scenarios. Additionally, when videos are decomposed into frame sequences, the resulting sequences tend to be quite long, leading to an imbalance in how models handle local and global information, which can impact the final summary results.

Using solely visual information is insufficient to differentiate between different video frames when the scene changes are minimal. In such cases, the frames may exhibit significant differences when considering the influence of audio information [21]. For long sequential video frames, some researchers have addressed the issue by constructing hierarchical structures [23] and employing attention mechanisms [3] to capture local or global dependencies. However, hierarchical structures can sometimes result in the loss of local information, while attention mechanisms tend to focus more on global relationships.

In this paper, we propose a multimodal local feature enhancement network. Our network comprises three components. The first component utilizes a shared-weight multi-head self-attention mechanism to capture semantic information among the same modality and explore the interrelationships between different modalities. In the second component, local attention is employed to capture the local dependencies between individual multimodal frames and their neighboring frames, while global attention explores the global dependencies among multimodal information. Finally, a regression network is utilized to predict the scores at the frame level.

Our work contributes and innovates in three aspects:

Integration of Audio Information: We introduce audio information into the video summarization task, enhancing the distinctiveness of video frames under minimal scene variations, making the process more akin to real-world video summarization.

Self-Attention-Based Network Architecture: We propose a method for fusing visual and audio features using weight-shared attention, and a prediction approach for frame-level scores using multimodal features.

Local Attention: We introduce a local attention mechanism to capture dependencies within a local range of video frames. This mechanism enhances its capability to capture fine-grained details and improve the overall summarization quality.

2 Related Work

2.1 Single Vision Modality

Traditional methods generate video summaries by considering the representativeness of the entire video and the diversity of frames or shots, using handcrafted shallow features such as color histogram [8], histograms of oriented gradient [4], and optical flow [7]. For instance, VSUMM [2] utilizes color features of video frames and the k-means clustering algorithm to select key video frames. In the work by Wang et al. [18], they introduce similarity suppression constraints into the dictionary selection model to ensure diversity among the selected key frames. This ensures that frames with a higher capacity of semantic information are more likely to be chosen for the summary.

In recent years, deep learning algorithms have been widely applied to video summarization tasks. For example, Zhang et al. [19] utilize Bi-LSTM to model video frames and incorporate the DPP (Determinantal Point Process) model to suppress similar shots. Zhao et al. [22] address the challenge of handling long-duration videos by proposing a two-layer RNN model that enhances the ability of RNN to capture long-term dependencies. Zhou et al. [24] employ reinforcement learning to construct a reward function based on the diversity and representativeness of generated summaries, enabling the prediction of frame scores in the video. Addressing the limitations of RNN models, VASNet [3] utilizes self-attention mechanisms to allocate different levels of attention to video frames, effectively handling video summarization tasks.

2.2 Multimodal Video Summarization

Indeed, different modalities of information exhibit semantic consistency, and they can complement and transform each other, aiding in better understanding and processing of relevant tasks. Multimodal data has also been introduced into video summarization tasks. For example, Zhao et al. [20] incorporated audio information into video summarization tasks by utilizing a Two-Stream LSTM architecture to explore the temporal dependencies of different modalities. They combined the multimodal information using gate units, and employed an Audio-Visual Fusion LSTM and self-attention encoding to analyze the dependencies between modalities. Li et al. [6] proposed a multimodal self-supervised learning framework that encodes both the textual and visual features of videos. They investigate the semantic consistency between coarse-grained and fine-grained representations and employ a progressively trained encoder for video summarization. CLIP-It [10] utilizes textual language guidance to generate video summaries by integrating video and text information using Transformer. It operates in both supervised and unsupervised settings for general video summarization and also explores query-focused video summarization.

3 Method

In this paper, we propose a multimodal self-attention network that captures both local and global dependencies of multimodal information for video summarization tasks. As shown in Fig. 1, MLFEN consists of three main components. The Fused Multi-Head Attention module integrates visual features and audio features together. The Local&Global Multi-Head Attention module captures the

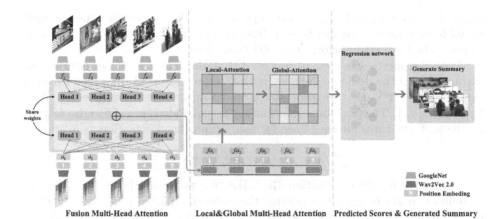

Fig. 1. Overview of the multimodal local feature enhancement network. The fusion multi-head attention is used to encode and integrate different modalities of information, while the local&global multi-head attention processes the fused multi-modal information. The regression network predicts frame-level scores, which are then used to generate the video summary.

local and global dependencies of multimodal information. The Regression Network predicts frame-level scores. In the following sections, each component will be described in detail.

3.1 Fused Multi-head Attention

Multi-head attention (MHA) was first proposed in [17] for machine translation tasks and has since been widely used in various fields of deep learning. Figure 2(a) shows the specific structure of MHA, which can explore the features of different subspaces of encoded information and capture global dependencies between information at a fine-grained level. In this section, we use MHA to process relevant unimodal information.

The video is decomposed into audio signals and frame images, and each modality information is encoded using MHA for the visual modality, a pretrained network is used to extract the deep visual features of the frames in the video as $F = \{f_1, f_2, \cdots, f_n\}$, where n represents the length of the frame sequence. In order to preserve the positional information of the sequence, we perform positional encoding on the sequence $F = \{f_1, f_2, \cdots, f_n\} + E_{pos}^f \in R^{d \times n}$, where E_{pos}^f is the positional encoding matrix, which has the same dimension as the visual features, and their sum is the new visual feature.

Then, in order to capture the global dependencies between visual features, multi-head attention is used to encode F, *i.e.*,

$$\bar{F} = F + MHA(LN(F)) \tag{1}$$

where LN represents the layer-norm operation and $\bar{F}$ represents the encoded visual feature matrix. As can be seen from the Eq. (1), our operation process is to place the layer-norm before the multi-head attention.

A similar method is used to process the audio features. Deep features are extracted from the audio and aligned with the dimensions of the visual features through a linear layer. The processed audio features are represented as $A = \{a_1, a_2, \cdots, a_n\}$. The positional information is embedded into the audio feature A, the audio features are then encoded using multi-head attention as follows,

$$\bar{A} = A + MHA(LN(A)) \tag{2}$$

In Fused Multi-Head Attention, the MHA that processes the visual and audio modalities shares weights. Considering the potential mutual promotion effect of visual and audio features, we fuse the two modalities of information as follows,

$$FA = \bar{F} + \bar{A} \tag{3}$$

where $FA = \{fa_1, fa_2, \cdots, fa_n\}$ are denoted as fused multimodal features. Multimodal features have richer representation capabilities, and simple processing may not be able to capture the dependencies between multimodal features. To this end, local and global encoding is performed on the multimodal features to further explore the local and global dependencies between multimodal features.

3.2 Local&Global Multi-head Attention

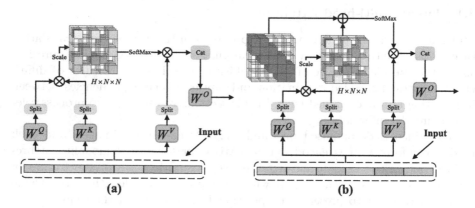

Fig. 2. Attention mechanisms used in this paper. (a) Multi-head attention structure. (b) Mask-based local multi-head attention structure.

Long videos are decomposed into longer frame sequences, and both attention-based and RNN-based models will reduce their focus on local information as the length of the processed sequence increases. We have developed a local&global multi-head attention to explore the dependencies between multimodal features.

To enhance the network's overall expressive power, we first employ the local multi-head attention (LMHA) to explore the influence weights within a local range for each frame. This allows us to capture the local interdependencies among the multimodal sequences, thereby supplementing the network's global representation capability. As shown in Fig. 2(b), LMHA has a similar structure to standard MHA, but differs in that LMHA uses a masking mechanism to only focus on sequences within N steps. In Fig. 2(b), the gray part represents the information that needs to be focused on, with a value of 0, and the white part is the information that does not need to be focused on, with a value of $-\infty$. The multimodal features after local encoding can be represented as,

$$LM = FA + LMHA(LN(FA)) \tag{4}$$

where LM is the feature after local encoding. Next, we further explore the global dependencies of LM, which can be expressed as,

$$GL = LM + MHA(LN(LM)) \tag{5}$$

where GL captures the local and global dependencies of multimodal features and serves as the input to the subsequent Regression Network.

3.3 Predicted Scores and Generated Summary

In order to calculate the frame-level importance scores for the video summarization task based on the local and global dependencies of multimodal features, we designed a Regression Network to complete this process, i.e.,

$$P = \text{ReLU}(W^1(LN(GL)) + b^1) \tag{6}$$

$$S = \text{Sigmoid}(W^2(LN(P)) + b^2) \tag{7}$$

where W^1 and W^2 are training weights, b^1 and b^2 are training biases, P is the output of the first layer network, and S is the frame-level predicted score. Specifically, the Regression Network mainly consists of two linear layers. The input information enters the layer normalization layer, linear layer and relu layer in turn to obtain the output P of the first layer network. The second layer network performs layer normalization, linear transformation and singmoid function activation on P in turn to obtain the frame-level score prediction.

We hope that the frame-level scores predicted by our method are as close as possible to the manual summaries, and use Mean Squared Error (MSE) as the loss function for optimization, i.e.,

$$\mathcal{L} = \frac{1}{n} \parallel S - G \parallel_2^2 \tag{8}$$

where G represents the ground truth scores annotated by humans, and $\mathcal{L}$ reflects the difference between the predicted scores and the ground truth scores.

Our method generates video summaries as collections of shots. We first segment the video into shots. The Kernel Temporal Segmentation (KTS) algorithm [13] segments a video into semantically coherent subshots by detecting general change points. The subshot scores are obtained by averaging the frame scores within each subshot. We set the length of the summary to be no more than 15% of the original video length. The selection of key shots is treated as a dynamic programming problem, where we use the knapsack algorithm to maximize the importance scores of the summary.

4 Experiments

4.1 Datasets and Metrics

Datasets. We evaluate our proposed method on two datasets: SumMe [4] and TVSum [15]. The SumMe dataset consists of 25 videos. Most of these videos have not undergone manual editing. The TVSum dataset comprises 50 videos sourced from YouTube. Additionally, we leverage the YouTube [2] and OVP [2] datasets for data augmentation and transfer learning during training.

Our model is evaluated using three different dataset settings. Canonical: Training is performed on 80% of the videos in one dataset, while the remaining 20% are used for testing. Augmented: The training set content is augmented with videos from three other datasets, building upon the Canonical setting. Transfer: Videos from three datasets are used for training, and videos from the remaining dataset are used for testing.

Metrics. The F-measure is utilized to evaluate the consistency between the predicted summary and the ground truth summary. A higher F-measure indicates a greater degree of overlap between the predicted and actual summaries.

In [11], it is pointed out that the F-measure is primarily influenced by the distribution of video segment lengths, and even random methods can achieve relatively high scores. In such cases, the F-measure alone cannot provide a comprehensive assessment of the system's performance. Therefore, it is recommended to compute the rank correlation coefficients, Kendall's τ and Spearman's ρ, between the predicted frame-level importance scores and the ground truth scores. These two rank correlation coefficients are used to comprehensively evaluate the quality of summary generation.

4.2 Implementation Details

To reduce temporal redundancy in videos and extract visual information, we employ a technique where features are extracted from every 15th frame of the video. We utilize GoogleNet [16], which has been pre-trained on ImageNet [14], to extract features from the video frames. The output vector of the penultimate layer (pool 5) with 1024-dim is used as the visual feature representation.

For audio information, we employ wav2vec 2.0 [1], pre-trained on the Librispeech [12] dataset, to extract audio features. We only utilize the output vector of the encoder layer with 512-dim as the audio feature representation. The audio is evenly divided every 0.5 s, ensuring that the sequence length of the audio features matches the sequence length of the video features. For videos in the SumMe dataset without any sound, we pad the audio features with zeros.

To ensure a fair comparison with existing methods, we perform 5 random splits of the training and testing sets and average the results. The proposed MLFEN is implemented in Python 3.8 and PyTorch 1.9. We employ the Adam optimizer for 75 epochs with a batch size of 1. The learning rate is set to 1e−4, and the decay rate is set to 1e−4. The number of heads for multi-head attention is 8 and the step size for local attention is 4.

4.3   Comparisons with the State of the Art

Comparing our method with previous approaches using the F-measure evaluation metric, as shown in Table 1, we can observe that our method achieves promising results across different training settings, particularly demonstrating significant improvements on the SumMe dataset. The SumMe dataset consists of videos with smooth scene transitions, where the involvement of audio information becomes crucial in selecting key frames. This observation indirectly highlights the necessity of incorporating audio information in our approach. It is worth noting that our method performs poorly in the transfer setting of the TVSum dataset, but outperforms other methods in the remaining settings. Further observations reveal that RNN-based methods demonstrate good transfer performance on the TVSum dataset. This could be attributed to the better handling capability of RNNs for transfer learning on small-scale datasets.

Figure 3 provides a visual representation of the summary results generated by MLFEN, demonstrating that our proposed method can effectively fit the score curve annotated by humans and select shots with higher scores.

From Table 2, it can be observed that our method also achieves favorable results in the rank-based evaluation, indicating that our method's predicted summaries are closer to the manually annotated summaries. HMT [21] outperforms MLFEN in terms of Kendall's τ score on the SumMe dataset, thanks to its utilization of a hierarchical structure for handling multimodal information. However, the varying lengths of shots in the hierarchical structure limit its concurrency. Taking all aspects into consideration, our method provides the best overall performance across both evaluation frameworks.

4.4   Ablation Study and Sensitivity Analysis

Ablation Study. To verify the effectiveness of our idea, we conducted ablation experiments on the proposed method. AV-MHA only performs Fused MHA and regression network, capturing the global dependencies between unimodal features without further exploring the local and global dependencies of multimodal features. MLFEN (w/o audio) does not consider audio information and the rest of the settings remain unchanged. MLFEN (w/o LMHA) removes the LMHA

Table 1. Comparison of F-measure for different methods on SumMe and TVSum datasets, with the best result highlighted in **bold**.

Methods	SumMe			TVSum		
	Canonical	Augmented	Transfer	Canonical	Augmented	Transfer
dppLSTM [19]	0.386	0.429	0.418	0.547	0.596	0.587
H-RNN [22]	0.421	0.438	-	0.579	0.619	-
HSA-RNN [23]	0.423	0.421	-	0.587	0.598	-
DR-DSN$_{sup}$ [24]	0.421	0.439	0.426	0.581	0.598	**0.589**
CAAN$_{sup}$ [9]	0.506	-	-	0.593	-	-
[25]	0.511	0.521	0.454	0.610	0.615	0.551
AVRN [20]	0.441	0.449	0.432	0.597	0.605	0.587
HMT [21]	0.441	0.448	-	0.601	0.603	-
SSPVS [6]	0.507	-	-	0.604	-	-
MLFEN	**0.528**	**0.540**	**0.465**	**0.612**	**0.623**	0.577

Table 2. Comparison of rank correlation coefficients (Kendall's τ and Spearman's ρ) for different methods, with the best result highlighted in **bold**.

Methods	SumMe		TVSum	
	Kendall's τ	Spearman's ρ	Kendall's τ	Spearman's ρ
VASNet [3]	0.054	0.058	0.082	0.088
HSA-RNN [23]	0.064	0.066	0.082	0.088
AVRN [20]	0.073	0.074	0.096	0.104
HMT [21]	**0.079**	0.080	0.096	0.107
MLFEN	0.076	**0.099**	**0.155**	**0.204**

part compared to the proposed MLFEN and only captures the global dependencies between multimodal features. MLFEN (w/o Fused MHA) does not use Fused MHA encoding for unimodal features and directly adds visual and audio features simply.

As shown in Table 3, when removing the audio features, LMHA, and Fusion MHA from the complete model, there is a varying degree of performance decline, which highlights the effectiveness of each component. It is worth noting that MLFEN (w/o LMHA) performs worse than AV-MHA on the SumMe dataset, indicating that the model with global attention performs worse than the one without it. This indirectly demonstrates the limitations of global attention in handling long sequential data and the need for complementary local attention.

Experiments on the TVSum dataset show that the addition of audio information does not significantly improve the performance. This is due to the rich scene transformation of video shots on the TVSum dataset, and the difference between shots can be well distinguished by using only visual information.

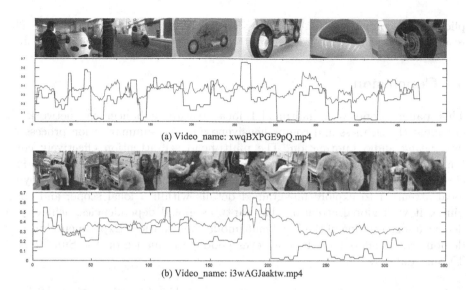

(a) Video_name: xwqBXPGE9pQ.mp4

(b) Video_name: i3wAGJaaktw.mp4

Fig. 3. Visualization of the summary generated by MLFEN. The red curve indicates the prediction made by MLFEN, while the blue curve represents the ground truth. (Color figure online)

Table 3. Results of ablation study. F-measure on different baselines.

Baseline	SumMe	TVSum
AV-MHA	0.520	0.606
MLFEN(w/o Audio)	0.461	0.606
MLFEN(w/o LMHA)	0.484	0.606
MLFEN(w/o Fused MHA)	0.508	0.604
MLFEN	**0.528**	**0.612**

Sensitivity Analysis. We also conducted a sensitivity analysis on the proposed method, mainly investigating the impact of different fusion methods for multimodal information on the final results. As shown in Table 4, we applied four operations to the features encoded by Fusion MHA: element-wise multi-

Table 4. Comparison of different modal fusion methods.

Fusion	SumMe	TVSum
Multiplication	0.473	0.608
Average	0.483	0.610
Maximum	0.504	0.610
Addition	**0.528**	**0.612**

plication, averaging, maximum value selection, and element-wise addition. The results indicate that adding different modalities together yields the best results.

5 Conclusion

This paper proposes a multimodal local feature enhancement network. It enhances the richness and comprehensiveness of the summarization process by integrating audio information. The multi-head self-attention effectively combines different modalities of information and captures their mutual dependencies, thereby improving the performance of video summarization. By employing local attention to explore fine-grained details within a local scope, and combining it with global attention that captures global dependencies, achieving a deeper understanding of the fused multimodal information. Experimental results demonstrate that our method achieves good performance on the SumMe and TVSum datasets.

Acknowledgement. This work was partly supported by the National Natural Science Foundation of China (61971272), and the research startup fund project of Shaanxi University of science and Technology (2020bj-01).

References

1. Baevski, A., Zhou, Y., Mohamed, A., Auli, M.: wav2vec 2.0: a framework for self-supervised learning of speech representations. Adv. Neural Inf. Process. Syst. **33**, 12449–12460 (2020)
2. De Avila, S.E.F., Lopes, A.P.B., da Luz Jr, A., de Albuquerque Araújo, A.: VSUMM: a mechanism designed to produce static video summaries and a novel evaluation method. Pattern Recogn. Lett. **32**(1), 56–68 (2011)
3. Fajtl, J., Sokeh, H.S., Argyriou, V., Monekosso, D., Remagnino, P.: Summarizing videos with attention. In: Carneiro, G., You, S. (eds.) ACCV 2018. LNCS, vol. 11367, pp. 39–54. Springer, Cham (2019). https://doi.org/10.1007/978-3-030-21074-8_4
4. Gygli, M., Grabner, H., Riemenschneider, H., Van Gool, L.: Creating summaries from user videos. In: Fleet, D., Pajdla, T., Schiele, B., Tuytelaars, T. (eds.) ECCV 2014. LNCS, vol. 8695, pp. 505–520. Springer, Cham (2014). https://doi.org/10.1007/978-3-319-10584-0_33
5. Ji, P., Yang, B., Zhang, T., Zou, Y.: Consensus-guided keyword targeting for video captioning. In: Yu, S., et al. (eds.) PRCV 2022, Part III. LNCS, vol. 13536, pp. 270–281. Springer, Cham (2022). https://doi.org/10.1007/978-3-031-18913-5_21
6. Li, H., Ke, Q., Gong, M., Drummond, T.: Progressive video summarization via multimodal self-supervised learning. In: Proceedings of the IEEE/CVF Winter Conference on Applications of Computer Vision, pp. 5584–5593 (2023)
7. Li, X., Zhao, B., Lu, X.: A general framework for edited video and raw video summarization. IEEE Trans. Image Process. **26**(8), 3652–3664 (2017)
8. Li, X., Zhao, B., Lu, X.: Key frame extraction in the summary space. IEEE Trans. Cybern. **48**(6), 1923–1934 (2017)

9. Liang, G., Lv, Y., Li, S., Zhang, S., Zhang, Y.: Video summarization with a convolutional attentive adversarial network. Pattern Recogn. **131**, 108840 (2022)

10. Narasimhan, M., Rohrbach, A., Darrell, T.: Clip-it! language-guided video summarization. Adv. Neural. Inf. Process. Syst. **34**, 13988–14000 (2021)

11. Otani, M., Nakashima, Y., Rahtu, E., Heikkila, J.: Rethinking the evaluation of video summaries. In: Proceedings of the IEEE/CVF Conference on Computer Vision and Pattern Recognition, pp. 7596–7604 (2019)

12. Panayotov, V., Chen, G., Povey, D., Khudanpur, S.: LibriSpeech: an ASR corpus based on public domain audio books. In: 2015 IEEE International Conference on Acoustics, Speech and Signal Processing (ICASSP), pp. 5206–5210. IEEE (2015)

13. Potapov, D., Douze, M., Harchaoui, Z., Schmid, C.: Category-specific video summarization. In: Fleet, D., Pajdla, T., Schiele, B., Tuytelaars, T. (eds.) ECCV 2014. LNCS, vol. 8694, pp. 540–555. Springer, Cham (2014). https://doi.org/10.1007/978-3-319-10599-4_35

14. Russakovsky, O.: ImageNet large scale visual recognition challenge. Int. J. Comput. Vision **115**, 211–252 (2015)

15. Song, Y., Vallmitjana, J., Stent, A., Jaimes, A.: TVSum: summarizing web videos using titles. In: Proceedings of the IEEE Conference on Computer Vision and Pattern Recognition, pp. 5179–5187 (2015)

16. Szegedy, C., et al.: Going deeper with convolutions. In: Proceedings of the IEEE Conference on Computer Vision and Pattern Recognition, pp. 1–9 (2015)

17. Vaswani, A., et al.: Attention is all you need. In: Advances in Neural Information Processing Systems, vol. 30 (2017)

18. Wang, S., et al.: Scalable gastroscopic video summarization via similar-inhibition dictionary selection. Artif. Intell. Med. **66**, 1–13 (2016)

19. Zhang, K., Chao, W.-L., Sha, F., Grauman, K.: Video summarization with long short-term memory. In: Leibe, B., Matas, J., Sebe, N., Welling, M. (eds.) ECCV 2016. LNCS, vol. 9911, pp. 766–782. Springer, Cham (2016). https://doi.org/10.1007/978-3-319-46478-7_47

20. Zhao, B., Gong, M., Li, X.: Audiovisual video summarization. IEEE Trans. Neural Netw. Learn. Syst. (2021)

21. Zhao, B., Gong, M., Li, X.: Hierarchical multimodal transformer to summarize videos. Neurocomputing **468**, 360–369 (2022)

22. Zhao, B., Li, X., Lu, X.: Hierarchical recurrent neural network for video summarization. In: Proceedings of the 25th ACM International Conference on Multimedia, pp. 863–871 (2017)

23. Zhao, B., Li, X., Lu, X.: HSA-RNN: hierarchical structure-adaptive RNN for video summarization. In: Proceedings of the IEEE Conference on Computer Vision and Pattern Recognition, pp. 7405–7414 (2018)

24. Zhou, K., Qiao, Y., Xiang, T.: Deep reinforcement learning for unsupervised video summarization with diversity-representativeness reward. In: Proceedings of the AAAI Conference on Artificial Intelligence, vol. 32 (2018)

25. Zhu, W., Lu, J., Han, Y., Zhou, J.: Learning multiscale hierarchical attention for video summarization. Pattern Recogn. **122**, 108312 (2022)

Asymmetric Attention Fusion for Unsupervised Video Object Segmentation

Hongfan Jiang, Xiaojun Wu$^{(\boxtimes)}$, and Tianyang Xu

School of Artificial Intelligence and Computer Science,
Jiangnan University, Wuxi, China
6213114014@stu.jiangnan.edu.cn, {wu_xiaojun,tianyang.xu}@jiangnan.edu.cn

Abstract. In this paper, we introduce a novel Asymmetric Attention Fusion Network (AAF-Net) based on an attention mechanism to complete unsupervised video object segmentation. Firstly, we propose an asymmetric attention fusion module (AAFM) to aggregate the two source inputs to exploit the complementary representations between optical flow and RGB images. The asymmetric attention structure of AAFM is equipped to enhance feature information. Next, we design a feature correction module (FCM) to balance the information ratio between motion features and appearance features. The results of experimental evaluation obtained on several well-known benchmarking datasets, including DAVIS16, FBMS, and SegTrack-V2, deliver outstanding performance compared to the other segmentation networks based on optical flow, reflecting the merit and advantage of the proposed approach.

Keywords: Video object segmentation · Convolutional neural network · Optical flow · Artificial intelligence

1 Introduction

Video object segmentation (VOS) is a crucial task to distinguish the pixels of the object from the background in sequential frames of a given video, which attracts widespread attention recently. VOS is divided into two main subtasks, semi-supervised video object segmentation and unsupervised video object segmentation (UVOS). Relatively, UVOS is more challenging compared with the semi-supervised one, which requires the network to adaptively predict the objects and extract them in the video.

Compared with semantic segmentation of images, UVOS requires additional temporal features. The core of temporal features is the difference between multiple frames, which represent objects' displacement and deformation. The video segmentation community, with the development of mathematical theory and modeling techniques in recent decades, has reported an impressive list of achievements. At present, there are two main frameworks for temporal feature extraction, which are based on memory mechanism and optical flow [15,28,31,32].

© The Author(s), under exclusive license to Springer Nature Singapore Pte Ltd. 2024
Q. Liu et al. (Eds.): PRCV 2023, LNCS 14430, pp. 170–182, 2024.
https://doi.org/10.1007/978-981-99-8537-1_14

The memory mechanism takes multiple different frames of the same video as the memory, assisting in segmenting the object in the current frame. The merit of this framework reflects in its high accuracy. However, it often entails a substantial computational burden, resulting in a diminished processing speed [15]. The other framework is to explore the optical flow images. The optical flow image is derived from two adjacent RGB images, which enable to reflect the displacement and deformation of the object. Therefore optical flow images can provide temporal features. We also call the temporal feature of optical flow images the motion feature, which reflects the motion of the object. This framework performs lower than the memory mechanisms, but with less computation load [8,28,32].

The discriminative features in RGB and optical flow images differ significantly. RGB images primarily capture the visual appearance of the target, whereas optical flow predominantly contains temporal features. Consequently, employing identical modules, as done in FS-Net [7], for processing RGB and optical flow images is deemed inappropriate. In light of these distinct modalities, it is more reasonable to employ distinct methods tailored to handle each modality accordingly. The identical method for RGB and optical flow images leads to two problems. On the one hand, temporal features only exist in optical flow images. It is difficult to extract this temporal information from optical flow images with the same method of RGB images. On the other hand, RGB images carry more visual features conducive to segmentation than optical flow. So how to keep a balance between appearance and motion also needs to be considered.

To solve the above problems, we design a segmentation network AAF-Net based on optical flow for UVOS. The effectiveness of AAF-Net has been evaluated on several benchmarks such as DAVIS16 [17], FBMS [16], SegTrack-V2 [11] and MCL [9]. The results demonstrate that AAF-Net outperforms the state-of-the-art segmentation networks.

The main innovations of AAF-Net are as follows:

- We design Asymmetric Attention Fusion Module to extract temporal features from motion, for interaction and complement between motion and appearance.
- We design Feature Correction Module. The module is to extract deeper features from motion. The serial connection structure enables finding the balance between two different modalities
- AAF-Net outperforms the state-of-the-art segmentation networks for UVOS in multiple datasets

The structure of the paper is as follows: In Sect. 2, we introduce the related work about AAF-Net. In Sect. 3, we introduce the specific structure of each module of AAF-Net in detail, and introduce the training method of the model. In Sect. 4, the experimental results are reported to verify the advantages of AAF-Net. In Sect. 5, the conclusions are drawn.

2 Related Work

We will review the pertinent technique related to our work briefly in this section.

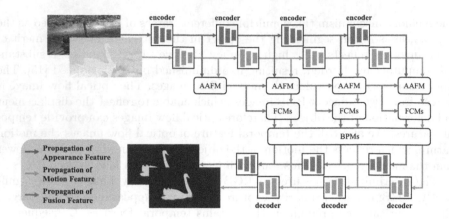

Fig. 1. Encoders are used to extract features with different sizes from RGB and optical flow images. The asymmetric Attention Fusion Module (AAFM) is equipped to extract temporal features from motion, passing to appearance. The output of the current-stage AAFM is passed into the next-stage AAFM and the current-stage Feature Correction Module (FCM). FCM is to extract deeper features from fusion and motion to balance between two different features. FCMs consist of multiple connecting FCM. Bidirectional Purification Module (BPM) is equipped to fuse multi-scale features. BPMs consist of multiple connecting BPM. The decoder enables generating the final mask.

2.1   UVOS Based on Optical Flow

Many existing methods for video object segmentation heavily rely on optical flow estimation [7,18,28,32]. Zhang et al. [32] extract dense local features from optical flow and appearance images. Then Wasserstein distance is used to calculate the global optimal flow for transferring feature information between different modalities. Yang et al. [28] take two continuous optical flow images as input for comparative learning. It also designed temporary consistency loss for training. Ji et al. [7] improve the mutual constraint scheme between motion and appearance in the fusion and decoding stages to enhance the robustness of the network.

3   Methodology

In this work, we propose an asymmetric attention fusion network (AAF-Net) for unsupervised video object segmentation. The overview of our AAF-Net framework is generally shown in Fig. 1. Our model adopts an encoder-decoder architecture.

3.1   Opital Flow and Encoder

Suppose there are T consecutive frames a in a video sequence $\{A^t\}_{t=1}^{T}$. FlowNet2.0 [6] is equipped as the optical flow field generator to generate $T-1$ optical flow images $\{M^t\}_{t=1}^{T-1}$, which are generated from two adjacent frames

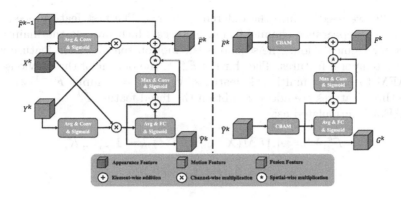

Fig. 2. Frameworks of AAFM (left) and FCM (right).

$(M^t = H[A^t, A^{t+1}])$, where $A^t \in \mathbb{R}^{H \times W \times 3}$, $M^t \in \mathbb{R}^{H \times W \times 1}$. The essence of optical flow estimation lies in the differential analysis of RGB videos, resulting in an excess of one RGB image compared to the corresponding optical flow images. In order to establish a rigorous one-to-one correspondence between the RGB optical flow images, we discard the last frame of the RGB video.

Then, we use two independent ResNet-50 branches [5] as the encoder, to extract $\{X^k\}_{k=1}^K$ and $\{Y^k\}_{k=1}^K$ from A^t and M^t, where K is the number of layers of the encoder. We set $K = 4$ in our network. We briefly express $\{X^k\}_{k=1}^K$ as X and $\{Y^k\}_{k=1}^K$ as Y, where $X, Y \in \mathbb{R}^{H \times W \times C}$.

3.2 Asymmetric Attention Fusion Module

As mentioned in Sect. 1, the motivation of AAFM is to align the two modalities while extracting motion features from optical flow. The detail of AAFM is shown in the left part of Fig. 2. Two inputs of AAFM are the appearance feature X and the motion feature Y, which are both from the kth layers of encoders.

The first step is to get the channel attentions of two different features. Using the feature vector of one modality to highlight the feature of the other modality. This operation closes the distance between appearance and motion modality before transferring motion features. First, We perform global average pooling (GAP), pointwise convolutions, and sigmoid function, to get two channel-weighted effective attention vectors. Perform channel-wise multiplication to get channel-fusion features $\hat{X}$ and $\hat{Y}$.

In the second step, we need to distill the spatial-fusion feature from the motion feature $\hat{Y}$ as temporal information to highlight the interesting regions of motion on the appearance modality. We use a two-layer structure to get the spatial-fusion feature, the structures of which are similar. In the first layer, the spatial vector is obtained by calculating the average value of each position from $\hat{Y}$, while using the max value in the second layer. Then use the full connection layer and the activation function to get spatial attention. Perform spatial-wise multiplication between spatial attention and $\hat{Y}$ to get the spatial-fusion feature.

Finally, we need to fuse the different features. The first feature need to be fused is the appearance channel-fusion features which provide the main visual feature for segmentation. The second is the motion spatial-fusion feature which provides temporal features. The third is $\hat{F}^{k-1}$, the output of the previous layer of AAFM to provide multi-scale features. When $k = 0$, input $\hat{F}^{k-1}$ will be 0. These three features are added to obtain the fused feature $\hat{F}^k$.

AAFM is formulated as:

$$\hat{F}^k, \hat{Y}^k = AAFM(X^k, Y^k, \hat{F}^{k-1}), k = 1, 2, ..., K, \tag{1}$$

$$\hat{F}^0 = AAFM(X^0, Y^0, 0). \tag{2}$$

3.3   Feature Correction Module

In order to further distill the effective information from the two features, we designed a feature correction module (FCM). The detail of FCM is shown in Fig. 2. The module takes the outputs $\hat{F}$ and $\hat{Y}$ of AAFM as inputs. Operate the convolutional block attention module (CBAM) [27] on the two features respectively. Then we use the two-stage spatial attention to extract more temporal features from motion, which is passed to fusion features. In each stage, we employ convolution to obtain a spatial attention matrix, which serves to highlight regions of interest within the features. Simultaneously, FCM is capable of being cascaded, the output of which serves as the input of another FCM. This unique serial connection structure allows for keeping a balance between two different modes by controlling the number of connecting multiple modules. The experimental results in Sect. 4 also prove it.

3.4   Bidirectional Purification Module and Decoder

We use the bidirectional purification module (BPM) proposed in [7] to get multi-scale features from fusion and motion. The module can also update the inconsistent features in spatiotemporal embedding. Similar to FCM, BPM employs the serial connection structure. The output of the final-layer BPM comprises 32 channels.

We choose the pyramid pool module (PPM) as the decoder. The decoded features undergo upsampling in order to match the size of the features from the subsequent stages, facilitating their concatenation. Subsequently, the channel dimension is reduced via pointwise convolution. The final output of the decoder is $S_t^A, S_t^M \in \mathbb{R}^{H \times W \times 1}$, activated by a sigmoid function.

3.5   Learning Objectives

Given a set of predicted $S_t \in S_t^A, S_t^M$ and the grand truth G_t corresponding to frame t. We use the standard binary cross entropy loss L_{BCE} and lovasz-hinge [1] L_{LH} to measure the difference between the output and the target jointly.

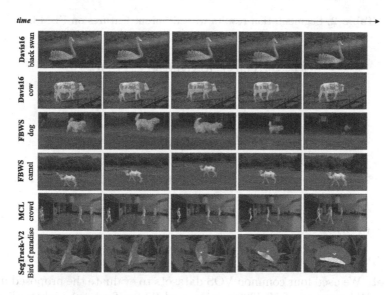

Fig. 3. Qualitative results on four datasets, including DAVIS16 [17], FBMS [16], MCL [9] and SegTrack-V2 [11].

The total loss is expressed as:

$$L_A(S_t^A, G_t) = L_{BCE}(S_t^A, G_t) + \alpha L_{LH}(S_t^A, G_t), \tag{3}$$

$$L_M(S_t^M, G_t) = L_{BCE}(S_t^M, G_t) + \alpha L_{LH}(S_t^M, G_t), \tag{4}$$

$$L_{total}(S_t, G_t) = L_A(S_t^A, G_t) + L_M(S_t^M, G_t). \tag{5}$$

α is set as 0.2. We take S_t^A as the final prediction because our experiments show that the segmentation results based on fusion features are better (Table 1).

Table 1. Segmentation results with different features on DAVIS16.

Method	Backbone	Mean-J	Mean-F	J&F
AFF-Net(Ours)	**ResNet-50** [5]	**83.1%**	**83.6%**	**83.4%**
CFAM [3]	ResNet101 [5]	83.5%	82.0%	82.7%
FS-Net [7]	ResNet-50 [5]	82.9%	83.0%	82.9%
MAT [34]	ResNet101 [5]	82.4%	80.7%	81.5%
WCS-Net [33]	EfficientNetv2 [20]	82.2%	81.5%	81.8%
AGNN [24]	DeepLabv3 [2]	80.7%	79.1%	79.9%
AD-Net [22]	ResNet101 [5]	81.7%	80.5%	81.1%
COS-Net [14]	DeepLabv3 [2]	80.5%	79.4%	79.9%
AGS [26]	ResNet101 [5]	79.7%	77.4%	78.5%

Table 2. Segmentation results with different features on FBMS.

Method	Backbone	Mean-J
AFF-Net(Ours)	**ResNet-50** [5]	**76.9%**
CFAM [3]	ResNet101 [5]	76.5%
AMC-Net [29]	ResNet-101 [5]	76.3%
FS-Net [7]	ResNet-50 [5]	75.9%
COS-Net [14]	DeepLabv3 [2]	75.6%
PDB [19]	ResNet-50 [5]	74.0%
IET [13]	DeepLabv2 [2]	71.9%

4   Experiments

4.1   Experimental Scheme

Data Set. We used four common VOS datasets to evaluate the proposed model. DAVIS16 [17] consists of 50 (30 training and 20 verification) high-quality and densely annotated video sequences. FBMS [16] includes 59 natural videos, 29 for training 30 for testing. SegTrack-v2 [11] consists of 13 segments, which is for testing in our work. MCL [9] contains 9 videos. Taking into account the limited content of MCL [9], we restrict its usage solely for visualization.

Training Strategy. According to the multi-task training setting like [12], we divide the training process into three steps:

We first use the well-known static significance dataset DUTS [23] to train appearance branches to avoid overfitting, such as [4, 25]. Under the same training settings mentioned in Sect. 3, this step lasts for 20 epochs, and the batch size is 4.

Then we train motion branches on the generated optical flow images. Under the same training settings mentioned in Sect. 3, This step lasts for 20 epochs, and the batch size is 4.

We finally loaded the pre-training weights on the two subtasks into the appearance and motion branches. Then, the whole network was trained end-to-end on the training sets of DAVIS16 [17] (30 clips) and FBMS [16] (29 clips). The last step takes about 10 h and converges after 20 epochs. The batch size is 4, which is the same as the training setting mentioned in Sect. 3.

Table 3. Segmentation results with different features on SegTrack-V2.

Method	**AFF-Net (Ours)**	CFAM [3]	AMC-Net [29]	FS-Net [7]	SAGE [30]	CUT [28]	FTS [21]	CIS [10]
Mean-J	**60.1%**	58.6%	57.3%	57.2%	57.6%	54.3%	47.8%	45.6%

Table 4. Ablation Study about Asymmetric Attention Fusion Module.

Method	Mean-J	Mean-F	J&F
w. AAFM	**83.1%**	**83.6%**	**83.4%**
w/o. AAFM	82.3%	82.6%	82.4%

4.2 Quantitative and Qualitative Results

Evaluation on DAVIS16. DAVIS16 [17] is a common video object segmentation dataset, which is used for single object segmentation task. AAF-Net obtains satisfactory results on this dataset. While the Mean-J of AAF-Net decreased by 4% compared with the latest method, CFAM [3], the Mean-F and J&F increased by 1.6% and 0.7%, respectively. Compared with FS-Net [7], which is also based on optical flow, our model shows significant improvement in all three indexes.

We believe that the reason for this is our model uses an asymmetric fusion strategy in multi-modalities fusion. This fusion strategy enables us to extract deeper temporal information from the optical flow images.

Evaluation on FBMS and SegTrack-v2. The result of the evaluation on FBMS is as shown in Table 2. The result of the evaluation on SegTrack-V2 is shown in Table 3. Compared with DAVIS16 [17], these two datasets are smaller but more challenging. AAF-Net performs well in both two datasets. In the realm of challenging tasks, we contend that the efficacy of FCMs is manifest. The serial connection structure of this module facilitates a dynamic control over the extent of information extraction and fusion. It is also through this structure that FCMs endow the system with remarkable resilience.

Because SegTrack-V2 [11] is a dataset just for testing. We only set DAVIS16 [17] and FBMS [16] for training. The result of this experience shows that AAF-Net exhibits strong adaptability, achieving excellent performance on different datasets.

Qualitative Results. We visualize the final masks of our network on the four datasets (DAVIS16 [17], FBMS [16], MCL [9] and SegTrack-V2 [11]). The results are shown in Fig. 3, showcasing the notable capabilities of AAF-Net in UVOS. For these challenging situations, e.g., background interference (1st and 2nd rows), fast-motion (3rd row), dynamic background (4th row), multi-object (5th row) and deformation (6th row) AAF-Net enables to infer the real target object.

4.3 Ablation Study

Effectiveness of AAFM. To validate the effectiveness of AAFM, we replace AAFM in the network with regular convolution. As expected (Table 4), the proposed AAFM performs consistently better than convolution on DAVIS16 [17].

Table 5. Ablation Study about the number N of Feature Correction Module

Method	Mean-J	Mean-F	J&F
N = 3	81.4%	81.8%	81.6%
N = 4	81.9%	82.5%	82.2%
N = 5	**83.1%**	**83.6%**	**83.4%**
N = 6	82.1%	82.8%	82.4%
N = 7	81.5%	82.2%	81.9%

Table 6. Ablation Study about the number M of Bidirectional Purification Module.

Method	Mean-J	Mean-F	J&F
M = 2	81.9%	82.5%	82.2%
M = 3	**83.1%**	**83.6%**	**83.4%**
M = 4	82.2%	82.9%	82.5%

Bidirectional channel attention transfer is positive for deeper features. one-way spatial attention transfer enable to extract more temporal information. We also observe a slight improvement (approximately 0.3%) in Mean-F when employing AAFM directly on FS-Net [7], indicating the generality of AAFM.

Visualization of AFMM. We set regular convolution as output head to visualiz the output of w. AAFM and w/o. AAFM. The results are shown in Fig. 4. Our findings indicate that AFMM not only serves to augment the temporal features present in the optical flow images, but also effectively mitigates unfavorable features within the optical flow modality. The disabled capability of optical flow images to accurately differentiate between reflections and objects can lead to misjudgment by the network, as observed in the first row of the results. Additionally, due to the inherent characteristics of the optical flow estimation algorithm, the same object in the optical flow image is consistently represented by various color blocks, adversely affecting the segmentation process, as evident in the second and third rows. By setting AFMM, we observe a noteworthy decrease in segmentation errors arising from the aforementioned limitations of optical flow images. This substantiates the advantage of AFMM in effectively purifying the noise present in optical flow.

Effectiveness of FCM. We hold the belief that FCM plays a crucial role in segmentation tasks. Nonetheless, we do not advocate for an indiscriminate increase in the number of FCMs. To investigate this further, we conduct ablation studies involving the variation of the number, denoted as N, of interconnected FCMs. The evaluation results, as depicted in Table 5, are obtained for $N = 3, 4, 5, 6, 7$. Analyzing the findings from this experiment, we enable to infer that the network performs optimally when N is set to 5. FCM employs spatial attention

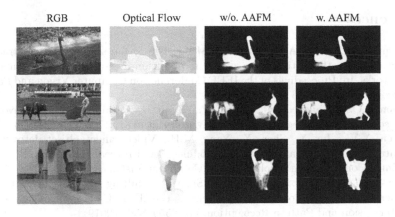

Fig. 4. Visualize experiments about AAFM. The masks without AAFM and with AAFM are the 3rd column and the 4th column. The masks with AAFM are obviously more accurate than masks without AAFM.

from motion, which includes temporal information, to rectify the fusion features. FCM also facilitates the preservation of a balance between two distinct modalities, thereby mitigating the risk of overemphasizing specific modalities. Changes in the number of interconnected FCMs enable to disrupt this equilibrium, resulting in an imbalance and subsequent decline in performance indicators.

Effectiveness of BPM. The optimal number, denoted as M, of interconnected BPMs also requires investigation. The results of the ablation study conducted on this aspect are presented in Table 6, with M values ranging from 3 to 5. Based on the result, AAF-Net achieves its best performance when M is set to 4. Furthermore, the inclusion of an excessive number of connecting BPMs (i.e., $M > 4$) leads to heightened model complexity, thereby increasing the risk of overfitting.

5 Conclusion

We propose an Asymmetric Attention Fusion Network based on optical flow for Unsupervised Video Object Segmentation. In our work, the asymmetric attention fusion module performs attention interaction between two modalities and extracts temporal information from motion features, which is then passed on to appearance features. The feature correction module further extracts deeper features while maintaining the balance between the two modalities. With the assistance of these two modules, our network achieves superior segmentation accuracy compared to other state-of-the-art segmentation models.

Acknowledgement. This work was supported in part by the National Natural Science Foundation of China (62020106012, 62106089).

References

1. Berman, M., Triki, A.R., Blaschko, M.B.: The lovász-softmax loss: a tractable surrogate for the optimization of the intersection-over-union measure in neural networks. In: Proceedings of the IEEE Conference on Computer Vision and Pattern Recognition, pp. 4413–4421 (2018)
2. Chen, L.C., Papandreou, G., Schroff, F., Adam, H.: Rethinking atrous convolution for semantic image segmentation. arXiv preprint arXiv:1706.05587 (2017)
3. Chen, Y.W., Jin, X., Shen, X., Yang, M.H.: Video salient object detection via contrastive features and attention modules. In: Proceedings of the IEEE/CVF Winter Conference on Applications of Computer Vision, pp. 1320–1329 (2022)
4. Fan, D.P., Wang, W., Cheng, M.M., Shen, J.: Shifting more attention to video salient object detection. In: Proceedings of the IEEE/CVF Conference on Computer Vision and Pattern Recognition, pp. 8554–8564 (2019)
5. He, K., Zhang, X., Ren, S., Sun, J.: Deep residual learning for image recognition. In: Proceedings of the IEEE Conference on Computer Vision and Pattern Recognition, pp. 770–778 (2016)
6. Ilg, E., Mayer, N., Saikia, T., Keuper, M., Dosovitskiy, A., Brox, T.: FlowNet 2.0: evolution of optical flow estimation with deep networks. In: Proceedings of the IEEE Conference on Computer Vision and Pattern Recognition, pp. 2462–2470 (2017)
7. Ji, G.P., Fu, K., Wu, Z., Fan, D.P., Shen, J., Shao, L.: Full-duplex strategy for video object segmentation. In: Proceedings of the IEEE/CVF International Conference on Computer Vision, pp. 4922–4933 (2021)
8. Jiang, Q., Wu, X., Kittler, J.: Insight on attention modules for skeleton-based action recognition. In: Ma, H., et al. (eds.) PRCV 2021. LNCS, vol. 13019, pp. 242–255. Springer, Cham (2021). https://doi.org/10.1007/978-3-030-88004-0_20
9. Kim, H., Kim, Y., Sim, J.Y., Kim, C.S.: Spatiotemporal saliency detection for video sequences based on random walk with restart. IEEE Trans. Image Process. **24**(8), 2552–2564 (2015)
10. Koh, Y.J., Kim, C.S.: Primary object segmentation in videos based on region augmentation and reduction. In: IEEE Conference on Computer Vision and Pattern Recognition (2017)
11. Li, F., Kim, T., Humayun, A., Tsai, D., Rehg, J.M.: Video segmentation by tracking many figure-ground segments. In: Proceedings of the IEEE International Conference on Computer Vision, pp. 2192–2199 (2013)
12. Li, H., Chen, G., Li, G., Yu, Y.: Motion guided attention for video salient object detection. In: Proceedings of the IEEE/CVF International Conference on Computer Vision, pp. 7274–7283 (2019)
13. Li, S., Seybold, B., Vorobyov, A., Fathi, A., Huang, Q., Kuo, C.C.J.: Instance embedding transfer to unsupervised video object segmentation. In: Proceedings of the IEEE Conference on Computer Vision and Pattern Recognition, pp. 6526–6535 (2018)
14. Lu, X., Wang, W., Ma, C., Shen, J., Shao, L., Porikli, F.: See more, know more: unsupervised video object segmentation with co-attention Siamese networks. In: Proceedings of the IEEE/CVF Conference on Computer Vision and Pattern Recognition, pp. 3623–3632 (2019)
15. Miao, J., Wei, Y., Yang, Y.: Memory aggregation networks for efficient interactive video object segmentation. In: Proceedings of the IEEE/CVF Conference on Computer Vision and Pattern Recognition, pp. 10366–10375 (2020)

16. Ochs, P., Malik, J., Brox, T.: Segmentation of moving objects by long term video analysis. IEEE Trans. Pattern Anal. Mach. Intell. **36**(6), 1187–1200 (2013)
17. Perazzi, F., Pont-Tuset, J., McWilliams, B., Van Gool, L., Gross, M., Sorkine-Hornung, A.: A benchmark dataset and evaluation methodology for video object segmentation. In: Proceedings of the IEEE Conference on Computer Vision and Pattern Recognition, pp. 724–732 (2016)
18. Rao, J., Xu, T., Song, X., Feng, Z.H., Wu, X.J.: Kitpose: Keypoint-interactive transformer for animal pose estimation. In: Yu, S., et al. (eds.) PRCV 2022. LNCS, vol. 13534, pp. 660–673. Springer, Cham (2022). https://doi.org/10.1007/978-3-031-18907-4_51
19. Song, H., Wang, W., Zhao, S., Shen, J., Lam, K.M.: Pyramid dilated deeper ConvLSTM for video salient object detection. In: Proceedings of the European conference on computer vision (ECCV), pp. 715–731 (2018)
20. Tan, M., Le, Q.: Efficientnetv2: smaller models and faster training. In: International Conference on Machine Learning, pp. 10096–10106. PMLR (2021)
21. Tokmakov, P., Alahari, K., Schmid, C.: Learning video object segmentation with visual memory. In: Proceedings of the IEEE International Conference on Computer Vision, pp. 4481–4490 (2017)
22. Tsai, Y.H., Hung, W.C., Schulter, S., Sohn, K., Yang, M.H., Chandraker, M.: Learning to adapt structured output space for semantic segmentation. In: Proceedings of the IEEE Conference on Computer Vision and Pattern Recognition, pp. 7472–7481 (2018)
23. Wang, L., et al.: Learning to detect salient objects with image-level supervision. In: Proceedings of the IEEE Conference on Computer Vision and Pattern Recognition, pp. 136–145 (2017)
24. Wang, W., Lu, X., Shen, J., Crandall, D.J., Shao, L.: Zero-shot video object segmentation via attentive graph neural networks. In: Proceedings of the IEEE/CVF International Conference on Computer Vision, pp. 9236–9245 (2019)
25. Wang, W., Shen, J., Shao, L.: Video salient object detection via fully convolutional networks. IEEE Trans. Image Process. **27**(1), 38–49 (2017)
26. Wang, W., ET AL.: Learning unsupervised video object segmentation through visual attention. In: Proceedings of the IEEE/CVF Conference on Computer Vision and Pattern Recognition, pp. 3064–3074 (2019)
27. Woo, S., Park, J., Lee, J.Y., Kweon, I.S.: CBAM: convolutional block attention module. In: Proceedings of the European conference on computer vision (ECCV), pp. 3–19 (2018)
28. Yang, C., Lamdouar, H., Lu, E., Zisserman, A., Xie, W.: Self-supervised video object segmentation by motion grouping. In: Proceedings of the IEEE/CVF International Conference on Computer Vision, pp. 7177–7188 (2021)
29. Yang, S., Zhang, L., Qi, J., Lu, H., Wang, S., Zhang, X.: Learning motion-appearance co-attention for zero-shot video object segmentation. In: Proceedings of the IEEE/CVF International Conference on Computer Vision, pp. 1564–1573 (2021)
30. Yang, Y., Loquercio, A., Scaramuzza, D., Soatto, S.: Unsupervised moving object detection via contextual information separation. In: Proceedings of the IEEE/CVF Conference on Computer Vision and Pattern Recognition, pp. 879–888 (2019)
31. Zhang, D., Wu, X.-J., Yu, J.: Discrete bidirectional matrix factorization hashing for zero-shot cross-media retrieval. In: Ma, H., et al. (eds.) PRCV 2021. LNCS, vol. 13020, pp. 524–536. Springer, Cham (2021). https://doi.org/10.1007/978-3-030-88007-1_43

32. Zhang, K., Zhao, Z., Liu, D., Liu, Q., Liu, B.: Deep transport network for unsupervised video object segmentation. In: Proceedings of the IEEE/CVF International Conference on Computer Vision, pp. 8781–8790 (2021)

33. Zhang, L., Zhang, J., Lin, Z., Měch, R., Lu, H., He, Y.: Unsupervised video object segmentation with joint hotspot tracking. In: Vedaldi, A., Bischof, H., Brox, T., Frahm, J.-M. (eds.) ECCV 2020. LNCS, vol. 12359, pp. 490–506. Springer, Cham (2020). https://doi.org/10.1007/978-3-030-58568-6_29

34. Zhou, T., Wang, S., Zhou, Y., Yao, Y., Li, J., Shao, L.: Motion-attentive transition for zero-shot video object segmentation. In: Proceedings of the AAAI Conference on Artificial Intelligence, vol. 34, pp. 13066–13073 (2020)

Flow-Guided Diffusion Autoencoder for Unsupervised Video Anomaly Detection

Aoni Zhu, Wenjun Wang, and Cheng Yan[✉]

Tianjin University, Tianjin, China
{zhu_aoni,wjwang,yancheng_work}@tju.edu.cn

Abstract. Video anomaly detection (VAD) aims to automatically detect abnormalities that deviate from expected behaviors. Due to the heavy reliance of mainstream one-class methods on labeled normal samples, some unsupervised VAD methods have emerged. However, these methods are unable to detect both appearance and motion anomalies in videos comprehensively. To address the above problem, we present for the first time a Flow-guided Diffusion AutoEncoder (FDAE) that generates objects of each frame to detect anomalous in an unsupervised manner. Our model takes foreground objects and motion information as inputs to train a conditional diffusion autoencoder for foreground reconstruction. To make our model concentrate on learning normal samples, we further design a sample refinement scheme and introduce a mixed Gaussian clustering network to enhance the capability of the diffusion model in capturing typical characteristics of normal samples. Comprehensive experiments on three public available datasets demonstrate that the proposed FDAE outperforms all competing unsupervised approaches.

Keywords: Video anomaly detection · Unsupervised learning · Diffusion model

1 Introduction

Video anomaly detection (VAD) is an important task of identifying abnormal events that deviate from the normality in real applications, e.g., intelligent surveillance systems. The task is challenging due to the unavailability of collecting all anomaly events, which are various in the real world. There have been many methods introduced for VAD over the years [1–11]. The majority of these methods belong to one-class VAD approach, which assumes that only normal samples are available during training and then learns patterns from normal samples to detect events that are different from these patterns as anomalies. Convolution neural networks based autoencoder [12] and U-Net [13] are popular frameworks in one-class VAD for learning such patterns with sufficient normal data. However, these methods face two problems: 1) labeling normal samples from massive data is extremely labor-intensive and 2) in natural scenes, the

© The Author(s), under exclusive license to Springer Nature Singapore Pte Ltd. 2024
Q. Liu et al. (Eds.): PRCV 2023, LNCS 14430, pp. 183–194, 2024.
https://doi.org/10.1007/978-981-99-8537-1_15

learned normal pattern may close to the abnormal one once the difference is slight, e.g., background changes from fallen leaves or shade maybe more notable than inconspicuous stealing action.

To overcome the annotating problem, a few unsupervised VAD methods have been proposed, such as [6–11]. Early methods [6–9] mainly design classifiers to distinguish the normal and abnormal video frames, while ignoring the influence from the background. The discrepancy from background usually affects the performance of these classifiers. To this end, object detection-based methods have been proposed [10] to concentrate the model on foreground learning. These models can effectively identify appearance anomalies, while ignoring action anomalies. Action-based methods [11] employ action recognition network to acquire motion features, while exhibiting a substantial dependence on the capacity of pre-trained action networks. How to effectively detect both appearance and motion anomalies in unsupervised VAD is still a challenge.

In this paper, we present a novel Flow-guided Diffusion AutoEncoder (FDAE) to learn the integration of appearance and motion information of the normal samples for unsupervised VAD. Our method is based on autoencoder architecture to capture spatio-temporal information of foreground objects by utilizing the consecutive video frames. Different from the previous works [6–9] that take the whole frame as input, we employ object detection and optical flows method to achieve a time series bounding boxes and optical flows as input samples. These appearance and motion features are used to train our diffusion autoencoder in which the optical flows serve as a condition for foreground object learning. To reduce the influence of the abnormal motion feature, we design a mixed gaussian clustering network that contains multiple Gaussian components to reconstruct optical flows for feature filter. Finally, we design a sample refinement scheme that alleviates the adverse effects of abnormal features and helps the model to discriminate normal from anomalies.

There are four contributions in our research summarized as follows:

- We propose a novel unsupervised VAD method, which for the first time introduces flow-guided diffusion model into unsupervised VAD to learn the spatio-temporal features of foreground objects in consecutive frames.
- We introduce a mixed gaussian clustering network to cluster important features and enhance the learning process.
- We design an efficient sample refinement scheme to concentrate our model on the normal features learning.
- Our experiment results show that the proposed method outperforms existing unsupervised VAD methods on standard benchmarks.

2   Related Work

2.1   Video Anomaly Detection

Unsupervised VAD. Benefiting from labeling free of training data, unsupervised VAD attracts more researchers' attention and becomes an important application in the video analysis field. Early unsupervised methods rely on the analysis of hand-crafted features extracted from each frame of videos, such as histogram of gradients (HoG) and histogram of flows (HoF) [6], then train a linear classifier to determine anomalies [7,8]. With the development of Convolution neural networks, an end-to-end deep unsupervised VAD method has been proposed [9]. The key step is to calculate an anomaly score for each frame as a label through other unsupervised models to improve the classifier's discriminatory power. As a result, the performance relies heavily on the performance of the extra model. Based on their anomaly score, recent methods [10,11] have emerged that utilize data pre-processing to remove and eliminate abnormal samples during training. This approach ensures that the model learns features solely from normal samples, even in unlabeled data.

Clustering-Based VAD. For better normal data learning, problem-specific memory modules have been incorporated into the VAD methods to store the significant features from the training set, which uses a cosine similarity measure to cluster the memory items. Gong et al. [1] developed a memory module into their deep model to store more relative features of normal samples, and in inference, the anomalies are hard to be reconstructed by these stored features. Zhian et al. [3] applied a memory module to store optical flows that provide video motion information for feature learning. Hyunjong et al. [2] compressed the memory module to keep the performance while reducing the model size. In addition, Yaxiang et al. [14] utilized a two-stream mixed Gaussian clustering network to learn video appearance and motion features and then fused the information over anomaly scores. However, all the above are in the field of one-class VAD, and no clustering network is proposed for unsupervised VAD.

2.2 Diffusion Models

Recently, the diffusion models have achieved SOTA performance on many generative tasks [15–18]. The diffusion model leverages the diffusion process to generate high-quality samples by denoising corrupted data. Denoising Diffusion Probabilistic Models (DDPM) [16] uses the denoising score matching objective, where the model is trained to estimate the score of the data distribution at each diffusion step. On the other hand, Denoising Diffusion Implicit Models (DDIM) [17] extend the concept of DDPM by introducing an implicit model formulation, allowing for more flexible and expressive modeling. In addition to applying it to image generation tasks, Diffusion Autoencoders [18] utilizes the DDIM process to gradually transform the input data into a more meaningful and decodable representation, which is achieved by adding high-level semantics to the latent space representation and encouraging the preservation of information during each diffusion step.

3   Method

The framework is shown in Fig. 1. The FDAE mainly consists of three modules: Mixed Gaussian Clustering Network (MGCN), Conditional Diffusion AutoEncoder (CDAE), and Refinement Schema (RS). In the training stage, following [4], we employ Cascade R-CNN [19] and FlowNet [20] as the foreground information extractor to obtain object bounding boxes of each frame and the corresponding optical flows from the previous continuous t_f video frames. The MGCN is designed to learn normal features from unlabeled videos. These optical flows are then fed into the CDAE together with the bounding box in $t_f + 1$ frame for diffusion learning. The RS guarantees CDAE to learn the normal samples. During the test phase, the model generates the bounding boxes in $t_f + 1$ frame and detects the bounding box with a large reconstruction error as anomalous.

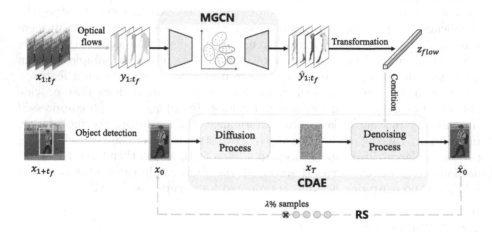

Fig. 1. The framework of our model FDAE. The optical flows $y_{1:t_f}$ of consecutive t_f video frames $x_{1:t_f}$ are extracted as input to the proposed Mixed Gaussian Clustering Network (MGCN) to obtain the reconstructed optical flows $\hat{y}_{1:t}$. Then, the foreground object x_{t_f+1} of the the $t_f + 1$ video frame is extracted by R-CNN as the input x_0 of Conditional Diffusion AutoEncoder (CDAE) for diffusion training. The Refinement Schema (RS) is used for normal sample selection.

3.1   Mixed Gaussian Clustering Network

The MGCN in our framework is inspired by [14], which supports a variational autoencoder to clustering. Through this network, an optical flow frame of each frame $y \sim \mathcal{N}\left(\mu_y, \sigma_y^2 \mathbf{I}\right)$ is clustered into K and reconstructed as $\hat{y} \sim \mathcal{N}\left(\mu_y, \sigma_y^2 \mathbf{I}\right)$ in the following process: 1) each component c_k of Mixed Gaussian clustering was initialized by K-Means. 2) Sample the implicit variable $\mathbf{z}$ from c. 3) $\mathbf{z}$ generates reconstructed optical flow $\hat{y}$ through decoder f parameterized by θ. It can be defined as follows:

$$p(c) = \text{Cat}(\boldsymbol{\pi}), \tag{1}$$

$$p(\mathbf{z} \mid c) = \mathcal{N}\left(\mathbf{z} \mid \mu_c, \sigma_c^2 \mathbf{I}\right), \tag{2}$$

$$f(\mathbf{z}; \theta) = \left[\mu_y; \log \sigma_y^2\right], \tag{3}$$

$$p(\hat{y} \mid \mathbf{z}) = \mathcal{N}\left(\mu_y, \sigma_y^2 \mathbf{I}\right), \tag{4}$$

where π_k is the prior probability that c_k belongs to cluster K, $\pi \in \mathbb{R}_+^K$, $\sum_{k=1}^K \pi_k = 1$. Similar to variational autoencoder [12], the encoder g parameterized by ϕ is utilized to model $q(\mathbf{z} \mid y)$:

$$g(y; \phi) = \left[\tilde{\mu}, \log \tilde{\sigma}^2\right], \tag{5}$$

$$q(\mathbf{z} \mid y) = \mathcal{N}\left(\mathbf{z}; \tilde{\mu}, \tilde{\sigma}^2 \mathbf{I}\right), \tag{6}$$

this yields the evidence lower bound (ELBO) for the log-likelihood rewritten as:

$$\begin{aligned}
\log p(y) &= \log \int_{\mathbf{z}} \sum_c p(y, \mathbf{z}, c) d\mathbf{z} \\
&\geq E_{q(\mathbf{z}, c|y)} \left[\log \frac{p(y, \mathbf{z}, c)}{q(\mathbf{z}, c \mid y)}\right] = \mathcal{L}_{\text{ELBO}}(y).
\end{aligned} \tag{7}$$

Further expansion obtains the following expression for ELBO:

$$\begin{aligned}
\mathcal{L}_{\text{ELBO}}(y) &= E_{q(\mathbf{z}, c|y)} \left[\log \frac{p(y, \mathbf{z}, c)}{q(\mathbf{z}, c \mid y)}\right] \\
&= E_{q(\mathbf{z}, c|y)}[\log p(y, \mathbf{z}, c) - \log q(\mathbf{z}, c \mid y)] \\
&= E_{q(\mathbf{z}, c|y)}[\log p(y \mid \mathbf{z}) + \log p(\mathbf{z} \mid c) \\
&\quad + \log p(c) - \log q(\mathbf{z} \mid y) - \log q(c \mid y)],
\end{aligned} \tag{8}$$

where $q(c \mid y)$ can be approximated by the Monte Carlo sampling method as follows:

$$\begin{aligned}
q(c \mid y) &= E_{q(\mathbf{z}|y)}[p(c \mid \mathbf{z})] \\
&= \frac{1}{L} \sum_{l=1}^L \frac{p(c)p\left(\mathbf{z}^{(l)} \mid c\right)}{\sum_{c'=1}^K p\left(c'\right)p\left(\mathbf{z}^{(l)} \mid c'\right)},
\end{aligned} \tag{9}$$

where $\mathbf{z}^{(l)} \sim \mathcal{N}\left(\tilde{\mu}, \tilde{\sigma}^2 \mathbf{I}\right)$, L is the number of Monte Carlo samples. By substituting the terms in Eq. 10 with Eqs. 2, 3, 5, 8 and 11, and using the SGVB estimator and the reparameterization trick, the loss function of the MGCN can be obtained as follows:

$$\mathcal{L}_{MGCN} = \alpha * \|y - \hat{y}\|_2^2$$
$$+ \beta * \left[\frac{1}{2} \sum_{c=1}^{k} \gamma_c \left(\log \sigma_c^2 + \frac{\tilde{\sigma}^2}{\sigma_c^2} + \frac{(\tilde{\mu} - \mu_c)^2}{\sigma_c^2} \right) \right.$$
$$\left. - \sum_{c=1}^{k} \gamma_c \log \frac{\pi_c}{\gamma_c} + \frac{1}{2} \left(1 + \log \tilde{\sigma}^2 \right) \right], \tag{10}$$

where the $\| \cdot \|_2^2$ is L_2 distance, α and β are hyperparameters to control balance, and γ_c denotes $q(c \mid \mathbf{x})$ for simplicity.

3.2   Conditional Diffusion AutoEncoder

Motivated by [18], motion features $y_{1:t_f}$ of continuous t_f frames are utilized to generate the latent variable $\mathbf{z}_{\text{flow}}$ as a condition for the CDAE parameterized by the U-net [13] architecture of noise prediction network ϵ_θ, thereby integrating appearance and motion features. Specifically, we add noise to the bounding boxes $\mathbf{x}_0$ of frame $t_f + 1$ to $\mathbf{x}_T$ through $t \sim \text{Uniform}(1, \ldots, T)$ step Gaussian diffusion process, which is formalized as follows:

$$q\left(\mathbf{x}_t \mid \mathbf{x}_{t-1}\right) = \mathcal{N}\left(\sqrt{1 - \beta_t}\mathbf{x}_{t-1}, \beta_t \mathbf{I}\right), \tag{11}$$

where $\{\beta_t\}_{t=1}^T$ is the variance used for each step. As a result, the noise image $\mathbf{x}_t$ at the moment t can be expressed in terms of $\mathbf{x}_0$ as follows:

$$q\left(\mathbf{x}_t \mid \mathbf{x}_0\right) = \mathcal{N}\left(\sqrt{\alpha_t}\mathbf{x}_0, (1 - \alpha_t)\mathbf{I}\right), \tag{12}$$

where $\alpha_t = \prod_{i=1}^{t}\left(1 - \beta_i\right)$. In turn, the denoising process and its distribution can be expressed as follows:

$$\mathbf{x}_{t-1} = \sqrt{\alpha_{t-1}}\left(\frac{\mathbf{x}_t - \sqrt{1 - \alpha_t}\epsilon_\theta\left(\mathbf{x}_t\right)}{\sqrt{\alpha_t}}\right) + \sqrt{1 - \alpha_{t-1}}\epsilon_\theta\left(\mathbf{x}_t\right), \tag{13}$$

$$q\left(\mathbf{x}_{t-1} \mid \mathbf{x}_t, \mathbf{x}_0\right) = \mathcal{N}\left(\sqrt{\alpha_{t-1}}\mathbf{x}_0 + \sqrt{1 - \alpha_{t-1}}\frac{\mathbf{x}_t - \sqrt{\alpha_t}\mathbf{x}_0}{\sqrt{1 - \alpha_t}}, 0\right). \tag{14}$$

The motion features of each set of consecutive t frames $\in \mathbb{R}^{c \times h \times w}$ are transformed by into a non-spatial vector $\mathbf{z}_{\text{flow}}$ by a simple linear transformation, where $\mathbf{z}_{\text{flow}} \in \mathbb{R}^d$, $d = 512$. As with [21], the addition of the condition is expressed using the adaptive group normalization layers (AdaGN) as follows:

$$\text{AdaGN}\left(\mathbf{h}, t, \mathbf{z}_{\text{flow}}\right) = \mathbf{y}_s\left(\mathbf{t}_s \text{ GroupNorm}(\mathbf{h}) + \mathbf{t}_b\right), \tag{15}$$

where $\mathbf{h} \in \mathbb{R}^{c \times h \times w}$ is the output feature map of $\mathbf{x}_t$ input to FDAE's encoder, GroupNorm$(\cdot)$ is group normalization [22], $\mathbf{y}_s \in \mathbb{R}^c = \text{Affine}(\mathbf{z}_{\text{flow}})$, and $(\mathbf{t}_s, \mathbf{t}_b) \in \mathbb{R}^{2 \times c} = \text{MLP}(\psi(t))$ is the output of a multilayer perceptron with a sinusoidal encoding function $(\psi(\cdot))$ of parameter $\mathbf{t}$. Through the involvement of $\mathbf{z}_{\text{flow}}$, the FDAE modeling $p_\theta\left(\mathbf{x}_{t-1} \mid \mathbf{x}_t, \mathbf{z}_{\text{flow}}\right)$ corresponds to the denoising distribution $q\left(\mathbf{x}_{t-1} \mid \mathbf{x}_t, \mathbf{x}_0\right)$ defined in Eq. 14, with the following procedure:

$$p_\theta\left(\mathbf{x}_{0:T} \mid \mathbf{z}_{\text{flow}}\right) = p\left(\mathbf{x}_T\right) \prod_{t=1}^{T} p_\theta\left(\mathbf{x}_{t-1} \mid \mathbf{x}_t, \mathbf{z}_{\text{flow}}\right). \tag{16}$$

The following loss function is optimized to train the FDAE:

$$L_{\text{FDAE}} = \sum_{t=1}^{T} \mathbb{E}_{\mathbf{x}_0, \epsilon_t}\left[\left\|\epsilon_\theta\left(\mathbf{x}_t, t, \mathbf{z}_{\text{flow}}\right) - \epsilon_t\right\|_2^2\right], \tag{17}$$

where $\epsilon_t \in \mathbb{R}^{3 \times h \times w} \sim \mathcal{N}(\mathbf{0}, \mathbf{I})$. And $\mathbf{x}_0$ was reconstructed from the noise ϵ_θ predicted by the FDAE as follows:

$$\mathbf{x}_0\left(\mathbf{x}_t, t, \mathbf{z}_{\text{flow}}\right) = \frac{1}{\sqrt{\alpha_t}}\left(\mathbf{x}_t - \sqrt{1 - \alpha_t}\epsilon_\theta\left(\mathbf{x}_t, t, \mathbf{z}_{\text{flow}}\right)\right). \tag{18}$$

Due to $\mathbf{z}_{\text{flow}}$'s limited ability to compress random details, in order to encourage $\mathbf{x}_T$ to encode only the information missed by $\mathbf{z}_{\text{flow}}$, during the testing phase, the noisy image $\mathbf{x}_{t+1}$ at time $t + 1$ can be obtained as follows:

$$\mathbf{x}_{t+1} = \sqrt{\alpha_{t+1}}\mathbf{x}_0\left(\mathbf{x}_t, t, \mathbf{z}_{\text{flow}}\right) + \sqrt{1 - \alpha_{t+1}}\epsilon_\theta\left(\mathbf{x}_t, t, \mathbf{z}_{\text{flow}}\right). \tag{19}$$

After generating the noise, the reconstructed image $\mathbf{x}_0$ is output using the inverse diffusion process of Eq. 13.

3.3   Sample Refinement

Inspired by the normality advantage from [10], we design a sample refinement to motivate the model to learn normal features in unlabeled videos. In the training process, our model discards the most suspected anomalous part of the detected samples. Specifically, for every n epochs, the model calculates the anomaly scores for each frame of the entire training set and then sorts the total anomaly scores from largest to smallest. The top λ percent of video frames are no longer involved in the subsequent n epochs training.

4   Experiments

4.1   Implementation Details

Dataset. Three datasets, UCSD Ped2 [23], CUHK Avenue [24], and Shang-haiTech [25] are used for experiments.

- UCSD Ped2 contains 16 training and 12 testing video clips, with a total of 4560 frames, which are from fixed shots. Among them, the video frames are all grayscale images, and most of the anomalies inside are abnormal appearances.

– CUHK Avenue contains 16 training and 21 testing video clips, with a total of 30,652 frames, which are from camera shake. Its abnormal events include the sprint of pedestrians and the appearance of pedestrians pushing bicycles in the street.
– ShanghaiTech contains 130 abnormal events and over 270, 000 training frames, among which abnormal events are mainly abnormal human movements. The dataset includes 13 scenes captured from various camera angles and under different lighting conditions.

Training Settings. Abnormal samples only exist in the testing sets of the above three datasets, so the training set and testing set are combined to participate in the training follows [9]. The condition latent variable $y_{1:t_f}$ is derived from the optical flow of 4 consecutive frames. The balancing parameters α and β of the loss function $\mathcal{L}_{MGCN}$ are set to 1.0 and 0.5, respectively. To reduce the impact of anomalous frames on model training, λ is set to 10%. The number of mixed Gaussian distribution components k was set to 0, 5, and 10 to train the model, and the experimental results are shown in Table 2. The batch size and training epochs are set to 256 and 60, respectively.

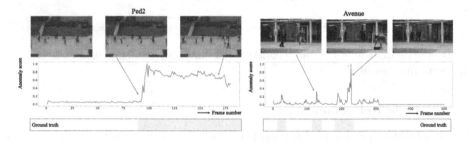

Fig. 2. The anomaly scores calculated by the model are given as ped2 and avenue, with the orange boxed area being the anomaly. (Color figure online)

4.2   Anomaly Score

The Anomaly score S of video frame at each time t_f is calculated by two parts: (1) Mean square error of the MGCN reconstructed optical flows: $S_{op} = \left\| \hat{y}_{t_f} - y_{t_f} \right\|_2^2$, (2) Mean square error of the FDAE reconstructed images: $S_{ap} = \left\| \hat{x}_{t_f} - x_{t_f} \right\|_2^2$. The combination of the two forms the anomaly score S shown below:

$$S = \frac{S_{op} - \mu_{op}}{\sigma_{op}} + \frac{S_{ap} - \mu_{ap}}{\sigma_{ap}}, \tag{20}$$

where $\mu_{op}, \sigma_{op}, \mu_{ap}, \sigma_{ap}$ are the mean and variance of S_{op} and S_{ap}, respectively. The larger the S, the larger possibility of the anomaly (see Fig. 2 for an example of anomaly score).

4.3   Comparison with Existing Methods

In Table 1, we compare our method with four existing state-of-the-art unsupervised VAD methods, including Ordinal Regression method (OR) [9], Classifier Two Sample method (CTS) [8], Unmasking the Abnormal Events (UAE) [7], Discriminative Framework (DF) [6], Self-paced Refinement (SR) [10], and Generative cooperative learning (GCL) [11]. We also present four groups of our method for ablation study. Following [1,2], Area Under Curve (AUC) is used to evaluate the performance. From Table 1, our method integrates appearance and motion features, resulting in superior performance compared to existing methods.

Table 1. AUC of different unsupervised VAD methods.

Method	Ped2	Avenue	ShanghaiTech
DF [6]	63.0%	78.3%	-
UAE [7]	82.2%	80.6%	-
CTS [8]	87.5%	84.4%	-
OR [9]	83.2%	-	-
SR [10]	97.2%	90.7%	72.6%
GCL [11]	-	-	72.4%
FDAE	**97.7%**	**91.1%**	**73.8%**

4.4   Ablation Study

Disturbance. Since the training samples have no annotation, intuitively, the Mixed Gaussian distribution retains key information about normal and abnormal features. However, anomalies are widely recognized as a minority in the real world [5,9,26]. Based on this fact, we assume that the Mixed Gaussian distribution to preserve the information of anomalies is much less than that of the normal samples. To verify the assumption, inspired by [26], we design an operation to disturb the component c'_k by:

$$c'_k + \mathcal{N}\left(0, 1 - \frac{\sum_{j \in Q} I_{U^{d'}}\left(q^j\right)}{Q}\right) \to c_k, \tag{21}$$

where Q is the feature map of encoder output, q^j is a feature vector in Q, $U^{d'}$ is the feature set whose cosine similarity with c'_k is the nearest, and c_k is the item after disturbance, $I_{U^{d'}}\left(\cdot\right)$ is indicator function as:

$$I_{U^{d'}}\left(q^j\right) = \begin{cases} 1 & \text{if } q^j \in U^{d'} \\ 0 & \text{if } q^j \notin U^{d'} \end{cases}. \tag{22}$$

In the experiment, we disturb all the original c'_k by Eq. 21 and set different numbers of Gaussian components. The results are summarized in Table 2, where it can be seen that the absence of the MGCN involvement in training and disturbing the Gaussian components leads to worse accuracy, which validates the MGCN has played an auxiliary role in learning normal sample features for the FDAE.

Table 2. AUC with different configurations of mixed Gaussian components. 0 is a training model without MGCN, 5 and 10 are mixed Gaussian components of size 5 and 10, respectively. * refers to using disturbance.

c_k	Ped2	Avenue	ShanghaiTech
0	96.5%	87.8%	69.2%
10	88.9%	89.1%	71.2%
5*	93.6%	88.8%	70.1%
5	**97.7%**	**91.1%**	**73.8%**

Content of Condition. To verify the guiding role of optical flow, we have carried out an experimental comparison of the content of the condition, which is either the bounding boxes of t_f video frames or the optical flows of t_f video frames, respectively. As seen in Table 3, the better experimental results obtained with optical flow as a condition confirmed the positive effect of optical flow on the model's learning of spatio-temporal information.

Transformation of Condition. In contrast to [18], our model employs a linear transformation to obtain the condition, as opposed to employing a parameterized network structure. We have carried out an experimental comparison of the way to obtain the transformation of the conditional hidden variable z_{flow} in Table 4, i.e. a Fully Connected Layer (FCL) or Linear Transformation (LT), respectively. The experimental findings demonstrate that a simple linear transformation yields superior performance compared to a parametric network, which maintains the linear nature of the optical flow space, thus retaining more information.

Table 3. AUC results for different content of condition. DAE is an unconditional diffusion autoencoder, BDAE is conditional on the bounding boxes, and FDAE is conditional on the optical flows.

	Ped2	Avenue	ShanghaiTech
DAE	93.1%	85.6%	66.2%
BDAE	94.6%	86.8%	68.1%
FDAE	**97.7%**	**91.1%**	**73.8%**

Table 4. AUC results for different transformation of condition. $FDAE_{FCL}$ adopts a fully connected layer, and $FDAE_{LT}$ adopts linear transformation.

	Ped2	Avenue	ShanghaiTech
$FDAE_{FCL}$	88.9%	89.1%	71.2%
$FDAE_{LT}$	**97.7%**	**91.1%**	**73.8%**

5 Conclusion

In this paper, we propose a flow-guided diffusion model for unsupervised VAD. We build a conditional diffusion autoencoder, a mixed gaussian clustering network, and a sample refinement scheme to guarantee that the model concentrates on normal feature learning. Through experiments, we find that the mixed gaussian clustering network can strengthen the process of normal feature learning. With the optical flow information, the conditional diffusion autoencoder can effectively learn spatio-temporal foreground object features. Experiments on three datasets confirm the effectiveness of the proposed method.

References

1. Gong, D., et al.: Memorizing normality to detect anomaly: memory-augmented deep autoencoder for unsupervised anomaly detection. In: Conference on Computer Vision and Pattern Recognition, pp. 1705–1714 (2019)
2. Park, H., Noh, J., Ham, B.: Learning memory-guided normality for anomaly detection. In: Conference on Computer Vision and Pattern Recognition, pp. 14372–14381 (2020)
3. Liu, Z., Nie, Y., Long, C., Zhang, Q., Li, G.: A hybrid video anomaly detection framework via memory-augmented flow reconstruction and flow-guided frame prediction. In: International Conference on Computer Vision, pp. 13588–13597 (2021)
4. Chen, C., et al.: Comprehensive regularization in a bi-directional predictive network for video anomaly detection. In: AAAI Conference on Artificial Intelligence, vol. 36, pp. 230–238 (2022)
5. Georgescu, M.I., Ionescu, R.T., Khan, F.S., Popescu, M., Shah, M.: A background-agnostic framework with adversarial training for abnormal event detection in video. IEEE Trans. Pattern Anal. Mach. Intell. **44**(9), 4505–4523 (2021)
6. Del Giorno, A., Bagnell, J.A., Hebert, M.: A discriminative framework for anomaly detection in large videos. In: Leibe, B., Matas, J., Sebe, N., Welling, M. (eds.) ECCV 2016. LNCS, vol. 9909, pp. 334–349. Springer, Cham (2016). https://doi.org/10.1007/978-3-319-46454-1_21
7. Tudor Ionescu, R., Smeureanu, S., Alexe, B., Popescu, M.: Unmasking the abnormal events in video. In: International Conference on Computer Vision, pp. 2895–2903 (2017)
8. Liu, Y., Li, C.L., Póczos, B.: Classifier two sample test for video anomaly detections. In: BMVC, p. 71 (2018)
9. Pang, G., Yan, C., Shen, C., Hengel, A.V.D., Bai, X.: Self-trained deep ordinal regression for end-to-end video anomaly detection. In: Conference on Computer Vision and Pattern Recognition, pp. 12173–12182 (2020)

10. Yu, G., Wang, S., Cai, Z., Liu, X., Xu, C., Wu, C.: Deep anomaly discovery from unlabeled videos via normality advantage and self-paced refinement. In: Conference on Computer Vision and Pattern Recognition, pp. 13987–13998 (2022)

11. Zaheer, M.Z., Mahmood, A., Khan, M.H., Segu, M., Yu, F., Lee, S.I.: Generative cooperative learning for unsupervised video anomaly detection. In: Conference on Computer Vision and Pattern Recognition, pp. 14744–14754 (2022)

12. Kingma, D.P., Welling, M.: Auto-encoding variational Bayes. arXiv preprint arXiv:1312.6114 (2013)

13. Ronneberger, O., Fischer, P., Brox, T.: U-Net: convolutional networks for biomedical image segmentation. In: Navab, N., Hornegger, J., Wells, W.M., Frangi, A.F. (eds.) MICCAI 2015. LNCS, vol. 9351, pp. 234–241. Springer, Cham (2015). https://doi.org/10.1007/978-3-319-24574-4_28

14. Fan, Y., Wen, G., Li, D., Qiu, S., Levine, M.D., Xiao, F.: Video anomaly detection and localization via gaussian mixture fully convolutional variational autoencoder. Comput. Vis. Image Underst. **195**, 102920 (2020)

15. Rombach, R., Blattmann, A., Lorenz, D., Esser, P., Ommer, B.: High-resolution image synthesis with latent diffusion models. In: Conference on Computer Vision and Pattern Recognition, pp. 10684–10695 (2022)

16. Ho, J., Jain, A., Abbeel, P.: Denoising diffusion probabilistic models. Adv. Neural. Inf. Process. Syst. **33**, 6840–6851 (2020)

17. Song, J., Meng, C., Ermon, S.: Denoising diffusion implicit models. arXiv preprint arXiv:2010.02502 (2020)

18. Preechakul, K., Chatthee, N., Wizadwongsa, S., Suwajanakorn, S.: Diffusion autoencoders: toward a meaningful and decodable representation. In: IEEE/CVF Conference on Computer Vision and Pattern Recognition, pp. 10619–10629 (2022)

19. Cai, Z., Vasconcelos, N.: Cascade R-CNN: delving into high quality object detection. In: Conference on Computer Vision and Pattern Recognition, pp. 6154–6162 (2018)

20. Ilg, E., Mayer, N., Saikia, T., Keuper, M., Dosovitskiy, A., Brox, T.: FlowNet 2.0: evolution of optical flow estimation with deep networks. In: Conference on Computer Vision and Pattern Recognition, pp. 2462–2470 (2017)

21. Dhariwal, P., Nichol, A.: Diffusion models beat GANs on image synthesis. Adv. Neural. Inf. Process. Syst. **34**, 8780–8794 (2021)

22. Wu, Y., He, K.: Group normalization. In: European Conference on Computer Vision, pp. 3–19 (2018)

23. Li, W., Mahadevan, V., Vasconcelos, N.: Anomaly detection and localization in crowded scenes. IEEE Trans. Pattern Anal. Mach. Intell. **36**(1), 18–32 (2013)

24. Lu, C., Shi, J., Jia, J.: Abnormal event detection at 150 FPS in MATLAB. In: International Conference on Computer Vision, pp. 2720–2727 (2013)

25. Luo, W., Liu, W., Gao, S.: A revisit of sparse coding based anomaly detection in stacked RNN framework. In: International Conference on Computer Vision, pp. 341–349 (2017)

26. Huyan, N., Quan, D., Zhang, X., Liang, X., Chanussot, J., Jiao, L.: Unsupervised outlier detection using memory and contrastive learning. IEEE Trans. Image Process. **31**, 6440–6454 (2022)

Prototypical Transformer for Weakly Supervised Action Segmentation

Tao Lin[1,4], Xiaobin Chang[2,4,5(✉)], Wei Sun[1], and Weishi Zheng[3,4]

[1] School of Electronics and Information Technology,
Sun Yat-sen University, Guangzhou, China
lint39@mail2.sysu.edu.cn, sunwei@mail.sysu.edu.cn
[2] School of Artificial Intelligence, Sun Yat-sen University, Guangzhou, China
[3] School of Computer Science and Engineering,
Sun Yat-sen University, Guangzhou, China
changxb3@mail.sysu.edu.cn, wszheng@ieee.org
[4] Key Laboratory of Machine Intelligence and Advanced Computing,
Ministry of Education, Beijing, China
[5] Guangdong Key Laboratory of Big Data Analysis and Processing,
Guangzhou 510006, People's Republic of China

Abstract. Weakly supervised action segmentation aims to recognize the sequential actions in a video, with only action orderings as supervision for model training. Existing methods either predict the action labels to construct discriminative losses or segment the video based on action prototypes. In this paper, we propose a novel Prototypical Transformer (ProtoTR) to alleviate the defects of existing methods. The motivation behind ProtoTR is to further enhance the prototype-based method with more discriminative power for superior segmentation results. Specifically, the Prediction Decoder of ProtoTR translates the visual input into action ordering while its Video Encoder segments the video with action prototypes. As a unified model, both the encoder and decoder are jointly optimized on the same set of action prototypes. The effectiveness of the proposed method is demonstrated by its state-of-the-art performance on different benchmark datasets.

Keywords: Video Segmentation · Action Recognition · Weak Supervision · Dynamic Time Warping

1 Introduction

Action segmentation aims to recognize different actions performed one after another in a video and localize their temporal spans by segmenting out the start and end timestamps. The fully supervised setting [6,14,17,25] requires frame-wise action labels for model training. Its labelling cost could be huge. On the contrary, the temporal orders of different actions in a video can be represented as the action ordering and obtained at a much lower price than the frame-wise labels. In this work, we focus on handling the weakly supervised action

© The Author(s), under exclusive license to Springer Nature Singapore Pte Ltd. 2024
Q. Liu et al. (Eds.): PRCV 2023, LNCS 14430, pp. 195–206, 2024.
https://doi.org/10.1007/978-981-99-8537-1_16

segmentation problem with such action orderings as weak supervision [4, 5, 9, 12, 15, 16, 18–22].

Many weakly supervised action segmentation methods [5, 15, 16, 19, 22] aim to recognize the action in each frame, as did under the fully supervised setting [6, 14, 17, 25]. However, the action orderings fail to provide direct supervision for the frame-wise model training. To construct discriminative learning objectives with the frame-wise predictions and action ordering weak supervision, recent approaches adopt the Viterbi and Hidden Markov Model (HMM) algorithms [12, 15, 18, 20], or calculate their mutual consistency [22]. Although competitive performance can be obtained by these methods, two shortcomings [22] are highlighted: (1) Over-segmentation, i.e., the adjacent frames may be classified into several different classes by error. (2) Such models rely on pseudo frame-wise labels for training. However, the generation process of pseudo labels could be unstable. In DP-DTW [4], a different paradigm was proposed. Instead of inferring the frame-wise labels, it treats the video segmentation problem as the alignment between video frames and action prototypes via Dynamic Time Warping (DTW) [1] algorithm, where each prototype is a learnable representative sequence of the corresponding action. The limitations of the previous paradigm are thus largely avoided. However, the visual model and action prototypes of DP-DTW are learned only with the hinge loss and distance loss. The direct optimization based on the action ordering supervision is not included. Therefore, the learned model could be less discriminative and leads to compromised performance.

In this paper, we aim to improve the prototype-based method with a novel Prototypical Transformer (ProtoTR), as shown in Fig. 1. Inspired by sequence-to-sequence (Seq2Seq) learning [8, 10, 24], a Prediction Decoder is designed in ProtoTR to translate the video into action ordering. As a unified model, the action prototypes are not only used to segment the video but also utilized as the input of the Prediction Decoder for action ordering prediction. Both the encoder and decoder are then jointly optimized with the same set of action prototypes. With the help of the proposed ProtoTR, more discriminative action prototypes and Video Encoder can be learned. Superior segmentation results are thus achieved by the proposed method. The main contributions of this work are summarized as follows:

1. A novel Prototypical Transformer (ProtoTR) is proposed to explicitly exploit the weak supervision by translating the video into action ordering. It helps to learn the more discriminative action prototypes and Video Encoder for superior weakly supervised action segmentation performance.
2. The proposed ProtoTR is a unified model. The action prototypes are not only used to segment the video but also utilized as the input of the Prediction Decoder to predict the action ordering. The whole model is jointly optimized in an end-to-end manner for superior results.
3. The proposed method achieves state-of-the-art performance on two large-scale weakly supervised action segmentation benchmark datasets and surpasses existing methods with clear margins.

2   Related Work

Most of the recent works for weakly supervised action segmentation are Viterbi and Hidden Markov Model (HMM) based methods [12,15,16,18–21]. Kuehne et al. [15] proposed an iteratively refined HMM model. It generates and refines frame-wise pseudo labels at each epoch. Following Kuehne et al. [15], the approaches, e.g., [12], using HMM with Recurrent Neural Network (RNN) are proposed. Different from the two-stage methods, Richard et al. [20] introduced the Neural Network Viterbi (NNV) to jointly generate and refine the pseudo labels. Based on NNV, the Constrained Discriminative Forward Loss (CDFL) [18,21] is proposed to boost the performance. Besides the Viterbi and HMM-based methods, the mutual consistency [22] between the frame-wise label prediction and the action ordering supervision is computed and used as a learning objective. However, these approaches can suffer from over-segmentation and unstable pseudo-label generation.

Following a different paradigm, DP-DTW [4] formulates video segmentation as the alignment between video frames and the concatenated action prototypes, where each prototype is a learnable representative sequence of the corresponding action. To find the optimal alignment, Dynamic Time Warping (DTW) [1] is used. As such alignment is monotonic and no pseudo label is required, the shortcomings of the previous paradigm are avoided. Although D^3TW [3] also utilizes DTW, it still belongs to the previous paradigm as its segmentation is based on frame-wise predictions and without action prototypes.

However, the visual model and action prototypes of DP-DTW are learned only with the hinge loss and distance loss. The direct optimization based on the action ordering supervision is not included. Therefore, the learned model could be less discriminative and leads to compromised performance. In this paper, a novel Prototypical Transformer (ProtoTR) is proposed to learn more discriminative action prototypes and Video Encoder with action ordering learning. As a unified model, the encoder and decoder of ProtoTR are jointly optimized with the same set of action prototypes and lead to superior segmentation performance.

3   Preliminaries

3.1   Notations

Let $X_{1:T} = [x_1, \ldots, x_T]$ denotes a video with T frames, the goal of action segmentation is to classify each frame of a given video $X_{1:T}$ into frame-wise label $\hat{Y}_{1:T} = [\hat{y}_1, \ldots, \hat{y}_T]$, where $\hat{y}_t \in C$ and C is the set of action classes. In weakly supervised action segmentation, only the action ordering (weak supervision) $A_{1:N} = [a_1, \ldots, a_N]$, $a_n \in C$, rather than the full supervision $Y_{1:T}$ is provided for training. In other words, N is usually less than T and the lengths of action ordering $L_{1:N} = [l_1, \ldots, l_N]$ are unknown. For each specific action class $c \in C$, there is a corresponding action prototype $p^c \in \mathbb{R}^{d_p \times \tau_p}$, where d_p and τ_p are the feature dimension and temporal length of prototype respectively.

3.2   Dynamic Time Warping

Dynamic Time Warping (DTW) [1] is widely used to measure the similarity between time series. Given two temporal sequences $s \in \mathbb{R}^{m \times \tau_1}$ and $s' \in \mathbb{R}^{m \times \tau_2}$ with the same m-dimensional space, DTW seeks the temporal alignment that minimizes Euclidean distance between s and s',

$$\mathrm{DTW}(s, s') = \min_{\pi \in \mathcal{W}(s,s')} \left(\sum_{(i,j) \in \pi} d(s_i, s'_j)^2 \right)^{\frac{1}{2}}, \tag{1}$$

where $\pi = [(i_1, j_1), \dots, (i_{\tau_1}, j_{\tau_2})]$ is the alignment path consisted of index pairs, and $\mathcal{W}(s, s')$ is the set of all the possible paths.

The outputs of DTW are denoted as

$$\hat{\pi}, \hat{d} = \mathrm{DTW}(s_1, s_2), \tag{2}$$

where $\hat{\pi}$ is the optimal alignment path between s_1 and s_2 with minimum Euclidean distance $\hat{d}$.

3.3   Attention Mechanism

Attention [23] is a fundamental mechanism used by ProtoTR. Given the keys K and values V, the attention output of query vector q is,

$$\mathrm{Attention}(q, K, V) = \mathrm{Softmax}(\frac{qK^\top}{\sqrt{d_k}})V, \tag{3}$$

where d_k is the feature dimension of the keys.

4   Prototypical Transformer

4.1   Video Encoder

The proposed ProtoTR is illustrated in Fig. 1. An input video $X_{1:T} = [x_1, ..., x_T]$ is first fed into the Video Encoder to extract the frame-wise features $F_{1:T} = [f_1, ..., f_T] \in \mathbb{R}^{d_f \times T}$,

$$F_{1:T} = \mathrm{Encoder}(X_{1:T}), \tag{4}$$

where d_f is the dimension of each frame feature, and it is identical to the feature dimension d_p of action prototype.

4.2   Action Ordering Prediction

Different from the traditional Seq2Seq model using individual embedding layers to convert the tokens to high-dimensional vectors, the action tokens in ProtoTR are converted to vectors of dimension d_{model} using corresponding action prototypes. Specifically, to predict the action ordering of an input video, the Prediction Decoder takes both the frame-wise features $F_{1:T}$ extracted from Video

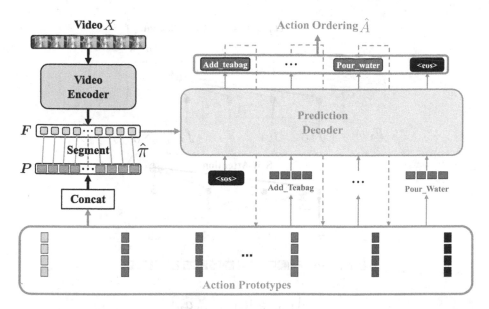

Fig. 1. Detailed illustration of our Prototypical Transformer (ProtoTR) architecture. A Video X is first fed into the Video Encoder and extracted as frame-wise feature $\boldsymbol{F}$. Then the Prediction Decoder takes both the tokens embedded by action prototypes and $\boldsymbol{F}$ as input and predicts the action ordering $\hat{A}$ of the input video. Finally, the optimal alignment path $\hat{\pi}$ between $\boldsymbol{F}$ and $\boldsymbol{P}$ is searched by DTW to segment the video. Different modules are centred on action prototypes and optimized jointly.

Encoder and action tokens as input. The action tokens can be used to index the corresponding action prototypes. The detailed procedure is depicted in Fig. 2. The first token for prediction is the start-of-sequence (sos) token. The prediction decoder aims to forecast the action $\hat{a}_{n+1}$ at step $n + 1$ based on the previous predictions $\hat{A}_{1:n} = [\hat{a}_1, \ldots, \hat{a}_n]$ till step n. $\hat{A}_{1:n}$ is used to index the corresponding action prototypes and obtain $\boldsymbol{E}_{1:n} = [\boldsymbol{p}^{<\text{sos}>}, \boldsymbol{p}^{\hat{a}_1}, \ldots, \boldsymbol{p}^{\hat{a}_n}] \in \mathbb{R}^{d_p \times \tau_p \times (n+1)}$, where $\boldsymbol{p}^{<\text{sos}>} \in \mathbb{R}^{d_p \times \tau_p}$ is the prototype of sos token. The average pooling is then applied to reduce the dimension of τ_p,

$$\overline{\boldsymbol{E}}_{1:n} = \text{AvgPooling}(\boldsymbol{E}_{1:n}), \tag{5}$$

$\overline{\boldsymbol{E}}_{1:n} = [\overline{\boldsymbol{p}^{<\text{sos}>}}, \overline{\boldsymbol{p}^{\hat{a}_1}}, \ldots, \overline{\boldsymbol{p}^{\hat{a}_n}}] \in \mathbb{R}^{d_p \times (n+1)}$. As the next action is closely related to the previous ones, the self-attention mechanism is thus used,

$$\$\left(\boldsymbol{p}^{\hat{a}_n}\right) = \text{Attention}\left(\overline{\boldsymbol{p}^{\hat{a}_n}}, \overline{\boldsymbol{E}}_{1:n}, \overline{\boldsymbol{E}}_{1:n}\right), \tag{6}$$

where $\overline{\boldsymbol{p}^{\hat{a}_n}}$ corresponds to the query vector and $\overline{\boldsymbol{E}}_{1:n}$ serves as the keys and values of Eq. (3). $\$(\boldsymbol{p}^{\hat{a}_n})$ encodes the self-attention information within $\hat{A}_{1:n}$. The cross-attention relations between the action ordering and visual content can be modelled as,

$$\mathbb{C}\left(\boldsymbol{p}^{\hat{a}_n}\right) = \text{Attention}\left(\$\left(\boldsymbol{p}^{\hat{a}_n}\right), \boldsymbol{F}_{1:T}, \boldsymbol{F}_{1:T}\right). \tag{7}$$

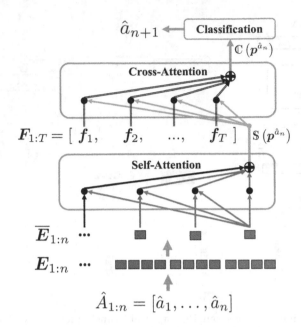

Fig. 2. Predicting the next action $\hat{a}_{n+1}$ based on the partially predicted action ordering $\hat{A}_{1:n}$ at previous step n. Common modules of the transformer like skip connection and feed-forward network are omitted here for clarity.

where $\$(p^{\hat{a}_n})$ serves as the query and the video feature $F_{1:T}$ services as the keys and values. $\mathbb{C}(p^{\hat{a}_n})$ is then fed into a classification layer to predict the next action label $\hat{a}_{n+1}$,

$$\hat{a}_{n+1} = \text{Classify}\left(\mathbb{C}\left(p^{\hat{a}_n}\right)\right). \tag{8}$$

With $\hat{a}_{n+1}$, the action ordering $\hat{A}_{n+1} = [\hat{a}_1, \ldots, \hat{a}_n, \hat{a}_{n+1}]$ at next step $n+1$ is predicted.

4.3   Segmentation

To segment a given video X, the video encoder is used to extract the frame-wise features $F_{1:T}$, as detailed in Sect. 4.1. The Prediction Decoder then takes both $F_{1:T}$ and action tokens as input and predicts the action orderings $\hat{A}_{1:N}$, as described in Sect. 4.2. According to the predicted action ordering $\hat{A}_{1:N} = [\hat{a}_1, \ldots, \hat{a}_N]$, the corresponding action prototypes are concatenated on the temporal dimension and resulting in $P_{1:N}$,

$$P_{1:N} = \text{Concat}([p^{\hat{a}_1}, \ldots, p^{\hat{a}_N}]). \tag{9}$$

The DTW (Eq. (2)) is then utilized to find the optimal alignment path $\hat{\pi}$ between the frame-wise feature $F_{1:T}$ and concatenated prototypes sequence $P_{1:N}$, where each prototype is a learnable representative sequence of the corresponding action.

As $\boldsymbol{F}_{1:T}$ is corresponding to the input video frames and $\boldsymbol{P}_{1:N}$ are concatenated one after another according to the predicted action ordering, $\hat{\pi}$ is the optimal alignment between input video frames and predicted action ordering. In other words, the predicted action labels are assigned to the video frames via $\hat{\pi}$. With such frame-wise action labels obtained, the video is segmented.

4.4 Training

The predicted action ordering $\hat{A}$ can be explicitly supervised by the ground truth ordering A via the cross-entropy loss,

$$\mathcal{L}_{CE} = \mathrm{CrossEntropy}(\hat{A}, A). \tag{10}$$

To learn the action segmentation, the hinge loss $\mathcal{L}_h$ and distance loss $\mathcal{L}_D$ in [4] is utilized. Given a training video, the ground truth action ordering is defined as positive ordering A^+, while all the other action orderings in the training set belong to negative ordering A^-. The corresponding positive and negative concatenated prototypes can then be obtained via Eq. (9), denoted as $\boldsymbol{P}^+$ and $\boldsymbol{P}^-$ respectively. The positive and negative distance $\hat{d}^+$, $\hat{d}^-$ between frame-wise features $\boldsymbol{F}$ and $\boldsymbol{P}^+$, $\boldsymbol{P}^-$ are calculated by DTW Eq. (2) respectively,

$$\hat{\pi}^+, \hat{d}^+ = \mathrm{DTW}(\boldsymbol{F}, \boldsymbol{P}^+), \tag{11}$$

$$\hat{\pi}^-, \hat{d}^- = \mathrm{DTW}(\boldsymbol{F}, \boldsymbol{P}^-). \tag{12}$$

The hinge loss $\mathcal{L}_h$ is designed to train the $\boldsymbol{F}$ to be closer to its positive concatenated prototype $\boldsymbol{P}^+$ than the negative ones with margin $\delta \geq 0$. Specifically,

$$\mathcal{L}_h = \sum_{q=1}^{Q} \max(0, \hat{d}^+ - \hat{d}^{-,q} + \delta), \tag{13}$$

where Q is the total number of negative samples used. The distance loss $\mathcal{L}_D$ is the positive distance,

$$\mathcal{L}_D = \hat{d}^+. \tag{14}$$

Therefore, the overall loss $\mathcal{L}_{w\_seg}$ of ProtoTR consists of three parts,

$$\mathcal{L}_{w\_seg} = \mathcal{L}_{CE} + \alpha\mathcal{L}_h + \beta\mathcal{L}_D, \tag{15}$$

where α, β are the balancing hyper-parameters.

5 Experiments

5.1 Datasets and Settings

Datasets. Two benchmark datasets for weakly supervised action segmentation are introduced. **Breakfast** [13] dataset contains $1,712$ videos of preparing breakfast at 18 different home kitchens, and the frames are annotated with 48 different action classes. On average, there are about 7 action instances per video. Four splits in the benchmark setting [13] are used for cross-validation and the averaged results are reported. **Hollywood Extended** [2] dataset contains 937 videos taken from Hollywood movies with 16 different action classes. On average, there are about 6 action instances per video. Ten splits in the benchmark setting [19] are used for cross-validation and the averaged results are reported.

Protocols. Weakly supervised action segmentation contains two benchmark sub-tasks, *segmentation* and *alignment*. During testing, the action ordering of a given test video is not available for *segmentation*, but it is provided for *alignment*. The performance is evaluated with three criteria [14,19]: frame accuracy (F-acc), intersection over union (IoU) and intersection over detection (IoD).

Implementation Details. The Prediction Decoder of ProtoTR follows the one-layer one-head transformer architecture for both the self-attention and cross-attention modules. The model is initialized following the Xavier procedures presented in [7] and trained from scratch. The hidden feature dimensions across different modules are fixed at 64. The action prototypes of ProtoTR are with temporal length t_p equals to 8 and 6 for Breakfast and Hollywood datasets respectively. Each prototype is implemented with an individual fully connected layer and initialized as the medoid of corresponding action class. The pre-computed video features in [19,20] are used for fair comparisons. The learning objective $\mathcal{L}_{w\_seg}$ of Eq. (15) is with $\alpha = 2.0$ and $\beta = 0.025$ for all experiments. The model is end-to-end trained by the Adam optimizer [11] with a learning rate of 0.001 and a mini-batch size of 64 for 6,000 iterations.

5.2 Main Results

Comparisons among different methods under the *segmentation* setting are shown in Table 1. The proposed ProtoTR consistently achieves the best performance.

Table 1. Comparisons among different methods under the *segmentation* setting.

Method	Breakfast			Hollywood		
	F-acc	IoU	IoD	F-acc	IoU	IoD
HMM+RNN [19]	33.3	-	-	-	11.9	-
NN-Viterbi [20]	43.0	-	-	-	-	-
D³TW [3]	45.7	-	-	33.6	-	-
MuCon [22]	48.5	-	-	41.6	-	-
CDFL [18]	50.2	33.7	45.4	45.0	19.5	25.8
TransAcT [21]	53.2	-	-	47.7	-	-
DP-DTW [4]	50.8	35.6	45.1	55.6	33.2	43.3
ProtoTR	**57.5**	**40.6**	**51.5**	**57.8**	**35.5**	**45.3**

It improves the state-of-the-art DP-DTW with more than 5.0% margins of all metrics on the Breakfast dataset. On the Hollywood dataset, ProtoTR can outperform its counterpart, DP-DTW, with non-trivial margins and achieves clearly better results than other SOTA methods. The qualitative segmentation results of ProtoTR and DP-DTW are shown and compared in Fig. 3. The action ordering from DP-DTW is flawed with one action missed. ProtoTR returns the matched action ordering and thus results in more accurate segmentation than that of DP-DTW.

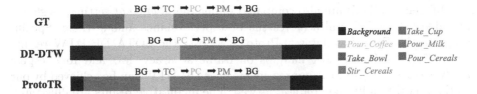

Fig. 3. Qualitative results on the Breakfast dataset under the *segmentation* setting. Texts in the first row represent the action ordering and the bars in the second row indicate the action segmentation. GT denotes the ground truth. Best viewed in colors. (Color figure online)

Table 2. Comparisons among different methods under the *alignment* setting.

Method	Breakfast			Hollywood		
	F-acc	IoU	IoD	F-acc	IoU	IoD
HMM+RNN [19]	-	-	47.3	-	-	46.3
NN-Viterbi [20]	-	-	-	-	-	48.7
D³TW [3]	57.0	-	56.3	59.4	-	50.9
MuCon [22]	-	-	66.2	-	-	52.3
CDFL [18]	63.0	45.8	63.9	64.3	40.5	52.9
TransAcT [21]	65.5	-	-	64.8	-	
DP-DTW [4]	67.7	50.8	66.5	66.4	46.8	61.7
ProtoTR	**74.4**	**55.3**	**70.5**	**68.8**	**49.4**	**64.5**

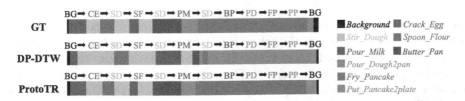

Fig. 4. Qualitative results on the Breakfast dataset under the *alignment* setting. Texts in the first row represent the action ordering and the bars in the second row indicate the action segmentation. GT denotes the ground truth. Best viewed in colors. (Color figure online)

Table 3. Ablation study of ProtoTR with Breakfast dataset.

Method	Segmentation			Alignment		
	F-acc	IoU	IoD	F-acc	IoU	IoD
DP-DTW	50.8	35.6	45.1	67.7	50.8	66.5
BasicTR	54.6	38.7	50.1	72.0	53.9	69.4
ProtoTR	**57.5**	**40.6**	**51.5**	**74.4**	**55.3**	**70.5**

Different methods are also compared under the *alignment* setting, as shown in Table 2. ProtoTR still achieves the best performance among them. Specifically, on the Breakfast dataset, ProtoTR boosts the SOTA DP-DTW results with 6.7%, 4.5% and 4.0% margins on frame accuracy, IoU and IoD respectively. Consistent and clear improvements over DP-DTW can be achieved by our method on the Hollywood dataset as well. The qualitative results are illustrated in Fig. 4. The performance by ProtoTR is more accurate than the DP-DTW one.

5.3   Ablation Study

As described in Sect. 4, ProtoTR is a unified model with the encoder and decoder jointly optimized with the same set of action prototypes. The basic transformer (BasicTR) is a competitor to ProtoTR. BasicTR is with additional independent action embeddings rather than the unified set of prototypes as in ProtoTR. Specifically, action tokens are converted to vectors by the additional independent action embeddings in BasicTR, while they are represented by corresponding action prototypes in ProtoTR. As shown in Table 3, ProtoTR consistently achieves superior results to BasicTR under all settings and metrics. The effectiveness of joint learning on the same action prototypes is demonstrated. Moreover, noticeable improvements are obtained by BasicTR over DP-DTW. It demonstrates that with the help of the Prediction Decoder, more discriminative prototypes and the Video Encoder can be learned to boost performance.

6   Conclusions

In this paper, we propose a novel Prototypical Transformer (ProtoTR) to exploit weak supervision by translating the video into action ordering. As a unified framework, different modules of ProtoTR are jointly optimized on the same set of action prototypes. With more discriminative action prototypes and visual encoder learned, superior action segmentation results are thus obtained by ProtoTR. The proposed method is evaluated on two weakly supervised benchmark datasets and achieves better results than existing methods with clear margins.

Limitation and future work: The proposed method is built upon the naive attention mechanism with high latency. As a future direction, the model efficiency can be improved by using more advanced transformer architectures.

Acknowledgement. This work was supported by the National Science Foundation for Young Scientists of China (62106289).

References

1. Berndt, D.J., Clifford, J.: Using dynamic time warping to find patterns in time series. In: KDD Workshop, Seattle, WA, USA, vol. 10, pp. 359–370. (1994)
2. Bojanowski, P., et al.: Weakly supervised action labeling in videos under ordering constraints. In: Fleet, D., Pajdla, T., Schiele, B., Tuytelaars, T. (eds.) ECCV 2014. LNCS, vol. 8693, pp. 628–643. Springer, Cham (2014). https://doi.org/10.1007/978-3-319-10602-1_41
3. Chang, C.Y., Huang, D.A., Sui, Y., Fei-Fei, L., Niebles, J.C.: D3TW: discriminative differentiable dynamic time warping for weakly supervised action alignment and segmentation. In: Proceedings of the IEEE/CVF Conference on Computer Vision and Pattern Recognition, pp. 3546–3555. IEEE (2019)
4. Chang, X., Tung, F., Mori, G.: Learning discriminative prototypes with dynamic time warping. In: Proceedings of the IEEE/CVF Conference on Computer Vision and Pattern Recognition, pp. 8395–8404. IEEE (2021)
5. Ding, L., Xu, C.: Weakly-supervised action segmentation with iterative soft boundary assignment. In: Proceedings of the IEEE/CVF Conference on Computer Vision and Pattern Recognition, pp. 6508–6516 (2018)
6. Farha, Y.A., Gall, J.: MS-TCN: multi-stage temporal convolutional network for action segmentation. In: Proceedings of the IEEE/CVF Conference on Computer Vision and Pattern Recognition, pp. 3575–3584 (2019)
7. Glorot, X., Bengio, Y.: Understanding the difficulty of training deep feedforward neural networks. In: Proceedings of the Thirteenth International Conference on Artificial Intelligence and Statistics, pp. 249–256. JMLR Workshop and Conference Proceedings (2010)
8. Guo, J., Xue, W., Guo, L., Yuan, T., Chen, S.: Multi-level temporal relation graph for continuous sign language recognition. In: Yu, S., et al. (eds.) PRCV 2022, Part III. LNCS, vol. 13536, pp. 408–419. Springer, Cham (2022). https://doi.org/10.1007/978-3-031-18913-5_32
9. Huang, D.-A., Fei-Fei, L., Niebles, J.C.: Connectionist temporal modeling for weakly supervised action labeling. In: Leibe, B., Matas, J., Sebe, N., Welling, M. (eds.) ECCV 2016. LNCS, vol. 9908, pp. 137–153. Springer, Cham (2016). https://doi.org/10.1007/978-3-319-46493-0_9
10. Ji, P., Yang, B., Zhang, T., Zou, Y.: Consensus-guided keyword targeting for video captioning. In: Yu, S., et al. (eds.) PRCV 2022, Part III. LNCS, vol. 13536, pp. 270–281. Springer, Cham (2022). https://doi.org/10.1007/978-3-031-18913-5_21
11. Kingma, D.P., Ba, J.: Adam: a method for stochastic optimization. arXiv preprint arXiv:1412.6980 (2014)
12. Koller, O., Zargaran, S., Ney, H.: Re-sign: re-aligned end-to-end sequence modelling with deep recurrent CNN-HMMS. In: Proceedings of the IEEE/CVF Conference on Computer Vision and Pattern Recognition, pp. 4297–4305. IEEE (2017)
13. Kuehne, H., Arslan, A.B., Serre, T.: The language of actions: recovering the syntax and semantics of goal-directed human activities. In: Proceedings of the IEEE/CVF Conference on Computer Vision and Pattern Recognition, pp. 780–787. IEEE (2014)

14. Kuehne, H., Gall, J., Serre, T.: An end-to-end generative framework for video segmentation and recognition. In: 2016 IEEE Winter Conference on Applications of Computer Vision (WACV), pp. 1–8. IEEE (2016)
15. Kuehne, H., Richard, A., Gall, J.: Weakly supervised learning of actions from transcripts. Comput. Vis. Image Underst. **163**, 78–89 (2017)
16. Kuehne, H., Richard, A., Gall, J.: A hybrid RNN-HMM approach for weakly supervised temporal action segmentation. IEEE Trans. Pattern Anal. Mach. Intell. **42**(4), 765–779 (2018)
17. Lea, C., Flynn, M.D., Vidal, R., Reiter, A., Hager, G.D.: Temporal convolutional networks for action segmentation and detection. In: Proceedings of the IEEE/CVF Conference on Computer Vision and Pattern Recognition, pp. 156–165 (2017)
18. Li, J., Lei, P., Todorovic, S.: Weakly supervised energy-based learning for action segmentation. In: IEEE/CVF International Conference on Computer Vision, pp. 6242–6250. IEEE (2019)
19. Richard, A., Kuehne, H., Gall, J.: Weakly supervised action learning with RNN based fine-to-coarse modeling. In: Proceedings of the IEEE/CVF Conference on Computer Vision and Pattern Recognition, pp. 754–763. IEEE (2017)
20. Richard, A., Kuehne, H., Iqbal, A., Gall, J.: NeuralNetwork-Viterbi: a framework for weakly supervised video learning. In: Proceedings of the IEEE/CVF Conference on Computer Vision and Pattern Recognition, pp. 7386–7395. IEEE (2018)
21. Ridley, J., Coskun, H., Tan, D.J., Navab, N., Tombari, F.: Transformers in action: weakly supervised action segmentation. arXiv preprint arXiv:2201.05675 (2022)
22. Souri, Y., Fayyaz, M., Minciullo, L., Francesca, G., Gall, J.: Fast weakly supervised action segmentation using mutual consistency. IEEE Trans. Pattern Anal. Mach. Intell. **44**(10), 6196–6208 (2021)
23. Vaswani, A., et al.: Attention is all you need. In: Advances in Neural Information Processing Systems, vol. 30 (2017)
24. Yang, B., Zhang, T., Zou, Y.: Clip meets video captioning: Concept-aware representation learning does matter. In: Yu, S., et al. (eds.) PRCV 2022, Part I. LNCS, vol. 13534, pp. 368–381. Springer, Cham (2022). https://doi.org/10.1007/978-3-031-18907-4_29
25. Zhao, Y., Xiong, Y., Wang, L., Wu, Z., Tang, X., Lin, D.: Temporal action detection with structured segment networks. In: Proceedings of the IEEE/CVF Conference on Computer Vision and Pattern Recognition, pp. 2914–2923 (2017)

Unimodal-Multimodal Collaborative Enhancement for Audio-Visual Event Localization

Huilin Tian, Jingke Meng$^{(\boxtimes)}$, Yuhan Yao, and Weishi Zheng

School of Computer Science and Engineering,
Sun Yat-sen University, Guangzhou, China
mengjke@gmail.com

Abstract. Audio-visual event localization (AVE) task focuses on localizing audio-visual events where event signals occur in both audio and visual modalities. Existing approaches primarily emphasize multimodal (*i.e.* audio-visual fused) feature processing to capture high-level event semantics, while overlooking the potential of unimodal (*i.e.* audio or visual) features in distinguishing unimodal event segments where only the visual or audio event signal appears within the segment. To overcome this limitation, we propose the Unimodal-Multimodal Collaborative Enhancement (UMCE) framework for audio-visual event localization. The framework consists of several key steps. Firstly, audio and visual features are enhanced by multimodal features, and then adaptively fused to further enhance the multimodal features. Simultaneously, the unimodal features collaborate with multimodal features to filter unimodal events. Lastly, by considering the collaborative emphasis on event content at both the segment and video levels, a dual interaction mechanism is established to exchange information, and video features are utilized for event classification. Experimental results demonstrate the significant superiority of our UMCE framework over state-of-the-art methods in both supervised and weakly supervised AVE settings.

Keywords: Audio-visual event · Action localization · Attention

1 Introduction

Audio-visual event localization (AVE) is a significant and challenging task introduced by Tian *et al.* [18]. Its objective is to develop a model capable of accurately localizing the position of audio-visual events and predicting their corresponding categories. An audio-visual event segment is defined as containing the event that is both visible and audible, exemplified by the first three segments in Fig. 1. As

Supported by National Science Foundation for Young Scientists of China (62206315), China Postdoctoral Science Foundation (2022 M 713574) and Fundamental Research Funds for the Central Universities, Sun Yat-sen University (23 ptpy 112).

Q. Liu et al. (Eds.): PRCV 2023, LNCS 14430, pp. 207–219, 2024.
https://doi.org/10.1007/978-981-99-8537-1_17

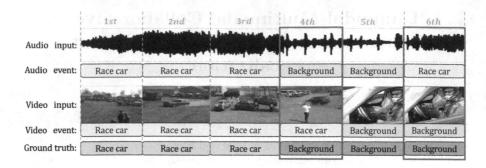

Fig. 1. An illustration example of the audio-visual event localization task.

illustrated in the figure, video segments are labeled as audio-visual events only when they exhibit concurrent visual and audio event signals. Over the past years, the AVE task has garnered growing attention, leading researchers to explore various methods [7,21,22,27] of inter-modal and intra-modal interaction between audio and visual content.

There are two types of segments labeled as background: pure background segments where neither visual nor audio modalities contain event information, such as the 5th segment in Fig. 1, and unimodal event segments where only the visual or audio event signal appears within the segment, indicated by the red regions in Fig. 1. Unimodal event segments, compared to pure background segments, contain event information and are harder to be identified as background, posing a significant challenge in AVE task. Previous studies [21–23] often struggle to distinguish unimodal events due to their reliance on audio-visual features for audio-visual event localization. We attribute this mislocalization to the inherent limitation of fused features in discerning the modality of event occurrence, as the visual and audio information within the fused features is extensively blended.

To overcome these limitations, we propose the **U**nimodal-**M**ultimodal **C**ollaborative **E**nhancement (UMCE) framework for audio-visual event localization. This framework promotes the synergistic use of unimodal and multimodal features. Given the independent nature of audio and visual features, we utilize them to identify unimodal events. Meanwhile, the audio-visual fused features exhibit higher robustness in predicting the multimodal event category at the video level. Recognizing the inadequate content of unimodal features and the comprehensive information contained in audio-visual fused features, we employ the latter as guidance to enhance the event information in the former. Moreover, we integrate audio/visual features to obtain fused features through multimodal adaptive fusion. Furthermore, by considering the collaborative emphasis on event content at both segment and video levels, we establish a dual interaction for exchanging information and employ video features for event classification.

The major contributions of this work are summarized as follows:

(1) To address the challenges posed by unimodal events, we introduce the unimodal-multimodal event-aware interaction module and collaborative uni-

modal event filter. These modules enhance unimodal and multimodal features and leverage them to achieve precise localization of audio-visual events.

(2) We introduce the dual-interactive multimodal event classification to enhance event information of video- and segment-level features for event classification.

(3) We integrate these modules and propose a unimodal-multimodal collaborative enhancement network, which outperforms the state-of-the-art methods by a large margin in both supervised and weakly-supervised AVE settings.

2   Related Works

2.1   Audio Visual Event Localization

In recent years, there have been extensive studies and improvements on the audio-visual event localization (AVE) task, which can be divided into two main directions. Most of methods explored the cross-modal feature fusion for better audio-visual representation. Xu et al. [23] proposed CMRA network, which interacts audio and visual feature for spatial-level and channel-level cross-modal attention. M2N proposed by Wang et al. [20] try to learn relations among audio, visual, and fusion features. Wu et al. [21] proposed dual attention matching (DAM) module. Yu et al. [27] used MPN framework to perceive global semantics and unmixed local information. Liu et al. [9] proposed BMFN to adjust unimodal features and fusion features with forward-backward fusion modules. Meanwhile, many other works focused on how to emphasize event information and suppress background noise. The positive sample propagation (PSP) [28] exploited relevant audio-visual paired features and prunes negative and weak connections. And Xia et al. [22] utilized cross-modal background suppression on the time-level and event-level. In contrast to existing models, our proposed UMCE method employs a collaborative approach between unimodal and fused features to identify unimodal events. Moreover, by capturing event-aware segment-level and video-level information, our method achieves precise localization.

2.2   Attention Mechanism

Attention mechanism is motivated by the human processing information method, which attends to the most relevant part of features and suppresses noise. Attention mechanism has many variants [1,6,11,19,25], which has been applied to several vision tasks like point clouds [12,26], action prediction [16], pose estimation [17] and etc. In addition to incorporating inter- and intra-modal attention mechanisms as employed by previous methods [21–23], our approach introduces attention between unimodal and audio-visual fused features. Specifically, we enhance the unimodal features by leveraging guidance from the fused features, thereby amplifying the event signals within each modality. Simultaneously, the fused features dynamically capture cross-modal event information, facilitating a comprehensive understanding of the audio-visual context.

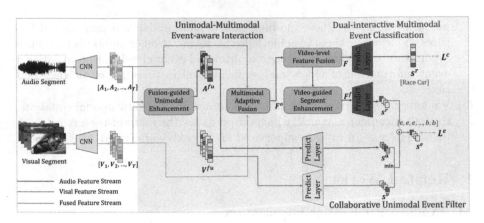

Fig. 2. Architecture of UMCE. Taking pre-extracted audio and visual features as inputs, the unimodal-multimodal event-aware interaction module is applied to enhance audio and visual features and fuse them adaptively for audio-visual features. Then, the collaborative unimodal event filter is used to identify unimodal events Finally, the dual-interactive multimodal event classification is deployed to exchange event information between segment and video for event classification.

3   Method

Our UMCE network comprises three components, as presented in Fig. 2. The detailed architecture is elaborated in the following subsections.

3.1   Problem Notations

Each video from the AVE dataset [18] is represented as: $S = (A_t, V_t)_{t=1}^{T}$, where T indicates the length of video, $A_t \in \mathbb{R}^{d_a}$ and $V_t \in \mathbb{R}^{H \times W \times d_v}$ respectively stand for the audio and visual features of the t-th segment, d_a and d_v are audio and video dimensions. In the supervised setting, the labels are segment-level on C event categories and a background category. And only the video-level labels are provided in the weakly-supervised setting. In the following formulations, σ denotes the sigmoid function, and $\odot$ means element-wise multiplication.

3.2   Unimodal-Multimodal Event-Aware Interaction

Fusion-Guided Unimodal Enhancement. Compared to fused multimodal features, the information provided by unimodal features is usually inadequate for event identification. In this section, the audio and visual features are capable of capturing information from the complementary modality to enhance the understanding of overall event semantics, while maintaining the independence of each modality.

Specifically, after audio-guided attention [22] and self-attention [8,19], we calculate the preliminary multimodal features: $F^{pre} = (A^{self} + V^{self})/2$, where

$A^{self}, V^{self} \in \mathbb{R}^{T \times d}$. Subsequently, we utilize F^{pre} as the event guidance to enhance A^{self} and V^{self} and obtain fusion-guided unimodal features V^{fu} and A^{fu}. For example, the details for calculating V^{fu} can be formulated as follows. First, we need to obtain the fusion-guided visual attention map $M_f^v \in \mathbb{R}^{T \times d}$, which indicates channel-wise feature weights of V^{self}:

$$M_f^v = \sigma(((F^{pre}U_f \odot V^{self}U_u)U_1)U_2), \tag{1}$$

where $U_f, U_u \in \mathbb{R}^{d \times d_h}$, $U_1 \in \mathbb{R}^{d_h \times d}$ and $U_2 \in \mathbb{R}^{d \times d}$ are fully-connected layers with ReLU And the fusion-guided visual features are calculated as:

$$V^{fu} = M_f^v \odot V^{self} + \alpha \cdot V^{self}, \tag{2}$$

where α is a hyperparameter. Similarly, A^{fu} can be performed in the same manner. By utilizing the guidance of fused features, fusion-guided visual and audio features $V^{fu}, A^{fu} \in \mathbb{R}^{T \times d}$ can enhance relevant event information and diminish the presence of irrelevant background noise for better localization.

Multimodal Adaptive Fusion. In order to enhance the fusion of robust event information derived from unimodal features, this module dynamically allocates attention to audio and visual modalities, prioritizing the relevant information in each segment. Referring to [22], the event importance scores are calculated based on the corresponding unimodal features A^{self} and V^{self}, respectively:

$$g^a = \sigma(W_g^a \cdot A^{self}), \quad g^v = \sigma(W_g^v \cdot V^{self}), \tag{3}$$

where $W_g^a, W_g^v \in \mathbb{R}^{d \times 1}$. g^a and g^v represent the degree of event information contained in the respective modalities. Hence, they can be utilized to determine the attention weights for the fusion of multimodal features: $w_t^a, w_t^v = Softmax(g_t^a, g_t^v)$. And the adaptive intra-segment fused features are calculated by weighted sum between w^a, w^v and unimodal features:

$$F_t^{ai} = A_t^{self} \odot w_t^a + V_t^{self} \odot w_t^v, \tag{4}$$

It is also essential to obtain the inter-segment event information. Following [23], we obtain the inter-segment fused features with cross-modal attention. The details of cross-modal attention (CMA) are formulated as:

$$\mathbf{CMA}(q, k) = Softmax(\frac{QK^T}{\sqrt{d_m}})V, \tag{5}$$

$$Q = qW^Q, \quad K = kW^K, \quad V = kW^V,$$

where W^Q, W^K and $W^V \in \mathbb{R}^{d \times d}$. Finally, the inter-segment and intra-segment features are combined as the segment-level fused features:

$$m_{av} = A^{self} \odot V^{self}, \quad c_{av} = Concat(A^{self}, V^{self}),$$
$$F^o = \mathbf{CMA}(m_{av}, c_{av}) + \beta \cdot F^{ai}, \tag{6}$$

Thus the fused features contain multi-scale audio-visual event information, enhancing the model's ability to comprehend events accurately and facilitate correct video classification. The significance of both inter- and intra-segment cross-modal features for localization is discussed in detail in the experimental section.

3.3 Collaborative Unimodal Event Filter

After fusion-guided unimodal enhancement, the enhanced unimodal features V^{fu}, A^{fu} can be utilized for the preliminary identification of the modality in which the events occur. In this subsection, we introduce a straightforward yet effective filter that leverages the collaboration between unimodal features and audio-visual fused features for localization.

First, we emphasize the event segments and suppress background segments with unimodal importance scores g^a, g^v calculated on Eq. (3), and obtain importance weighted unimodal features V^g and A^g, which are formulated as follows: $A^g = A^{fu} + (g^a + g^v) \cdot A^{fu}$, $V^g = V^{fu} + (g^a + g^v) \cdot V_t^{fu}$. In order to filter out unimodal events, we predict the audio and visual event scores as follows:

$$s^a = \sigma(A^g \cdot W_{uni}^a), \quad s^v = \sigma(V^g \cdot W_{uni}^v), \tag{7}$$

where $W_{uni}^a, W_{uni}^v \in \mathbb{R}^{d \times 1}$. Unimodal event scores reflect the likelihood of event occurrence in their respective modalities. Therefore, by getting the smaller value between s^a and s^v, we can effectively reduces the score of audio-visual events.

Meanwhile, we utilize the segment-level fused features from Eq. (12) to calculate the fused event scores:

$$s^r = \sigma(F^f W^r), \tag{8}$$

where $W^r \in \mathbb{R}^{d \times 1}$. Finally $s^r \in \mathbb{R}^T$ and $s^{av} \in \mathbb{R}^T$ collaborate to calculate the eventness scores s^e, which is formulated as:

$$s^e = (1 - \gamma) \cdot s^r + \gamma \cdot min(s^a, s^v), \tag{9}$$

where γ is a hyperparameter. The eventness score represents the probability that an audio-visual event exists in the segment, which is used for localization.

3.4 Dual-Interactive Multimodal Event Classification

Cross-Modal Video-Level Feature Fusion. In addition to segment-level eventness prediction, video-level event classification is also essential for precise audio-visual event localization. Hence we integrate event information from all segments within the video to obtain comprehensive video-level features. Specifically, we first calculate the segment-level multimodal importance scores: $g^f = \sigma(W_g^f \cdot F^o)$, where $W_g^f \in \mathbb{R}^{1 \times d}$. And we use them as weights to obtain video-level audio-visual features with weighted sum: $F = \sum_{t=1}^{T}(F_t^o \cdot g_t^f)$. The final video-level audio-visual features focus on event-related information cross segments and are employed to predict video event category scores $s^c \in \mathbb{R}^C$:

$$s^c = Softmax(FW_c), \tag{10}$$

Video-Guided Segment Enhancement. Compared to audio-visual features at the segment level, the video-level features F are the aggregation of all segment features within each video, thereby providing a richer context of event information.

Therefore, we employ video-level features to provide guidance for segment features, directing their attention toward event-related information. We give the details as follows:

$$M_t^g = \sigma(((FU_g \odot F_t^o U_s)U_3)U_4), \tag{11}$$

where $U_g, U_s \in \mathbb{R}^{d \times d_h}$, $U_3 \in \mathbb{R}^{d_h \times d}$ and $U_4 \in \mathbb{R}^{d \times d}$ are fully-connected layers with ReLU. M_t^g adaptively emphasizes event information in t th segment. The final segment-level fused features are calculated as:

$$F_t^f = F_t^o \odot M_t^g, \tag{12}$$

and F^f is used for the calculation of eventness in Eq. (8).

3.5   Localization and Objective Function

After the above modules, we obtain two outputs: video-level event category scores s^c (Eq. (9)) and segment-level eventness scores s^e (Eq. (10)). These scores are crucial for the final prediction and model optimization.

To train our model on the supervised setting, first we decouple the ground truth of video Y^f to two labels: video-level event category label $c = \{c^k, c^k \in \{0, 1\}, k = 0, ..., C, \sum_{k=1}^{C} c^k = 1\}$ and segment-level eventness label $e \in \mathbb{R}^T$. Based on these two labels, the overall objective function is formulated as follows:

$$L_{fully} = L^c + \frac{1}{T} \sum_{t=1}^{T} L_t^e, \tag{13}$$

where L_c denotes the cross entropy loss between category scores s^c and GT label c, and L_t^e is binary cross entropy loss between the eventness score s_t^e with GT label e_t. During the inference phase, if $s_t^e > 0.5$, the t-th video segment is classified as belonging to the s^c class. Conversely, if the value falls below the threshold, the t-th video segment is labeled as background.

In the weakly-supervised setting, considering only video-level event category labels are available during training, we cannot supervise the eventness scores directly. So we combine the eventness scores with segment-level event category scores for training, which is formulated as:

$$s^{wc} = F^o W_{wc}, \quad s^w = Softmax(\sum_{t=1}^{T}(s_t^{wc} \odot s_t^e)), \tag{14}$$

$$L_{weakly} = L^c + L^{wc},$$

where $W_{wc} \in \mathbb{R}^{d \times C}$ and L^{wc} is cross entropy loss of between s^w and GT label c. After training, we generate segment-level pseudo labels on the training set, which is used to retrain our model with the supervised setting.

4   Experiments and Analysis

4.1   Experiment Setup

The Audio-Visual Event Dataset. The Audio-Visual Event Dataset, proposed by Tian *et al.* [18], is the sole publicly accessible dataset designed for the audio-visual event localization task. It is derived from AudioSet [2] and comprised of 4143 videos, each spanning a duration of 10 s. These videos encompass 28 diverse event categories, such as Guitar, Frying food, Church bell, *etc.*

Evaluation Metrics. Following previous works [5,18,21,23,28], we exploit the overall segment classification accuracy on C events and one background categories as the evaluation metric. we utilize the integration of the final segment-level eventness s^e and the video-level event category scores s^c as the event prediction for evaluation.

Implementation Details. We adopt the VGG-19 [15] model pre-trained on ImageNet [4] to extract segment-level visual features. Similarly, the 128-dimensional audio features are obtained by VGG-like [3] model pre-trained on AudioSet [2]. The training batch size is set to 64, while the test batch size is 32. Following the approach in previous works [18,22], we initialize the learning rate to 7×10^{-4} initially and gradually decay it by multiplying with 0.5 at epochs 10, 20 and 30.

Table 1. Comparisons with state-of-the-arts in supervised and weakly-supervised manners on AVE dataset.

Models	Reference	Supervised	Weakly-supervised
AVEL (audio) [18]	ECCV 2018	59.5	53.4
AVEL (visual) [18]	ECCV 2018	55.3	52.9
AVSDN [5]	ICASSP 2019	72.6	66.8
AVEL [18]	ECCV 2018	72.7	66.8
CMAN [24]	AAAI 2020	73.3	66.7
DAM [21]	ICCV 2019	74.5	70.4
AVRB [14]	WACV 2020	74.8	68.9
AVIN [13]	ICASSP 2020	75.2	69.4
AVT [7]	ACCV 2020	76.8	70.2
CMRAN [23]	ACMMM 2020	77.4	72.9
PSP [28]	CVPR 2021	77.8	73.5
BMFN [9]	ICASSP 2022	78.7	74.0
CMBS [22]	CVPR 2022	79.3	74.2
DCMR [10]	TMM 2022	79.6	74.3
UMCE		**80.1**	**75.1**

4.2   Comparisons with State-of-the-Art Methods

As presented in Table 1, we conduct a comprehensive comparison between our proposed UMCE network and several previous methods. The results demonstrate that our approach achieves superior performance in both supervised and weakly-supervised settings. Specifically, in the supervised setting, when compared to the top two AVE methods, UMCE surpasses CMBS [22] and DCMR [10] by 0.8% and 0.5%, respectively. Moreover, in the weakly-supervised setting, our method achieves the highest performance of 75.1%, surpassing CMBS [22] by 0.9% and DCMR [10] by 0.8%.

Table 2. Ablation study of different modules on AVE dataset

Models	Supervised	Weakly-supervised
w/o FGUE	78.13	73.93
w/o MAF	79.45	73.55
w/o CUEF	79.65	74.92
w/o VFF	79.08	74.62
w/o VSE	78.61	74.57
UMCE	**80.10**	**75.07**

4.3   Ablation Studies

We validate the effectiveness of each component of our method in Table 2, including fusion-guided unimodal enhancement (FGUE), multimodal adaptive fusion (MAF), collaborative unimodal event filter (CUEF), cross-modal video-level feature fusion (VFF) and video-guided segment enhancement (VSE). The removal of any module leads to a decrease in localization accuracy, indicating the significant impact of each module on localization capability. Moving forward, we will delve into a discussion on the effectiveness of the structural design of certain modules.

Impact of Fusion-Guided Unimodal Enhancement. As illustrated in Table 3, we compare our method with several different settings of fusion-guided unimodal enhancement module, where "w/o FGUE" refers to deleting this module entirely, and "audio attention" and "visual attention" present making fusion-guided attention only on audio or visual features respectively. The effect of fusion-guided enhancement to multimodal features is better than using one modal features alone. We confirm that these two pieces of attention are indispensable for the UGFA.

The Effectiveness of the Multimodal Adaptive Fusion. In the multimodal adaptive fusion module, fused feature F^o is comprised of two components: one is the intra-segment feature derived from the cross-modal attention [23]; the other

Table 3. Ablation studies on unimodal-multimodal event-aware interaction. (a) is on fusion-guided unimodal enhancement (FGUE). (b) is on multimodal adaptive fusion (MAF).

Models	Supervised	Weakly-supervised
w/o FGUE	78.13	73.93
audio attention	79.23	74.05
visual attention	78.51	74.77
FGUE	**80.10**	**75.07**

(a)

Models	Supervised	Weakly-supervised
w/o MAF	79.45	73.55
w/o CMRA[23]	78.56	74.23
MAF+mean	**80.10**	**75.07**

(b)

Table 4. Ablation study on collaborative unimodal event filter.

Models	Supervised	Weakly-supervised
unimodal branch	78.56	74.90
fusion branch	79.65	74.92
unimodal+fusion branch	**80.10**	**75.07**

is the inter-segment feature obtained from the multimodal adaptive fusion. In Tabel 3, The "w/o MAF" denotes that the model directly assigns features from the cross-modal attention [23] to F^o, while "w/o CMRA" means utilize MAF only. The above-conducted ablation experiments demonstrate the necessity of both intra- and inter-segment information in audio-visual event localization.

Influence of Unimodal and Fusion Filter Branch. As for collaborative unimodal event filter, we conduct an ablation study to verify the necessity of unimodal and fused event scores, which compares with following different settings: only use unimodal event scores ("unimodal branch"), only use fused event scores(fusion branch) and use both scores ("unimodal + fusion branch"). As illustrated in Tabel 4, utilize both scores yields significantly better performance compared to using either unimodal event scores or fused event scores alone. This demonstrates the indispensability of both unimodal and fused event scores for accurate event localization.

Influence of Hyper-parameters. Figure 3 shows the experimental results obtained by varying the hyperparameters, namely α from Eq. (2), β from Eq. (6), and γ from Eq. (9). The optimal performance for supervised setting is achieved when $\alpha = 0.7$, $\beta = 0.3$ and $\gamma = 0.9$. Similarly, for weakly-supervised event localization, the best performance is achieved when $\alpha = 0.7$, $\beta = 0.2$ and $\gamma = 0.4$.

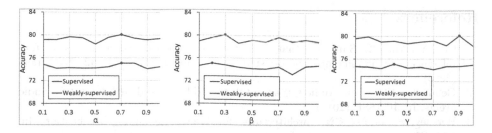

Fig. 3. Comparison of different hyperparameter values.

Audio input:										
Audio event scores:	0.01	0.31	1.00	1.00	1.00	1.00	1.00	1.00	0.90	0.86
Video input:										
Video event scores:	0.00	0.00	0.13	0.00	1.00	1.00	0.47	0.21	0.01	0.01
Fused event scores:	0.00	0.00	1.00	0.00	1.00	1.00	1.00	0.00	0.00	0.00
Eventness scores:	0.00	0.00	0.32	0.00	0.99	1.00	0.41	0.02	0.00	0.00
Ground truth:	BG	BG	BG	BG	Bark	Bark	BG	BG	BG	BG

Fig. 4. Qualitative results of our model on dog bark event. The red region stands for the result we predict. (Color figure online)

4.4   Qualitative Analysis

Figure 4 presents a qualitative example that showcases the efficacy of our proposed UMCE network. Although the fused event scores are capable of identifying audio-visual events, they are prone to misclassifying unimodal events as audio-visual events. Conversely, the unimodal event scores accurately pinpoint event occurrences within each modality and filter out unimodal event segments. By considering both the unimodal event scores and fused event scores, the final eventness scores mitigate interference caused by unimodal events and enable precise localization of audio-visual events.

5   Conclusion

In this paper, we propose the Unimodal-Multimodal Collaborative Enhancement (UMCE) network to address the mislocalization of unimodal events. Our approach focuses on enhancing event information in both audio/visual and fused features through unimodal-multimodal event-aware interaction. Then we leverage the event semantics of unimodal features to identify unimodal events and assist fused features in the localization. Additionally, our network facilitates information exchange between segment- and video-level fused features to improve event classification. Experimental results demonstrate the superiority of our UMCE approach in both supervised and weakly-supervised settings.

References

1. Cao, Y., Min, X., Sun, W., Zhai, G.: Attention-guided neural networks for full-reference and no-reference audio-visual quality assessment. TIP **32**, 1882–1896 (2023)
2. Gemmeke, J.F., et al.: Audio set: an ontology and human-labeled dataset for audio events. In: ICASSP (2017)
3. Hershey, S., et al.: CNN architectures for large-scale audio classification. In: ICASSP (2017)
4. Krizhevsky, A., Sutskever, I., Hinton, G.E.: ImageNet classification with deep convolutional neural networks. Commun. ACM **60**, 84–90 (2017)
5. Lin, Y., Li, Y., Wang, Y.F.: Dual-modality seq2seq network for audio-visual event localization. In: ICASSP (2019)
6. Lin, Y.B., Tseng, H.Y., Lee, H.Y., Lin, Y.Y., Yang, M.H.: Exploring cross-video and cross-modality signals for weakly-supervised audio-visual video parsing. NIPS (2021)
7. Lin, Y.B., Wang, Y.C.F.: Audiovisual transformer with instance attention for audio-visual event localization. In: ACCV (2020)
8. Lin, Z., et al.: A structured self-attentive sentence embedding. In: ICLR (2017)
9. Liu, S., Quan, W., Liu, Y., Yan, D.: Bi-directional modality fusion network for audio-visual event localization. In: ICASSP (2022)
10. Liu, S., Quan, W., Wang, C., Liu, Y., Liu, B., Yan, D.M.: Dense modality interaction network for audio-visual event localization. TMM (2022)
11. Mercea, O.B., Riesch, L., Koepke, A., Akata, Z.: Audio-visual generalised zero-shot learning with cross-modal attention and language. In: CVPR (2022)
12. Qin, S., Li, Z., Liu, L.: Robust 3D shape classification via non-local graph attention network. In: CVPR (2023)
13. Ramaswamy, J.: What makes the sound?: A dual-modality interacting network for audio-visual event localization. In: ICASSP (2020)
14. Ramaswamy, J., Das, S.: See the sound, hear the pixels. In: WACV (2020)
15. Simonyan, K., Zisserman, A.: Very deep convolutional networks for large-scale image recognition. In: ICLR (2015)
16. Stergiou, A., Damen, D.: The wisdom of crowds: temporal progressive attention for early action prediction. In: CVPR (2023)
17. Tang, Z., Qiu, Z., Hao, Y., Hong, R., Yao, T.: 3D human pose estimation with spatio-temporal criss-cross attention. In: CVPR (2023)
18. Tian, Y., Shi, J., Li, B., Duan, Z., Xu, C.: Audio-visual event localization in unconstrained videos. In: ECCV (2018)
19. Vaswani, A., et al.: Attention is all you need. In: NIPS (2017)
20. Wang, H., Zha, Z., Li, L., Chen, X., Luo, J.: Multi-modulation network for audio-visual event localization. CoRR (2021)
21. Wu, Y., Zhu, L., Yan, Y., Yang, Y.: Dual attention matching for audio-visual event localization. In: ICCV (2019)
22. Xia, Y., Zhao, Z.: Cross-modal background suppression for audio-visual event localization. In: CVPR (2022)
23. Xu, H., Zeng, R., Wu, Q., Tan, M., Gan, C.: Cross-modal relation-aware networks for audio-visual event localization. In: ACM MM (2020)
24. Xuan, H., Luo, L., Zhang, Z., Yang, J., Yan, Y.: Discriminative cross-modality attention network for temporal inconsistent audio-visual event localization. TIP **30**, 7878–7888 (2021)

25. Xuan, H., Zhang, Z., Chen, S., Yang, J., Yan, Y.: Cross-modal attention network for temporal inconsistent audio-visual event localization. In: AAAI (2020)
26. Yang, J., et al.: Modeling point clouds with self-attention and gumbel subset sampling. In: CVPR (2019)
27. Yu, J., Cheng, Y., Feng, R.: MPN: multimodal parallel network for audio-visual event localization. In: ICME (2021)
28. Zhou, J., Guo, D., Wang, M.: Contrastive positive sample propagation along the audio-visual event line. TPAMI (2023)

Dual-Memory Feature Aggregation
for Video Object Detection

Diwei Fan[1,2,3], Huicheng Zheng[1,2,3(✉)], and Jisheng Dang[1,2,3]

[1] School of Computer Science and Engineering,
Sun Yat-Sen University, Guangzhou, China
zhenghch@mail.sysu.edu.cn
[2] Key Laboratory of Machine Intelligence and Advanced Computing,
Ministry of Education, Guangzhou, China
[3] Guangdong Province Key Laboratory of Information Security Technology,
Guangzhou, China

Abstract. Recent studies on video object detection have shown the advantages of aggregating features across frames to capture temporal information, which can mitigate appearance degradation, such as occlusion, motion blur, and defocus. However, these methods often employ a sliding window or memory queue to store temporal information frame by frame, leading to discarding features of earlier frames over time. To address this, we propose a dual-memory feature aggregation framework (DMFA). DMFA simultaneously constructs a local feature cache and a global feature memory in a feature-wise updating way at different granularities, i.e., pixel level and proposal level. This approach can partially preserve key features across frames. The local feature cache stores the spatio-temporal contexts from nearby frames to boost the localization capacity, while the global feature memory enhances semantic feature representation by capturing temporal information from all previous frames. Moreover, we introduce contrastive learning to improve the discriminability of temporal features, resulting in more accurate proposal-level feature aggregation. Extensive experiments demonstrate that our method achieves state-of-the-art performance on the ImageNet VID benchmark.

Keywords: video object detection · feature aggregation · temporal information · global memory · local feature cache

1 Introduction

As the basic task of computer vision, image object detection has achieved remarkable progress. Recent research efforts have focused on video tasks in real-world scenarios, which often face appearance degradation caused by occlusion, motion blur, and defocus. To overcome these challenges, existing video object detection (VOD) methods have attempted to aggregate temporal information for feature enhancement.

© The Author(s), under exclusive license to Springer Nature Singapore Pte Ltd. 2024
Q. Liu et al. (Eds.): PRCV 2023, LNCS 14430, pp. 220–232, 2024.
https://doi.org/10.1007/978-981-99-8537-1_18

Specifically, the local temporal information is crucial in improving the localization performance, while the global temporal information is beneficial for augmenting the feature semantic representation. FGFA [28] and STMN [26] enhance the key frame by propagating pixel-level features within a local range. SELSA [25] and LRTR [21] leverage the long-range temporal relationship between the proposals to aggregate features. MEGA [1] is the first to model both local and global temporal information across proposal-level features, significantly boosting detection accuracy. However, these methods only use a single granularity of temporal information for feature enhancement. Without pixel-level temporal alignment, frame variations would lead to erroneous proposals that decrease the proposal-level feature aggregation performance. On the other hand, ignoring proposal-level enhancement would prevent the establishment of fine-grained spatio-temporal correlations between objects. Besides, none of the aforementioned methods fully exploit the temporal information from all previous frames, as they rely on a frame-wise first-in-first-out (FIFO) updating structure that discards all information from earlier frames due to GPU memory limitations.

In this paper, we propose a dual-memory feature aggregation framework (DMFA) for pixel-level and proposal-level feature enhancement. Concretely, dual memory consists of a local feature cache that stores spatio-temporal contexts from nearby frames and a global feature memory that captures temporal semantic features from all previous frames. To accommodate comprehensive temporal information, we adopt a feature-wise updating strategy rather than the frame-wise updating approach used in FIFO structures, which enables the partial preservation of features across frames. Through these designs, we extract sufficient temporal information for effective aggregation at different granularities, leading to high-quality spatio-temporal consistent feature representations that benefit subsequent detection. Besides, we employ contrastive learning to explore the temporal correlations among object features, enhancing their discriminability without introducing additional computational complexity. This approach makes the features of the same objects closer in the embedding space and those of different objects farther away. The contributions can be summarized as follows:

1) We propose a dual-memory feature aggregation framework that simultaneously captures local spatio-temporal contexts and global temporal information from all previous frames to enhance pixel-level and proposal-level features.
2) We introduce contrastive learning into the VOD task to enhance the discriminability of features and further improve detection performance.
3) Without bells and whistles, our DMFA achieves state-of-the-art performance on the commonly-adopted ImageNet VID benchmark.

2 Related Work

2.1 Image Object Detection

Current image object detection methods can be divided into one-stage and two-stage methods. One-stage methods are generally faster but less accurate, as they directly predict object category probabilities and position coordinates. The corresponding representative works include SSD [17], RetinaNet [16], and YOLO [18], etc. Two-stage methods prioritize accuracy over speed. They first generate candidate regions and refine them to achieve accurate detection results. Related works include Fast R-CNN [4], Faster R-CNN [20], and Mask R-CNN [10], etc.

2.2   Video Object Detection

Although video object detection is sensitive to appearance degradation, e.g., partial occlusion, motion blur, and defocus, abundant temporal information from other video frames may help improve detection performance. Existing VOD methods can be categorized into box-level and feature-level methods.

Box-level methods use additional tracking models or post-processing techniques to model the temporal associations between detected bounding boxes. For example, T-CNN [15] incorporates a tracker to model object motion and refines the results through temporal convolution. Seq-NMS [7] re-scores boxes belonging to the same object sequence by computing the intersection over union (IoU) between the bounding boxes. However, these methods introduce high computational costs or cannot serve as an end-to-end framework.

In contrast, feature-level methods attempt to aggregate temporal features from other frames to enhance the current frame features end-to-end. They generally associate features between frames in one of two ways: optical flow-based and attention-based methods. Specifically, DFF [29] and FGFA [28] introduce the optical flow predicted by FlowNet [3] to align motion features across frames. Due to the heavy optical flow computational cost, a series of attention-based methods have been proposed. SELSA [25] exploits a multi-frame semantic similarity strategy to aggregate high-level proposal features. RDN [2] and MEGA [1] leverage relation modules [13] to capture spatial-temporal contexts across frames to augment object features. MAMBA [22] introduces a unified feature enhancement module to aggregate multi-level features. QueryProp [8] speeds up the aggregation process through a lightweight framework. Moreover, TransVOD [12] applies transformer architecture to VOD and achieves remarkable results.

3   Method

We first present an overview of the notations commonly used in VOD and the relation module [13] in Sect. 3.1. In Sect. 3.2, we explain the architecture of our dual memory and its pivotal role in feature aggregation. Additionally, we illustrate how to employ contrastive learning to improve the temporal correlations of object features in Sect. 3.3.

3.1   Preliminary

As illustrated in Fig. 1, we denote the current t-th frame and its corresponding pixel-level features as I_t and F_t, respectively. Through RPN and ROI pooling

[20], we obtain the proposal-level features P_t. Our goal is to enhance F_t and P_t by our dual memory, thereby improving the performance of subsequent classification and regression.

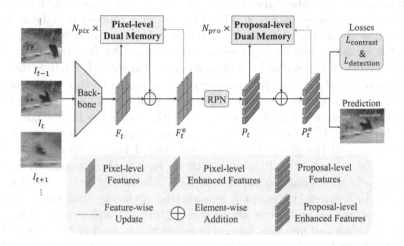

Fig. 1. Overview of our DMFA framework. Given the current frame I_t, dual memories enhance both pixel-level and proposal-level features. During training, the standard detection loss, i.e., $L_{\text{detection}}$, and our proposed contrastive embedding loss, i.e., L_{contrast}, are combined to update the model parameters.

Relation Module. We employ the relation module to model the relationship between features. Given a set of features $\mathbf{F} = \{f_i\}_{i=1}^{N_c}$ (N_c denotes the number of features), a relation module is applied to enhance each f_i by computing relation features as a weighted sum of features from other features with N multi-attention heads [23]. Specifically, the relation features of f_i from the n-th head are computed as:

$$f_R^{n,*}(f_i, \mathbf{F}) = \sum_j \omega_{ij}^{n,*}(W_V^n f_j), \quad n = 1, \ldots, N \tag{1}$$

where W_V^n denotes a linear transformation matrix. $\omega_{ij}^{n,*}$ is the relation weight between f_i and f_j, which indicates the appearance and optional geometric similarity. According to [1], $* \in \{Z, U\}$ denotes whether the geometric similarity is considered, where Z indicates inclusion and U indicates exclusion. Finally, we obtain the enhanced features of f_i by adding the concatenated relation features from N heads:

$$f_{rn}^*(f_i, \mathbf{F}) = f_i + [f_R^{n,*}(f_i, \mathbf{F})_{n=1}^N]. \tag{2}$$

where $[\cdot]$ and $+$ mean the channel-wise concatenation and element-wise addition.

3.2   Dual-Memory Feature Aggregation

In this part, we discuss how dual memory captures sufficient temporal information for feature aggregation at various granularities.

Dual-Memory Architecture. We provide an overview of our dual memory and its updating procedure in Fig. 2. The dual memory consists of a local feature cache and a global feature memory. The local feature cache $\mathbf{M}_L = \{m_i^L\}_{i=1}^{N_l}$ stores the semantic feature representations from nearby frames and preserves box information for proposal-level enhancement. On the other hand, the global feature memory $\mathbf{M}_G = \{m_i^G\}_{i=1}^{N_g}$ captures rich semantic feature representations from all previous frames. To reduce computational cost while ensuring feature diversity, we randomly sample features from $\mathbf{M}_G$ to form a lightweight feature subset $\mathbf{S}_G = \{s_i^G\}_{i=1}^{N_s}$ for feature aggregation. Here, we denote N_s, N_g, and N_l as the sizes of the global feature subset, global feature memory, and local feature cache, respectively. Previous methods update the memory by a framewise approach, which only considers temporal information from a fixed number of frames. In contrast, we adopt a feature-wise updating strategy to partially preserve features across frames, which can capture a broader range of temporal information across all previous frames. Suppose the amount of the current frame features to be updated is K, we replace the earliest K features in $\mathbf{M}_L$ with them. To accommodate these displaced local features, $\mathbf{M}_G$ randomly removes an equal amount of existing global features.

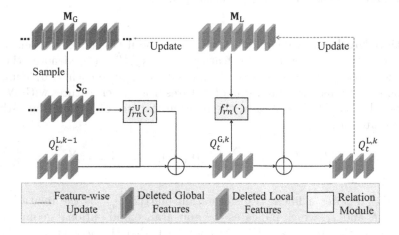

Fig. 2. The architecture of our dual memory. It takes pixel-level features F_t or proposal-level features P_t as input to generate the enhanced features. For convenience, we uniformly define the input as $Q_t^{L,k-1}$, globally-enhanced features as $Q_t^{G,k}$ and locally-enhanced features as $Q_t^{L,k}$. k denotes the k-th iteration.

Feature Aggregation. As illustrated in Fig. 1, dual memories are constructed on the pixel level and proposal level to enhance the features of the current frame.

Given a set of features Q_t (F_t or P_t), we apply the relation module to aggregate the global and local temporal information captured in the dual memories. Specificly, Q_t aggregates sampled global features from $\mathbf{S_G}$ and afterwards, these globally-enhanced features integrate local information from $\mathbf{M_L}$. Please note that a single relation module is challenging to model complex temporal contexts. To fully exploit the potential of temporal information, the enhancement process can be recursively performed for N_p (N_{pix} or N_{pro}) times, which is formulated as follows:

$$Q_t^{G,k} = f_{rn}^U(Q_t^{L,k-1}, \mathbf{S_G}), \quad k = 1, \ldots, N_p \tag{3}$$

$$Q_t^{L,k} = f_{rn}^*(Q_t^{G,k}, \mathbf{M_L}), \quad k = 1, \ldots, N_p \tag{4}$$

where $Q_t^{L,0}$ ($Q_t^{L,k-1}, k = 1$) denotes the input at the first iteration. $Q_t^{G,k}$ and $Q_t^{L,k}$ represent the global and local features enhanced at the k-th iteration, respectively. Additionally, $f_{rn}^U(\cdot)$ and $f_{rn}^Z(\cdot)$ refer to the geometry-free and geometry-based relation modules, respectively, defined in Eq. 2. When performing local feature aggregation in Eq. 4, we utilize the geometry-free relation module at the pixel level and the geometry-based relation module at the proposal level. This is because leveraging the box information preserved in $\mathbf{M_L}$ can further improve the ability to locate the objects. Finally, the output of the N_p-th local feature aggregation is taken as F_t^e or P_t^e.

With the support of iterative dual-memory enhancement, the pixel-level features establish fine-grained temporal consistent representations, enabling the generation of high-quality proposals. Meanwhile, proposal-level features perceive sufficient spatio-temporal information, thus improving the robustness of object features against appearance degradation.

3.3 Contrastive Temporal Correlation

More discriminative features can help distinguish various objects, boosting the quality of temporal correlations. To achieve this, we adopt contrastive learning to encourage gathering features of the same objects in the embedding space while pushing those of different objects farther away. Before updating the enhanced proposal-level features into dual memory, we calculate a contrastive embedding loss, which is denoted as follows:

$$L_{contrast} = \sum_{P_t^e} \log[1 + \sum_{\mathbf{M_L^+}} \sum_{\mathbf{M_L^-}} \frac{\exp(P_t^e \cdot \mathbf{M_L^-})}{\exp(P_t^e \cdot \mathbf{M_L^+})}], \tag{5}$$

where $\exp(a \cdot b)$ measures the similarity between feature embeddings of a and b, P_t^e denotes the enhanced proposal-level feature embeddings, $\mathbf{M_L^+}$ and $\mathbf{M_L^-}$ are positive and negative feature embeddings from the local feature cache. Specifically, we select positive and negative samples by a predefined similarity threshold ε, which is empirically set as 0.6. If the similarity between the feature embeddings is greater than ε, we label them as $\mathbf{M_L^+}$; Otherwise, they are regarded as $\mathbf{M_L^-}$. By minimizing the contrastive embedding loss, our framework enhances the discriminability of feature representations and temporal consistency without adding additional computational complexity.

4 Experiments

4.1 Dataset and Metric

We conduct experiments on the ImageNet VID dataset [19], which is a large-scale public benchmark for VOD. This dataset comprises 3,862 training videos and 555 validation videos. Following the previous protocols in [1,2,22,25], we train our model on the intersection of 30 classes shared by the ImageNet VID and DET datasets. Then we evaluate the performance of our method using the mean average precision (mAP) metric.

4.2 Implementation Details

Backbone and Detection Network. We adopt ResNet-101 [11] as the backbone. Following the standard practice in [1,2,25,28], we modify the stride of the first conv block in the conv5 stage from 2 to 1 in order to increase the resolution of feature maps. To maintain the receptive field, all the 3×3 conv layers are modified by the dilated convolutions with a dilation rate 2. We use Faster R-CNN [20] as the baseline detector. We place the RPN on the top of the conv4 stage. In RPN, we utilize 4 scales $\{64^2, 128^2, 256^2, 512^2\}$ and 3 aspect ratios $\{1:2, 1:1, 2:1\}$ to generate 12 anchors for each spatial location. Besides, non-maximum suppression is applied with an IoU threshold of 0.7 to generate 300 proposals per frame.

Training and Inference Details. During both training and testing, we resize images to a shorter side of 600 pixels. Our models are trained on two Titan X GPUs with SGD. Each GPU processes a mini-batch, including the key frame and a randomly selected frame. The training consists of two stages. In the first stage, we train the pixel-level enhancement for 60K iterations with an initial learning rate 0.001. After 40K iterations, the learning rate is reduced to 0.0001. In the second stage, we train the entire model for 120K iterations, and the learning rate is 0.001 for the first 80K iterations and 0.0001 for the remaining 40K iterations. After detection, the proposal-level dual memory is updated using the top $K = 75$ proposal features with the highest object scores. For the pixel-level dual memory update, $K = 50$ pixel features are randomly sampled within each proposal.

4.3 Main Results

Quantitative Results. Table 1 compares our method with state-of-the-art methods using the ResNet-101 backbone. Our DMFA achieves the best accuracy (mAP). Specifically, our DMFA achieves 83.5% mAP, 8.9% higher than that of the baseline Faster R-CNN [20]. Compared with optical flow-based methods like FGFA [28] and MANet [24], our DMFA performs better, showing improvements of 7.2% and 5.4%, respectively. The reason is that they only adopt local temporal information to align pixel-level features while ignoring the broader range of temporal information. Besides, RDN [2], SELSA [25], and TROI [5] utilize limited temporal information to enhance proposal-level features, resulting in inferior

Table 1. Comparison with state-of-the-art methods on the ImageNet VID validation set. "*" denotes the methods that we re-implemented.

Methods	Publication	Backbone	Detector	mAP	FPS
Faster R-CNN [20]	NIPS-2015	ResNet-101	–	74.6	15.6
FGFA [28]	ICCV-2017	ResNet-101	R-FCN	76.3	1.4
MANet [24]	ECCV-2018	ResNet-101	R-FCN	78.1	4.9
THP [27]	CVPR-2018	ResNet-101&DCN	R-FCN	78.6	13
SELSA [25]	ICCV-2019	ResNet-101	Faster R-CNN	80.3	3.2
RDN [2]	ICCV-2019	ResNet-101	Faster R-CNN	81.8	8.9
MEGA [1]	CVPR-2020	ResNet-101	Faster R-CNN	82.9	5.5
TROI [5]	AAAI-2021	ResNet-101	Faster R-CNN	82.1	–
QueryProp [8]	AAAI-2022	ResNet-101	Sparse R-CNN	82.3	**26.8**
GMLCN [6]	TMM-2022	ResNet-101	Faster R-CNN	78.6	25.2
TSFA [9]	PR-2022	ResNet-101	Faster R-CNN	81.9	7.2
BoxMask [14]	WACV-2023	ResNet-101	Faster R-CNN	82.3	–
TransVOD [12]	PAMI-2023	ResNet-101	DETR	81.9	2.9
MEGA* [1]	CVPR-2020	ResNet-101	Faster R-CNN	81.8	4.5
Ours	–	ResNet-101	Faster R-CNN	**83.5**	6.8

performance than ours. As two strong competitors, MEGA [1] aggregates both global and local information to improve accuracy, while QueryProp [8] uses a sparse query-based framework to achieve a trade-off between accuracy and speed. Our DMFA outperforms MEGA by 0.6% mAP and QueryProp by 1.2% mAP. This is due to our effective utilization of sufficient temporal information at various granularities. Furthermore, we re-implement MEGA on our device for a fair comparison. Our DMFA shows a gain of 1.7% mAP over MEGA, which is a compelling evidence of our state-of-the-art performance.

Qualitative Results. We present the visualization results of MEGA [1] and our DMFA in Fig. 3. Our DMFA outperforms MEGA in handling complex sce-

Table 2. Ablation study of each component in DMFA. "Pixel" denotes pixel-level dual memory, "Proposal" means proposal-level dual memory, "Feature-wise update" denotes the feature-wise updating strategy and "Contrast" indicates the contrastive embedding loss. We use Faster R-CNN as the baseline and default to the frame-wise updating approach.

Methods	Pixel	Proposal	Contrast	Feature-wise update	mAP (%)	FPS
a					74.6	15.6
b		✓			81.5	9.1
c		✓		✓	82.5	9.4
d	✓				80.6	9.5
e	✓	✓		✓	83.1	6.8
f	✓	✓	✓	✓	83.5	6.8

narios involving motion blur and partial occlusion. Specifically, while MEGA fails to detect the lizard under occlusion and assigns lower confidence scores to the rapidly moving cat, DMFA provides accurate classification and localization results. This is attributed to the comprehensive mining and effective utilization of spatio-temporal information by our DMFA.

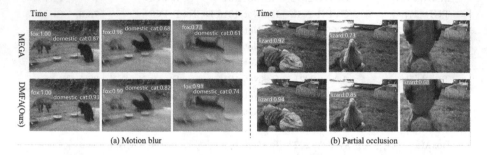

(a) Motion blur   (b) Partial occlusion

Fig. 3. Visual comparisons of two cases between the MEGA [1] and our DMFA. Each bounding box labels a detected object with the class name and corresponding confidence score. Overall, our DMFA exhibits superior robustness against (a) motion blur and (b) partial occlusion.

4.4   Ablation Study

Effectiveness of Each Component. We conduct experiments to analyze the contributions of each component in our DMFA, and the results are presented in Table 2. Method (a) is the baseline detector Faster R-CNN using ResNet-101, which achieves 74.6% overall mAP. For proposal-level enhancement, Method (b) incorporates the frame-wise updating dual memory, while Method (c) introduces the feature-wise updating dual memory. Their performances are improved to 81.5% mAP and 82.5% mAP, respectively, with a 1% mAP gap. This is because the feature-wise updating dual memory can model a wider range of temporal information from all previous frames, leading to more robust proposal-level features. Method (d) introduces the feature-wise updating dual memory at the pixel level into (a), significantly improving 6% mAP. Method (e) utilizes pixel-level and proposal-level dual memory, achieving 83.1% mAP with an improvement of 8.2% mAP over Method (a). This demonstrates the superiority of leveraging sufficient temporal information for feature aggregation at different granularities. Method (f) integrates contrastive embedding loss into (e), which brings a 0.4% mAP improvement. This gain is because contrastive learning can further enhance the discriminability of features. These results demonstrate the effectiveness of each component in DMFA.

Number of Iterations in the Pixel-Level Enhancement. In Table 3a, we adjust the number of iterations in the pixel-level enhancement N_{pix} from 0 to 3. Our method degenerates to the baseline when $N_{\text{pix}} = 0$. With $N_{\text{pix}} = 1$, the accuracy is improved to 80.6% mAP. As the N_{pix} increases, the accuracy reaches a saturation point, and the FPS drops significantly. Therefore, we set N_{pix} to 1 by default.

Number of Iterations in the Proposal-Level Enhancement. Table 3b shows the effect of using the different number of iterations in proposal-level enhancement. As N_{pro} increases from 0 to 3, our method improves accuracy from 74.6% to 82.6% in mAP. To balance the speed-accuracy trade-off, $N_{pro} = 2$ is the default setting.

Table 3. Influence of the number of iterations in the pixel-level enhancement N_{pix}, number of iterations in the proposal-level enhancement N_{pro}, local feature cache size N_l, and global feature subset size N_s.

(a) Number of iterations in the pixel-level enhancement N_{pix}				
N_{pix}	0	1	2	3
mAP (%)	74.6	80.6	80.5	80.5
FPS	15.6	9.5	6.8	5.3
(b) Number of iterations in the proposal-level enhancement N_{pro}				
N_{pro}	0	1	2	3
mAP (%)	74.6	80.7	82.5	82.6
FPS	15.6	11.8	9.4	7.8

(c) Size of local feature cache N_l					
N_l	0	500	1000	1500	2000
mAP (%)	74.6	80.8	81.5	81.2	80.9
FPS	15.6	9.6	9.5	9.5	9.4
(d) Size of global feature subset N_s					
N_s	0	500	1000	2000	5000
mAP (%)	81.5	82.3	82.5	82.5	82.6
FPS	9.5	9.5	9.4	9.2	8.7

Size of the Local Feature Cache. Table 3c&d show the results of experiments using the proposal-level enhancement with the default setting. First, we explore how the size of the local feature cache N_l affects the performance. As shown in Table 3c, we vary N_l from 0 to 2000. When $N_l = 1000$, the accuracy is increased to 81.5% mAP. However, when N_l exceeds 1000, the performance has a downward trend since modeling position information between distant proposals in the temporal dimension may harm the performance. Therefore, we use $N_l = 1000$ by default.

Size of the Global Feature Subset. Regarding the global feature memory, we customize its capacity N_g to 20,000 based on the size of our GPU memory. As discussed in Sect. 3.1, we randomly sample global features to form a feature subset for lightweight aggregation. In Table 3d, we incorporate the default local feature cache mentioned above to test the size of the global feature subset N_s. As the number of sampled features increases, the detection accuracy is improved. For efficient aggregation, we choose $N_s = 1000$ as the default configuration.

5   Conclusion

In this paper, we propose a dual-memory feature aggregation framework, termed DMFA. Specifically, DMFA employs a feature-wise updating strategy to capture sufficient temporal information at various granularities. Then, the dual memory allows accurately utilizing these temporal cues to enhance features. Moreover, the contrastive embedding loss can enhance the discriminability of features without introducing additional computational complexity. Experiments show the state-of-the-art performance of our DMFA on the ImageNet VID dataset.

Acknowledgments. This work was supported in part by the National Natural Science Foundation of China under Grant 61976231 and in part by the Guangdong Basic and Applied Basic Research Foundation under Grant 2019A1515011869 and Grant 2023A1515012853.

References

1. Chen, Y., Cao, Y., Hu, H., Wang, L.: Memory enhanced global-local aggregation for video object detection. In: IEEE Conference on Computer Vision and Pattern Recognition, pp. 10334–10343 (2020)
2. Deng, J., Pan, Y., Yao, T., Zhou, W., Li, H., Mei, T.: Relation distillation networks for video object detection. In: IEEE International Conference on Computer Vision, pp. 7022–7031 (2019)
3. Dosovitskiy, A., et al.: FlowNet: learning optical flow with convolutional networks. In: IEEE International Conference on Computer Vision, pp. 2758–2766 (2015)
4. Girshick, R.B.: Fast R-CNN. In: IEEE International Conference on Computer Vision, pp. 1440–1448 (2015)
5. Gong, T., et al.: Temporal ROI align for video object recognition. In: AAAI Conference on Artificial Intelligence, pp. 1442–1450 (2021)
6. Han, L., Yin, Z.: Global memory and local continuity for video object detection. IEEE Trans. Multimed. (2022). https://doi.org/10.1109/TMM.2022.3164253
7. Han, W., et al.: Seq-NMS for video object detection. arXiv:1602.08465 (2016)
8. He, F., Gao, N., Jia, J., Zhao, X., Huang, K.: QueryProp: object query propagation for high-performance video object detection. In: AAAI Conference on Artificial Intelligence, pp. 834–842 (2022)
9. He, F., Li, Q., Zhao, X., Huang, K.: Temporal-adaptive sparse feature aggregation for video object detection. Pattern Recogn. **127**, 108587 (2022)
10. He, K., Gkioxari, G., Dollár, P., Girshick, R.B.: Mask R-CNN. In: IEEE International Conference on Computer Vision, pp. 2980–2988 (2017)

11. He, K., Zhang, X., Ren, S., Sun, J.: Deep residual learning for image recognition. In: IEEE Conference on Computer Vision and Pattern Recognition, pp. 770–778 (2016)
12. He, L., et al.: TransVOD: end-to-end video object detection with spatial-temporal transformers. IEEE Trans. Pattern Anal. Mach. Intell. **45**, 7853–7869 (2023)
13. Hu, H., Gu, J., Zhang, Z., Dai, J., Wei, Y.: Relation networks for object detection. In: IEEE Conference on Computer Vision and Pattern Recognition, pp. 3588–3597 (2018)
14. Hashmi, K. A., Pagani, A., Stricker, D., Afzal, M. Z.: BoxMask: revisiting bounding box supervision for video object detection. In: IEEE Winter Conference on Applications of Computer Vision, pp. 2029–2039 (2023)
15. Kang, K., et al.: T-CNN: tubelets with convolutional neural networks for object detection from videos. IEEE Trans. Circuits Syst. Video Technol. **28**, 2896–2907 (2018)
16. Lin, T.Y., Goyal, P., Girshick, R.B., He, K., Dollár, P.: Focal loss for dense object detection. IEEE Trans. Pattern Anal. Mach. Intell. **42**, 318–327 (2020)
17. Liu, W., et al.: SSD: single shot multibox detector. In: Leibe, B., Matas, J., Sebe, N., Welling, M. (eds.) ECCV 2016. LNCS, vol. 9905, pp. 21–37. Springer, Cham (2016). https://doi.org/10.1007/978-3-319-46448-0_2
18. Redmon, J., Divvala, S.K., Girshick, R.B., Farhadi, A.: You only look once: unified, real-time object detection. In: IEEE Conference on Computer Vision and Pattern Recognition, pp. 779–788 (2016)
19. Russakovsky, O., et al.: ImageNet large scale visual recognition challenge. Int. J. Comput. Vision **115**, 211–252 (2014)
20. Shaoqing Ren, Kaiming He, R.B.G., Sun, J.: Faster R-CNN: towards real-time object detection with region proposal networks. In: Advances in Neural Information Processing Systems, pp. 91–99 (2015)
21. Shvets, M., Liu, W., Berg, A.C.: Leveraging long-range temporal relationships between proposals for video object detection. In: IEEE International Conference on Computer Vision, pp. 9755–9763 (2019)
22. Sun, G., Hua, Y., Hu, G., Robertson, N.M.: MAMBA: multi-level aggregation via memory bank for video object detection. In: AAAI Conference on Artificial Intelligence, pp. 2620–2627 (2021)
23. Vaswani, A., et al.: Attention is all you need. In: Advances in Neural Information Processing Systems, pp. 5998–6008 (2017)
24. Wang, S., Zhou, Y., Yan, J., Deng, Z.: Fully motion-aware network for video object detection. In: Ferrari, V., Hebert, M., Sminchisescu, C., Weiss, Y. (eds.) ECCV 2018. LNCS, vol. 11217, pp. 557–573. Springer, Cham (2018). https://doi.org/10.1007/978-3-030-01261-8_33
25. Wu, H., Chen, Y., Wang, N., Zhang, Z.: Sequence level semantics aggregation for video object detection. In: IEEE International Conference on Computer Vision, pp. 9216–9224 (2019)
26. Xiao, F., Lee, Y.J.: Video object detection with an aligned spatial-temporal memory. In: Ferrari, V., Hebert, M., Sminchisescu, C., Weiss, Y. (eds.) ECCV 2018. LNCS, vol. 11212, pp. 494–510. Springer, Cham (2018). https://doi.org/10.1007/978-3-030-01237-3_30
27. Zhu, X., Dai, J., Yuan, L., Wei, Y.: Towards high performance video object detection. In: IEEE Conference on Computer Vision and Pattern Recognition, pp. 7210–7218 (2018)

28. Zhu, X., Wang, Y., Dai, J., Yuan, L., Wei, Y.: Flow-guided feature aggregation for video object detection. In: IEEE International Conference on Computer Vision, pp. 408–417 (2017)
29. Zhu, X., Xiong, Y., Dai, J., Yuan, L., Wei, Y.: Deep feature flow for video recognition. In: IEEE Conference on Computer Vision and Pattern Recognition, pp. 4141–4150 (2017)

Going Beyond Closed Sets: A Multimodal Perspective for Video Emotion Analysis

Hao Pu[1], Yuchong Sun[1], Ruihua Song[1(✉)], Xu Chen[1], Hao Jiang[2], Yi Liu[2], and Zhao Cao[2]

[1] Gaoling School of Artificial Intelligence, Renmin University of China, Beijing, China
{puhao,ycsun,xu.chen}@ruc.edu.cn, songruihua_bloon@outlook.com
[2] Huawei Technologies, Beijing, China
{jianghao66,liuyi139,caozhao1}@huawei.com

Abstract. Emotion analysis plays a crucial role in understanding video content. Existing studies often approach it as a closed set classification task, which overlooks the important fact that the emotional experiences of humans are so complex and difficult to be adequately expressed in a limited number of categories. In this paper, we propose **MM-VEMA**, a novel **M**ulti**M**odal perspective for **V**ideo **EM**otion **A**nalysis. We formulate the task as a crossmodal matching problem within a joint multimodal space of videos and emotional experiences (e.g. emotional words, phrases, sentences). By finding experiences that closely match each video in this space, we can derive the emotions evoked by the video in a more comprehensive manner. To construct this joint multimodal space, we introduce an efficient yet effective method that manipulates the multimodal space of a pre-trained vision-language model using a small set of emotional prompts. We conduct experiments and analyses to demonstrate the effectiveness of our methods. The results show that videos and emotional experiences are well aligned in the joint multimodal space. Our model also achieves state-of-the-art performance on three public datasets.

Keywords: Multimodal · Video · Emotion

1 Introduction

With the rapid development of video social media platforms, there has been a tremendous number of user-generated videos being shared and viewed on the internet every day. This makes video content analysis emerge as a hot research topic. In particular, video emotion analysis is essentially important because it has great research value in investigating human's emotional reaction to what they see and thus makes a robot more natural and humanized. Also, it is helpful in many real applications, such as video recommendations [22] and video management on social platforms [37].

Supplementary Information The online version contains supplementary material available at https://doi.org/10.1007/978-981-99-8537-1_19.

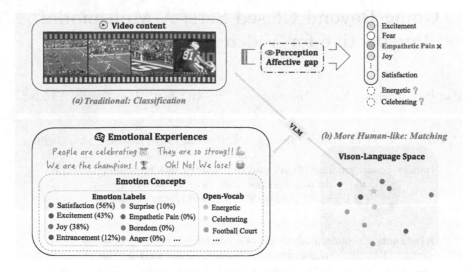

Fig. 1. Illustrating two pipelines of VEA. (*a*) depicts the traditional 1-N classification scheme, which lacks the understanding of emotion labels as language and is difficult to expand categories. In contrast, we introduce a **Multimodal Matching Perspective** in (*b*) that is more in line with the human emotional reaction process and can leverage the power of language models.

There is a long line of research on video emotion analysis (VEA) that regards this task as a classification task. Early studies [9,28] categorize audience emotion into a few categories using emotion models like Plutchik's [18] and Ekman's [7] to annotate emotional experiences on social media. The number of categories is expanded to 27 recently [5]. However, these works still have limitations of describing complex human feelings with a closed set of category labels, making it challenging to express richer emotional experiences using open vocabularies.

The paradigm of previous works maps the visual features to emotion labels [19,29,36], but it regards labels as IDs and thus ignores the meaning of emotion labels. In this way, it is challenging to fill the huge affective gap [8] between video content and emotion. Actually, as shown in Fig. 1, when viewers perceive a video of football games, they first have experiences such as concepts (i.e. celebrating) or comments (i.e. we are the champions!) that come to mind, which are then matched to one or several emotion labels (satisfaction, excitement, etc.). Language serves as the carrier of this process. A few recent works start to use language as an additional signal for training emotion classifiers [6,17]. However, there is still large room to improve the alignment between videos and languages from the perspective of emotion.

To overcome the above challenges, we propose **MM-VEMA**, a novel Multi-Modal perspective for **V**ideo **EM**otion **A**nalysis. Different from previous works that classify videos into several emotion categories, we reformulate this task as a crossmodal matching problem. Given videos and emotional experiences, we project them into a joint space and align them. Then the emotion of each video can be matched to the nearby experiences. These emotional experiences include a large number of emotional words, phrases, and even sentences, thus can adequately convey the emotions of audiences. Moreover, the surrounding emotional experiences of each video reflect the process of humans producing emotional responses. Constructing such a joint space is not trivial because of the affective gap [8] between videos and emotional experiences. We propose to utilize a multimodal space (e.g., CLIP [20]), where videos and texts are well-aligned, then we manipulate this space with a number of emotional prompts to obtain an emotion-oriented space. We conduct extensive experiments to verify the effectiveness of our methods. As our analysis showed, the multimodal representation space after training has a strong ability to align video and emotional experiences. Moreover, our model achieves state-of-the-art performance on three VEA datasets.

We summarize our contributions as follows: (1) We propose a novel multimodal perspective for video emotion analysis, instead of classifying videos into closed sets of categories, we formulate this task as a crossmodal matching problem. (2) We propose an emotional prompt-guided multimodal space manipulating method, which can effectively construct a well-aligned joint multimodal space of videos and emotional experiences. (3) We achieve state-of-the-art performance on three public VEA datasets.

2   Related Work

Emotion Modeling for Video Emotion Analysis. Two emotion models are widely used by affective computing researchers in video emotion analysis area: dimensional emotion space (DES) and categorical emotion states (CES). Baveye et al. [2] propose a large-scale movie affective analysis dataset LIRIS-ACCEDE annotated with VA emotional states, enabling researchers to study the impact of videos on viewers using deep learning method for the first time. VideoEmotion-8 [9] collects videos from social media and using Plutchik's [18] eight basic categories to manually annotate. YouTube/Flickr-EkmanSix [28] leverage Ekman's [7] emotion model to present annotations. Both datasets provide a limited yet valuable resource for recognizing emotions in video content. Mazeika et al. [16] suggest a dataset that encompasses a significantly wider range and diversity of videos, with annotations on the intensity levels of 27 self-reported emotional categories [5]. Despite the finer categorization of emotions, these categories remain a closed set. We argue that complex emotional responses should not be constrained by fixed categories, but rather can be conveyed through language as any subjective feeling. Few works have been done from this perspective. Thus we try to explore open-vocab video emotion analysis in this paper.

Methods for Video Emotion Analysis. Similar to the semantic gap in computer vision [37], one main challenge for video emotion analysis is the affective gap between low-level pixel and high-level abstract emotion concepts. To fill the gap, a lot of efforts have been made to find discriminative features that can effectively capture emotion information [1,4,15,30]. SentiBank [4] is a large visual

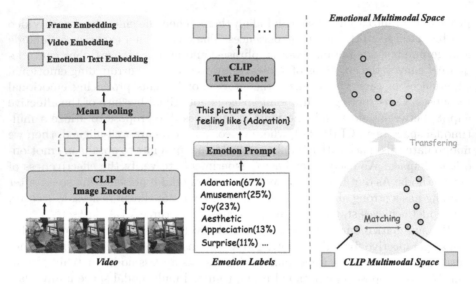

Fig. 2. Overview of our proposed MM-VEMA framework. Videos and emotional texts are projected into a joint CLIP-initialized representation space. To efficiently transfer the space to a more emotional one, emotion prompts are designed to make the emotion labels more descriptive. A matching strategy is proposed to smoothly manipulate the space.

sentiment ontology that contains 1,200 concepts and provides rich emotional semantic representation. Many works [11,12,31,32,34,35] study how to use deep learning methods to obtain better emotional visual representation. Deng et al. [6] introduce a language-supervised method, which can combine the features of language and visual emotion to drive the visual model to gain stronger emotional discernment with language prompts. Pan et al. [17] utilize time-synchronized comments (TSCs) as auxiliary supervision due to their accessibility and rich emotional cues. Both of them show that language plays a key essential role in visual emotion perception. However, while previous efforts attempt to use language to bridge the affective gap, they were still restricted to closed-set video classification tasks. In contrast, we first propose a novel multimodal approach that incorporates language models to understand emotional concepts and treats closed-set emotion classification as an open-vocabulary video-emotional experience matching task.

3   Method

The goal of Video Emotion Analysis is to figure out the emotion evoked by videos. Previous methods typically regard the task as simply assigning a 1-N label to a given video, ignoring the semantic information of emotion labels, and cannot be effectively expanded to retrieving emotional experiences. To break this limitation, we propose a multimodal matching framework MM-VEMA, which views video emotion recognition as a special case of a general matching task between videos and emotional experiences. Section 3.1 outlines the overall architecture of our proposed framework called MM-VEMA. In Sect. 3.2, we present our implementation of utilizing efficient emotion prompts to transfer the CLIP representation space to a new emotion-oriented multimodal space. Section 3.3 introduces the training objective of our method.

3.1 Multimodal Matching Framework

As Fig. 2 shows, our multimodal matching framework consists of three parts: (1) **Video Encoding**, which includes an image encoder, denoted as E_i; and a temporal modeling module, denoted as M_v; (2) **Text Encoding**, which consists of an emotional prompt generator, denoted as P_t; and a text encoder, denoted as E_t; (3) **Matching Strategy**, which utilizes cosine similarity to measure the distance between the two embeddings, denoted as the function $S(\cdot, \cdot)$. Given a frame set $F = \{f_1, f_2, \cdots, f_T\}$ extracted from video v and its candidate emotion labels $L = \{l_1, l_2, \cdots, l_C\}$, where T is the number of sampled frames and C is the number of categories, we obtain the video embedding $\mathbf{e_v}$ and the text embedding $\mathbf{e_t}$ as follows:

$$\mathbf{e_v} = M_v(\{E_i(f_1), E_i(f_2), \cdots, E_i(f_T)\}), \mathbf{e_t} = E_t(P_t(l)). \tag{1}$$

Through such encoding process, we map the information from both modalities onto a shared representation space. This unified perspective allowed the model to effectively perceive and understand visual and textual information jointly. Then, by computing the cosine similarity between the video embedding and the text embedding, where

$$s_v = S(\mathbf{e_v}, \mathbf{e_t}) = \frac{\mathbf{e_v} \cdot \mathbf{e_t}^\mathbf{T}}{\|\mathbf{e_v}\|\|\mathbf{e_t}\|} \tag{2}$$

Thus we can rank texts in terms of the similarities. The video's predicted labels are determined by selecting the most similar labels or texts. To initialize the entire training framework, we start with CLIP's [20] parameter as CLIP can provide an initial well-aligned multimodal representation space.

Text Encoding. Before encoding a label l as embedding, they undergo a process of generating emotional experiences through the prompt generator, such as "this picture evokes feeling like l". Then we use CLIP's text encoder, which is a transformer [25] based module, to encode the sentence. We finally use the [EOS] embedding generated by E_t as the embedding for the entire emotional experience.

Video Encoding. The process of video encoding can be split into two distinct stages. In the first stage, the video is sampled into T frames $f_1, f_2, ..., f_T$, with each frame being encoded by the same CLIP's image encoder E_i to obtain T frame embeddings. As for temporal modeling, we employ mean pooling, a widely used simple but effective method, to aggregate frame embeddings and generate the final video representation.

Matching Strategy. The similarity score between e_v and e_t serves as a confidence measure. A higher similarity score between e_v and e_t suggests a closer semantic relationship between them. During training, the objective is to maximize $s(v, l)$ when v and l are matched, and minimize it in all other instances. During inference, the label with the highest similarity to the video score is regarded as the predicted category. In addition, any word or sentence, regardless of category, can be used as a query to determine its emotional similarity to the video. This score indicates how closely the emotion involved in the query aligns with that evoked by the video.

3.2 Emotion Prompt

While CLIP exhibits a well-aligned visual-language space that can capture emotional semantics to some extent from texts, its text representation space tends to be more biased towards comprehending descriptive sentences rather than individual abstract emotional labels. To accommodate this characteristic, we design a number of prompt templates for emotion labels, converting them into more descriptive emotional experiences. This approach facilitates a smooth transition from CLIP's original representation space, which favors descriptive comprehension, to a representation space that prioritizes emotional understanding through training.

Following [27], we design three kinds of emotion prompts: prefix prompt, cloze prompt, and suffix prompt. Prefix prompt refers to a fixed prompt added before the label, such as "This picture evokes feeling like {}". Cloze prompt inserts label in the middle of the prompt, such as "Feeling {} after viewing this image". Finally, suffix prompt refers to a prompt added after the label, for instance, "{} is the emotion the photo evoked". It is notable that we use "image" or "photo" rather than "video" in the experience description since CLIP pre-training is based on image-text pairs, which makes it more effective in representing the visual modality. *See Supplementary for the complete set of emotion prompts.*

3.3 Model Training

During training, the video encoder and text encoder are optimized jointly to construct an emotional video-text representation space by increasing the similarity between the video representation e_v and its paired emotional experience representation e_t, while decreasing the similarity between unpaired ones.

Training Loss. For the task of predicting a single emotion label for each video, each video sample v corresponds to a ground truth label $c \in C$. We adopt the common implementation [10,27] of infoNCE loss to maximize the similarity between matched video-category pairs while minimizing the similarity for non-matching pairs.

$$\mathcal{L}_{single} = -\frac{1}{B} \sum_i^B \sum_{c \in \mathcal{C}(i)} \log \frac{\exp(s(\mathbf{e}_{vi}, \mathbf{e}_{tc})/\tau)}{\sum_j^B \exp(s(\mathbf{e}_{vi}, \mathbf{e}_{tj})/\tau)}, \quad (3)$$

Note that τ refers to the temperature hyper-parameter for scaling and B is the batch for training.

For the task of multi-label classification of C emotion labels for each video, such as classification for VCE dataset, we optimize Kullback-Leibler (KL) divergence to better fit the distribution of target emotion intensity distribution. Unlike contrastive learning within a batch, we minimize the distance of two C-dimensional intensity distribution vectors. One is the similarity vector of the video sample v to all candidate labels, the other is the ground truth intensity distribution $\mathbf{gt} \in \mathbb{R}^C$.

$$\mathcal{L}_{multi} = KL(s(\mathbf{e_v}, \mathbf{e_t}), \mathbf{gt}) \quad (4)$$

Moreover, in order to address the issue of uneven label distribution in emotion classification, we utilized FOCAL Loss [13] to mitigate the spatial learning bias.

4 Experiments

4.1 Experiment Setup

We use the following three public datasets in our experiments. **Video Cognitive Empathy** (VCE) dataset contains 61,046 videos with annotations for the emotional response of human viewers. The dataset is divided into a training set of 50,000 videos and a test set of 11,046 videos, with each video lasting an average of 14.1 s. Each video is annotated with 27 descriptive emotion states' [5] intensity distribution. **VideoEmotion-8** (VE-8) dataset [9] is composed of 1,101 videos from YouTube and Flickr, with an average duration of 107 s. Each video is labeled with one of eight emotions in Plutchik's, with a minimum of 100 videos per category. We do ten-fold cross validation as previous works did [9,17,36]. We report the average results across the ten runs. **YouTube/Flickr-EkmanSix** (YF-6) dataset [28] consists of 1,637 videos collected from YouTube and Flickr, with an average duration of 112 s. The videos are labeled with one of Ekman's six basic emotion categories, with a minimum of 221 videos per category. There are 819 videos for training and 818 for testing [28].

We adopt the top-3 accuracy metric used by [16] which measures the proportion of test samples where the predicted maximum emotion falls within the top three emotions of the ground-truth distribution. For VE-8 and YF-6, the model is trained with a learning rate of 1×10^{-5} and batch size of 32 for 50 epochs. The standard top-1 accuracy is used for the metric.

For vision encoding, we use CLIP's ViT-B/32 models, which is a 12-layer vision transformer with input patch sizes of 32. We extract the [CLS] token from their last layers' outputs. The text encoder is based on the 12-layer, 512-wide Transformer used in CLIP, with 8 attention heads and the highest layer's

Table 1. Comparison with SOTA methods on VCE.

Method	Performance
Majority Emotion [16]	35.7
R(2+1)D [24]	65.6
STAM [21]	66.4
TimeSformer [3]	66.6
VideoMAE [23]	68.9
MM-VEMA	**73.3**

Table 2. Comparision with SOTA method on VE-8 and YF-6.

Method	VE-8	YF-6
CSS [29]	51.48	55.62
VAANet [36]	54.50	55.30
Dual [19]	53.34	57.37
FAEIL [33]	57.63	60.44
TAM [17]	57.53	61.00
MM-VEMA	**59.38**	**62.75**

activation at the [EOS] token serving as the feature representation. The input frames have a spatial resolution of 224×224. We use the same frame sampling and data augmentation strategy as [26] to sample 16 frames. For VCE, the model is trained with a learning rate of 1×10^{-5} and a batch size of 16 for 10 epochs. We train our models using the AdamW optimizer, with a weight decay of 0.2. The learning rate is warmed up for 10% of the total training epochs and then decayed to zero following a cosine schedule for the remaining training. We set the temperature hyperparameter τ to 0.01.

4.2 Results on Video Emotion Classification

We evaluate our approach against state-of-the-art methods (SOTAs) on three video emotion classification datasets: VCE, VE-8, YF-6. As shown in Table 1 and Table 2, our approach outperforms other video emotion analysis methods while requiring only visual modality from the video. Our approach achieves significant improvement on the VCE dataset compared to the best previous method (+4.4%), contributing to the rich language annotation. VCE annotate 27 emotion labels with intensity for each video, providing richer language information for our method to utilize, compared to VE-8's single annotation of 8 emotion labels and YF-6's single annotation of 6 emotion labels. Our method also achieve improvements on the VE-8 and YF-6 datasets. Unlike previous methods that utilized several modalities of videos (i.e. audio, optical flow) and introduced auxiliary training data, our method only utilizes the visual information of videos. As we use the same visual encoder (ViT) and pre-training parameters (CLIP) with TAM [17], which demonstrates the importance of introducing a text encoder.

We do an ablation study on the impact of our multimodal framework and emotional prompt on the three datasets. The method "CLIP-Uni" in Table 3 refers to the way of only using the visual branch in our framework. Following [6], we feed the output of the video encoder into a multi-layer perceptron(MLP). The method "CLIP-Mul" refers to the way of using the multimodal framework without any prompt. The results show that incorporating the understanding of text modality can significantly enhance the model's classification effectiveness, and our emotion prompt further improves CLIP's emotion perception capability.

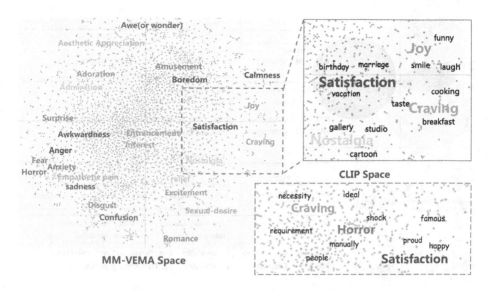

Fig. 3. Visualization of the space of MM-VEMA in comparison to CLIP. Emotional words in our representation space exhibit a more reasonable distribution, with surrounding words that are more relevant and concrete.

Table 3. Ablation study of network structure on three datasets.

Method	Prompt	VCE	VE-8	YF-6
CLIP-Uni	✗	71.80	51.94	57.63
CLIP-Mul	✗	72.77	58.52	62.25
MM-VEMA	✔	**73.29**	**59.38**	**62.75**

4.3 Visualization of the Representation Space

To determine if MM-VEMA provides a video-language representation space with a more comprehensive understanding of emotions, we visualize and compare the representation space of our model trained on VCE with that of CLIP using t-SNE [14]. We project the 27 emotion words in VCE and 5000 high-frequency words onto a two-dimensional plane and show the space of MM-VEMA in Fig. 3. Then we zoom out and display details of a part on both spaces, i.e., **MM-VEMA Space** and **CLIP Space**, for comparison.

We have two observations from Fig. 3: (1) In our MM-VEMA Space, the words with the same emotion are gathered and the word clusters transit smoothly from one emotion to another related emotion. For instance, "aesthetic appreciation" is closely related to "admiration" and "awe" as beautiful things often deserve people's appreciation. Emotional progression like "excitement" → "relief" → "satisfaction" is highly intuitive for humans. It is noteworthy that "amusement" and "boredom" are closely related, as many user-generated videos aim to entertain viewers. When these videos fail to meet viewers' expectations of

	Closed-set Classification		
Ground Truth	Empathetic pain / Amusement / Awe(or Wonder)		
TimeSformer	Surprise ✕ / Amusement / Interest		
MM-VEMA (Ours)	**Empathic Pain** ✓ / Amusement / Surprise		
	Beyond Closed-set Experiences		
CLIP Open-vocab	Slide / Skill / Flip / Brutal / Technique		
CLIP comments	That dog did a net backflip		
MM-VEMA Open-vocab	**Pain / Injury / Ankle / Jump / Punishment**		
MM-VEMA Comments	**OMG! see also can feel the pain ready**		

	Closed-set Classification		
Ground Truth	Awe(or Wonder) / Fear / Boredom		
TimeSformer	Aesthetic Appreciation ✕ / Surprise / Awe(or Wonder)		
MM-VEMA (Ours)	**Awe(or Wonder)** ✓ /Surprise /Aesthetic Appreciation		
	Beyond Closed-set Experiences		
CLIP Open-vocab	Explosion / Launch / Bombing / Video / Missile		
CLIP Comments	I love your videos of nature, how beautiful		
MM-VEMA Open-vocab	**Astonish/Surprise/Scientist/Spectacular/Impressive**		
MM-VEMA Comments	**WOW!!! Amazing. That's really pretty.**		

Fig. 4. Two cases of **Video-Emotional Experience Matching**. MM-VEMA not only has superior classification accuracy but also allows for more diverse and nuanced experiences of emotional reactions through open-vocabulary retrieval.

amusement, they are often labeled as "boredom". *See Supplementary for CLIP Space visualization.* (2) Our MM-VEMA Space provides a more concrete and comprehensive understanding of emotional words compared to CLIP Space. Take the category of "craving", which means a strong desire for something, as an example, the nearby words in CLIP Space, such as "ideal" and "necessity", are either synonyms or abstract words that are less related; whereas, the nearby words in MM-VEMA Space are related to food, such as "cooking" and "taste", which often triggers a strong appetite. For another instance, the words near to "satisfaction" have transferred from synonyms like "happy" in CLIP Space to more concrete words like "birthday" and "marriage" that can bring satisfaction in MM-VEMA Space. These indicate that after training, the model has a more specific understanding of emotions, effectively bridging the "affective gap".

4.4   Video-Emotional Experience Matching

In addition to achieving superior performance in closed-set emotion classification, our model also leverages matching methods to obtain more diverse and nuanced emotional experiences. We utilized the 5000 high-frequency words used in visualizing MM-VEMA space as an open-vocab retrieval library, and 5000 randomly selected video comments from a short video platform as a comment retrieval library. As shown in Fig. 4, our model successfully matches the emotion label "Empathetic pain" for a video depicting a child falling off a skateboard, and captured related emotional concepts such as "injury" and "punishment". In contrast to descriptive comments retrieved by CLIP matching, our approach yielded more direct and expressive comments that better reflect the audience's emotional reaction to the video. *See Supplementary for more qualitative results.*

5   Conclusion

In this paper, we introduce a novel multimodal perspective for video emotion analysis, which formulates this task as a crossmodal matching problem in a joint multimodal space of videos and emotional descriptions. To construct this space, we start from a common space of vision and language (e.g., CLIP), and propose an efficient and effective method of using emotion prompts to manipulate CLIP space to make it more emotion-oriented. We analyze our emotional multimodal space and find it can align videos with rich emotional experiences. We also achieve state-of-the-art performance on three video emotion analysis datasets. In the future, we will further explore using more forms of emotional experiences (e.g., comments) to make our model better understand video emotion.

Acknowledgments. This work was supported by the Fundamental Research Funds for the Central Universities, and the Research Funds of Renmin University of China (21XNLG28), National Natural Science Foundation of China (No. 62276268) and Huawei Technology. We acknowledge the anonymous reviewers for their helpful comments.

References

1. Ali, A.R., et al.: High-level concepts for affective understanding of images. In: WACV, pp. 679–687. IEEE (2017)
2. Baveye, Y., et al.: LIRIS-ACCEDE: a video database for affective content analysis. TAC **6**(1), 43–55 (2015)
3. Bertasius, G., et al.: Is space-time attention all you need for video understanding? In: ICML, vol. 2, p. 4 (2021)
4. Borth, D., et al.: Large-scale visual sentiment ontology and detectors using adjective noun pairs. In: ACM MM, pp. 223–232 (2013)
5. Cowen, A.S., et al.: Self-report captures 27 distinct categories of emotion bridged by continuous gradients. PNAS **114**(38), E7900–E7909 (2017)
6. Deng, S., et al.: Simple but powerful, a language-supervised method for image emotion classification. TAC (2022)
7. Ekman, P.: An argument for basic emotions. Cogn. Emot. **6**(3–4), 169–200 (1992)
8. Hanjalic, A.: Extracting moods from pictures and sounds: towards truly personalized tv. SPM **23**(2), 90–100 (2006)
9. Jiang, Y.G., et al.: Predicting emotions in user-generated videos. In: AAAI, vol. 28 (2014)
10. Ju, C., et al.: Prompting visual-language models for efficient video understanding. In: Avidan, S., Brostow, G., Cissé, M., Farinella, G.M., Hassner, T. (eds.) ECCV 2022. LNCS, vol. 13695, pp. 105–124. Springer, Cham (2022). https://doi.org/10.1007/978-3-031-19833-5_7
11. Lee, J., et al.: Context-aware emotion recognition networks. In: ICCV, pp. 10143–10152 (2019)

12. Li, Y., et al.: Decoupled multimodal distilling for emotion recognition. In: CVPR, pp. 6631–6640 (2023)
13. Lin, T.Y., et al.: Focal loss for dense object detection. In: ICCV, pp. 2980–2988 (2017)
14. Van der Maaten, L., et al.: Visualizing data using t-SNE. J. Mach. Learn. Res. **9**(11) (2008)
15. Machajdik, J., et al.: Affective image classification using features inspired by psychology and art theory. In: ACM MM, pp. 83–92 (2010)
16. Mazeika, M., et al.: How would the viewer feel? Estimating wellbeing from video scenarios. arXiv preprint arXiv:2210.10039 (2022)
17. Pan, J., et al.: Representation learning through multimodal attention and time-sync comments for affective video content analysis. In: ACM MM, pp. 42–50 (2022)
18. Plutchik, R.: Emotions: a general psychoevolutionary theory. Approaches Emot. **1984**(197–219), 2–4 (1984)
19. Qiu, H., et al.: Dual focus attention network for video emotion recognition. In: ICME, pp. 1–6. IEEE (2020)
20. Radford, A., et al.: Learning transferable visual models from natural language supervision. In: ICML, pp. 8748–8763. PMLR (2021)
21. Sharir, G., et al.: An image is worth 16×16 words, what is a video worth? arXiv preprint arXiv:2103.13915 (2021)
22. Stray, J., et al.: What are you optimizing for? Aligning recommender systems with human values. arXiv preprint arXiv:2107.10939 (2021)
23. Tong, Z., et al.: VideoMAE: masked autoencoders are data-efficient learners for self-supervised video pre-training. arXiv preprint arXiv:2203.12602 (2022)
24. Tran, D., et al.: A closer look at spatiotemporal convolutions for action recognition. In: CVPR, pp. 6450–6459 (2018)
25. Vaswani, A., et al.: Attention is all you need. In: NeurIPS, vol. 30 (2017)
26. Wang, L., et al.: Temporal segment networks for action recognition in videos. TPAMI **41**(11), 2740–2755 (2018)
27. Wang, M., et al.: ActionCLIP: a new paradigm for video action recognition. arXiv preprint arXiv:2109.08472 (2021)
28. Xu, B., et al.: Heterogeneous knowledge transfer in video emotion recognition, attribution and summarization. TAC **9**(2), 255–270 (2016)
29. Xu, B., et al.: Video emotion recognition with concept selection. In: ICME, pp. 406–411. IEEE (2019)
30. Yanulevskaya, V., et al.: Emotional valence categorization using holistic image features. In: ICIP, pp. 101–104. IEEE (2008)
31. Yu, W., et al.: CH-SIMS: a Chinese multimodal sentiment analysis dataset with fine-grained annotation of modality. In: ACL, pp. 3718–3727 (2020)
32. Yu, W., et al.: Learning modality-specific representations with self-supervised multi-task learning for multimodal sentiment analysis. In: AAAI, vol. 35, pp. 10790–10797 (2021)
33. Zhang, H., et al.: Recognition of emotions in user-generated videos through frame-level adaptation and emotion intensity learning. TMM (2021)
34. Zhang, Z., et al.: Temporal sentiment localization: listen and look in untrimmed videos. In: ACM MM, pp. 199–208 (2022)
35. Zhang, Z., et al.: Weakly supervised video emotion detection and prediction via cross-modal temporal erasing network. In: CVPR, pp. 18888–18897 (2023)
36. Zhao, S., et al.: An end-to-end visual-audio attention network for emotion recognition in user-generated videos. In: AAAI, vol. 34, pp. 303–311 (2020)
37. Zhao, S., et al.: Affective image content analysis: two decades review and new perspectives. TPAMI **44**(10), 6729–6751 (2021)

Temporal-Semantic Context Fusion for Robust Weakly Supervised Video Anomaly Detection

Yuan Zeng, Yuanyuan Wu$^{(\boxtimes)}$, Jing Liang, and Wu Zeng

College of Computer Science and Cyber Security, Chengdu University of Technology, Chengdu 610059, Sichuan, China
{2021020855,2021020874,zengwu}@stu.cdut.edu.cn, wuyuanyuan@cdut.edu.cn

Abstract. Video anomaly detection (VAD) has emerged as a vital and challenging task in computer vision, driven by the rapid advancement of surveillance videos. Weakly supervised learning is a critical branch in this field, with Multiple Instance Learning (MIL) standing out as the prevalent approach. In the weakly supervised VAD task, video context information provides crucial cues. However, the majority of studies in this field have primarily focused on the temporal context, neglecting the emphasis of semantic context in videos, which provides a deeper understanding of the observed video content. Therefore, this paper proposes a MIL-based framework for efficiently capturing video context information in weakly supervised VAD task. A GCN-based Temporal-Semantic context fusion module is employed to comprehensively capture both temporal and semantic context. Additionally, in order to enhance feature discrimination and learning robustness, our approach integrates highly effective anomaly classification learning and feature magnitude learning loss functions. Extensive experiments on two challenging datasets (i.e., ShanghaiTech and UCF-Crime) outperform some recent methods and demonstrate our approach's effectiveness.

Keywords: Video Anomaly Detection · Context Information · MIL-based Framework

1 Introduction

Video anomaly detection (VAD) is a is a challenging task in computer vision, receiving increasing attention from researchers. Previous studies [10,18] have defined anomalies as patterns that deviate from the normal patterns in videos. This led to a one-class classifier (OCC) method that trains only on normal samples has been proposed. However, collecting comprehensive real-world normal samples poses significant challenges. To overcome this, researchers in [13] propose treating VAD as a weakly supervised task, leveraging training samples annotated with video-level labels. This approach achieves higher accuracy with less human annotation effort compared to the OCC method. Multi-Instance

Q. Liu et al. (Eds.): PRCV 2023, LNCS 14430, pp. 245–256, 2024.
https://doi.org/10.1007/978-981-99-8537-1_20

Learning (MIL) is widely used in computer vision as it has shown ability in handling weakly supervised learning problems. It forms the basis for addressing the weakly supervised VAD task introduced by [13]. It treats each video as a bag, with video snippets as instances. Bags holding all normal instances as negative bags, and bags containing abnormal instances as positive bags. This paper aims to describe anomaly detection as a weakly supervised learning problem trained with MIL approaches.

The weakly supervised VAD task poses two main challenges. Firstly, the subjective nature of the anomaly demands expertise to determine video abnormality. Secondly, abnormal videos contain both normal and abnormal instances, making it difficult to distinguish due to their similarity.

Regarding the first challenge, context information is crucial in video processing. Incorporating context into videos has been applied to various video processing tasks [11,17,18], yielding significant outcomes. In weakly supervised VAD task, methods [5,14,18] explore aggregating context into extracted features. These methods effectively capture insights from the context information, which contribute to anomaly identification. However, their focus solely on the temporal context, overlooking the emphasis of semantic context. By fusing semantic context, a deeper understanding of the observed content can be achieved, facilitating the identification of anomalies beyond low-level visual features. For instance, certain actions may be considered abnormal in a specific scene, but normal in a different scene. Through the fusion of semantic context, more informed judgments can be made. Drawing inspiration from G-TAD [17], we propose fusing both temporal and semantic context into features for the weakly supervised VAD task, which leads to encouraging results and enhances its effectiveness.

For challenge 2, various loss functions have been designed to enhance the distinguishability between normal and abnormal instances in abnormal videos [3,13–15]. As MIL-based weakly supervised anomaly detection can be transformed into a binary-class classification problem, common loss is sigmoid cross-entropy loss, employing video-level labels for instances. However, it may be affected by label noise from normal instances in positive bags. Therefore, the top-K select method is introduced to improve feature separability. Moreover, feature magnitude learning is introduced to enhance the robustness of top-K selection. This work integrates highly effective anomaly classification and feature magnitude learning loss functions to enhance feature discrimination and learning robustness.

To summarize, our contributions can be described as follows:

(1) We propose a MIL-based framework for weakly supervised VAD task that efficiently captures video context information and integrates highly effective anomaly classification learning and feature magnitude learning loss functions to enhance feature discrimination and learning robustness.
(2) A novel GCN-based temporal-semantic context fusion module called GCNeXt-T is designed to utilize both temporal and semantic context more sufficiently.

(3) The effectiveness of our approach has been validated through extensive experiments on two challenging datasets (i.e., ShanghaiTech and UCF-Crime). A frame-level AUC of 96.41% is achieved on ShanghaiTech, and a frame-level AUC of 82.27% is achieved on UCF-Crime.

2   Related Work

In this paragraph, we introduce some works that most relevant to our research briefly. AR-Net [15] maps feature representations to anomaly scores by a anomaly regression layer. MCR [5] introduces differential contextual information for anomaly scores refinement. TCA-VAD [18] uses multi-scale attention to establish video temporal dependencies. However, these methods are limited by their inadequate utilization of context information. WAGCN [1] and STGCNs [12] employ Graph Convolutional Network(GCN) [7] for complicated context relationship modeling. Some methods employ transformer to capture temporal and semantic context. MSL [9] utilizes a Transformer-based network to encode snippet features, and [6] proposes a spatio-temporal transformer framework for encoding well-structured skeleton features. These methods highlight context's role in anomaly detection. Our method uses a graph to connect both temporal and semantic context, capturing intricate dependencies between instances effectively.

To improve the performance of MIL-based weakly supervised VAD method, AR_Net [15] proposes dynamic multiple-instance learning (DMIL) Loss and center Loss to improve feature separability, and dynamically selects top-K instances for DMIL Loss. RTFM [14] tackles unstable top-K selection due to label noise, using feature magnitude learning for accurate positive instance detection, enhancing robustness. MGFN [3] proposes magnitude contrastive (MC) Loss for scene adaptive feature magnitudes learning. We extend these advancements by integrating DMIL, center, and MC Loss, demonstrating effective and robust performance.

3   Proposed Method

3.1   Problem Statement

The input of our network consists of N videos, each containing L_i snippets. In our approach, we define a snippet as a sequence of 16 consecutive frames from videos. The network is constructed based on features extracted from the I3D network [2]. These features are denoted as $X = \{x_i \mid x_i \in \mathbb{R}^{L_i \times C}, i \in (1, 2, \ldots, N)\}$, where C denotes the dimension of each snippet, and L_i denotes the number of snippets in video i. The video-level label of videos can be represented as $Y = \{y_i \mid i \in (1, 2, \ldots, N)\}$, where $y_i \in \{0, 1\}$. And the predicted anomaly scores vector of videos can be expressed as $S = \{s_i \mid s_i \in [0, 1]^{L_i \times 1}, i \in (1, 2, \ldots, N)\}$. The feature magnitudes vector of videos can be expressed as $M = \{m_i \mid m_i \in [0, 1]^{L_i \times 1}, i \in (1, 2, \ldots, N)\}$.

3.2   Architecture

The structure of our proposed framework is shown in Fig. 1. The video features (X) are inputted into a GCN-based temporal-semantic context fusion module called GCNeXt-T. Through n iterations, this module produces fused feature representations denoted as $H^{(n)}$. Then, these feature representations are utilized to generate feature magnitudes (M) and anomaly scores (S). Finally, the top-K feature magnitudes are selected from bags to calculate feature magnitude learning loss (MC Loss), and the selected top-K indexes are employed in anomaly classification learning to calculate the DMIL Loss. Additionally, the center Loss is calculated within normal bags.

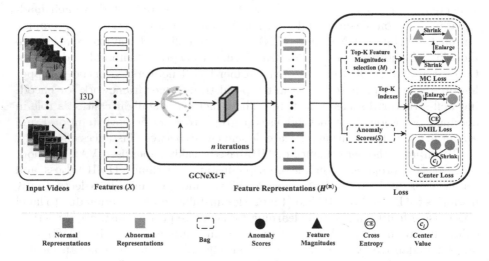

Fig. 1. Overview of our proposed network architecture. Snippet-level features are extracted using the I3D network. Subsequently, a graph is constructed based on feature similarity. Then the context information is captured using the GCN-based context fusion module GCNeXt-T. Finally, model training is performed by combining anomaly classification learning and feature magnitude learning.

3.3   Temporal-Semantic Context Fusion Module

In weakly supervised VAD task, most methods focus only on temporal context. Inspired by G-TAD [17], which proposed GCNeXt to capture both temporal and semantic context effectively for temporal behavior detection, we design a GCN-based temporal-semantic context fusion module called GCNeXt-T to utilize context information more sufficiently in weakly supervised VAD task (see Fig. 2). Initially, a video is represented as a graph, and graph convolution network is applied separately to aggregate temporal and semantic context information. Then, the processed context information and the input are fused together in a single iteration. After n iterations, the final fused features are obtained.

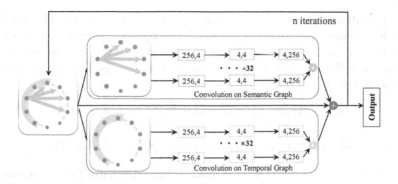

Fig. 2. The GCN-based temporal-semantic context fusion module GCNeXt-T. It has two streams, one aggregates temporal context, and the other aggregates semantic context. The output is the fused context information and input after n iterations.

Construction of the Graph. To model contextual dependencies between video snippets, we conceptualize a video as a graph, where each video snippet represents a node. Temporal edges connect temporally adjacent temporal contexts, and semantic edges connect semantic contexts that share similar semantic content (see Fig. 3). Therefore, a context dependency graph $G = \{V, E\}$ is built, where $V = \{v_i \mid i \in (1, 2, \ldots, L_i)\}$ represents the node set of the graph, with each v_i denoting a video snippet, and $E = E_t \cup E_s$ represents the set of edges, containing temporal edges E_t and semantic edges E_s.

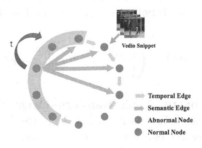

Fig. 3. The context dependency graph of a video. A video is viewed as a graph. The nodes represent video snippets, and the edges contain temporal edges and semantic edges. A sliding window is introduced to temporal edges on each node to capture time dependencies more effectively, depicted by a blue shaded area in this figure, with the current node highlighted in green. (Color figure online)

Temporal Edges. Temporal edges are utilized to capture relationships between temporally adjacent nodes. Each node v_i is connected to a forward edge E_t^f and a backward edge E_t^b in G-TAD [17] proposed GCNeXt. However, it only focuses on the two closest nodes, neglecting the potential information provided by several adjacent nodes. Recognizing this limitation, we propose GCNeXt-T, which incorporates a sliding window of size $2l+1$ for temporal edges on each node. This is intended to enhance the utilization of temporal context information. The equations for forward edges E_t^f and backward edges E_t^b are presented as Eq. (1) and Eq. (2) respectively.

$$E_t^f = \{(v_i, v_{i+1}) \ldots (v_i, v_{i+l}) | i \in (1, 2, \ldots, L_i - l)\}. \tag{1}$$

$$E_t^b = \{(v_{i-l}, v_i) \ldots (v_{i-1}, v_i) | i \in (l+1, l+2, \ldots, L_i)\}. \tag{2}$$

Semantic Edges. The concept of semantic edges originates from dynamic edge convolution (DGCNN) [19]. In determining the anomalies of a video snippet, semantically relevant information provides important clues for weakly supervised VAD task. Therefore, semantic edges (E_s) (see Eq. (3)) are dynamically constructed within the graph to capture semantic context information between video snippets. Here, $v_{ni}(k)$ denotes the node index of the k-th nearest neighbor of node v_i. The K Nearest Neighbors (KNN) algorithm is employed to establish these semantic edges, selecting them based on the Euclidean distance between node features.

$$E_s = \left\{ (v_i, v_{ni}(k)) \, | i \in (1, 2, \ldots, L_i), k \in (1, 2, \ldots, K) \right\}. \tag{3}$$

Graph Convolution. To effectively aggregate local neighborhoods and learn global information, DGCNN is employed for implementing the graph convolution network. Equation (4) presents the graph convolution equation, where $W^{(n)}$ is a trainable weight vector and $A^{(n)}$ represents the adjacency matrix without self-connection. [] denotes concatenation operation, and n denotes the number of iterations.

$$F\left(x_i^{(n)}, A^{(n)}, W^{(n)}\right) = \left(\left[x_i^{(n)^T}, A^{(n)} x_i^{(n)^T} - x_i^{(n)^T} \right] W^{(n)} \right)^T. \tag{4}$$

Multi-stream Incorporate. A split-transform-merge strategy is used for temporal and semantic streams. Each stream has 32 paths for transformation diversity, enabling comprehensive capture of context information from multiple perspectives. Each path includes a graph convolution and two 1-by-1 convolutions represented as F'. Information from streams and input is fused to build the updated graph. GCNeXt-T obtains fused information after n iterations (see Eq. (5)). The adjacency matrices $A = \left\{ A_t^{f\,(n)}, A_t^{b\,(n)}, A_s^{(n)} \right\}$ correspond the forward edges E_t^f, backward edges E_t^b, semantic edges E_s respectively. The trainable weights $W = \left\{ W_t^{f\,(n)}, W_t^{b\,(n)}, W_s^{(n)} \right\}$ are associated with these edges.

$$\begin{aligned} H^{(n)}\left(x_i^{(n)}, A^{(n)}, W^{(n)}\right) = ReLU \left(F'\left(x_i^{(n)}, A_t^{f\,(n)}, W_t^{f\,(n)}\right) \right. \\ \left. + F'\left(x_i^{(n)}, A_t^{f\,(n)}, W_t^{b\,(n)}\right) + F'\left(x_i^{(n)}, A_s^{(n)}, W_s^{(n)}\right) + x_i^{(n)} \right). \end{aligned} \tag{5}$$

3.4   Model Training

Multiple-Instance Learning Loss (L_{DMIL}). In the L_{DMIL} formulation (see Eq. (6)), cross-entropy is calculated between these selected top-K anomaly scores and their corresponding labels, expanding the inter-class distance between normal and abnormal instances.

$$L_{DMIL} = \frac{1}{k_i} \sum_{j=1}^{k_i} \left[-y_i log\left(s_i^j\right) + (1 - y_i) log\left(1 - s_i^j\right) \right]. \tag{6}$$

Center Loss (L_c). AR_Net introduces the L_c (see Eq. (7)) to reduce intra-class distance among normal instances by learning the feature centre, addressing similarity between maximum anomaly scores in normal and anomaly videos.

$$L_c = \begin{cases} \frac{1}{L_i} \sum_{j=1}^{L_i} \left\| s_i^j - c_i \right\|^2, & if \ y_i = 0 \\ 0, & otherwise \end{cases}, \tag{7}$$

where c_i represents center of the anomaly score vector s_i of the i-th video, which can be obtained by calculating the average value.

Magnitude Contrastive Loss (L_{mc}). MGFN introduces the L_{mc} to facilitate the learning of scene-adaptive feature magnitudes and enhance the leaning robustness of the MIL approach. Equation (8) shows the expression, in which D(.,.) refers the distance between selected top-K feature magnitudes ($m_i^{k_i}$). p, q denote indices of normal video instances, and u, v represent indices of abnormal instances. Further details can be found in [3].

$$L_{mc} = \sum_{p,q=0}^{B/2} (1 - l)\left(D\left(m_n^p, m_n^q\right)\right) + \sum_{u,v=B/2}^{B} (1 - l)\left(D\left(m_a^u, m_a^v\right)\right)$$
$$+ \sum_{p=0}^{B/2} \sum_{u=B/2}^{B} l\left(Margin - D\left(m_n^p, m_a^u\right)\right). \tag{8}$$

Overall Loss Function. In the proposed method, we introduce a integration of highly effective anomaly classification learning and feature magnitude learning loss functions. Considering the complexity of video duration and incorrect label assignment during the early training stages, we employ the L_{DMIL} and L_c to increase the distinguishability of features. In order to improve the learning robustness of the MIL-based approach, the L_{mc} is employed for scene-adaptive feature magnitude learning. By integrating these loss functions, we aim to enhance feature discrimination and robustness of our proposed approach. The total loss in the model training is defined as Eq. (9):

$$L = L_{DMIL} + \lambda L_c + \alpha L_{mc}, \tag{9}$$

where λ and α are parameters representing the loss weights that balance the impact of each factor.

4 Experiments

4.1 Datasets and Evaluation Measure

Our method is evaluated on two challenging datasets for weakly supervised VAD tasks: ShanghaiTech [10] and UCF-Crime [13]. The ShanghaiTech dataset comprises 437 videos from 13 scenes. For binary classification, we utilize the split method from [20]. The training set includes 238 videos, and the test set contains 199 videos. The UCF-Crime dataset is gathered from real-world surveillance videos, encompassing 13 anomalies. The dataset consists of 1900 videos, with 1610 in the training set and 290 in the test set. We evaluate the performance of our method by calculating the frame-level area under the ROC curve (AUC).

4.2 Implementation Details

Following previous works [15,16,18], we combine the 1024-dimensional $I3D^{RGB}$ and $I3D^{Optical\ Flow}$ features as input. We conducted relevant experiments[1], and the best results are achieved with a batch size of 60 and balanced sampling. The size of sliding window is 11. The GCNeXt-T module undergoes 4 iterations ($n = 4$). We set the center Loss weight $\lambda = 20$ and the MC Loss weight $\alpha = 0.1$. The training process utilizes the Adam optimizer with weight decay of 0.00001 and learning rate of 0.0001.

4.3 Experiment Results and Discussions

We compare our method with state-of-the-art approaches on the ShanghaiTech and UCF-Crime datasets (see Table 1 and Table 2). On the ShanghaiTech dataset, our method demonstrates superior performance compared to MIST and BN-SVP, with improvements of 1.58% and 0.41%, respectively. Additionally, our method outperforms AR_Net, MSAF, MCR, and TCA-VAD, achieving improvements of 3.17%, 2.95%, 1.49%, and 2.81% respectively. These results demonstrate the advantage of utilizing semantic context in anomaly detection, as opposed to methods that merely rely on temporal context. Notably, in comparison with other GCN-based methods, our method outperforms GCN-Anomaly, STGCNs, and WAGCN by 11.97%, 0.36%, and 2.11%, respectively. Furthermore, our method also performs excellent on the UCF-Crime dataset. Specifically, our method outperforms MIL, AR_Net, MCR by 6.86%, 3.12%, and 1.27%, respectively and improves upon the GCN-based method [8] by 0.15%. These results highlight the superior performance of our approach in addressing the weakly supervised VAD task.

The visualized results obtained from our method and AR_Net are shown in Fig. 4. In video 01-0035, the anomaly corresponds to a car breaking into the pavement. Our method better aligns with the ground truth, featuring more prominent boundaries. AR_Net fails to detect the anomaly around the 150 frame,

---

[1] https://github.com/ycolourful/video-anomaly-detection.

Table 1. AUC (%) performance comparison on ShanghaiTech.

Method	GCN used	Feature	ShanghaiTech
MIL (2018, CVPR) [13]	False	I3D^{RGB}	86.30
AR_Net (2021, ICME) [15]		I3D^{RGB}	85.38
AR_Net (2021, ICME) [15]		I3D$^{RGB+Optical\ Flow}$	93.24
MIST (2021, CVPR) [4]		I3D^{RGB}	94.83
MSAF (2022) [16]		I3D$^{RGB+Optical\ Flow}$	93.46
TCA-VAD (2022, ICME) [18]		I3D$^{RGB+Optical\ Flow}$	93.60
MCR (2022, ICME) [5]		I3D$^{RGB+Optical\ Flow}$	94.92
MGFN (2022, AAAI) [3]		I3D^{RGB}	96.02
GCN-Anomaly (2019, CVPR) [20]	True	I3D^{RGB}	84.44
WAGCN (2022) [1]		I3D^{RGB}	94.30
STGCNs (2022) [12]		I3D^{RGB}	96.05
Ours		I3D^{RGB}	91.36
Ours		I3D$^{RGB+Optical\ Flow}$	**96.41**

Table 2. AUC (%) performance comparison on UCF-Crime.

Method	GCN used	Feature	UCF
MIL (2018, CVPR) [13]	False	I3D^{RGB}	75.41
AR_Net (2021, ICME) [15]		I3D^{RGB}	78.96
AR_Net (2021, ICME) [15]		I3D$^{RGB+Optical\ Flow}$	79.15
MCR (2022, ICME) [5]		I3D^{RGB}	81.0
GCN (2022, Neurocomputing) [8]	True	C3D	81.08
GCN (2022, Neurocomputing) [8]		TSN^{RGB}	82.12
Ours		I3D^{RGB}	80.07
Ours		I3D$^{RGB+Optical\ Flow}$	**82.27**

while our method successfully identifies it. In normal video 04-007, our method obtains a predicted score of 0 to all frames, while AR_Net does not. Based on these results, our method demonstrates superior effectiveness.

In video 01-0064, an anomaly occurs involving a skateboard passing through the pedestrian sidewalk. Both our method and AR_Net fail to detect it. The main reasons are as follows: 1) Before frame 130, the skateboarder is partially visible. This highlights the challenge of judging anomalies from localized object parts. 2) After frame 130, despite the anomaly object appearing, the skateboard still closely resembles normal walking in the video. Therefore, video anomaly detection in this scenario remains notably challenging.

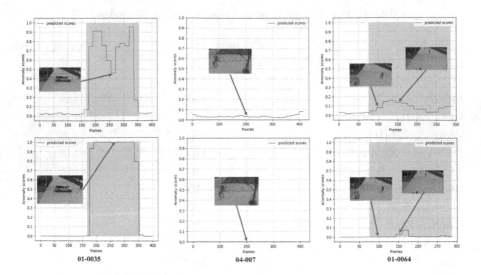

Fig. 4. Visualization of the testing results for ShanghaiTech. The first line corresponds to the visualization results of AR_Net, and the second line represents the visualization results of our method. The red blocks in the graph indicate the ground truth annotations of abnormal events. (Color figure online)

4.4 Ablation Study

To evaluate our proposed method, we conduct ablation experiments on the ShanghaiTech dataset using the baseline model AR_Net (see Table 3). The initial AR_Net model achieved an AUC of 93.25% on the frame-level AUC. The introduction of the GCNeXt, as proposed by G-TAD, the performance increased to 94.56%. Moreover, the employment of the proposed GCNeXt-T further enhances the results to 94.68%. Regarding the loss function, combining these highly effective loss functions results in an AUC of 93.83%. Notably, when integrating GCNeXt-T with the new loss function, a performance gain of 3.16% is observed.

Table 3. AUC (%) Performance Comparison of Baseline and Our Method.

Model	GCNeXt	GCNeXt-T	MC Loss	DMIL loss+Center loss	AUC (%)
AR_Net				✓	93.25
AR_Net			✓	✓	93.83
AR_Net	✓			✓	94.56
AR_Net	✓		✓	✓	96.28
AR_Net		✓		✓	94.68
AR_Net		✓	✓	✓	**96.41**

To investigate the impact of the number of iterations on the GCNeXt-T, we conduct experiments with varying iterations (see Fig. 5). The AUC demonstrates an increasing trend as iterations (n) increases, peaking at 4. Increasing n facilitates context learning and captures more valuable context information. However, it can also introduce noise and result in higher computational costs.

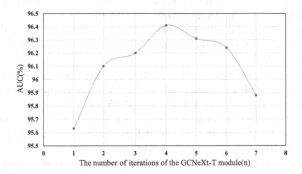

Fig. 5. The frame-level AUC on shanghaiTech dataset with different iterations of GCNeXt-T module.

5   Conclusion

In this paper, we propose a novel approach for weakly supervised video anomaly detection. Our method captures both temporal context and semantic context, distinguishing it from existing methods. Additionally, we introduce a integration of highly effective anomaly classification learning and feature magnitude learning loss functions to enhance the model's learning capability. Extensive experiments are conducted on the ShanghaiTech and UCF-Crime datasets to validate our approach, demonstrating the efficacy of our method.

References

1. Cao, C., Zhang, X., Zhang, S., Wang, P., Zhang, Y.: Adaptive graph convolutional networks for weakly supervised anomaly detection in videos. IEEE Signal Process. Lett. **29**, 2497–2501 (2022)
2. Carreira, J., Zisserman, A.: Quo vadis, action recognition? A new model and the kinetics dataset. In: Proceedings of the IEEE Conference on Computer Vision and Pattern Recognition, pp. 6299–6308 (2017)
3. Chen, Y., Liu, Z., Zhang, B., Fok, W., Qi, X., Wu, Y.C.: MGFN: magnitude-contrastive glance-and-focus network for weakly-supervised video anomaly detection (2022)
4. Feng, J.C., Hong, F.T., Zheng, W.S.: MIST: multiple instance self-training framework for video anomaly detection. In: Proceedings of the IEEE/CVF Conference on Computer Vision and Pattern Recognition, pp. 14009–14018 (2021)
5. Gong, Y., Wang, C., Dai, X., Yu, S., Xiang, L., Wu, J.: Multi-scale continuity-aware refinement network for weakly supervised video anomaly detection. In: 2022 IEEE International Conference on Multimedia and Expo (ICME), pp. 1–6. IEEE (2022)

6. Huang, C., Liu, Y., Zhang, Z., Liu, C., Wen, J., Xu, Y., Wang, Y.: Hierarchical graph embedded pose regularity learning via spatio-temporal transformer for abnormal behavior detection. In: Proceedings of the 30th ACM International Conference on Multimedia, pp. 307–315 (2022)

7. Kipf, T.N., Welling, M.: Semi-supervised classification with graph convolutional networks. arXiv preprint arXiv:1609.02907 (2016)

8. Li, N., Zhong, J.X., Shu, X., Guo, H.: Weakly-supervised anomaly detection in video surveillance via graph convolutional label noise cleaning. Neurocomputing **481**, 154–167 (2022)

9. Li, S., Liu, F., Jiao, L.: Self-training multi-sequence learning with transformer for weakly supervised video anomaly detection. In: Proceedings of the AAAI Conference on Artificial Intelligence, vol. 36, pp. 1395–1403 (2022)

10. Liu, W., Luo, W., Lian, D., Gao, S.: Future frame prediction for anomaly detection-a new baseline. In: Proceedings of the IEEE Conference on Computer Vision and Pattern Recognition, pp. 6536–6545 (2018)

11. Long, F., Yao, T., Qiu, Z., Tian, X., Luo, J., Mei, T.: Gaussian temporal awareness networks for action localization. In: Proceedings of the IEEE/CVF Conference on Computer Vision and Pattern Recognition, pp. 344–353 (2019)

12. Mu, H., Sun, R., Wang, M., Chen, Z.: Spatio-temporal graph-based CNNs for anomaly detection in weakly-labeled videos. Inf. Process. Manage. **59**(4), 102983 (2022)

13. Sultani, W., Chen, C., Shah, M.: Real-world anomaly detection in surveillance videos. In: Proceedings of the IEEE Conference on Computer Vision and Pattern Recognition, pp. 6479–6488 (2018)

14. Tian, Y., Pang, G., Chen, Y., Singh, R., Verjans, J.W., Carneiro, G.: Weakly-supervised video anomaly detection with robust temporal feature magnitude learning. In: Proceedings of the IEEE/CVF International Conference on Computer Vision, pp. 4975–4986 (2021)

15. Wan, B., Fang, Y., Xia, X., Mei, J.: Weakly supervised video anomaly detection via center-guided discriminative learning. In: 2020 IEEE International Conference on Multimedia and Expo (ICME), pp. 1–6. IEEE (2020)

16. Wei, D., Liu, Y., Zhu, X., Liu, J., Zeng, X.: MSAF: multimodal supervise-attention enhanced fusion for video anomaly detection. IEEE Signal Process. Lett. **29**, 2178–2182 (2022). https://doi.org/10.1109/LSP.2022.3216500

17. Xu, M., Zhao, C., Rojas, D.S., Thabet, A., Ghanem, B.: G-TAD: sub-graph localization for temporal action detection. In: Proceedings of the IEEE/CVF Conference on Computer Vision and Pattern Recognition, pp. 10156–10165 (2020)

18. Yu, S., Wang, C., Xiang, L., Wu, J.: TCA-VAD: temporal context alignment network for weakly supervised video anomly detection. In: 2022 IEEE International Conference on Multimedia and Expo (ICME), pp. 1–6 (2022). https://doi.org/10.1109/ICME52920.2022.9859607

19. Yue, W., Yongbin, S., Ziwei, L., Sarma, S.E., Bronstein, M.M., Solomon, J.M.: Dynamic graph CNN for learning on point clouds. ACM Trans. Graph. (TOG) **38**(5), 1–12 (2019)

20. Zhong, J.X., Li, N., Kong, W., Liu, S., Li, T.H., Li, G.: Graph convolutional label noise cleaner: train a plug-and-play action classifier for anomaly detection. In: Proceedings of the IEEE/CVF Conference on Computer Vision and Pattern Recognition, pp. 1237–1246 (2019)

A Survey: The Sensor-Based Method for Sign Language Recognition

Tian Yang[1,2], Cong Shen[1,2(✉)], Xinyue Wang[1,2], Xiaoyu Ma[1,2], and Chen Ling[3]

[1] School of Computer Science and Engineering, Tianjin University of Technology,
Tianjin, People's Republic of China
{tyang,zhujue1998}@stud.tjut.edu.cn
[2] Engineering Research Center of Learning-Based Intelligent System,
Ministry of Education, Tianjin, People's Republic of China
congshen@email.tjut.edu.cn
[3] R&D Center (China), Intel-Mobileye, Shanghai, People's Republic of China
lingc1@mobileye.com

Abstract. Sign language is a crucial communication carrier among deaf people to express and exchange their thoughts and emotions. However, ordinary individuals cannot acquire proficiency in sign language in the short term, which leads to deaf people facing huge barriers with the sound community. Regarding this conundrum, it is valuable to investigate Sign Language Recognition (SLR) equipped with sensors which collect data for the following computer vision processing. This study has reviewed the sensor-based SLR methods, which can transform heterogeneous signals from various underlying sensors into high-level motion representations. Specifically, we have summarized current developments in sensor-based SLR techniques from the perspective of modalities. Addtionally, we have also distilled the sensor-based SLR paradigm and compared the state-of-the-art works, including computer vision. Following that, we have concluded the research opportunities and future work expectations.

Keywords: Sign Language Recognition · Sensor · Computer Vision

1 Introduction

Compared with traditional sound-mediated language, sign language has an extraordinary visual perception mode for communication among deaf individuals, which owns almost 489 million worldwide [28]. Along with the support of facial expressions and mouth movements, it incorporates various gesture forms, locations, and movement trajectories [41]. It is challenging for hearing people to communicate with the deaf sign language always performs diverse representation in contrast with natural language in grammar and syntax. Therefore, Sign

This work was supported in part by the Postgraduate Scientific Research Innovation Practice Program of Tianjin University of Technology (YJ2247).

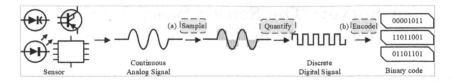

Fig. 1. The operation of an Analog-Digital converter. (a) The analog signals are sampled at a given frequency. (b) The various discrete waves can be utilized to approximate the quantized digital signals with binary code.

Language Recognition (SLR) will provide great aid in removing the hurdles to communication faced by hearing-impaired individuals [31].

Most designed methods in SLR are heavily relying on sophisticated visual data collection processes and peripheral input devices. Among vision-based scenarios, cameras are set up to capture and record hand motions for the sign language dataset construction [43]. Although several algorithms and models have been created to process the captured images and transform them into the corresponding text, the application is still constrained under several conditions.

1. The surrounding elements, such as lighting and shadows, must be carefully dealt with.
2. Noisy signals should be removed during the generation process of the dataset.
3. The recognition accuracy is highly likely to be affected by the complexity of the environment [51].
4. Conventional cameras can only capture information in a foreground perspective towards the performer [19].

Sensor-based SLR approaches can fix this problem by sampling, quantization, and encoding. Particularly, an Analogue Digital (AD) converter can transform the continuous analog signal obtained from sensors into a discrete digital output, which has been illustrated in Fig. 1. Once the motion data has been converted to a digital value, it is convenient for the computer to gain the normalized features and feed them into the model to recognize gestures.

In contrast with the images, the data obtained from sensors exclude cameras containing more physical quantities, which can reflect the hand and finger movement, location, and posture in multiple views. The sensor is robust since it can resolve SLR in various illumination, background, and occlusion conditions. While only depending on video is more constrained by the surrounding influences, which undermined recognition accuracy. Low-power sensors and communication modules are frequently utilized in sensor devices for their economy, contrasted with extensive computational devices and considerable energy consumption [22].

The related works on sensor-based devices for SLR have received extensive attention. We have summarized the prior research to elucidate its implementation and future challenges. The rest of this survey is arranged as follows. We have illustrated the existing work by classifying them according to the data modality

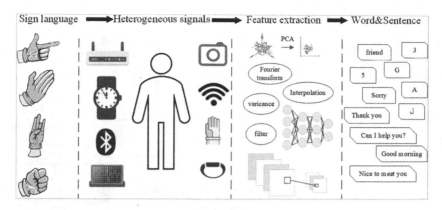

Fig. 2. The paradigm of sensor-based SLR. The heterogeneous signals obtained from sensors mainly consist of WiFi signals, myoelectric signals, vision-based information, and data recorded from hand sensors.

of the sensors captured in Sect. 2. Section 3 has classified research from the perspectives of the dataset, evaluation metrics, and device appearance. Section 4 has highlighted opportunities for future approaches. We have concluded in Sect. 5.

2 Sensor Modality

The state-of-the-art SLR works, including computer vision-based analysis in understanding analog signals captured by sensors such as cameras, are not only affected by the development of feature engineering in the vision domain but also require choosing the appropriate machine learning strategies based on the type of signals. We have summarized the paradigm of cutting-edge research based on the activity signals and feature extractor in Fig. 2. The feature extraction operation can be considered the means of processing digital signals using computer vision.

From the view of modalities, we compared various sensors from the cost, valid range, merits and cons, including Kinect, flex sensor, Force Myography (FMG), Electromyographic EMG, Inertial sensor, Photoplethysmography PPG and Radio Frequency RF sensor, which has been shown in Table 1. Moreover, to deal with the motion signals obtained from the sensors, we classify the cutting-edge works according to the type of sensors, which has been shown in Table 2.

Table 1. The Sensors Modalities.

Modalities	Cost ($)	Valid Range	Merits	Cons
Kinect sensor	100–200	meter	Depth perception	Less effective for long distance tracking
Flex sensor	1–10	centimeter	High flexibility	For close range measurements only
FMG sensor	100–500	centimeter	Non-invasive	Susceptible to external interference
EMG sensor	50–200	centimeter	High precision	Sensitive to muscle mass and position
Inertial sensor	10–50	centimeter	Motion capturing	Large cumulative error
PPG sensor	1–20	millimeter	Heart rate measurement	Highly disturbed by ambient light
RF sensor	100–1000	meter	Long distances	Expensive cost

Table 2. The Sensor-based SLR Works.

Work	Year	Kinect	Flex	FMG	EMG	Inertial	PPG	RF	Appearance	Classifiers	Size	Class	Country	Metrics
[46]	2015	✓							-	CRF		24	American	ACC
[14]	2016				✓				glove	HMM	2000	40		ACC
[45]	2016			✓	✓				-	DT, LibSVM NN, NV		80	American	ACC
[23]	2017	✓							-	HMM,BiLSTM	7500	50	Indian	ACC
[47]	2017			✓	✓				-	Tree-Structure	5500	150	Chinese	ACC
[4]	2020		✓						Wrist-worn	ELM	1350	9	American	ACC
[1]	2020				✓				glove	RNN	750	50	Malaysian	ACC, F1
[17]	2020							✓	-	-	25		American	ACC
[40]	2020	✓							glove	-		23		ACC Response Time
[27]	2020	✓							glove	-			American	
[30]	2021	✓							glove	-			American	ACC
[16]	2021			✓	✓				-	Ensemble	10000	50	Indian	ACC
[39]	2021	✓			✓				-	-		20	Dzongkha	Response Time
[25]	2021						✓		-	DTW		9	Japanese	ACC
[12]	2021	✓			✓				glove	-		28	Arabic	ACC
[10]	2021	✓			✓				glove	SVM,DTW	700	7	Japanese	ACC
[50]	2021					✓			Wrist-worn	Gradient Boost Tree ResNet		9	American	Recall Precision
[49]	2022	✓							glove	Hierarchical Fusion		26		ACC
[13]	2022				✓				Armband	BiLSTM, CNN	6000	15	American	ACC
[20]	2022	✓			✓				glove	SVM, DT, KNN Random Forest	2000	20	Japanese	ACC
[21]	2022	✓							glove	-	520	26	Polish	ACC
[36]	2022				✓				-	LD, Fine Tree, SDE WNN, NV, SVM	780	30	Italian	ACC
[29]	2022	✓							glove	-		29		
[6]	2022				✓				Armband	CNN, LSTM	9350	28	Arabic	ACC
[48]	2022						✓		-	ResNet		40	American	ACC
[8]	2022						✓		-	GRU			Chinese	ACC
[32]	2023						✓		-	BiLSTM	5200	26	American	ACC
[3]	2023	✓				✓			glove	Random Forest LR, SVM, MLP		26 29	American Arabic	ACC, F1, Precision Recall

Note: FMG: Force Myography, EMG: Electromyographic, PPG:Photoplethysmography, RF: Radio Frequency. SVM: Support Vector Machine, KNN: K-Nearest Neighbor, LR: Logistic Regression, DTW: Dynamic Time Wrapping, DT: Decision Tree, ELM: Extreme Learning Machine, NV: NaiveBayes, SDE: Subspace Discriminant Ensemble, WNN: Wide Neural Network, LD: Linear Discriminant, ACC: Accuracy.

2.1 Kinect

Kinect is a depth sensor platform that integrates a traditional RGB camera, an infrared camera, and a projector together [11]. Due to its strong ability to track moving objects, Kinect sensors are frequently applied to record sign language video to generate the corresponding datasets [33]. Kumar *et al.* utilized Leap Motion and Kinect sensors to capture finger and palm positions from two separate viewpoints, respectively [23]. Yang *et al.* acquired 3D depth information of hand movements from the Kinect sensor and employed hierarchical conditional random fields (CRF) as a gesture classifier [46].

A number of Kinect-based computer vision algorithms and applications have been proposed [18]. The manufacturer of Kinect, Microsoft^{TM} Cooperation, has provided software development kits to customize the behavior of the sensor. However, Kinect does not equip more adaptability and flexibility compared with the exchangeable sensor in the traditional infrastructure layer. Moreover, the updated iteration of the device limited the continued availability of hardware and influenced Kinect-based SLR research to some extent.

2.2 Flexible Sensor

Contrasted with Kinect, the flex sensor is made of flexible material with excellent ductility, which adapts to more stringent criteria [38]. By means of changing

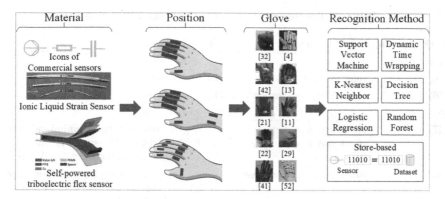

Fig. 3. The SLR works based on flex sensors. (The images of ionic liquid and self-powered triboelectric material are derived from [30] and [27], respectively).

the amount of charge, flex sensors utilize resistor and capacitor to respond to mechanical deformation [35]. Due to the adaptability of its structure, it offers various promising applications, including wearable equipment, environmental monitoring, and medical electronics, which make it easier to assess the movement of the body and skin-related parameters [34].

Flex sensors always served as the glove element to measure the signal of gesture deformation [12,40,49], which has been demonstrated in Fig. 3. Obviously, the flexible and off-the-shelf commercial existing is the main consideration of SLR researchers, which is determined by the specific materials [21,29,39]. For instance, Pukar *et al.* designed a self-powered triboelectric flex sensor with randomly distributed microstructures to detect hand movement [27]. Moreover, Nhu *et al.* utilized an ionic liquid strain sensor to assess the deformation data of the hand, which has superior sensitivity and longevity compared to traditional flexible sensors [30]. Flex sensors have emerged in the high-end mobile market to measure the degree of joint bendings, such as the fingers and wrists. Although various studies have employed machine learning approaches to build classifiers [3,10,20], most of the SLR works align sensor data with records in the database.

2.3 Electromyographic Sensor

EMG can measure the electrical activity of muscles during relaxation and contraction via surface sensors, which deliver a more stable analog signal than flex sensors. It can be applied to assess the muscular movements of fingers or arms in SLR [6,13,16], as demonstrated in Table 3. We have summarized the number of sensors, locations, signal processing algorithms, and the dimension of the obtained feature. Despite its excellent performance, EMG sensors are difficult to promote among the general public in an invasive manner. Moreover, the sensitivity to external noise of the surface EMG is also a factor that influences the SLR accuracy.

Table 3. The Summarization of EMG-based SLR Works.

Work	Amount	Signal Processing	Dimension	Placement
[13]	5	–	26	arm
[16]	3	interpolation	312	arm
[36]	8	infinite impulse response	11	arm
[6]	8	–	400	arm
[47]	4	two-stage amplifier band-pass filter	–	extensor digiti minimi palmaris longus extensor carpi radialis longus extensor carpi ulnaris
[45]	4	infinite impulse response	19	extensor digitorum flexor carpi radialis longus extensor carpi radialis longus extensor carpi ulnaris

2.4 Force Myography Sensor

FMG Sensor is a flex sensor-based electromyogram that can measure muscle squeeze and deformation produced during muscle contraction, to predict muscle activity [42], such as the connection between FSR sensors and the bracelet to extract signals from Bariou *et al.* [4]. Information about movement and deformation can be obtained through the Force Sensing Resistor (FSR) corresponding to the muscle deformation. An existing study has shown that FMG performs better than EMG in human action recognition and intention detection [15].

2.5 Inertial Sensor

Inertial sensors can measure the acceleration and angular velocity of an object, including posture measurement, motion monitoring [16,39], navigation and positioning, posture recognition, and human-computer interaction [2]. To enable functions like gesture control and virtual reality interaction, it can be employed to assess hand position and movement [3,12].

As the most explored sensor, accelerometers use tiny springs or piezoelectric materials to measure changes in the acceleration of an object [1,16,24]. Several researchers have utilized it to calculate the rotation of hand components and variations in speed [14,47].

Gyroscopes track changes in the angular velocity of object rotation, which can detect rotational inertia or angular displacement using suspended or rotating structures [1]. Gyroscopes are suitable for precisely capturing the angular velocity or improving the accuracy of gesture detection [39,47].

Inertial Measurement Unit (IMU) is an integrator of various inertial sensors, including accelerometers, gyroscopes, and magnetometers [5]. Magnetometers can access the strength and direction of a magnetic field, sense changes in the magnetic field, and measure direction and position. The merits of IMUs make them popular in motion control and precise calculation, such as displacements on gloves, which can capture motion with higher accuracy [1,45].

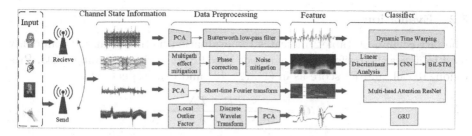

Fig. 4. The SLR work based on WiFi signals. The four different colors from the top to the bottom of the job are from [8, 25, 32, 48], respectively. The CSI can be converted into visual image information, such as spectrograms, by signal analysis operations.

2.6 Photoplethysmography Sensor

PPG sensors can monitor physiological parameters, such as blood flow and heart rate, using optoelectronic measuring technology [7]. It has been widely employed in wearable devices, including smart bracelets, smart watches, and smart glasses, to accomplish real-time monitoring and analysis of human physiological data [9]. Mehdi *et al.* extracted meaningful characteristics to identify human activities [7]. Furthermore, Zhao *et al.* applied the PPG data to investigate the variations in blood flow induced during gestural movements [50]. Although distinct gesture expressions can be captured by the fine-grained PPG signal, the color of skin and wetness should be considered since they can both affect light emission.

2.7 Radio Frequency Sensor

RF sensors can measure physical quantities by assessing reflection, scattering, and interference of RF signals [37]. It can be applied to measure data such as distance, speed, position, object form in industrial production, vital signs, and lesion locations in medical diagnostics. In addition, RF data can be exploited to recognize motions after data processing procedures such as signal transformation [17, 25]. Gurbuz *et al.* recorded an ASL sign language dataset by combining three RF sensors and two Xethru sensors [17]. Liu *et al.* have collected the Channel State Information (CSI) from WiFi signals to capture hand movement [25].

The general paradigm of WiFi-based SLR works in Fig. 4 shows the acquired CSI information can be fed into the classifiers for gesture identification [32]. Data preprocessing aims to eliminate the noise and retrieve the signal with the most distinctive features [48]. To increase SLR accuracy, feature classifiers must extract contextual data from the collected temporal vectors [8].

3 Other Categories

3.1 Dataset

We summarize the existing datasets in terms of size, category, and country, as shown in Table 2. According to these items, existing works have not used identical datasets, so that required us to establish a consistent assessment standard to facilitate conducting comparison.

3.2 Evaluation

The last column of Table 2 indicates the evaluation metrics that are commonly used to assess the performance of sensor-based SLR works, including ACC, Precision, Recall, and F1 score as in (1), (2), (3) and (4), respectively.

$$ACC = \frac{TP + TN}{TP + TN + FP + FN} \tag{1}$$

$$Recall = \frac{TP}{TP + FN} \tag{2}$$

$$Precision = \frac{TP}{TP + FP} \tag{3}$$

$$F1 = \frac{2 \cdot (Precision \cdot Recall)}{Precision + Recall} \tag{4}$$

where TP, TN, FP, FN denote True Positive, True Negative, False Positive, and False Negative, respectively. In addition, response time are also utilized as an evaluation metric to measure the sensitivity of the sensor components.

3.3 Appearance

Although Table 2 has shown that gloves are currently the most dominant form of sensor-dependent appearance in SLR, several of them appear uncomfortable and unpopular due to the bulky design and expensive cost. Recently, researchers have focused on integrating sensors into a compact, transportable bracelet or armband to overcome these drawbacks.

4 Discussion

4.1 Material

We have illustrated that semiconductor materials have occupied the predominated position in sensor-based SLR. The development themes of the future are prone to miniaturization, flexibility, and intelligence towards the novel materials design in contrast with the traditional utilization. Specifically, sensitivity and portability are the chief focusing points of the wearable SLR equipment explored by the researchers [44]. As an interdisciplinary task, novel sensor materials can boost SLR work in computer vision, which has a broad application prospect [30].

4.2 Heterogeneous SLR

Despite the rapid growth of sensor-based gesture recognition, which can be resumed from the mentioned introduction, device-based solutions have partly addressed the problem of isolated word recognition. There is still a challenge in continuous SLR work concerning reality. Two reasons should be taken into account.

1. The sensor can only be used for large-scale motion capture due to the noise and sample frequency, which means that the tracking of fine movements still requires visual information to assist in the computation.
2. The fluency and coherence of gestures in continuous sign language means that the construction of device-based SLR datasets cannot rely on the combination of isolated word datasets. The absence of relevant datasets and the alignment of annotation should be considered.

4.3 Combination of Sensor and Vision

Computer vision-based and device-based SLRs have achieved remarkable effects but still have their limits because of the lack of a panorama view. It is necessary to fix the problems in the processing of visual data, including the denoising of complex backgrounds in RGB images and the decreasing computational cost concerning massive data. Although various works have employed multi-sensor fusion techniques to facilitate the collaboration of devices from a few views, it is always perplexing that without sufficient kinds of sensors in continuous SLR because the technicality of the sensor determines the mode of captured features about sign language.

In virtue of the multimodal fusion across the data from vision and sensors, the complementarity of information can be conducive to increasing recognition accuracy. However, additional research challenges, such as the growth in data volume and the increasing cost of computational power, must be dealt with. Besides, overfitting phenomena will be more likely to occur under the unbalanced data distribution. Furthermore, data mapping from different spaces will cause computational bias, which brings trouble in the model converging. The semantic discrepancies caused by various tasks can be reduced by learning shared information from several domains [26].

5 Conclusion

SLR is a crucial research topic in human-computer interaction and computer vision. In this paper, SLR methods motivated by the taxonomy of sensor modalities have been investigated in depth. We have summarized and discussed the most representative sensor-based SLR. In the end, current challenges pointing out several promising future research directions have been proposed. Finally, we anticipate that this survey will offer a panorama of the SLR technology, which is conducive to prompting researchers in future works.

References

1. Abdullah, A., Abdul-Kadir, N.A., Che Harun, F.K.: An optimization of IMU sensors-based approach for Malaysian sign language recognition. In: ICCED, pp. 1–4 (2020)

2. Alaoui, F., Fourati, H., Kibangou, A., Robu, B., Vuillerme, N.: Kick-scooters identification in the context of transportation mode detection using inertial sensors: methods and accuracy. J. Intell. Transport. Syst. (2023). https://doi.org/10.1080/15472450.2022.2141118

3. Alosail, D., Aldolah, H., Alabdulwahab, L., Bashar, A., Khan, M.: Smart glove for bi-lingual sign language recognition using machine learning. In: IDCIoT, pp. 409–415 (2023)

4. Barioul, R., Ghribi, S.F., Ben Jmaa Derbel, H., Kanoun, O.: Four sensors bracelet for American sign language recognition based on wrist force myography. In: CIVEMSA, pp. 1–5 (2020)

5. Barraza Madrigal, J.A., Contreras Rodríguez, L.A., Cardiel Pérez, E., Hernández Rodríguez, P.R., Sossa, H.: Hip and lower limbs 3D motion tracking using a double-stage data fusion algorithm for IMU/MARG-based wearables sensors. Biomed. Signal Process. Control **86**, 104938 (2023)

6. Ben Haj Amor, A., El Ghoul, O., Jemni, M.: Deep learning approach for sign language's handshapes recognition from EMG signals. In: ITSIS, pp. 1–5 (2022)

7. Boukhechba, M., Cai, L., Wu, C., Barnes, L.E.: ActiPPG: using deep neural networks for activity recognition from wrist-worn photoplethysmography (PPG) sensors. Smart Health **14**, 100082 (2019)

8. Chen, H., Feng, D., Hao, Z., Dang, X., Niu, J., Qiao, Z.: Air-CSL: Chinese sign language recognition based on the commercial WiFi devices. Wirel. Commun. Mob. Comput. **2022** (2022). https://doi.org/10.1155/2022/5885475

9. Choi, J., Hwang, G., Lee, J.S., Ryu, M., Lee, S.J.: Weighted knowledge distillation of attention-LRCN for recognizing affective states from PPG signals. Expert Syst. Appl. 120883 (2023)

10. Chu, X., Liu, J., Shimamoto, S.: A sensor-based hand gesture recognition system for Japanese sign language. In: LifeTech, pp. 311–312 (2021)

11. DiFilippo, N.M., Jouaneh, M.K.: Characterization of different Microsoft Kinect sensor models. IEEE Sens. J. **15**(8), 4554–4564 (2015)

12. Dweik, A., Qasrawi, H., Shawar, D.: Smart glove for translating Arabic sign language "SGTArSL". In: ICCTA, pp. 49–53 (2021)

13. Fouts, T., Hindy, A., Tanner, C.: Sensors to sign language: a natural approach to equitable communication. In: ICASSP, pp. 8462–8466 (2022)

14. Galka, J., Masior, M., Zaborski, M., Barczewska, K.: Inertial motion sensing glove for sign language gesture acquisition and recognition. IEEE Sens. J. **16**(16), 6310–6316 (2016)

15. Godiyal, A.K., Singh, U., Anand, S., Joshi, D.: Analysis of force myography based locomotion patterns. Measurement **140**, 497–503 (2019)

16. Gupta, R., Bhatnagar, A.S.: Multi-stage Indian sign language classification with sensor modality assessment. In: ICACCS, vol. 1, pp. 18–22 (2021)

17. Gurbuz, S.Z., et al.: ASL recognition based on Kinematics derived from a multi-frequency RF sensor network. IEEE Sens. J. 1–4 (2020)

18. Han, J., Shao, L., Xu, D., Shotton, J.: Enhanced computer vision with Microsoft Kinect sensor: a review. IEEE T. Cybern. **43**(5), 1318–1334 (2013)

19. Hu, H., Wang, W., Zhou, W., Zhao, W., Li, H.: Model-aware gesture-to-gesture translation. In: CVPR, pp. 16423–16432 (2021)

20. Ji, L., Liu, J., Shimamoto, S.: Recognition of Japanese sign language by sensor-based data glove employing machine learning. In: LifeTech, pp. 256–258 (2022)

21. Kania, M., Korzeniewska, E., Zawiślak, R., Nikitina, A., Krawczyk, A.: Wearable solutions for the sign language. In: MEES, pp. 1–4 (2022)

22. Kudrinko, K., Flavin, E., Zhu, X., Li, Q.: Wearable sensor-based sign language recognition: a comprehensive review. IEEE Rev. Biomed. Eng. **14**, 82–97 (2021)

23. Kumar, P., Gauba, H., Roy, P.P., Dogra, D.P.: A multimodal framework for sensor based sign language recognition. Neurocomputing **259**(SI), 21–38 (2017)

24. Kwon, J., Nam, H., Chae, Y., Lee, S., Kim, I.Y., Im, C.H.: Novel three-axis accelerometer-based silent speech interface using deep neural network. Eng. Appl. Artif. Intell. **120**, 105909 (2023)

25. Liu, C., Liu, J., Shimamoto, S.: Sign language estimation scheme employing Wi-Fi signal. In: SAS, pp. 1–5 (2021)

26. Ma, Y., Zhao, S., Wang, W., Li, Y., King, I.: Multimodality in meta-learning: a comprehensive survey. Knowl.-Based Syst. **250**, 108976 (2022)

27. Maharjan, P., et al.: A human skin-inspired self-powered flex sensor with thermally embossed microstructured triboelectric layers for sign language interpretation. Nano Energy **76**, 105071 (2020)

28. Mitra, S., Acharya, T.: Gesture recognition: a survey. IEEE Trans. Syst. Man Cybern.-Syst. **37**(3), 311–324 (2007)

29. Muralidharan, N.T., Rohidh, M.R., Harikumar, M.E.: Modelling of sign language smart glove based on bit equivalent implementation using flex sensor. In: WiSP-NET, pp. 99–104 (2022)

30. Nhu, C.T., Dang, P.N., Thanh, V.N.T., Thuy, H.T.T., Thanh, V.D., Thanh, T.B.: A sign language recognition system using ionic liquid strain sensor. In: ISMEE, pp. 263–267 (2021)

31. Qahtan, S., Alsattar, H.A., Zaidan, A.A., Deveci, M., Pamucar, D., Martinez, L.: A comparative study of evaluating and benchmarking sign language recognition system-based wearable sensory devices using a single fuzzy set. Knowl.-Based Syst. **269**, 110519 (2023)

32. Qin, Y., Pan, S., Zhou, W., Pan, D., Li, Z.: WiASL: American sign language writing recognition system using commercial WiFi devices. Measurement **218**, 113125 (2023)

33. Rakun, E., Andriani, M., Wiprayoga, I.W., Danniswara, K., Tjandra, A.: Combining depth image and skeleton data from Kinect for recognizing words in the sign system for Indonesian language (SIBI [Sistem Isyarat Bahasa Indonesia]). In: ICACSIS, pp. 387–392 (2013)

34. Rashid, A., Hasan, O.: Wearable technologies for hand joints monitoring for rehabilitation: a survey. Microelectron. J. **88**, 173–183 (2019)

35. Saggio, G., Riillo, F., Sbernini, L., Quitadamo, L.R.: Resistive flex sensors: a survey. Smart Mater. Struct. **25**(1), 013001 (2016)

36. Saif, R., Ahmad, M., Naqvi, S.Z.H., Aziz, S., Khan, M.U., Faraz, M.: Multi-channel EMG signal analysis for Italian sign language interpretation. In: ICETST, pp. 1–5 (2022)

37. Sarkar, B., Takeyeva, D., Guchhait, R., Sarkar, M.: Optimized radio-frequency identification system for different warehouse shapes. Knowl.-Based Syst. **258**, 109811 (2022)

38. Sharma, A., Ansari, M.Z., Cho, C.: Ultrasensitive flexible wearable pressure/strain sensors: parameters, materials, mechanisms and applications. Sens. Actuat. A **347**, 113934 (2022)

39. Subedi, B., Dorji, K.U., Wangdi, P., Dorji, T., Muramatsu, K.: Sign language translator of Dzongkha alphabets using Arduino. In: i-PACT, pp. 1–6 (2021)

40. Suri, A., Singh, S.K., Sharma, R., Sharma, P., Garg, N., Upadhyaya, R.: Development of sign language using flex sensors. In: ICOSEC, pp. 102–106 (2020)

41. Sze, F.: From gestures to grammatical non-manuals in sign language: a case study of polar questions and negation in Hong Kong sign language. Lingua **267**, 103188 (2022)
42. Ul Islam, M.R., Bai, S.: A novel approach of FMG sensors distribution leading to subject independent approach for effective and efficient detection of forearm dynamic movements. Biomed. Eng. Adv. **4**, 100062 (2022)
43. Venugopalan, A., Reghunadhan, R.: Applying deep neural networks for the automatic recognition of sign language words: a communication aid to deaf agriculturists. Expert Syst. Appl. **185**, 115601 (2021)
44. Wang, Z., et al.: Hear sign language: a real-time end-to-end sign language recognition system. IEEE Trans. Mob. Comput. **21**(7), 2398–2410 (2022)
45. Wu, J., Sun, L., Jafari, R.: A wearable system for recognizing American sign language in real-time using IMU and surface EMG sensors. IEEE J. Biomed. Health Inform. **20**(5, SI), 1281–1290 (2016)
46. Yang, H.D.: Sign language recognition with the Kinect sensor based on conditional random fields. IEEE Sens. J. **15**(1), 135–147 (2015)
47. Yang, X., Chen, X., Cao, X., Wei, S., Zhang, X.: Chinese sign language recognition based on an optimized tree-structure framework. IEEE J. Biomed. Health Inform. **21**(4), 994–1004 (2017)
48. Zhang, N., Zhang, J., Ying, Y., Luo, C., Li, J.: Wi-phrase: deep residual-multihead model for WiFi sign language phrase recognition. IEEE Internet Things J. **9**(18), 18015–18027 (2022)
49. Zhang, Y., Xu, W., Zhang, X., Li, L.: Sign annotation generation to alphabets via integrating visual data with somatosensory data from flexible strain sensor-based data glove. Measurement **202**, 111700 (2022)
50. Zhao, T., Liu, J., Wang, Y., Liu, H., Chen, Y.: Towards low-cost sign language gesture recognition leveraging wearables. IEEE Trans. Mob. Comput. **20**(4), 1685–1701 (2021)
51. Zhou, H., Zhou, W., Zhou, Y., Li, H.: Spatial-temporal multi-cue network for sign language recognition and translation. IEEE Trans. Multimed. **24**, 768–779 (2022)

Utilizing Video Word Boundaries and Feature-Based Knowledge Distillation Improving Sentence-Level Lip Reading

Hongzhong Zhen, Chenglong Jiang, Jiyong Zhou, Liming Liang,
and Ying Gao[✉]

School of Computer Science and Engineering, South China University of Technology,
Guangzhou, China
{cshzzhen,csjiangcl_gx,202121044812,csliangliming}@mail.scut.edu.cn,
gaoying@scut.edu.cn

Abstract. Lip reading is to recognize the spoken content from silent video of lip movement. There is a general problem in sentence-level lip reading that the length of predicted text is inconsistent with actual text. To alleviate this problem, we introduce video word boundary information into sentence-level lip reading and propose TLiM-VWB model. Besides, to deal with the situation that video word boundaries can not be obtained in wild environment, we propose LiM-VWB-KD method with two knowledge distillation strategies utilizing video word boundary information implicitly. We evaluate our model and method on CMLR and LRS2 datasets with metrics of CER/WER and our proposed length difference rate (LDR). We verify the effectiveness of video word boundary information to improve sentence-level lip reading accuracy through the results of TLiM-VWB model. We also show the effectiveness of LiM-VWB-KD method especially with feature-based strategy. Our LiM-VWB-KD method achieves the best result on Chinese sentence-level lip reading among methods using Transformer architecture and achieves the new state-of-the-art performance in speaker-independent setting on CMLR.

Keywords: Lip reading · Deep learning · Video word boundary · Knowledge distillation

1 Introduction

Lip reading, also known as visual speech recognition, is to recognize spoken content from silent video of lip movement. In recent years, the performance of lip reading is gradually improving with the development of deep learning. Previous researches [1,10,19] on lip reading mainly paid attention to improving the model structure, but few researches try to improve the accuracy of lip reading model from the perspective of text. We find a frequent problem in sentence-level lip reading that the length of predicted text is inconsistent with the actual text, which is illustrated in Fig. 1(a). This problem greatly reduces the accuracy

© The Author(s), under exclusive license to Springer Nature Singapore Pte Ltd. 2024
Q. Liu et al. (Eds.): PRCV 2023, LNCS 14430, pp. 269–281, 2024.
https://doi.org/10.1007/978-981-99-8537-1_22

because missing or redundant words will be considered as wrong words and even will result in the wrong recognition of other words in the context. This problem shows current lip reading models have not grasped the video duration of each word so that they don't accurately predict the number of words in the video.

In order to alleviate this problem, we use video word boundary information to strengthen the lip reading model. We propose Transformer-based Lip Reading Model Merging Video Word Boundaries (TLiM-VWB). To the best of our knowledge, we are the first to introduce video word boundary information to sentence-level lip reading. On the other hand, we can not directly obtain the video word boundaries in the wild scene. In order to utilize the video word boundary information implicitly, we propose Lip Reading Model Training Method Utilizing Video Word Boundaries by Knowledge Distillation (LiM-VWB-KD).

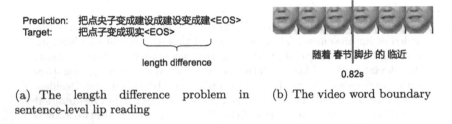

(a) The length difference problem in sentence-level lip reading

(b) The video word boundary

Fig. 1. The length difference problem and the video word boundary.

We conduct LiM-VWB-KD method with response-based strategy and feature-based strategy. To evaluate the accuracy of sentence-level lip reading, we not only use conventional metrics character error rate (CER) and word error rate (WER) but also propose to use length difference rate (LDR) to measure the length accuracy of the model to predict sentences. We conduct experiments on CMLR and LRS2 datasets and verify the effectiveness of TLiM-VWB model, which shows video word boundary information can effectively help to improve accuracy of sentence-level lip reading. We also verify the effectiveness of LiM-VWB-KD method. On CMLR, our LiM-VWB-KD method achieves CER of 18.06% in speaker-dependent setting and 28.67% in speaker-independent setting. Among methods using Transformer architecture, our method achieves the best result on Chinese sentence-level lip reading. More importantly, we update the state-of-the-art result in speaker-independent setting on CMLR.

In general, the contributions of our work are as follows:

- We introduce video word boundary information into sentence-level lip reading model for the first time and propose TLiM-VWB model. We verify utilizing video word boundary information can effectively improve the accuracy of sentence-level lip reading.
- To improve the accuracy of lip reading in the wild scene, we propose LiM-VWB-KD method utilizing video word boundary information implicitly. Our

method achieves the best result among Chinese sentence-level lip reading models with Transformer architecture and achieves the new state-of-the-art performance in speaker-independent setting on CMLR dataset.

- We propose to use length difference rate (LDR) to measure the length accuracy of the model to predict sentences. Using this metric, we show the improvement of our model and method on length accuracy of lip reading.

2   Related Work

2.1   Sentence-Level Lip Reading

Sentence-level lip reading recognizes an entire sentence from a lip movement video and is a complex sequence-to-sequence task. Afouras et al. [1] introduced Transfomer architecture to English sentence-level lip reading and proposed TM-CTC model. In Chinese field, Zhao et al. [19] proposed a specialized CSSMCM model and released CMLR dataset. After that, CTCH-LipNet [10] model utilized Transformer architecture to perform Chinese sentence-level lip reading and achieved better result on CMLR. Recently, LipSound2 [11] proposed a extra speaker-independent task baseline for Chinese sentence-level lip reading.

2.2   Video Word Boundary

The video word boundary refers to the video frame position corresponding to the boundary between words in the lip movement video, which is illustrated in Fig. 1(b). In lip reading datasets, video word boundaries are usually indicated by the start and end times of the words. In word-level lip reading field, Stafylakis et al. [14] used video word boundaries to differentiate the video mapped to one word into proper clip and redundant clips and improved the accuracy of lip reading on LRW [3] dataset. Feng et al. [4] used the same method on LRW-1000 [17] dataset and also gain the effect. Following [4,14], Ma et al. [9] also showed using video word boundaries is a beneficial strategy for word-level lip reading.

2.3   Knowledge Distillation

Knowledge distillation [7] transfers the output layer knowledge from the teacher model into the student model traditionally. In the past few years, feature-based knowledge distillation was also widely used, and the losses on intermediate representation were proposed [13,18]. There are also some methods [8,12,20] utilizing knowledge distillation to improve the accuracy of lipreading. For example, LIBS [20] distilled multi-granularity knowledge from the speech recognizer to enhance the lip reader.

3   Methods

3.1   Transformer-Based Lip Reading Model Merging Video Word Boundaries (TLiM-VWB)

Sentence-Level Lip Reading Pipeline. We employ a cascaded lip reading pipeline to perform sentence-level lip reading task. Firstly, the video of lip movement is input in a visual front-end model described in [15] to generate visual feature sequence. Then our proposed model uses this visual feature sequence to generate character probabilities that are used to predict the sentence.

Overall Architecture. The overall architecture of TLiM-VWB is showed in Fig. 2(a). To merge the video word boundaries when dealing visual features, we set the video word boundary embedding module between the encoder and decoder. We design the structure of the encoder and decoder according to TM-CTC [1] model. The main component of encoder and the decoder is a stack of Transformer encoder layers [16]. The video word boundary embedding module utilizes video word boundaries to enhance the encoding result and assist the decoding process. The output of decoder are the probabilities that are used to calculate the CTC loss [5] in the model training process.

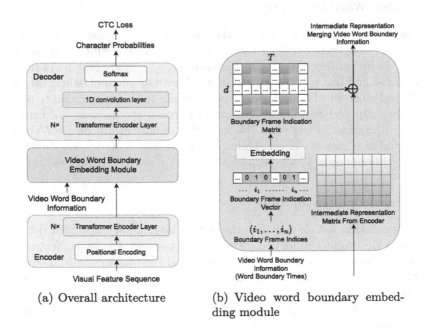

(a) Overall architecture   (b) Video word boundary embedding module

Fig. 2. The architecture of TLiM-VWB.

Video Word Boundary Embedding. Video word boundary embedding module combines video word boundary information with intermediate representation from encoder and forms intermediate representation merging video word boundary information. The video word boundary information is often recorded in the form of word boundary times. Therefore, we carry out a series of transformations to make the video word boundary information can be added with the intermediate representation matrix. Firstly, we use a sequence $S_{et} = (t_1, ..., t_n)$ to denotes the end times (boundary times) of words in the video, where n is the number of words. Then we multiply the time in S_{et} by the video frame rate to get the boundary frame index sequence S_{bfi}, i.e.

$$S_{bfi} = (i_1, ..., i_n), \ i_k = t_k \cdot f, \ t_k \in S_{et}, \tag{1}$$

where f is the video frame rate. We set a boundary frame indication vector $V_b = (b_1, ..., b_T)$, where T denotes the number of frames in the video. As a result, V_b is in the same length as the video frame sequence. We set the element value of V_b to 1 at the boundary frame index and 0 at the rest, i.e.

$$b_i = \begin{cases} 1, \ i \in S_{bfi} \ and \ i \in [1, T], \\ 0, \ i \notin S_{bfi} \ and \ i \in [1, T]. \end{cases} \tag{2}$$

Then the boundary frame indication vector V_b passes through the embedding layer to form the the boundary frame indication matrix M_b, i.e.

$$M_b = Embedding(V_b). \tag{3}$$

In this process, every element in V_b is extended to the d dimensional vector, where d is the dimension of intermediate representation vector corresponding to a frame. As a result, the size of M_b is consistent with that of the intermediate representation matrix M_{inter} processed by the encoder for the whole video. After that, the corresponding positional values of M_b and M_{inter} are added to form the intermediate representation merging video word boundary information M_{imb}, i.e.

$$M_{imb} = M_{inter} + M_b. \tag{4}$$

The illustration of video word boundary embedding module is showed in Fig. 2(b).

3.2 Lip Reading Model Training Method Utilizing Video Word Boundaries by Knowledge Distillation (LiM-VWB-KD)

Overview. In the inference process of wild scene, the video word boundaries are unavailable. To improve the accuracy of lip reading in this situation, we propose a two-stage knowledge distillation based lip reading model training method utilizing video word boundary information implicitly. In the first stage, we pre-train TLiM-VWB model. In the second stage, we take the pre-trained TLiM-VWB model as the teacher model and the similar model without video word boundary embedding module as the student model for knowledge distillation.

Knowledge Distillation Strategies. In the first stage of our method, we pre-trained the teacher model using CTC loss. In the second stage, i.e. knowledge distillation stage, the pre-trained weights of teacher model are frozen. In this process, the teacher model and studnet model receive the same visual feature sequence as input, and only the teacher model receives video word boundary information. We design a feature-based knowledge distillation strategy that is related to video word boundary information more closely. In contrast, we also design the response-based knowledge distillation strategy. The illustration of two strategies is showed in Fig. 3.

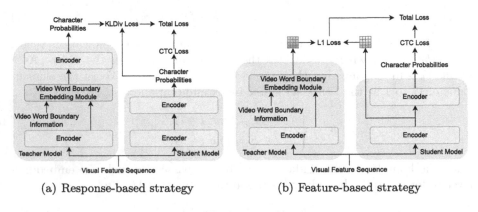

(a) Response-based strategy   (b) Feature-based strategy

Fig. 3. Two knowledge distillation strategies used in LiM-VWB-KD.

Response-Based Strategy. We distill the output layer knowledge from the teacher to the student in response-based strategy. We calculate the KL-divergence loss between the probability distributions that output by decoders of the student and the teacher. Then we add it with CTC loss of the student model weightedly, i.e.

$$L_{stu1} = L_{CTC} + \lambda_1 L_{KD1}, \tag{5}$$

$$L_{KD1} = \sum_{i=1}^{T} \sum_{c \in C} p_i^t(c) \cdot \log \frac{p_i^t(c)}{p_i^s(c)}, \tag{6}$$

where L_{CTC} is the CTC loss and T is the number of frames in the video. C denotes the character set that we use. The $p_i^t(c)$ denotes the probability that teacher model recognize the ith frame of video as character c and $p_i^s(c)$ denotes the similar meaning for the student model.

Feature-Based Strategy. In feature-based strategy, we use the intermediate representation merging video word boundary information as the knowledge. For the input visual feature sequence, the teacher model calculates and outputs the intermediate representation merging video word boundary information M_{imb}^t. The student model output the intermediate representation M_{inter}^s calculated by the encoder for the same visual feature sequence. We calculate the L1 loss between the M_{inter}^s and M_{imb}^t. The student model also uses the decoder to get the character probabilities and calculate the CTC loss. The two losses are added together weightedly to get the total target loss of the student model, i.e.

$$L_{stu2} = L_{CTC} + \lambda_2 L_{KD2}, \tag{7}$$

$$L_{KD2} = \frac{1}{Td} \sum_{i=1}^{T} \left\| M_{inter\,i}^s - M_{imb\,i}^t \right\|_1, \tag{8}$$

where $\|\cdot\|_1$ is the ℓ_1-norm. $M_{inter\,i}^s$ denotes the ith vector in M_{inter}^s corresponding to the feature of the ith frame in the video and $M_{imb\,i}^t$ denotes the similar meaning. The d is the dimension of the vector $M_{inter\,i}^s$ and $M_{imb\,i}^t$. The student model is optimized by minimizing the L_{stu2}. In this strategy, the output of the student's encoder will be close to the output of teacher's video word boundary embedding module. In this way, the encoder of the student model implicitly learns about the video word boundary information to help the decoder decode.

4   Experiment

4.1   Dataset

CMLR   [19] is the largest public Chinese sentence-level lip reading dataset. It includes 102072 spoken sentences by 11 speakers from a Chinese national news program "News Broadcast". The boundary time of each word is included in the transcription. Its training, validation and test set contain 71448, 10206 and 20418 sentences respectively.

LRS2   [1] is a large-scale English sentence-level lip reading dataset. It includes 144482 spoken sentences from BBC television. Its pre-training, training, validation and test set contain 96318, 45839, 1082 and 1243 sentences respectively. In its pre-training set, the boundary time of each word is included.

4.2   Evaluation Metrics

We use CER and WER to measure the accuracy of sentence-level lip reading. Here, WER is only used to evaluate English sentence-level lip reading.

$$CER/WER = \frac{S + D + I}{N}, \tag{9}$$

where S, D, I are the number of substitution, deletion and insertion from predicted sentence to target sentence. N is the number of characters or words in target sentence.

In addition, we propose to use length difference rate (LDR) to measure the degree of length difference between the predicted sentence and the target sentence.

$$LDR = \frac{|\hat{N} - N|}{N},\tag{10}$$

where $\hat{N}$ is the number of characters or words in predicted sentence.

We calculate the global value of these metrics by using the corresponding total numbers in all sentences that are evaluated.

4.3 Implementation

Preprocess. We clip the lip regions of video frames in the dataset and adjust them to the size of 112×112. Then we use the pre-trained visual front-end model in [1] to extract the visual feature sequences.

Experiment Schedule. On CMLR dataset, we pretrain the TLiM-VWB model and carry out the LiM-VWB-KD method in two settings. In speaker-dependent setting, we use the original dataset division as in [19]. In speaker-independent setting, we need to make the speakers of training set and test set are different. So we divide the dataset in the same way with [11]. S1 and S6 parts of CMLR serve as test set, and other parts serve as training set and validation set. On LRS2 dataset, we use the pre-training set to pretrain the TLiM-VWB model with the curriculum learning and carry out the LiM-VWB-KD method. We divide the pre-validation set from the pre-training set for validation. In the curriculum learning, we train the model 12 times iteratively with the number of words in the sample increasing from 1 to all in the sentence.

Text Generation. In the inference process, we use the greedy search and the prefix beam search [6] with language model to get the predicted sentence. We use the swig_decoders python package that implements the transcription score calculation method in [2] to carry out the prefix beam search.

Training Details. The CMLR vocabulary has 3517 Chinese characters and the end mark <EOS>, while the LRS2 vocabulary includes 26 letters, 10 Arabic numerals, space, apostrophe and <EOS>. We train 14290 samples in a step (virtual epoch) for CMLR and 16384 for LRS2. The batch size is 32 for CMLR and 16 for LRS2. We train the models using Adam optimizer with defult parameters execpt $\beta_2 = 0.98$ in knowledge distillation stage for CMLR. The learning rate starts at 10^{-4} and decrease as 70% untill 10^{-7} and 50% untill 10^{-9} every time the validation set accuracy plateaus for 10 steps in the pre-training and knowledge distillation stage for CMLR. For LRS2, the learning rate wait 25 steps for accuracy plateauing and decrease as 50% untill 10^{-6}.

4.4   Results

Effectiveness of TLiM-VWB Model. To compare our TLiM-VWB model, we also implement TM-CTC model as a baseline on CMLR dataset in both speaker-dependent and speaker-independent settings. Table 1 shows the results of two models on the test set of CMLR, where +LM means using prefix beam search with language model. On CMLR dataset, the lip reading accuracy of the TLiM-VWB model exhibits significant improvement over that of TM-CTC. In all conditions, CER experiences a reduction of about a third. Particularly in the speaker-independent setting, there is a more pronounced decrease in CER, indicating that TLiM-VWB model improve the generalization performance more obviously. In addtion, LDR of TLiM-VWB model decreases more than a half compared to TM-CTC and has become very small. This result demonstrates that the TLiM-VWB model can achieve more accurate sentence length prediction by leveraging video word boundary information.

Table 1. Comparison of results between TLiM-VWB and TM-CTC on the test set of CMLR.

Medel	Spk-Dep				Spk-Indep			
	CER		LDR		CER		LDR	
		+LM		+LM		+LM		+LM
TM-CTC	32.31%	22.61%	4.81%	4.18%	47.01%	34.56%	6.16%	5.64%
TLiM-VWB	**22.67%**	**14.89%**	**1.63%**	**1.27%**	**31.12%**	**19.96%**	**2.40%**	**2.00%**
Variation	↓9.64%	↓7.72%	↓3.18%	↓2.91%	↓15.89%	↓14.60%	↓3.86%	↓3.64%

Table 2. Comparison of results between TLiM-VWB and TM-CTC on the pre-validation set of LRS2.

Model	N = 2			N = 5		
	CER	WER	LDR	CER	WER	LDR
TM-CTC	48.26%	82.52%	2.80%	51.11%	85.82%	8.10%
TLiM-VWB	**39.84%**	**71.99%**	**0.83%**	**39.80%**	**69.64%**	**1.29%**
Variation	↓8.42%	↓10.53%	↓1.97%	↓11.31%	↓16.18%	↓6.81%
Model	N = 9			N = 17		
	CER	WER	LDR	CER	WER	LDR
TM-CTC	50.59%	83.72%	8.94%	49.77%	82.53%	10.08%
TLiM-VWB	**36.45%**	**62.73%**	**1.24%**	**34.55%**	**58.24%**	**1.08%**
Variation	↓14.14%	↓20.99%	↓7.70%	↓15.22%	↓24.29%	↓9.00%
Model	N = 29			N=all		
	CER	WER	LDR	CER	WER	LDR
TM-CTC	49.33%	81.69%	11.23%	48.13%	79.48%	12.59%
TLiM-VWB	**33.58%**	**55.36%**	**0.94%**	**32.59%**	**53.83%**	**1.07%**
Variation	↓15.75%	↓26.33%	↓10.29%	↓15.54%	↓25.65%	↓11.52%

For LRS2 dataset, Table 2 shows the results of TLiM-VWB and TM-CTC model on the pre-validation set during partial iterations of curriculum learning when pre-training, where N denotes the number of words in the sentences. For all evaluation metrics, the performance of TLIM-VWB model is obviously better than TM-CTC. Especially, compared with TM-CTC, the LDR of TLIM-VWB is very small. Besides, as the number of words in sentences increases in the curriculum learning process, the improvement of TLIM-VWB model also becomes greater. This is because the video word boundary information plays a bigger role when the sentence is longer. The results on CMLR and LRS2 dataset proves the effectiveness of TLiM-VWB model, which shows that the accuracy of sentence-level lip reading can be effectively improved with video word boundary information.

Effectiveness of LiM-VWB-KD Method. We carry out two strategies of LiM-VWB-KD method in speaker-dependent setting on CMLR dataset. Table 3 shows the results of two strategies with different knowledge distillation loss weights. We compare these with the results of TM-CTC, which are equivalent to the results of training the student model alone without the guidance of the teacher. From this table, we can see that both two strategies can improve the accuracy of lip reading with appropriate knowledge distillation loss weight. But the feature-based strategy has better performance. The reason for this is that the key factor of teacher model's better performance is the participation of video word boundary information. Through feature-based knowledge distillation strategy, the student model can implicitly learn the key features brought by video word boundary information and have a more similar computational process to the teacher model in the decoder. In contrast, response-based strategy only makes the student learn the output of the teacher and ignore the process.

Table 3. Results of two LiM-VWB-KD strategies in speaker-dependent setting on CMLR.

Method	Weight	CER	+LM	LDR	+LM
LiM-VWB-KD (Response-based)	$\lambda_1 = 0.002$	29.62%	20.68%	3.36%	3.87%
	$\lambda_1 = 0.01$	29.45%	20.98%	3.79%	5.79%
	$\lambda_1 = 0.05$	29.71%	21.75%	6.42%	8.06%
	$\lambda_1 = 0.10$	31.78%	22.57%	6.59%	7.45%
LiM-VWB-KD (Feature-based)	$\lambda_2 = 0.5$	29.77%	21.24%	3.80%	3.89%
	$\lambda_2 = 1.0$	**26.73%**	18.18%	**2.88%**	2.65%
	$\lambda_2 = 2.0$	26.89%	**18.06%**	2.89%	**2.63%**
	$\lambda_2 = 5.0$	27.30%	18.63%	3.00%	2.75%
TM-CTC	–	32.31%	22.61%	4.81%	4.18%

We also carry out LiM-VWB-KD method in speaker-independent setting on CMLR. The results of our method and other state-of-the-art methods on CMLR are showed in Table 4. Using the feature-based strategy, our LiM-VWB-KD method can achieve the CER of 18.06% in speaker-dependent setting and 28.67% in speaker-independent setting. Compared with other methods that use Transformer architecture such as CTCH-LipNet, our method achieve the best result. It is worth noting that our method achieve the new state-of-the-art performance in speaker-independent setting on CMLR.

Table 4. Comparison of results with other existing methods.

Method	CMLR		LRS2	
	Spk-Dep	Spk-Indep		
	CER	CER	CER	WER
WAS [20]	38.93%	–	48.28%	68.19%
LipCH-Net-seq [19]	34.07%	–	–	–
CSSMCM [19]	32.48%	–	–	–
LIBS [20]	31.27%	–	45.53%	65.29%
TM-CTC [1]	32.31%	47.01%	37.21%	65.00%
TM-CTC + LM [1]	22.61%	34.56%	35.57%	54.70%
LipSound2 [11]	25.03%	36.56%	–	–
LipSound2 + LM [11]	22.93%	33.44%	–	–
CTCH-LipNet [10]	22.01%	–	–	–
LiM-VWB-KD (Response-based)	29.45%	43.86%	36.90%	62.46%
LiM-VWB-KD (Response-based) + LM	20.68%	30.27%	35.69%	54.58%
LiM-VWB-KD (Feature-based)	26.73%	41.44%	35.73%	61.52%
LiM-VWB-KD (Feature-based) + LM	**18.06%**	**28.67%**	**34.72%**	**52.84%**

On LRS2, we use LiM-VWB-KD method only on the pre-training set because of the absence of video word boundaries in the training set. We continue to train the student model without the teacher on the training set. Nevertheless, we also see the improvement of lip reading accuracy compared with baseline using this method, which is also showed in Table 4.

5   Conclusion

In this paper, we propose TLiM-VWB model to introduce the video word boundary information for sentence-level lip reading. Through carrying out experiments on CMLR and LRS2 dataset, we show the effect of video word boundary information to improve sentence-level lip reading accuracy through our TLiM-VWB model. We also propose LiM-VWB-KD method to utilize video word boundary information implicitly. Through comparing our method with other state-of-the-art methods on CMLR, we show the advancement of our method in Chinese sentence-level lip reading field.

References

1. Afouras, T., Chung, J.S., Senior, A., et al.: Deep audio-visual speech recognition. IEEE Trans. Pattern Anal. Mach. Intell. **44**(12), 8717–8727 (2018)
2. Amodei, D., Ananthanarayanan, S., Anubhai, R., et al.: Deep speech 2: end-to-end speech recognition in English and Mandarin. In: International Conference on Machine Learning, pp. 173–182. PMLR (2016)
3. Chung, J.S., Zisserman, A.: Lip reading in the wild. In: Lai, S.-H., Lepetit, V., Nishino, K., Sato, Y. (eds.) ACCV 2016. LNCS, vol. 10112, pp. 87–103. Springer, Cham (2017). https://doi.org/10.1007/978-3-319-54184-6_6
4. Feng, D., Yang, S., Shan, S.: An efficient software for building lip reading models without pains. In: 2021 IEEE International Conference on Multimedia & Expo Workshops (ICMEW), pp. 1–2. IEEE (2021)
5. Graves, A., Fernández, S., Gomez, F., et al.: Connectionist temporal classification: labelling unsegmented sequence data with recurrent neural networks. In: Proceedings of the 23rd International Conference on Machine Learning, pp. 369–376 (2006)
6. Hannun, A.Y., Maas, A.L., Jurafsky, D., et al.: First-pass large vocabulary continuous speech recognition using bi-directional recurrent DNNs. arXiv preprint arXiv:1408.2873 (2014)
7. Hinton, G., Vinyals, O., Dean, J.: Distilling the knowledge in a neural network. arXiv preprint arXiv:1503.02531 (2015)
8. Ma, P., Martinez, B., Petridis, S., et al.: Towards practical lipreading with distilled and efficient models. In: ICASSP 2021-2021 IEEE International Conference on Acoustics, Speech and Signal Processing (ICASSP), pp. 7608–7612. IEEE (2021)
9. Ma, P., Wang, Y., Petridis, S., et al.: Training strategies for improved lip-reading. In: ICASSP 2022-2022 IEEE International Conference on Acoustics, Speech and Signal Processing (ICASSP), pp. 8472–8476. IEEE (2022)
10. Ma, S., Wang, S., Lin, X.: A transformer-based model for sentence-level Chinese mandarin lipreading. In: 2020 IEEE Fifth International Conference on Data Science in Cyberspace (DSC), pp. 78–81. IEEE (2020)
11. Qu, L., Weber, C., Wermter, S.: LipSound2: self-supervised pre-training for lip-to-speech reconstruction and lip reading. IEEE Trans. Neural Netw. Learn. Syst. (2022)
12. Ren, S., Du, Y., Lv, J., et al.: Learning from the master: distilling cross-modal advanced knowledge for lip reading. In: Proceedings of the IEEE/CVF Conference on Computer Vision and Pattern Recognition, pp. 13325–13333 (2021)
13. Romero, A., Ballas, N., Kahou, S.E., et al.: FitNets: hints for thin deep nets. In: International Conference on Learning Representations (2015)
14. Stafylakis, T., Khan, M.H., Tzimiropoulos, G.: Pushing the boundaries of audio-visual word recognition using residual networks and LSTMs. Comput. Vis. Image Underst. **176**, 22–32 (2018)
15. Stafylakis, T., Tzimiropoulos, G.: Combining residual networks with LSTMs for lipreading. In: Interspeech (2017)
16. Vaswani, A., Shazeer, N., Parmar, N., et al.: Attention is all you need. In: Advances in Neural Information Processing Systems, vol. 30 (2017)

17. Yang, S., Zhang, Y., Feng, D., et al.: LRW-1000: a naturally-distributed large-scale benchmark for lip reading in the wild. In: 2019 14th IEEE International Conference on Automatic Face & Gesture Recognition (FG 2019), pp. 1–8. IEEE (2019)
18. Yu, L., Yazici, V.O., Liu, X., et al.: Learning metrics from teachers: compact networks for image embedding. In: Proceedings of the IEEE/CVF Conference on Computer Vision and Pattern Recognition, pp. 2907–2916 (2019)
19. Zhao, Y., Xu, R., Song, M.: A cascade sequence-to-sequence model for Chinese mandarin lip reading. In: Proceedings of the ACM Multimedia Asia, pp. 1–6 (2019)
20. Zhao, Y., Xu, R., Wang, X., et al.: Hearing lips: improving lip reading by distilling speech recognizers. In: Proceedings of the AAAI Conference on Artificial Intelligence, vol. 34, pp. 6917–6924 (2020)

Denoised Temporal Relation Network for Temporal Action Segmentation

Zhichao Ma and Kan Li[✉]

Beijing Institute of Technology, Beijing 100081, China
likan@bit.edu.cn

Abstract. Temporal relations among action segments play a crucial role in temporal action segmentation. Existing methods tend to employ the graph neural network to model the temporal relation. However, the performance is unsatisfactory and exhibits serious over-segmentation due to the generated noisy features. To solve the above issues, we present an action segmentation framework, termed a denoised temporal relation network (DTRN). In DTRN, a temporal reasoning module (TRM) models inter-segment temporal relations and conducts feature denoising jointly. Specifically, the TRM conducts an uncertainty-gated reasoning mechanism for noise-immune and utilizes a cross-attention-based structure to combine the informative clues from the discriminative enhance module which is trained under Selective Margin Plasticity (SMP) to ensure informative clues, SMP adjusts the decision boundary adaptively by changing specific margins in real-time. Our framework is demonstrated to be effective and achieves state-of-the-art performance of accuracy, edit score, and F1 score on the challenging 50Salads, GTEA, and Breakfast benchmarks.

Keywords: Denoised Temporal Relation Network · Temporal Action Segmentation · Selective Margin Plasticity

1 Introduction

Temporal action segmentation [8] involves temporal segmenting and recognizing human actions from untrimmed videos, which is an essential task for analyzing the ability of understanding human actions [10,24] and with wide applications [6,8]. Modeling temporal relation among action segments is beneficial for temporal action segmentation [12,24], since sequential human behavior in daily life always forms a meaningful event and both the correlation and anti-correlation among actions provide informative clues for recognizing actions [12], which conduces to reasoning the actions whose objects are blurred or blocked. For example, if we see the taking cabbage action followed by taking a knife, the cutting cabbage action would be correctly inferred although the cabbage is blocked by the hand. Most existing methods [6,8,24] focus on capturing frame-level temporal dependencies or introducing some complementary techniques [6,8,24] to improve the perfromance through temporl modeling. They [8,24] are inadequate

Q. Liu et al. (Eds.): PRCV 2023, LNCS 14430, pp. 282–294, 2024.
https://doi.org/10.1007/978-981-99-8537-1_23

to capture temporal relations among action segments due to lacking explicit segment-level relation modeling. Some few works [12,21] explore the graph convolution networks (GCNs) to model the segment-level temporal relations, they first transform the initial segmentation predicted by the module of the previous stage to temporally ordered node representations according to the boundaries obtained from initial segmentation, then update the node representations through message passing on the graph for refining the segmentation. However, their performance is very unsatisfactory triggered by the generated noisy features. Specifically, they [12,21] directly perform message passing on the graph using heuristic graphs from the initial segmentation to update node representations and do not pay attention to process noise nodes caused by inaccurate initial segmentation.

In this paper, we propose a novel action segmentation framework, called denoised temporal relation network (DTRN), to overcome the above issues. To be specific, a temporal reasoning module in DTRN first conducts an uncertainty-gated reasoning mechanism for noise-immune, then conducts feature denoising (FD) by utilizing a cross-attention-based structure to combine the clues from the discriminative enhance module (DEM). To provide informative clues, DEM is trained under Selective Margin Plasticity (SMP). In summary, our contributions are four-fold:

- We propose a novel action segmentation framework, called DTRN, which overcomes incorrect action boundaries and noisy features to improve action segmentation.
- In DTRN, a temporal reasoning module (TRM) conducts feature denoising which first controls inter-node message passing by uncertainty-guided reasoning mechanism for noise-immune, and utilize the cross-attention-based structure to integrate informative clues for feature denoising.
- In DTRN, SMP adjusts the decision boundary adaptively to provide informative clues for FD by changing specific margins in real time during training.
- Our framework is demonstrated to be effective and efficient by extensive ablation studies, and achieve new state-of-the-art performance on the 50Salads, GTEA, and Breakfast benchmarks.

2   Related Work

2.1   Action Segmentation

Given a video of human daily activities, temporal action segmentation methods predict human actions at every frame [5]. There are some representative early works to realize action segmentation, such as sequential models [3] and multi-stream architectures, but they suffer from information forgetting [8] and computational redundancy respectively due to the iterative way and the way of integrating multiple streams. To ensure temporal smoothness, many approaches [21] apply probabilistic models over framewise features. Among these models, the transformer-based method [24] is competitive but not efficient due to the full connection and self-attention operation on long video sequences, TCN [6,8] is more efficient than the transformer where long-term temporal dependency can be easily captured by simply stacking. However, TCN usually suffers from over-segmentation [19] due to ignoring the temporal relations among action segments. Although GCNs-based methods [12,21] are devoted to model segment-level temporal relations, they have unsatisfying performance in frame recognition. Our DTRN is inspired by the above phenomena to explore effective structures to integrate the advantages of different models.

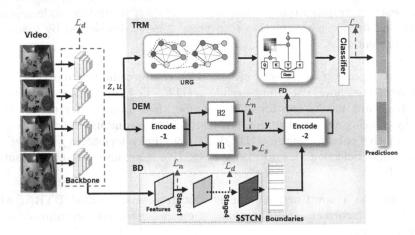

Fig. 1. The architecture of DTRN. The backbone module generates an initial segmentation z and uncertainties u, and u and detected boundaries are fed to TRM to refine z. Both Encoder-1 and Ecoder-2 are the sub-networks of SSTCN [8]. H1 and H2 are two different 1×1 temporal convolutions.

2.2   Uncertainty and Evidence Theory

Various uncertainty learning theories [14,17] have been proposed, such as the subjective logic [14] and the RBF distances. The epistemic uncertainty learning is our focus, where epistemic uncertainty [14] denotes the uncertainty of model parameters induced by the training and framework, and it can be reduced with an increasing amount of training data. Typical methods include the bayesian network, TCP [7], and ensemble methods [17]. Bayesian network replaces the deterministic parameters with the distribution of stochastic characteristics, but it takes a huge computation cost, TCP [7] pursues a more efficient way that generates uncertainty by predicting confidence with an extra module. Ensemble method [17] utilizes the difference in prediction provided by multiple predictors to determine the uncertainty for further reducing the computation cost. However, the above methods are not efficient due to relying on extra modules. The Dempster-Shafer Evidence Theory (DST) can be applied to model the epistemic uncertainty and achieve the Bayesian inference based on the subjective logic [14] without extra modules. Our DTRN learns the uncertainty based on the DST.

3   Technical Approach

3.1   Overview

Given an untrimmed video of T frames where each frame is with size $H \times W$ in RGB color, the goal of DTRN is to predict the action categories of every frame. The pipeline of DTRN is shown in Fig. 1, the video is first fed to an off-the-shelf backbone module for generating an initial segmentation, then TRM models inter-segment temporal relation and conducts FD jointly to refine the segmentation basing the boundaries awarded by the boundary detector (BD). The backbone model first extracts the features f by I3D [3] usually, then feeds f to the unique temporal models. Our BT receives f as input and outputs the predicted segmentation first with SSTCN [8], then obtains the boundaries according to the segmentation. Thus the outputs of BD and TRM are supervised under function $\mathcal{L}_n$ during training:

$$\mathcal{L}_n(p_{t,c}) = -\frac{1}{T}\sum_{c=1}^{C}(\beta_{h_t}\log p_{t,c}) + \lambda\mathcal{L}_m(p_{t,c}), \tag{1}$$

$$\mathcal{L}_m(p) = \mathbb{E}_{(h_t,t)}(\kappa_p^2(1-\iota(\kappa-\tau))+\tau^2\iota(\kappa_p-\tau)), \kappa_p = |\log(p_{t,c})-\log(p_{t-1,c})|. \tag{2}$$

where $\mathcal{L}_m$ is the truncated mean squared error to smooth predictions, $p_{t,c}$ is the cth (ground truth category) element of the softmax output of the tth frame, and $\iota(\cdot)$ would output one if the input is true, otherwise output zero. The backbone outputs uncertainties $u = (u_1, \cdots, u_T)$ for its predictions $z = (z_1, \cdots, z_T)$, where u_i is the uncertainty for $z_i \in \mathbb{R}^C$ (the predicted action likelihood for the ith frame). To achieve it, we keep the structure unchanged and make z obtained from the e exponent function and supervised by the objective function $\mathcal{L}_d$ (introduced in Sect. 3.2) during training. The TRM contains two branches where DEM provides informative clues to another branch for feature denoising (FD), another branch conducts FD and the uncertainty-gated reasoning on the graph (URG) to update features, and the updated features are fed to a linear classifier to obtain the final segmentation. To ensure informative clues, DEM is trained under SMP and $\mathcal{L}_n$. By jointly optimizing the backbone, BD, DEM, and TRM, the segmentation z can be effectively refined.

3.2   Epistemic Uncertainty Prediction

To obtain the uncertainties of the inference, z are supervised by $\mathcal{L}_d$ based on the Empirical Baye [13], where an adjusted Dirichlet distribution (4) that assigns belief masses to every class is as a prior to polynomial likelihood to obtain the marginal likelihood for the frame i:

$$\mathcal{L}_d = -\log\left[\int \prod_{c=1}^{C} p_{i,c}^{y_{ic}} \frac{1}{B(\boldsymbol{\alpha}_i)} \prod_{c=1}^{C} p_{i,c}^{\alpha_c^i-1} d\boldsymbol{p}_i\right] + \lambda\mathcal{L}_m(\hat{e})$$

$$= \sum_{c=1}^{C} y_{ic}\left[\log(\sum_{c=1}^{C} \alpha_c^i) - \log(\alpha_c^i)\right] + \lambda\mathcal{L}_m(\hat{e}) \tag{3}$$

$$D(\boldsymbol{p}|\boldsymbol{\alpha}) = \begin{cases} \dfrac{1}{B(\boldsymbol{\alpha})} \prod_{c=1}^{C} p_c^{\alpha_c^t-1} & \text{for } \boldsymbol{p} \in S_t \\ 0 & \text{otherwise} \end{cases}, \tag{4}$$

where $B(\cdot)$ is the beta function, $\boldsymbol{\alpha}$ is the distribution parameter, and $\mathcal{L}_m$ has the same form as (2), but replaces p in (2) to $\hat{e}$ to smooth the evidence, thus κ_p converts to $\kappa_{\hat{e}}$:

$$\kappa_{\hat{e}} = |\log(\frac{e_t}{\sum_{c=1}^{C} e_{tc}}) - \log(\frac{e_{t-1}}{\sum_{c=1}^{C} e_{tc}})|, \tag{5}$$

In this way, u_t and the evidence e^t of the tth frame can be interpreted by α_c^t:

$$u_t = \frac{C}{\sum_{c=1}^{C} \alpha_c^t}, \ \alpha_c^t = e_c^t + I, \ e_c^t = [e_1^t, \cdots, e_C^t] \in \mathbb{R}^C, I = \{1, \cdots, 1\} \tag{6}$$

3.3 Temporal Reasoning Module

Uncertainty-gated Reasoning on Graph. TRM conducts URG with matrix $\check{D}$ for noise immune, where the graph can be denoted by $\mathcal{G}(\mathcal{V}, \mathcal{E})$ and $\mathcal{V}$ is a node-set, and $e(i,j) \in \mathcal{E}$ denotes the edge between node i and node j. To construct the graph, input features are first encoded to features o by a TDC, then node representations $\mathbf{h}$ and $\mathbf{h}_i$ (the features of node i or segment i) are given by

$$\mathbf{h} = \{\mathbf{h}_1, \cdots, \mathbf{h}_N\}, \mathbf{h}_i = \frac{1}{t_e^i - t_s^i} \sum_{t=t_s^i}^{t_e^i} o_t, o = (o_1, \cdots, o_T), \mathbf{h}_i \in \mathbb{R}^D. \tag{7}$$

where segment i starts from the frame t_s^i to t_e^i decided by the detected boundaries. The edges of the graph model the action relations and are constituted by the inter-node similarity, thus given kernel size k, $e(i,j)$ is given by

$$e(i,j) = \begin{cases} \dfrac{h_i \cdot h_j}{\max\{\|h_i\|_2 \cdot \|h_j\|_2, \epsilon\}} & \text{if } |i-j| \leq k \ (j \in N_i) \\ 0 & \text{otherwise.} \end{cases}$$

Given the graph, we label every node with an uncertainty to adjust the message passing adaptively. If a segment contains many high-uncertainty frames, the node would be labeled a noise node. Thus, the uncertainties $v = \{v_1, \cdots, v_N\}$ of every nodes, where v_i is the uncertainty of node i, are generated with

$$v_i = \iota(\hat{v}_i - \frac{\sum_{t=1}^{T} \hat{v}_i}{T}), \hat{v}_i = \frac{\sum_{t=t_s^i}^{t_e^i} \bar{u}_t}{t_e^i - t_s^i + 1}, \bar{u}_t = \iota(\hat{u}_t - \frac{\sum_{t=1}^{T} \hat{u}_t}{T}), \tag{8}$$

$$\hat{u}_t = \frac{u_t - \min(u)}{\max(u) - \min(u) + \epsilon}, \iota(x) = \min\{1, \max\{0, x\} \times exp(\max\{0, x\})\}. \quad (9)$$

where $\min(\cdot)$ and $\max(\cdot)$ are the min and max values of all elements of a vector. $\check{D}$ is built based on v and the adjacency matrix $\mathbf{A}$ by following the rule reducing the influence of nodes with high uncertainties. Thus URG is given by

$$\mathbf{h}_i^{(m)} = PRelu(\sum_{j \in N_i} \check{\mathbf{D}}(i,j)\mathbf{h}_i^{(m-1)}\mathbf{w}_{i,j}^{(m)}),$$

$$\check{\mathbf{D}}(i,j) = \mathbf{A}(i,j) - \mathbf{A}(i,j) \times \beta * v_j, \mathbf{A}(i,j) = \frac{exp(e(i,j))}{\sum_{k \in N_i} exp(e(i,k))},$$

where $\mathbf{h}_i^{(m)}$ is the representation of node i in the mth layer, $\mathbf{w}_{i,j}^{(m)}$ are the parameters, $\beta \in [0,1]$ is the hyper-parameter for weighting, $exp(\cdot)$ and $PRelu(\cdot)$ are the e exponent and PRelu function respectively.

Feature Denoising. We assume that the features from DEM is informative ensured by SMP and the frame-level temporal relation reasoning, thus is beneficial for FD. The features updated by URG are further updated by (10) where a cross-attention summarizes the relations between a node and the nearby features from DEM, then are transformed to frame-level representations.

$$\mathbf{h}^{(m)} = \eta + \hat{\mathbf{h}}^{(m)}, \eta = PRelu(W_\eta * \mathbf{h}^*), \quad (10)$$

$$\hat{\mathbf{h}}_t^{(m)} = \mathbf{h}_i^{(m)} + Softmax(W_h \mathbf{h}_{l_1:l_2}^* \mathbf{K}_{l_1^t:l_2^t}) \mathbf{V}_{l_1^t:l_2^t}, \text{ if } t \in \{t_s^i, \cdots, t_e^i\}. \quad (11)$$

We denote that $\mathbf{K}, \mathbf{V}$ are the features output from two 1×1 convolutions by feeding $\mathbf{h}^*$ (the features from the DEM), and $X_{l_1:l_2}$ are the vectors from the l_1^{th} to the l_2^{th} frame of the variable matrix X, thus $\mathbf{K}_{l_1^t:l_2^t}, \mathbf{V}_{l_1^t:l_2^t} \in \mathbb{R}^{(l_2^t - l_1^t)*D}$, and l_1^t, l_2^t stimulate the associated frames with the hyper-paramter ς:

$$l_1^t = t - l, l_2^t = t + l, l = \min\{\lfloor (t_e^i - t_s^i)/2 \rfloor, \varsigma\}, \quad (12)$$

Selective Margin Plasticity. SMP builds a loss function $\mathcal{L}_s$ to ensure DEM prvoide informative clues. During training, the outputs from H1 are supervised under $\mathcal{L}_s$ and the weights of H1 are normalized ($\|w_i\| = 1$, w_i is the weight of the ith convolution kernel in H1). SMP considers the easily confused actions where similar motions exist usually, we build a zero-initialized matrix $\lceil$ which labels every easily confused action j for some action i with 1 through semantic induction, marked as $\lceil_j^i = 1$. Given $\lceil$, the barrier vector $b_{g_t} = [b_{g_t}^1, \cdots, b_{g_t}^C]$ (g_t is the ground truth class of the frame t) for the tth frame can be obtained to constitute the $\mathcal{L}_s$, which is to adjust the angular margin between the classifier of the ground truth category and sample, also the strength of being close to the classifier for each sample. b_{g_t} reduces the corresponding angles by hindering the optimization, thus, whose items indicate easily confused categories will be assigned a large value. $b_{g_t}^j$ is changed during training based on the discriminativenss of y in real time. Thus, $b_{g_t}^j$ is given by

$$b_{g_t}^j = \lceil^{(g_t,j)} \times \iota(j \neq g_t, j \in \Xi), \quad (13)$$

where Ξ is the category set within the K highest probability in y_i. We define $U = [1, \cdots, C]$, the set $\Phi_{g_t} = \complement_U\{g_t\}$, and the loss for the tth frame is given by

$$\mathcal{L}_s(x, t) = -\frac{1}{T} \sum_{t=1}^{T} \log \frac{e^{s \cdot (\cos \theta_{g_t} - m)}}{e^{s \cdot (\cos \theta_{g_t} - m)} + J_t}, J_t = \sum_{j=1, j \in \Phi_{g_t}}^{c} e^{s \cdot (\cos \theta_j + b_{g_t}^j)}, \quad (14)$$

where $\cos \theta_j$ is the cosine value between w_j and the normalized feature of the tth frame x_t ($\|x_t\| = 1$), and the hyper-parameter m is to keep a measure of feature discriminativeness of the current frame, $s = 1$. Training with $\mathcal{L}_d, y$ would be forced to be discriminative among easily confused actions, which makes the featurs of DEM informative.

4   Experiments

4.1   Datasets and Evaluation Metrics

The **50Salads** [20] dataset contains 50 videos of 17 action categories, where each video involving thousands of frames and 20 (average) action instances of various temporal durations is 6.4 min long on average and records salad preparation activities achieved by 25 actors. The **GTEA** [9] dataset contains 28 egocentric videos of 11 action categories, involving 7 kinds of daily activities conducted by 4 actors and 20 (average) action instances in each video. The **Breakfast** [15,16] dataset involves 1712 videos of 48 different actions, recording the breakfast activities in 18 different kitchens. On average, three are 6 action instances in each video. Following the default setting [8], we set 5-fold, 4-fold, and 4-fold cross-validation to record their average values for evaluation on 50Salads, GTEA, and breakfast respectively. For quantitative evaluation, we adopt the following **evaluation metrics** as in [18]: frame-wise accuracy (Acc), segmental edit score (Edit), and F1 score at overlapping thresholds 0.1, 0.25, 0.5 (F1@0.1, F1@0.25, F1@0.5). Edit and F1 metrics consider the temporal relations between ground truth and inference, which can reflect the degree of over-segmentation and measure the quality of inference.

4.2   Implementation Details

Consistent with [8,22], we adopt the temporal resolution of 15 frames per second (fps) to sample the videos for all datasets. We set λ and τ to 0.15 and 4 respectively, they control the importance of $\mathcal{L}_m$, we set β to 0.1 and ς to 23. All models are trained with Adam optimizer and implemented by using PyTorch. The learning rates of all models are 0.005.

4.3   Compared with State-of-the-Art Methods

To evaluate the overall performance of our framework, our DTRN is compared with recent state-of-the-art methods as shown in Table 1 and Table 2, where

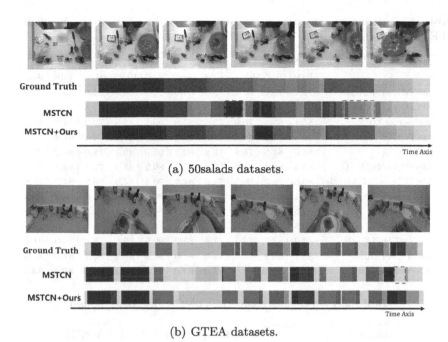

(a) 50salads datasets.

(b) GTEA datasets.

Fig. 2. The visualization of action segmentation for different methods with color-coding (Different colors represent different types of actions).

DTGRM [21] and GTRM [12] are graph-based models for modeling temporal relations among segments, and ASFormer [24] is a competitive transformer-based method and is inefficient due to huge computing and memory cost, and other methods are built based on MSTCN [8], where various complementary techniques with high computation costs are used for improving the performance, such as temporal domain and adaption structure search, thus not efficient enough. Our MSTCN+Ours indicates that MSTCN is the backbone model in our DTRN, thus compare our structure with the above models is more illustrative. In Table 1 and Table 2, we can see that compared with recent state-of-the-art methods, our method achieves the best performance at all metrics, that is improved significantly compared to Baseline. For example, on the 50salads dataset, MSTCN+Ours outperforms the baseline by 5.3% for the framewise accuracy. We also visualize some segmentation results as shown in Fig. 2, they illustrate the effecitve of our method. Moreover, our MSTCN+Ours only requires 1.87 M model parameters, while ASRF, SSTDA, and BCN requires 2.31 M, 4.60 M, and 5.72 M parameters respectively, the number of parameters for BCN approximately even 3 times our temporal model, but our DTRN can still obtain better performance. Thus, MSTCN+Ours can maintain both efficiency and effectiveness.

Table 1. Performance comparisons (in percentage units (%)) on the 50Salads and GTEA.

Method	50 Salads					GTEA				
	F1@{10,25,50}			Edit	Acc	F1@{10,25,50}			Edit	Acc
MSTCN+DTGRM [21]	79.1	75.9	66.1	72.0	80.0	87.8	86.6	72.9	83.0	77.6
MSTCN+GTRM [12]	75.4	72.8	63.9	67.5	82.6	–	–	–	–	–
ASFormer [24]	85.1	83.4	76.0	79.6	85.6	90.1	88.8	79.2	84.6	79.7
MSTCN [8]	76.3	74.0	64.5	67.9	80.7	87.5	85.4	74.6	81.4	79.2
SSTDA [5]	83.0	81.5	73.8	75.8	83.2	90.0	89.1	78.0	86.2	79.8
MSTCN+HASR [1]	83.4	81.8	71.9	77.4	81.7	89.2	87.3	73.2	85.4	77.4
MSTCN++ +UARL [4]	80.8	78.7	69.5	74.6	82.7	90.1	87.8	76.5	84.9	78.8
BUIMS-TCN [6]	81.1	79.8	72.4	74.0	83.9	89.4	86.6	76.6	85.0	80.6
AU-TCN [2]	74.8	72.7	66.1	–	86.3	84.9	83.1	74.5	–	83.5
MSTCN+RF [11]	80.3	78.0	69.8	73.4	82.2	89.9	87.3	75.8	84.6	78.5
BCN+TACHA [23]	84.0	82.7	74.2	77.4	85.2	90.6	89.5	78.2	86.0	81.2
MSTCN+Ours	85.1	83.5	77.1	79.6	86.0	90.6	88.8	80.1	87.1	83.0
ASFormer+Ours	**86.0**	**85.2**	**81.9**	**81.4**	**87.1**	**91.8**	**90.0**	**82.5**	**87.3**	**83.5**

Table 2. Performance comparison on the Breakfast dataset.

Method	F1@{10,25,50}(%)			Edit(%)	Acc(%)
DTGRM	68.7	61.9	46.6	68.9	68.3
GTRM	57.5	54.0	43.3	58.7	65.0
MSTCN	52.6	48.1	37.9	61.7	66.3
BCN	68.7	65.5	55.0	66.2	70.4
SSTDA	75.0	69.1	55.2	73.7	70.2
BUIMS-TCN	71.0	65.2	50.6	70.2	68.7
MSTCN++ + UARL	65.2	59.4	47.4	66.2	67.8
MSTCN+RF	74.9	69.0	55.2	73.3	70.7
BCN+TACHA	70.8	67.7	57.9	68.0	72.2
MSTCN+Ours	**71.5**	**68.7**	**60.0**	**73.1**	**72.4**

4.4 Ablation Study

Analysis of SMP. SMP provides a auxiliary training to DEM for outputing informative features. We performed an ablation study on SMP by adding it and removing it from DEM, whose results are shown in Table 3. It shows the SMP is effective in improving the performance, the Act value in Table 3 illustrates the function of SMP in generating targetedly discriminative features (in this paper, we select the easily confused actions as the target), they illustrate that the SMP is effective in ensuring DEM generate informative features.

Analysis of URG. We conduct an ablation study of URG in TRM by adding and removing URG from TRM, whose testing results are shown in Table 4 where

Table 3. Performance comparison (in percentage units (%)) on the 50Salads and Breakfast datasets.

SMP	50 Salads						Breakfast				
	F1@{10,25,50}			Edit	Acc	Act	F1@{10,25,50}			Edit	Acc
−	85.0	83.3	75.3	79.5	85.0	4512	68.4	58.1	48.1	72.0	71.9
✓	**85.1**	**83.5**	**77.1**	**79.6**	**86.0**	5291	**71.5**	**68.7**	**60.0**	**73.1**	**72.4**

the results of the first row are from the method removing the URG, where $\check{D}(i,j)$ is replaced with $A(i,j)$ for updating the features. Table 4 shows that the method with URG is effective for improving the qualitiy of segmentation, especially in improving ACC. The URG conduces to correct some predictions that also illustrates the function of reducing the impact of data noise.

Table 4. Performance comparison (in percentage units) on the 50Salads and GTEA datasets.

URG	50 Salads					GTEA				
	F1@{10,25,50}			Edit	Acc	F1@{10,25,50}			Edit	Acc
−	85.0	83.5	75.2	79.5	85.0	90.4	**89.1**	79.2	86.0	82.8
✓	**85.1**	**83.5**	**77.1**	**79.6**	**86.0**	**90.6**	88.8	**80.1**	**87.1**	**83.0**

Table 5. Testing the effect of TRM, the results are in percentage units (%), and the backbone we adopted is MSTCN.

Method	50 Salads					GTEA				
	F1@{10,25,50}			Edit	Acc	F1@{10,25,50}			Edit	Acc
TGraph	83.3	81.9	75.5	78.1	84.2	88.9	85.7	76.1	83.5	80.2
AGraph	84.0	82.3	75.5	79.0	85.0	90.3	87.9	78.5	86.3	81.9
TRM	**85.1**	**83.5**	**77.1**	**79.6**	**86.0**	**90.6**	**88.8**	**80.1**	**87.1**	**83.0**

Analysis of Cross-Attention Based Structure for FD. To test the function of the cross-cascaded based structure, we conduct some variants of TRM for comparison as shown in Table 5. The first one is the TGraph where the cross-cascaded structure in TRM is removed, which means that operations (11) and (10) in TRM are removed. The second one is the AGraph where operation (11) is removed and operation (10) is kept. Table 5 indicates the positive effect of (11) and (10) respectively, it shows that (10) helps to reduce over-segment since it improves Edit score hugely, and (11) helps to correct many wrong predictions. For example, the TRM outperforms the AGraph by 1.6% for F1@50, and 1.1% for the frame-wise accuracy, and the performance outperforms that of the backbone hugely. It also validates the advantage of TRM in understanding human actions.

5   Conclusions

In our work, we propose a novel action segmentation framework, called DTRN to overcome the impact of noisy features. The TRM in DTRN utilizes the URG for noise-immune and utilizes a cross-attention-based structure to realize feature denoising. The DEM in TRM provides informative clues for FD ensured by the proposed SMP and frame-level temporal relation reasoning. The SMP adjusts adaptively the decision boundary of key neural by changing specific margins in real-time during training to provide informative clues for FD. Extensive ablation studies illustrated that our superiority over existing methods both in efficiency and effectiveness and our DTRN outperforms the state-of-the-art methods on challenging 50Salads, GTEA, and Breakfast datasets.

Acknowledgments. This work is supported by Beijing Natural Science Foundation (No. 4222037, L181010) and National Natural Science Foundation of China (No. 61972035).

References

1. Ahn, H., Lee, D.: Refining action segmentation with hierarchical video representations. In: 2021 IEEE/CVF International Conference on Computer Vision (ICCV), pp. 16282–16290. IEEE, Montreal, QC, Canada (2021). https://doi.org/10.1109/iccv48922.2021.01599
2. Cao, J., Xu, R., Lin, X., Qin, F., Peng, Y., Shao, Y.: Adaptive receptive field u-shaped temporal convolutional network for vulgar action segmentation. Neural Comput. Appl. **35**(13), 9593–9606 (2023). https://doi.org/10.1007/s00521-022-08190-5
3. Carreira, J., Zisserman, A.: Quo vadis, action recognition? A new model and the kinetics dataset. In: 2017 IEEE Conference on Computer Vision and Pattern Recognition (CVPR), pp. 4724–4733. IEEE Computer Society, Honolulu, HI, USA (2017). https://doi.org/10.1109/cvpr.2017.502
4. Chen, L., Li, M., Duan, Y., Zhou, J., Lu, J.: Uncertainty-aware representation learning for action segmentation. In: Proceedings of the Thirty-First International Joint Conference on Artificial Intelligence, IJCAI-22, pp. 820–826. ijcai.org, Vienna, Austria (2022). https://doi.org/10.24963/ijcai.2022/115
5. Chen, M.H., Li, B., Bao, Y., AlRegib, G., Kira, Z.: Action segmentation with joint self-supervised temporal domain adaptation. In: 2020 IEEE/CVF Conference on Computer Vision and Pattern Recognition (CVPR), pp. 9451–9460. Computer Vision Foundation/IEEE, Seattle, WA, USA (2020). https://doi.org/10.1109/cvpr42600.2020.00947
6. Chen, W., et al.: Bottom-up improved multistage temporal convolutional network for action segmentation. Appl. Intell. **52**(12), 14053–14069 (2022). https://doi.org/10.1007/s10489-022-03382-x

7. Corbière, C., Thome, N., Bar-Hen, A., Cord, M., Pérez, P.: Addressing failure prediction by learning model confidence. In: Advances in Neural Information Processing Systems, pp. 2898–2909. Vancouver, BC, Canada (2019)

8. Farha, Y.A., Gall, J.: MS-TCN: multi-stage temporal convolutional network for action segmentation. In: 2019 IEEE/CVF Conference on Computer Vision and Pattern Recognition (CVPR), pp. 3570–3579. Computer Vision Foundation/IEEE, Long Beach, CA, USA (2019). https://doi.org/10.1109/cvpr.2019.00369

9. Fathi, A., Ren, X., Rehg, J.M.: Learning to recognize objects in egocentric activities. In: CVPR 2011. pp. 3281–3288. IEEE Computer Society, Colorado Springs, CO, USA (2011). DOI: 10.1109/cvpr.2011.5995444

10. Gao, S.H., Han, Q., Li, Z.Y., Peng, P., Wang, L., Cheng, M.M.: Global2local: efficient structure search for video action segmentation. In: 2021 IEEE/CVF Conference on Computer Vision and Pattern Recognition (CVPR), pp. 16800–16809. Computer Vision Foundation/IEEE, virtual event (2021). https://doi.org/10.1109/cvpr46437.2021.01653

11. Gao, S., Li, Z.Y., Han, Q., Cheng, M.M., Wang, L.: RF-Next: efficient receptive field search for convolutional neural networks. IEEE Trans. Pattern Anal. Mach. Intell. 45(3), 2984–3002 (2023). https://doi.org/10.1109/TPAMI.2022.3183829

12. Huang, Y., Sugano, Y., Sato, Y.: Improving action segmentation via graph-based temporal reasoning. In: 2020 IEEE/CVF Conference on Computer Vision and Pattern Recognition (CVPR), pp. 14021–14031. Computer Vision Foundation/IEEE, WA, USA, June 2020. https://doi.org/10.1109/cvpr42600.2020.01404

13. Jamil, T., Braak, C.: Selection properties of type ii maximum likelihood (empirical Bayes) in linear models with individual variance components for predictors. Pattern Recognit. Lett. 33(9), 1205–1212 (2012)

14. Josang, A., Hankin, R.: Interpretation and fusion of hyper opinions in subjective logic. In: 15th International Conference on Information Fusion (FUSION), pp. 1225–1232. IEEE, Singapore (2012)

15. Kuehne, H., Arslan, A., Serre, T.: The language of actions: Recovering the syntax and semantics of goal-directed human activities. In: 2014 IEEE Conference on Computer Vision and Pattern Recognition, pp. 780–787. IEEE Computer Society, Columbus, OH, USA (2014). https://doi.org/10.1109/CVPR.2014.105

16. Kuehne, H., Gall, J., Serre, T.: An end-to-end generative framework for video segmentation and recognition. In: 2016 IEEE Winter Conference on Applications of Computer Vision (WACV), pp. 1–8. IEEE Computer Society, Lake Placid, NY, USA (2016). https://doi.org/10.1109/WACV.2016.7477701

17. Lakshminarayanan, B., Pritzel, A., Blundell, C.: Simple and scalable predictive uncertainty estimation using deep ensembles. In: Advances in Neural Information Processing Systems, pp. 6402–6413. Long Beach, CA, USA (2017)

18. Lea, C., Flynn, M.D., Vidal, R., Reiter, A., Hager, G.D.: Temporal convolutional networks for action segmentation and detection. In: 2017 IEEE Conference on Computer Vision and Pattern Recognition (CVPR), pp. 1003–1012. IEEE Computer Society, Honolulu, HI, USA (2017). https://doi.org/10.1109/cvpr.2017.113

19. Li, S., Farha, Y.A., Liu, Y., Cheng, M.M., Gall, J.: MS-TCN++: multi-stage temporal convolutional network for action segmentation. IEEE Trans. Pattern Anal. Mach. Intell. 1 (2020). https://doi.org/10.1109/tpami.2020.3021756

20. Stein, S., Mckenna, S.J.: Combining embedded accelerometers with computer vision for recognizing food preparation activities. In: The 2013 ACM International Joint Conference on Pervasive and Ubiquitous Computing, vol. 33, pp. 3281–3288. ACM, Zurich, Switzerland (2013)

21. Wang, D., Hu, D., Li, X., Dou, D.: Temporal relational modeling with self-supervision for action segmentation. In: Thirty-Fifth AAAI Conference on Artificial Intelligence, pp. 2729–2737. AAAI Press, Virtual Event (2021)
22. Wang, Z., Gao, Z., Wang, L., Li, Z., Wu, G.: Boundary-aware cascade networks for temporal action segmentation. In: Vedaldi, A., Bischof, H., Brox, T., Frahm, J.-M. (eds.) ECCV 2020. LNCS, vol. 12370, pp. 34–51. Springer, Cham (2020). https://doi.org/10.1007/978-3-030-58595-2_3
23. Yang, D., Cao, Z., Mao, L., Zhang, R.: A temporal and channel-combined attention block for action segmentation. Appl. Intell. **53**(3), 2738–2750 (2023). https://doi.org/10.1007/s10489-022-03569-2
24. Yi, F., Wen, H., Jiang, T.: Asformer: transformer for action segmentation. In: The British Machine Vision Conference, p. 236. BMVA Press, Online (2021)

Vision Applications and Systems

3D Lightweight Spatial-Spectral Attention Network for Hyperspectral Image Classification

Ziyou Zheng[1] ⓘ, Shuzhen Zhang[1,2](✉) ⓘ, Hailong Song[1], and Qi Yan[1]

[1] College of Communication and Electronic Engineering, Jishou University, People's South Road, Jishou 416000, Hunan, China
shuzhen_zhang@hnu.edu.cn
[2] Key Laboratory of Visual Perception and Artificial Intelligence, Hunan University, Lushan Road, Changsha 410000, Hunan, China

Abstract. Compared with the convolutional neural network (CNN), a 3D lightweight network (3D-LWNet) can successfully perform hyperspectral image (HSI) classification using fewer network parameters. However, results of these methods can not reflect the effect of each of the different features used in the classification process. To address this problem, we propose a 3D lightweight spatial-spectral attention network (3D-LSSAN), which adopts a 3D large kernel attention (3D-LKA) mechanism to better determine the effect of each feature in HSI classification. 3D-LKA directly operates on spatial-spectral features of HSI, which can reduce the number of trainable parameter and overcome the network overfitting. Moreover, 3D-LKA disassembles the convolution kernel, which can capture local information and long-range information in HSI using a small amount of computational power. Specifically, the HSI is first sent through a convolutional network layer to generate shallow spatial-spectral features. Second, these shallow features are input into six spatial-spectral attention (SSA) units based on 3D-LKA to emphasize the importance of different parts of features. Finally, output features of the SSA units are fed into a fully connected layer to obtain the classification result. Experimental results on two publicly available data sets demonstrate that the proposed 3D-LSSAN achieves better classification performance than the other techniques.

Keywords: Convolution neural network · Hyperspectral image classification · Attention mechanism · Lightweight network · Large kernel attention

1 Introduction

Hyperspectral images (HSI) include hundreds of contiguous narrow spectral bands that reveal small spectral differences among ground objects, and make them suitable for remote sensing applications, including agriculture [1], mineralogy [2], and astronomy [3]. Nonetheless, high-dimensional data is susceptible to

Q. Liu et al. (Eds.): PRCV 2023, LNCS 14430, pp. 297–308, 2024.
https://doi.org/10.1007/978-981-99-8537-1_24

the Hughes phenomenon [4], in which classification accuracy initially increases and subsequently decreases with each increase in the number of bands [5]. Therefore, efficient feature extraction method is necessary to process high-dimensional HSI.

In early studies, most methods relies on hand-crafted features. However, these features required extensive a priori knowledge. Therefore, a more powerful feature extraction method is needed. Deep learning (DL) has been gained significant attention in vision tasks. End-to-end training is a distinct feature of DL models, which provides the capability to directly learn from the input data without requiring any manual feature engineering or other human intervention for the intermediate layers. In recent years, DL methods have been implemented for HSI classification with remarkable success. CNN architectures such as 2D-CNN [6], 3D-CNN [7], and ResNet [8] have been utilized for HSI classification. The classification performance of a CNN is directly related to its model size. Due to the small number of labeled HSI samples, the use of CNN to classify HSI often leads to overfitting. Zhang et al. [9] proposed a 3D-Lightweight Network (3D-LWNet) to mitigate this issue by utilizing depth-separable convolution to extend network depth and without increasing the parameter count. The 3D-LWNet demonstrates an improved classification performance with fewer parameters.

Moreover, CNN method lacks feature significance recognition as the convolving kernel within a receptive field without prioritizing vital information. To address this issue, attention mechanisms has been introduced into the CNN. The attention mechanism is a selection process that can automatically identify crucial features in a set of input data. Mei et al. [10] integrated both RNN and CNN with attention mechanisms in am attempt to determine both spectral and spatial relationships. Ma et al. [11] employed a similar structure to the one proposed by Mei et al. the squeeze-and-excitation network (SENet) [12] is utilized as a substitute for RNN to identify spectral relationships. This result in lower overall computational complexity and better classification performance. Furthermore, these attention mechanisms all have a distinct shortcoming: they operate discretely in the spectral and spatial dimensions, which increases the complexity of the model.

Self-attention [13,14] has been proven to be a notable method for learning the relationship between different parts. For example, Vaswani et al. [15] utilized several self-attention mechanisms to create an impressive Transformer model, which does not require traditional network structures such as RNN and CNN, relying solely on attention mechanisms for machine translation. Self-attention is limited to unidimensional sequences and cannot incorporate the two-dimensional images. Hong et al. proposed SpectralFormer [16], which applies the transformer to the HSI classification task. Experiments proved that using self-attention for HSI can achieve better classification performance. Furthermore, using larger convolution kernels (e.g., 15×15, 21×21) [17,18] can expand the perception field area and, to some extent, capture long-range feature relationships similar to self-attention methods. Zhong et al. proposed a lightweight criss-cross large kernel convolutional neural network [19], where the lightweight module of the network consists

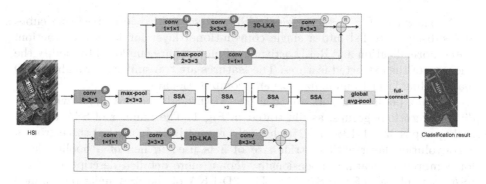

Fig. 1. Structure of the 3D-LSSAN. The six SSA units are divided into four groups with 32, 64, 128 and 256 channels. Ⓡ represents ReLU function, Ⓑ represents the batch normalization layer and ⊕ represents the addition operations.

of two 1D convolutions with self attention in orthogonal directions, and long range context features are aggregated by setting large kernel convolution in the 1D convolutional layers. According to the above analysis, this study employed the large kernel convolution.

The main contributions of this paper are as follows:

- We propose a 3D lightweight spatial-spectral attention network (3D-LSSAN) and introduce a 3D large kernel attention (3D-LKA) in the network for enhancing the details of the extracted features of HSI.
- The 3D-LKA used in the network structure is a novel spatial-spectral attention mechanism, which can simultaneous process spatial and spectral dimensions.

The rest of this paper is arranged as follows. Section 2 gives the details of the proposed method. Section 3 describes the datasets used in the study and the comparison experiments with other HSI classification techniques. Section 4 summarises the conclusions of the study and gives the outlook for the future.

2 Proposed Method

This section presents the 3D-LSSAN method with the 3D-LKA attention mechanism, which aims to improve the HSI classification accuracy by using CNN. A detailed explanation of the method is provided as below.

2.1 3D-LSSAN Structure

As previously noted, CNN cannot assess different effects of differernt extracted features on the HSI classification. Thus, we propose a 3D-LSSAN for HSI classification in this study. Figure 1 illustrates the network model framework for this study.

As shown in Fig. 1, the original HSI is first divided into fixed-size cubes and subsequently fed into a single convolutional layer for feature extraction. Batch normalization and ReLU activation functions are employed to modify the linearity of the extracted features. The features are transmitted through the 3D max-pooling layer to minimize size and computational load.

3D max pooling outputs are then passed through six SSA units, which are divided into four groups, as illustrated in Fig. 1. The number of SSA units in each group is 32, 64, 128, and 256 channels, respectively. In the last three groups, a convolution layer with a stride size of 2 is used in lieu of a pooling layer for compressing feature dimensionality. For a more detailed description of the SSA unit, please refer to Sect. 2.2. The 3D-LKA of the SSA units determined weight distribution by capturing spatial-spectral local and long-range features. A detailed introduction to 3D-LKA is provided in Sect. 2.3.

Finally, the features are compressed to a fixed size after passing through the adaptive averaging pooling layer and are subsequently sent to the fully connected layer for classification.

2.2   Spatial-Spectral Attention Unit

To increase model depth without gradient vanishing issues, ResNet is used with shortcut connections and bottleneck structures. However, directly applying ResNet to HSI still faces gradient vanishing due to limited labeled samples.

Grouped convolution reduces parameters by decomposing the kernel while maintaining scale and adaptability. It's commonly used in deep separable convolution, where groups, input channels, and output channels are kept equal. Depthwise separable convolution is also used in SSA units in this study.

Figure 1 displays the structure of the SSA unit. This study employs the two SSA structures. Each SSA structure is comprised of 2 point-wise convolution (PW-conv) layers, a depth-wise convolution (DW-conv) layer ($3 \times 3 \times 3$ kernel size), and an 3D-LKA.

The key difference between the two SSA structures is that the former lacks extra operations in the shortcut path, while the latter includes downsampling. This involves an average pooling layer and a PW-conv layer that reduce feature map dimensions before the addition, guided by shortcuts. The second SSA has a stride of 2, causing feature size misalignment without downsampling. The initial PW-conv increases channels by a factor of t. Following the method proposed in reference [9], t is set at 4. For D input channels, the bottleneck holds $32D^2$ parameters, while the SSA unit has $5D^2 + 27D$ parameters. Given input size $S \times S \times L$, the bottleneck requires $32 \times S^2 \times L \times D^2$ floating point operations (FLOPs), while the SSA unit needs $S^2 \times L \left(5D^2 + 27D\right)$ FLOPs.

2.3   3D Large Kernel Attention Model

Self-attention can capture distant dependencies and enhance relationships between inputs. For images, it refines details via local and global pixel relations and aids contextual understanding.

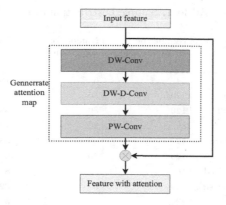

Fig. 2. Structure of 3D the Large Kernel Attention(3D-LKA) model. ⊗ represents the multiplication operation.

Large kernel convolution differs from standard 3×3 or 5×5 kernels. A larger size widens the receptive field, capturing complex features over a broader range. In images, using large kernels enhances feature info, aligning with attention mechanisms.

Large Kernel Attention (LKA) [20] merges self-attention and large kernel convolution benefits. It uses large kernels for distant relationships and integrates self-attention for detail image features. LKA's ability to capture both local and global relations makes it potent for image analysis.

The LKA algorithm is originally devised to tackle problems with nature images. However, it is not directly applicable to HSI due to the dissimilar dimensions between the two data types. Typically, natural images comprise only three bands, R, G, and B. In contrast, HSI data may consist of hundreds of bands. Consequently, we reconstruct a new model that employ LKA to create 3D-LKA. We deploy the 3D convolutional kernel to enable spatial-spectral feature processing in three dimensions. Specifically, we decompose a kernel of size $K \times K \times K$ into a $\lceil \frac{K}{d} \rceil \times \lceil \frac{K}{d} \rceil \times \lceil \frac{K}{d} \rceil$ depth wise dilation convolution (DW-D-conv) with dilation rate d, a $(2d-1) \times (2d-1) \times (2d-1)$ DW-conv, and a $1 \times 1 \times 1$ convolution. DW-D-conv can determine long-range relationships in HSI, similar to prior self-attention methods, while DW-conv captures local HSI information. The $1 \times 1 \times 1$ convolution (PW-conv) manage relationships between the different channels. The flowchart for the 3D-LKA model is shown in Fig. 2.

3   Experimental Results

3.1   Description of Datasets

In this study, we perform experiments on two datasets entitled Indian Pines(IP) and Pavia University(PU). We provide a concise overview of each dataset as below. The IP dataset was collected by an AVIRIS sensor in 1992. It is comprised of 145×145 pixels with a spatial resolution of 20 m. The dataset contains 224 bands within the 0.4–2.5 $\times 10^{-6}$ meters wavelength range. It contains of 16

Table 1. Sample Distribution of IP.

No	Class_name	Number of training samples	Number of validation samples	Number of test samples
1	Alfalfa	24	6	16
2	Corn notill	120	30	1278
3	Corn mintill	120	30	680
4	Corn	80	20	137
5	Grass pasture	120	30	333
6	Grass trees	120	30	580
7	Pasture mowed	16	4	8
8	Hay windrowed	120	30	328
9	Oats	12	3	5
10	Soybean notill	120	30	822
11	Soybean mintill	120	30	2305
12	Soybean clean	120	30	443
13	Wheat	120	30	55
14	Woods	120	30	1115
15	Buildings Grass Trees	40	10	336
16	Stone Steel Towers	40	10	43
total		1412	353	8484

categories. The number of bands was reduced to 200 by removing the bands covering the absorption area. The 16 main landcover types are listed in the Table 1, along with the number of samples used for training, validation and testing in the classification task.

The PU dataset was acquired using a ROSIS sensor during a flight over Pavia University in northern Italy in 2001. The number of bands in the dataset is 103, with a 0.43–0.86×10^{-6} meters wavelength range. The spatial resolution of the dataset is $1.3\,m$ per pixel and it contains 610×340 pixels. The image contains nine landcover categories. The specific category names and the division of the training, validation and test sets are shown in Table 2.

3.2   Discussion of Hyperparameters

In terms of hyperparameter settings, we compare experimental results with different patch size, kernel size K and dilation rate d. The patch size used for comparison is not less than 15, because a patch size under 15 would result in the disappearance of features during the network forward propagation process. We choose values of 15, 17, 19, 21, 23, 25, 27 and 30 for this study. As show in Fig. 3(a), our model performs best with a patch size of 27. Therefore, in subsequent experiments, the patch size is set to 27.

As describe previously by Guo *et al.*, values of (7, 2), (14, 3), (21, 3), and (28, 4) are chosen for K, and d. The results of this analysis are show in Fig. 3(b). The results are best when K and d are 21 and 3. If the (K, d) value are too small,

Table 2. Sample Distribution of PU.

No	Class_name	Number of training samples	Number of validation samples	Number of test samples
1	Asphalt	160	40	6431
2	Meadows	160	40	18449
3	Gravel	160	40	1899
4	Trees	160	40	2899
5	Metal sheets	160	40	1145
6	Bare Soil	160	40	4829
7	Bitumen	160	40	1130
8	Bricks	160	40	3482
9	Shadows	160	40	747
total		1440	360	40976

it prevented 3D-LKA from capturing long-range feature relations. If the (K, d) value are too large, the 3D-LKA capture features that did not belong to this category, thus reducing classification accuracy. According to the above results, we choose (21, 3) for (K,d).

3.3 Classification Results

To demonstrate the effectiveness of 3D-LSSAN, we compare its performance against five models: spectralNet [21], 3D-hyper generative adversarial minority oversampling (3D-hyperGAMO) [22], double branch multi-attention mechanism network (DBMA) [11], double branch dual-attention mechanism network (DBDA) [23], and the pyramidal residual network (pResNet) [24].

The classification performance is evaluated using four indicators, which include overall accuracy (OA), average accuracy (AA), category accuracy(CA), and the kappa coefficient (Kappa). Tables 3 and 4 present the CA, AA, OA, and Kappa of each of the test methods for both the IP and PU datasets. Additionally, Figs. 4 and 5 provide the corresponding classification maps.

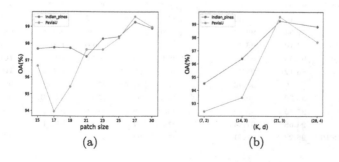

(a)   (b)

Fig. 3. OA values corresponding to different patch size and (K, d).

Table 3. Classification Result of Indian Pines(IP) Dataset.

Class	SpectralNet	3D-HyperGAMO	DBDA	DBMA	pResNet	3D-LWNet	Proposed
Alfalfa	95.12	100.0	98.70	98.03	100.0	100.0	**100.0**
Corn notill	92.68	**100.0**	96.14	94.80	99.61	95.33	96.67
Corn mintill	97.05	97.50	98.06	95.07	99.33	99.33	**100.0**
Corn	96.24	100.0	96.52	97.55	98.59	100.0	**100.0**
Grass pasture	95.17	100.0	97.12	95.55	98.16	100.0	**100.0**
Grass trees	96.19	99.02	98.27	98.63	98.63	100.0	**100.0**
Pasture mowed	100.0	95.18	84.02	80.76	92.00	100.0	**100.0**
Hay windrowed	100.0	100.0	99.94	98.31	100.0	100.0	**100.0**
Oats	66.67	95.42	84.77	97.46	88.89	100.0	**100.0**
Soybean notill	**98.97**	96.86	94.21	89.07	96.57	94.67	98.67
Soybean mintill	98.50	**99.36**	97.27	96.65	99.23	92.67	98.00
Soybean clean	94.38	96.36	97.19	96.15	95.69	**100.0**	98.67
Wheat	100.0	97.85	98.59	98.45	100.0	100.0	**100.0**
Woods	99.12	99.54	99.30	98.65	98.42	100.0	**100.0**
Buildings Grass Trees	93.08	95.11	97.92	91.58	93.37	100.0	**100.0**
Stone Steel Towers	90.47	98.01	93.13	90.55	98.81	100.0	**100.0**
OA	96.83	97.63	97.56	96.33	98.42	98.34	**99.26**
AA	94.60	98.14	95.79	95.27	97.33	98.88	**99.50**
Kappa(×100)	96.39	97.30	97.22	95.82	98.20	98.19	**99.20**

Findings reveal that spectralNet, a 2D method, demonstrated poor classification performance, with an OA of 96.83% and 97.15% in the two datasets analyses. Conversely, pResNet, a 3D structure, achieve better classification performance. It attain an OA of 98.42% and 98.30%, in two dataset analyses as the 3D structure can directly extract spatial-spectral features. Moreover, 3D-hyperGAMO produce high-quality data for training through generative means, resulting in better classification performance. Remarkably, compare to Spectral-Net, both DBMA and DBDA exhibit enhance classification performance in the PU dataset due to the incorporation of an attention mechanism, achieving 0.97% and 1.05% improvement, respectively. The 3D-LSSAN method outperform all

Table 4. Classification Result of Pavia University (PU) Dataset.

Class	SpecialNet	3D-HyperGAMO	DBDA	DBMA	pResNet	3D-LWNet	Proposed
Asphalt	97.59	99.88	98.91	97.99	97.35	96.88	**100.0**
Meadows	99.03	96.5	**99.84**	98.94	99.82	98.75	98.75
Gravel	88.16	100.0	97.33	97.32	86.83	98.13	**100.0**
Trees	97.34	90.53	98.08	98.13	97.30	98.13	**98.75**
Metal sheets	94.10	96.43	99.83	99.50	100.0	100.0	**100.0**
Bare Soil	97.58	95.01	98.65	99.56	100.0	100.0	**100.0**
Bitumen	86.12	98.47	95.58	99.22	93.48	100.0	**100.0**
Bricks	96.89	98.42	91.69	92.75	98.03	95.00	**99.38**
Shadows	95.32	99.91	97.69	96.98	100.0	**100.0**	99.38
OA	97.15	98.28	98.20	98.12	98.30	98.54	**99.58**
AA	94.68	97.24	97.51	97.82	96.98	98.54	**99.58**
Kappa(×100)	96.23	97.71	97.62	97.50	97.75	98.36	**99.53**

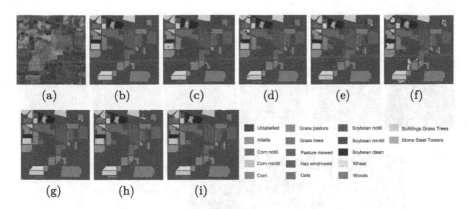

Fig. 4. Classification maps of the IP dataset. (a) Pseudocolor image. (b) Ground-truth. (c) SpectralNet, OA = 96.83%. (d) 3D-HyperGAMO, OA = 97.63%. (e) DBDA, OA = 97.56%. (f) DBMA. OA = 96.33%. (g) pResNet, OA = 98.42%. (h) 3D-LWNet, OA = 98.34%. (i) proposed method, OA = 99.26%.

others quantitatively. Our approach produces cleaner classification maps with less noise. For example, in the Soybean mintill section of the IP dataset, DBDA and DBMA maps has considerable noise, while 3D-LSSAN maps are clearer. Similarly, in the Meadows section of the PU dataset, 3D-LSSAN generate maps with significantly less noise compare to other methods.

We compare different model performance using various sample sizes to show 3D-LLSAN's superiority. For the PU dataset, we get classification results for sample sizes of 50, 80, 110, and 160 per class. Due to the imbalance nature of the IP dataset, we use a portion of the total sample size for training: 5%, 8%, 12%, and 14%. These results are in Tables 6 and 5. The data in these tables indicate that as training data increase, all methods improve. Notably, 3D-LSSAN consistently outperformed other methods with larger sample sizes. These findings highlight our approach's effectiveness.

Table 5. Classificaton OA(%) of PU Dataset with different training sample number.

Methods	50	80	110	160
SpectralNet	90.66	94.60	96.27	97.15
3D-HyperGMAO	95.21	97.55	97.77	97.94
DBDA	96.96	97.40	97.44	98.20
DBMA	93.75	96.79	95.88	98.12
pResNet	94.59	95.94	97.00	98.30
3D-LWNet	96.44	97.11	97.56	98.54
Proposed	**97.56**	**97.78**	**98.43**	**99.58**

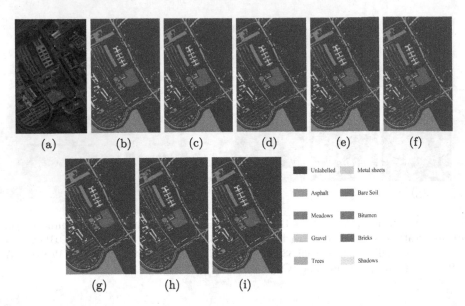

Fig. 5. Classification maps of the PU dataset. (a) Pseudocolor image. (b) Ground-truth. (c) SpectralNet, OA = 97.15%. (d) 3D-HyperGAMO, OA = 98.28%. (e) DBDA, OA = 98.20%. (f) DBMA. OA = 98.12%. (g) pResNet, OA = 98.30%. (h) 3D-LWNet, OA = 98.54%. (i) proposed method, OA = 99.58%.

Table 6. Classification OA(%) of IP dataset with different training sample percent.

Methods	5%	8%	12%	14%
SpectralNet	74.24	95.23	97.51	98.71
3D-HyperGMAO	90.36	93.7	95.29	95.06
DBDA	93.95	95.79	94.71	97.56
DBMA	86.86	91.51	94.71	94.06
pResNet	94.48	95.38	98.13	98.42
3D-LWNet	93.43	95.33	97.61	98.34
Proposed	**95.02**	**95.96**	**98.35**	**99.26**

4   Conclusion

In this study, we introduce a novel attention mechanism 3D-LKA for HSI classification that applies the attention mechanisim to spatial and spectral features simultaneously. Our approach uses 3D-LKA to capture local and long-range information, enhancing feature extraction accuracy. We also present 3D-LSSAN, a network incorporating 3D-LKA into the bottleneck structure in ResNet. Unlike lightweight CNNs, 3D-LSSAN prevents the neglect of critical information within receptive fields. Comparative experiments on real-world datasets show the superiority of the proposed method over recent approaches. Our future plans involve combining 3D-LSSAN with traditional clustering method to address difficulties by the limited labeled samples and enhance the HSI classification accuracy.

Acknowledgments. This work is funded by the Research Foundation of Education Department of Hunan Province of China under Grant No. 22A0371; the Graduate Research Project of Jishou University under Grant No. jdy22024.

References

1. Lacar, F., Lewis, M., Grierson, I.: Use of hyperspectral imagery for mapping grape varieties in the Barossa valley, South Australia. In: 2001 International Geoscience and Remote Sensing Symposium, vol. 6, pp. 2875–2877 (2001)
2. Van Der Meer, F.: Analysis of spectral absorption features in hyperspectral imagery. Int. J. Appl. Earth Obs. Geoinf. 5(1), 55–68 (2004)
3. Hege, E.K., O'Connell, D., Johnson, W., Basty, S., Dereniak, E.L.: Hyperspectral imaging for astronomy and space surveillance. In: Imaging Spectrometry IX, vol. 5159, pp. 380–391 (2004)
4. Hughes, G.: On the mean accuracy of statistical pattern recognizers. IEEE Trans. Inf. Theory 14(1), 55–63 (1968)
5. Chen, Y., Lin, Z., Zhao, X., Wang, G., Gu, Y.: Deep learning-based classification of hyperspectral data. IEEE J. Sel. Top. Appl. Earth Obs. Remote Sens. 7(6), 2094–2107 (2014)
6. Sharma, V., Diba, A., Tuytelaars, T., Van Gool, L.: Hyperspectral CNN for image classification & band selection, with application to face recognition. Technical report KUL/ESAT/PSI/1604, KU Leuven, ESAT, Leuven, Belgium (2016)
7. Hamida, A.B., Benoit, A., Lambert, P., Amar, C.B.: 3-D deep learning approach for remote sensing image classification. IEEE Trans. Geosci. Remote Sens. 56(8), 4420–4434 (2018)
8. Zhong, Z., Li, J., Luo, Z., Chapman, M.: Spectral-spatial residual network for hyperspectral image classification: A 3-d deep learning framework. IEEE Trans. Geosci. Remote Sens. 56(2), 847–858 (2017)
9. Zhang, H., Li, Y., Jiang, Y., Wang, P., Shen, Q., Shen, C.: Hyperspectral classification based on lightweight 3-d-cnn with transfer learning. IEEE Trans. Geosci. Remote Sens. 57(8), 5813–5828 (2019)
10. Mei, X., et al.: Spectral-spatial attention networks for hyperspectral image classification. Remote Sens. 11(8), 963 (2019)
11. Ma, W., Yang, Q., Wu, Y., Zhao, W., Zhang, X.: Double-branch multi-attention mechanism network for hyperspectral image classification. Remote Sens. 11(11), 1307 (2019)
12. Hu, J., Shen, L., Sun, G.: Squeeze-and-excitation networks. In: 2018 IEEE/CVF Conference on Computer Vision and Pattern Recognition, pp. 7132–7141 (2018)
13. Wang, X., Girshick, R., Gupta, A., He, K.: Non-local neural networks. In: Proceedings of the IEEE Conference on Computer Vision and Pattern Recognition, pp. 7794–7803 (2018)
14. Yuan, Y., Huang, L., Guo, J., Zhang, C., Chen, X., Wang, J.: Ocnet: object context network for scene parsing. arXiv preprint arXiv:1809.00916 (2018)
15. Vaswani, A., et al.: Attention is all you need. Adv. Neural Inf. Process. Syst. 30 (2017)

16. Hong, D., et al.: Spectralformer: rethinking hyperspectral image classification with transformers. IEEE Trans. Geosci. Remote Sens. **60**, 1–15 (2022)
17. Woo, S., Park, J., Lee, J.Y., Kweon, I.S.: CBAM: convolutional block attention module. In: Proceedings of the European Conference on Computer Vision, pp. 3–19 (2018)
18. Wang, F., et al.: Residual attention network for image classification. In: Proceedings of the IEEE Conference on Computer Vision and Pattern Recognition, pp. 3156–3164 (2017)
19. Zhong, C., Gong, N., Zhang, Z., Jiang, Y., Zhang, K.: Litecclknet: a lightweight criss-cross large kernel convolutional neural network for hyperspectral image classification. IET Computer Vision (2023)
20. Guo, M.H., Lu, C.Z., Liu, Z.N., Cheng, M.M., Hu, S.M.: Visual attention network. arXiv preprint arXiv:2202.09741 (2022)
21. Chakraborty, T., Trehan, U.: Spectralnet: exploring spatial-spectral waveletcnn for hyperspectral image classification. arXiv preprint arXiv:2104.00341 (2021)
22. Roy, S.K., Haut, J.M., Paoletti, M.E., Dubey, S.R., Plaza, A.: Generative adversarial minority oversampling for spectral-spatial hyperspectral image classification. IEEE Trans. Geosci. Remote Sens. **60**, 1–15 (2021)
23. Li, R., Zheng, S., Duan, C., Yang, Y., Wang, X.: Classification of hyperspectral image based on double-branch dual-attention mechanism network. Remote Sens. **12**(3), 582 (2020)
24. Paoletti, M.E., Haut, J.M., Fernandez-Beltran, R., Plaza, J., Plaza, A.J., Pla, F.: Deep pyramidal residual networks for spectral-spatial hyperspectral image classification. IEEE Trans. Geosci. Remote Sens. **57**(2), 740–754 (2018)

Deepfake Detection via Fine-Grained Classification and Global-Local Information Fusion

Tonghui Li, Yuanfang Guo[✉], and Yunhong Wang

Laboratory of Intelligent Recognition and Image Processing, School of Computer Science and Engineering, Beihang University, Beijing 100191, China
{lthlth,andyguo,yhwang}@buaa.edu.cn

Abstract. In response to the increasing amount of deepfake content on the internet, a large number of deepfake detection methods have been recently developed. To our best knowledge, existing methods simply perform binary classification, i.e., they simply consider the deepfake images generated from different forgery methods as a single category. Unfortunately, different deepfake forgery methods usually generate deepfake images with different artifacts/appearances. Under such circumstance, a simple binary classification mechanism may limit the learning ability of the detection models, i.e., they may ignore certain forgery traces. Therefore, we propose a novel deepfake detection method via fine-grained classification and global-local information fusion. Specifically, we improve the binary classification task with a fine-grained classification mechanism, such that the deepfake detection model can learn more precise features for fake images from different forgery methods. Besides, we construct a global-local information fusion architecture to emphasize the important information in certain local regions and fuse them with global semantic information. In addition, we design a global center loss, which makes the real images features more cohesive and enlarge the distance between real and fake images features, to further enhance the generalization ability of the detection model. Extensive experiments demonstrate the effectiveness and superiority of our method.

Keywords: Deepfake detection · Fine-grained classification · Global-local information fusion · Global center loss

1 Introduction

With the continuous development of generative techniques such as Generative Adversarial networks [5] and Variational AutoEncoders [8], people can easily produce high-quality manipulated (deepfake) images which are difficult to be identified by human perception. Unfortunately, these technologies may be abused for various malicious purposes, such as political slander, economic fraud, fake news, etc., which usually induce huge negative impacts on social trust, public

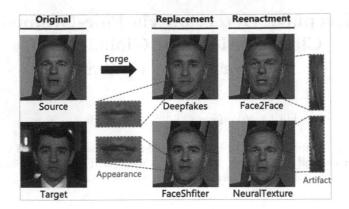

Fig. 1. The results of different deepfake forgery methods when forging the same target and source images. Note that different methods tend to generate different artifacts/appearances.

trust, political trust, etc. Therefore, it is of great significance to develop effective counterfeit detection methods.

To alleviate these risks, many deepfake detection methods have been proposed [4,10,13,22]. Some of the early work focused on exploiting biological inconsistencies in fake images [20]. Recently, Liu et al. achieved promising results in [10] by mining the artifact information of blending boundaries. Dong *et al.* proposed an Identity Consistency Transformer [4] by identifying the identity consistency of the inner and outer regions of each face. Liu *et al.* utilized the upsampling artifacts in the frequency domain to detect forgery in [13]. Masi *et al.* utilized a dual-stream network in [15] to fuse RGB information and high-frequency information. Zhao *et al.* suggested multiple attention maps, which focus on different regions, by introducing a region loss in [22].

Currently, most of the existing methods regard the deepfake detection problem as a binary classification problem [13,17,20,22]. A typical deepfake detection pipeline utilizes a backbone network to extract features from input images, and then classify them as real or fake.

Unfortunately, this binary classification mechanism may possess certain problems. Since there exists a variety of forging methods, their correspondingly generated fake images tend to possess various types of artifacts and different data distributions. As shown in Fig. 1, though the same images are forged, the fake images from different forging methods tend to show differences in many details. Therefore, if the model is trained to perform a binary classification on the real images and the fake images from different forging methods, the model learning type may not be conductive and the model may ignore certain forgery traces. Besides, certain local regions, such as eyes, nose and lip, usually contain more information than other facial regions. As shown in Fig. 1, different forgery methods tend to vary more obviously when forging these regions.

Inspired by the above observations, in this paper, we propose a deepfake detection method via fine-grained classification and global-local information fusion. Specifically, we improve the binary classification task of deepfake detection via a fine-grained classification mechanism. This mechanism divides the fake category into multiple fake categories corresponding to different forging methods, such that the detection model can learn more precise features for the images from different fake categories. Besides, we construct a global-local information fusion architecture, which extracts local features with shallow networks and fuses them with the global features. These extracted local features can better describe the local forgery artifacts at the eyes, nose and lip regions, to further boost our fine-grained classification mechanism. In addition, to alleviate the overfitting issue which exists widely in existing methods, we add a global center loss into our training loss to make the real image features more cohesive and enlarge the distance between the real and fake image features. Benefiting from the global center loss, the classification boundary learned by our model tends to be more explicit.

Our major contributions can be summarized as follows:

- We propose a novel deepfake detection method, by improving the conventional binary classification to a fine-grained classification.
- We propose a global-local information fusion architecture, which effectively extracts and combines global deep semantic information and local shallow texture information.
- We design a global center loss to further enhance the generalization ability of our model.

2   Methodology

To solve the problems induced by the simple binary classification mechanism in conventional deepfake detection methods, we propose a deepfake detection method via fine-grained classification mechanism and global-local information fusion.

The framework of our method is shown in Fig. 2. In the global branch, a pretrained backbone network is adopted to extract deep global features. In the local branch, multiple shallow convolutional neural networks are constructed to extract shallow features from three informative local regions, i.e., nose, eyes and lip. The shallow features are then concatenated and enhanced via a channel-wise attention module to obtain the local features. At last, the global deep features and the local shallow features are fused and a fine-grained classification is performed. Meanwhile, a global center loss is calculated with respect to the fused features, to further expand the margin between the real and fake images, such that the detection performance and generalization ability of the model can be improved.

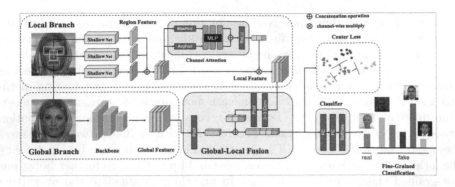

Fig. 2. Overview of our method.

Algorithm 1 Pseudocode of FGC in a PyTorch-like style.

```
# input: input image (N, C, H, W).
# output: pred (N, 2).
# backbone: the model on which the fine-grained classification mechanism is based
# S: number of fake categories
# y: ground truth label of fine-grained categories
# category_weight: loss weight for differnet categories

# extract features from the input image (N, C, H, W) -> (N, C_out)
features = backbone(features)

# fine-grained full connected layer (N, C_out) -> (N, S)
output = softmax(linear(features), dim=1)

# during training phase, fine-grained classification mechanism calculate loss
if phase == "train":
    loss = CrossEntropyLoss((y, output), category_weight, class=S)
    Optimize()

# during test phase, fine-grained classification mechanism predict the label
elif phase == "test":
    pred = cat((output[:, 0], sum(output[:, 1:])), dim=1)
    pred = argmax(pred, dim=1)
```

2.1 Fine-Grained Classification Mechanism

As mentioned above, images generated by different forgery methods tend to present different data distributions in the feature space. The previous methods simply consider the forged images as a single category, which may misguide the detection model to miss certain forgery cues. Therefore, we improve the binary classification performance with a fine-grained classification mechanism in our method.

Specifically, we construct the forgery categories according to the forgery methods utilized in the training process. Then, we constrain our method to further classify the fake images into its corresponding forgery category. With the help of the fine-grained classification mechanism, the model can learn more precise features for the fake images generated from different forgery methods, such that the real images can also be better identified. In the actual detection (testing) process, we directly convert the results of fine-grained classification into the corresponding real or fake categories. The calculation process of the fine-grained mechanism is summarized in Algorithm 1.

2.2 Global-Local Information Fusion Architecture

In our method, we employ deep neural network as the backbone to extract features from the input images. With the help of the backbone network, the output features are correlated to the entire input image and contain high-level semantic information. Unfortunately, the extracted deep features tend to contain less local information, which can better describe the local forgery artifacts. Based on the above observation, we propose a global-local information fusion architecture, which can effectively exploit the complementary global deep information and local shallow information.

Intuitively, typical deepfake methods tend to have much more difficulties in forging the local regions with more details, such as eyes, nose and lip. Then, these deepfake methods tend to induce more forgery artifacts/cues within these local regions. Therefore, the local branch is constructed to emphasize the local information in these regions. Specifically, the local branch utilizes the key point information of face detection to identify three local regions, i.e., eyes, nose and lip, in the input image. These local regions are denoted as R^p, where $p \in \{eyes, nose, lip\}$.

Then, the local branch input the local regions into their corresponding shallow convolutional neural network f_p for shallow feature extraction. The extracted features will be downsampled to $F^p \in \mathbb{R}^{H_l \times W_l \times C_l}$ by average pooling. Then, the output of the local branch, $F^r \in \mathbb{R}^{H_l \times W_l \times (C_l \times 3)}$, can be obtained via a concatenation operation, as

$$F^r = Concatenate(Pooling_p(f_p(R^p))). \tag{1}$$

Although the architectures of the shallow networks employed in the above local feature extraction process are identical, the importance and the amount of information contained in the above three local regions are different in the actual deepfake detection process. Thus, the channel attention module in [19] is adopted as

$$A(F^r) = \sigma(W_1\delta(W_0 F^r_{Avg}) + W_1\delta(W_0 F^r_{Max})), \tag{2}$$

$$F^{local} = A(F^r) \cdot F^r. \tag{3}$$

Here, F^r_{Max} and F^r_{Avg} correspond to F^r after the maximum and average pooling operation, respectively. W_0 and W_1 correspond to the parameters of the two layers in the channel attention module. σ and δ denote the Sigmoid and ReLU activation functions. After obtaining the enhanced local feature, we fuse the local feature F^{local} with the global feature F^{global} to generate the final feature F^{gl}. The fusion operation is denoted as $Fuse(\cdot, \cdot)$. In this paper, it is achieved by using a concatenation operation.

$$F^{gl} = Fuse(F^{global}, F^{local}) \tag{4}$$

2.3   Global Center Loss

To further improve the generalization ability of the model, a global center loss is designed based on the single center loss in [9] and added to the loss function of our model. The global center loss constrains the features of real images to be more cohesive, while maintains a certain distance between the features of fake images and the cluster center of the features of real images. By adopting the global center loss, our model can learn a more explicit classification boundary.

The existing single center loss in [9] calculates the cluster center of the features of real images within each batch, which may be obviously deviated from the global center and induces degradation to the learning of our model. On the contrary, we initialize the cluster center C_g of the features of real images globally on the entire training dataset. In each training iteration, the cluster center C_b of the features of real images is calculated on the entire data batch. Then, the global center C_g is updated accordingly.

$$C_b = \frac{\sum_{i=0}^{n-1} F_i^{gl} \cdot \mathbb{I}\{y_i = 0\}}{\sum_{i=0}^{n-1} \mathbb{I}\{y_i = 0\}} \tag{5}$$

$$C_g = C_g + \alpha \cdot (C_b - C_g) \tag{6}$$

where α controls the updating strength, y_i is the ground truth label of the i-th sample in the batch, and $\mathbb{I}$ is an indicator function. This indicator function takes the value 1 if the internal condition is satisfied and 0 otherwise. After updating the global center, we calculate the Euclidean distances from the real and fake image features to the global center, respectively.

$$D_m = \frac{\sum_{i=0}^{n-1} \|F_i^{gl} - C_g\|_2 \cdot \mathbb{I}\{y_i = m\}}{\sum_{i=0}^{n-1} \mathbb{I}\{y_i = m\}} \tag{7}$$

Then, our global center loss L_{center} is calculated according to the Euclidean distances D_0 and D_1, as

$$L_{center} = D_0 + ReLU(\beta \cdot \sqrt{N} - D_1), \tag{8}$$

where β controls the expected distance from the features of fake images to the global center, and N represents the dimension of the feature. Note that this global center loss constrains the forged features to prevent them from unlimited pushbacks, which tends to force their distributions to be overly scattered.

At last, we compute the overall loss function L_{total} by combining the center loss L_{center} with the cross-entropy loss L_{ce}, as

$$L_{total} = L_{ce} + \lambda \cdot L_{center}, \tag{9}$$

where λ controls the importance of the center loss.

After adding the global center loss, a larger distance between the real and fake image features will be maintained, and the classification boundary will be more explicit. This assigns our model a better generalization ability to detect the fake images generated from unknown forgery methods.

Table 1. Architecture of the shallow network in the local branch.

layer name	parameters
conv1	$3 \times 3, 32$, stride 2
conv2	$5 \times 5, 64$
conv3	$3 \times 3, 128$
conv4	$1 \times 1, 128$

3 Experiments

3.1 Experimental Settings

Datasets. To facilitate the comparisons with the existing methods, the proposed model is trained on the most widely used benchmark FF++ [17]. The tests are performed on FF++ [17], Celeb-DF [12] and DeepFakeDetection (DFD) [2].

Metrics. In this paper, Accuracy Rate (ACC) and Area Under the ROC Curve (AUC), which are commonly applied in deepfake detection tasks [13,15,17], are employed as the evaluation metrics.

Implementation Details. In our experiments, EfficientNet-B4 [18] is utilized as the backbone network in the global branch. In the local branch, we construct a 4-layered convolutional neural network as the shallow network to extract shallow features from important local regions. Note that the details of the shallow network is provided in Table 1. The Adam [7] optimizer is adopted to update the parameters of the model. We set the learning rate to 0.001 and decay to half of itself every 5 epochs, until the validation loss has not decreased in 5 consecutive epochs. The relevant hyperparameters of the center loss are set as $\alpha = 0.3$, $\beta = 0.3$, $\lambda = 1$. All the input images are preprocessed by RetinaFace for face and keypoint detection. Note that since FF++, which contains five forgery methods, i.e., Deepfakes, Face2Face, FaceSwap, NeuralTexture and Faceshifter, are utilized as the training dataset, the fake categories in our fine-grained classification mechanism are constructed accordingly.

3.2 Performance Evaluations

Intra-Dataset Evaluations. The intra-dataset evaluations are performed on the FF++ dataset. The results in Table 2 demonstrate that our method outperforms all the existing methods in the low-quality (LQ) setting. According to the results in the low-quality (LQ) setting, our method obtains remarkable improvements compared to the current state-of-the-art methods, e.g., 3.99% in terms of the ACC score against PD [21] and 8.53% against SPSL [13]. Although our ACC score in the high-quality (HQ) setting is slightly inferior to MADD [22], MADD requires to repeatedly mask images with different shapes during training, which makes it cost approximately $3\times$ time of ours to process a single epoch in the training process.

Table 2. Intra-dataset evaluation results on FaceForensics++.

Methods	HQ		LQ	
	ACC(%)	AUC(%)	ACC(%)	AUC(%)
MesoNet [1]	83.10	–	70.47	–
Face X-ray [10]	–	87.40	–	61.60
Xception [3]	95.73	96.30	86.86	89.30
Xception-ELA [6]	93.86	94.80	79.63	82.90
Two-Branch [15]	–	98.70	–	86.59
MADD [22]	**97.60**	**99.29**	88.69	90.40
SPSL [13]	91.50	95.32	81.57	82.82
PD [21]	95.43	98.71	86.11	89.29
EfficientNet-B4 [18]	96.63	99.18	88.15	90.71
Ours	97.28	**99.29**	**90.10**	**93.07**

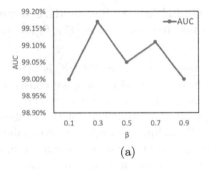
     (a)

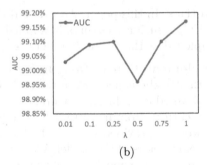
     (b)

Fig. 3. The impacts of varying different hyperparameters β, λ. (a) Varying λ when β is fixed at 0.3. (b) Varying β when λ is fixed at 1.

Cross-Dataset Evaluations. The cross-dataset evaluations are conducted on the Celeb-DF dataset to evaluate the generalization ability of the proposed method. Since the transform-based data augmentation strategy is a commonly utilized cross-dataset performance boosting strategy, we also adopt this strategy in the training phase for fair comparison. According to the results presented in Table 3, our method outperforms the existing methods and achieves the overall best performance, when utilizing the transformation-based data augmentation strategy. Even without employing the data augmentation strategy, our method surpasses other approaches in such situation. This unequivocally demonstrates the superiority of our proposed method in terms of generalization ability.

3.3 Analysis

Hyperparameter Analysis. We analyze the impacts of varying the hyperparameters β and λ on the FF++(HQ). The results are shown in Fig. 3. According

Table 3. Cross-dataset evaluation results (AUC) on Celeb-DF. † indicates that the transform-based data augmentation strategy is applied during the training process.

Methods	Celeb-DF
Meso4 [1]	54.80
MesoInception4 [1]	53.60
HeadPose [20]	54.60
Xception-c23 [12]	65.30
DSP-FWA [11]	64.60
F^3-Net [16]	65.17
EfficientNet-B4 [18]	64.29
Face X-ray [10]†	74.76
MADD [22]†	67.44
SPSL [13]†	76.88
PD [21]†	74.22
Ours	70.75
Ours†	**77.62**

Table 4. Ablation results (AUC(%)) of our method. The complete method achieves the best results. FGC represents the fine-grained classification mechanism, GLF denotes the global-local information fusion architecture and GCL stands for the global center loss.

FGC	GCL	GLF	FF++(HQ)	DFD	Celeb-DF
–	–	–	99.04	80.16	64.29
✓	–	–	99.12	79.29	67.27
✓	✓	–	99.16	85.14	70.28
✓	✓	✓	**99.29**	**86.10**	**70.75**

to the results, except for the abnormality at $\lambda = 0.5$, the performance of the model increases with the increase of λ, and reaches the optimum when $\lambda = 1$. Regarding β, it controls the expected lower bound of the distance between the fake image features and the global center of real features. When β is too large, the model can hardly converge. When β is too small, the model may not be able to distinguish between the real and fake image features. According to the results, when $\beta = 0.3$, the model achieves a good balance in terms of discriminability and convergence, and exhibits best performance. Thus, $\beta = 0.3$ and $\lambda = 1$ are selected in other experiments. Since we observed that α possesses little impact on the performance of our model after convergence, we set $\alpha = 0.3$.

Ablation Study. Table 4 shows the ablation results of our proposed work. As can be observed, each of our key component, i.e., fine-grained classification mechanism, global-local information fusion architecture and global center loss, can contribute solidly to the final method.

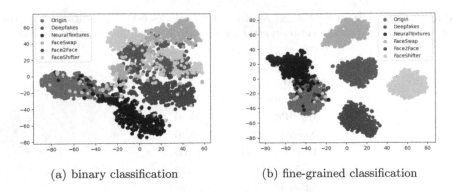

(a) binary classification   (b) fine-grained classification

Fig. 4. The t-SNE result of the image features extracted from the FF++ HQ testing dataset. (a) The results of EfficientNet-B4. (b) The results of EfficientNet-B4 with fine-grained classification mechanism.

Table 5. The fine-grained classification performances of different backbone networks with our fine-grained classification mechanism on FF++.

Backbone	HQ		LQ	
	ACC(%)	AUC(%)	ACC(%)	AUC(%)
Xception	95.20	99.54	75.71	94.86
EfficientNet-B4	96.86	99.64	83.44	97.04

Effectiveness of the Fine-grained Classification Mechanism. To verify the effectiveness of the fine-grained classification mechanism, we utilize Efficient-Net [18] and Xception [3] as the backbone networks, and apply the fine-grained classification mechanism to them respectively.

The fine-grained classification and binary classification results are shown in Table 5 and Table 6, respectively. Table 5 reveals that the backbone networks can well distinguish the deepfake images generated by different forgery methods, when the fine-grained classification mechanism is applied. For the binary classification task in Table 6, the fine-grained classification mechanism can obviously improve the binary classification performance of the model. Meanwhile, this mechanism can actually make the model to converge faster and thus reduces the training time. In addition, the fine-grained classification mechanism can be easily applied to the existing models to enhance their detection capabilities.

To further verify the effectiveness of the fine-grained classification mechanism, we employ t-SNE [14] to visualize the intermediate features extracted from the testing dataset. As shown in Fig. 4(a), although the binary classification model treats the deepfake images from different forgery methods as a single fake category, the extracted features still intend to form multiple clusters related to the forgery methods. After applying our fine-grained classification mechanism, the model can extract features which can be clustered in a more precise manner. As shown in the Fig. 4(b), the explicit clusters verify that the data distributions

Table 6. Binary forgery classification results of the backbone networks with/without the fine-grained classification mechanism (FGC).

Backbone	HQ		LQ	
	ACC(%)	AUC(%)	ACC(%)	AUC(%)
Xception w/o FGC	95.73	96.30	86.86	89.30
EfficientNet-B4 w/o FGC	96.66	99.04	88.15	90.71
Xception w/ FGC	95.84	98.60	87.68	89.50
EfficientNet-B4 w/ FGC	**97.06**	**99.12**	**89.68**	**93.01**

Table 7. Comparison of different shallow network designs.

Shallow Network Designs	ACC(%)	AUC(%)
4-layered CNN	**97.19**	**99.28**
4-layered CNN + dense connections	97.12	99.16
ROI + 4-layered CNN	97.15	99.17

of fake images from different forgery methods exist differently. By adopting our fine-grained classification mechanism, the model can better discriminate between real and fake images.

Comparison of Different Shallow Network Designs. In this analysis, we verify the design of the shallow network in the local branch. Here, three designs are compared: 1) extracting features via our designed 4-layered convolutional neural network, 2) extracting features via the designed 4-layered convolutional neural network with dense connections, 3) obtaining the ROI features corresponding to the eyes, nose and lip regions from the shallow layer of the backbone network, and then extracting features via the designed 4-layered convolutional neural network. According to Table 7, our 4-layered convolutional neural network gives the best performance and is utilized in other experiments.

4   Conclusion

In this paper, we propose a new deepfake detection method via fine-grained classification and global-local information fusion. Our method improves the binary classification performance with fine-grained classification mechanism. Meanwhile, our method also effectively exploit the complementary global deep information and local shallow information. These improvements allow our deepfake detection model to learn finer forgery traces and better identify the deepfake images from different forgery methods within the feature space. In addition, we design a global center loss to effectively enlarge the margin between the real and fake images, which can make the classification boundary more explicit. Experimental results have demonstrated the effectiveness and superiority of our method.

References

1. Afchar, D., Nozick, V., Yamagishi, J., Echizen, I.: Mesonet: a compact facial video forgery detection network. In: IEEE WIFS, pp. 1–7 (2018)
2. AI, G.: Contributing data to deepfake detection research (2019). https://ai.googleblog.com/2019/09/contributing-data-to-deepfake-detection.html. Accessed 09 Apr 2022
3. Chollet, F.: Xception: deep learning with depthwise separable convolutions. In: IEEE CVPR, pp. 1251–1258 (2017)
4. Dong, X., et al.: Protecting celebrities from deepfake with identity consistency transformer. In: IEEE/CVF CVPR, pp. 9468–9478 (2022)
5. Goodfellow, I., et al.: Generative adversarial nets. In: Advances in Neural Information Processing Systems, pp. 2672–2680 (2014)
6. Gunawan, T.S., Hanafiah, S.A.M., Kartiwi, M., Ismail, N., Za'bah, N.F., Nordin, A.N.: Development of photo forensics algorithm by detecting photoshop manipulation using error level analysis. Indones. J. Electr. Eng. Comput. Sci. $7(1)$, 131–137 (2017)
7. Kingma, D.P., Ba, J.: Adam: a method for stochastic optimization. arXiv preprint arXiv:1412.6980 (2014)
8. Kingma, D.P., Welling, M.: Auto-encoding variational bayes. arXiv preprint arXiv:1312.6114 (2013)
9. Li, J., Xie, H., Li, J., Wang, Z., Zhang, Y.: Frequency-aware discriminative feature learning supervised by single-center loss for face forgery detection. In: IEEE/CVF CVPR, pp. 6458–6467 (2021)
10. Li, L., et al.: Face x-ray for more general face forgery detection. In: IEEE/CVF CVPR, pp. 5001–5010 (2020)
11. Li, Y., Lyu, S.: Exposing DeepFake videos by detecting face warping artifacts. arXiv preprint arXiv:1811.00656 (2018)
12. Li, Y., Yang, X., Sun, P., Qi, H., Lyu, S.: Celeb-DF: a large-scale challenging dataset for DeepFake forensics. In: IEEE/CVF CVPR, pp. 3207–3216 (2020)
13. Liu, H., et al.: Spatial-phase shallow learning: rethinking face forgery detection in frequency domain. In: IEEE/CVF CVPR, pp. 772–781 (2021)
14. der Maaten, L.V., Hinton, G.: Visualizing data using t-SNE. JMLR $9(86)$, 2579–2605 (2008)
15. Masi, I., Killekar, A., Mascarenhas, R.M., Gurudatt, S.P., AbdAlmageed, W.: Two-branch recurrent network for isolating deepfakes in videos. In: Vedaldi, A., Bischof, H., Brox, T., Frahm, J.-M. (eds.) ECCV 2020. LNCS, vol. 12352, pp. 667–684. Springer, Cham (2020). https://doi.org/10.1007/978-3-030-58571-6_39
16. Qian, Y., Yin, G., Sheng, L., Chen, Z., Shao, J.: Thinking in frequency: face forgery detection by mining frequency-aware clues. In: Vedaldi, A., Bischof, H., Brox, T., Frahm, J.-M. (eds.) ECCV 2020. LNCS, vol. 12357, pp. 86–103. Springer, Cham (2020). https://doi.org/10.1007/978-3-030-58610-2_6
17. Rossler, A., Cozzolino, D., Verdoliva, L., Riess, C., Thies, J., Nießner, M.: Face-forensics++: learning to detect manipulated facial images. In: IEEE/CVF ICCV, pp. 1–11 (2019)
18. Tan, M., Le, Q.: Efficientnet: rethinking model scaling for convolutional neural networks. In: ICML, pp. 6105–6114. PMLR (2019)
19. Woo, S., Park, J., Lee, J.-Y., Kweon, I.S.: CBAM: convolutional block attention module. In: Ferrari, V., Hebert, M., Sminchisescu, C., Weiss, Y. (eds.) ECCV 2018. LNCS, vol. 11211, pp. 3–19. Springer, Cham (2018). https://doi.org/10.1007/978-3-030-01234-2_1

20. Yang, X., Li, Y., Lyu, S.: Exposing deep fakes using inconsistent head poses. In: IEEE ICASSP, pp. 8261–8265 (2019)
21. Zhang, B., Li, S., Feng, G., Qian, Z., Zhang, X.: Patch diffusion: a general module for face manipulation detection. AAAI **36**(3), 3243–3251 (2022)
22. Zhao, H., Zhou, W., Chen, D., Wei, T., Zhang, W., Yu, N.: Multi-attentional deepfake detection. In: IEEE/CVF CVPR, pp. 2185–2194 (2021)

Unsupervised Image-to-Image Translation with Style Consistency

Binxin Lai and Yuan-Gen Wang[✉]

School of Computer Science and Cyber Engineering, Guangzhou University, Guangzhou, China
laibinxin@e.gzhu.edu.cn, wangyg@gzhu.edu.cn

Abstract. Unsupervised Image-to-Image Translation (UNIT) has gained significant attention due to its strong ability of data augmentation. UNIT aims to generate a visually pleasing image by synthesizing an image's content with another's style. However, current methods cannot ensure that the style of the generated image matches that of the input style image well. To overcome this issue, we present a new two-stage framework, called Unsupervised Image-to-Image Translation with Style Consistency (SC-UNIT), for improving the style consistency between the image of the style domain and the generated image. The key idea of SC-UNIT is to build a style consistency module to prevent the deviation of the learned style from the input one. Specifically, in the first stage, SC-UNIT trains a content encoder to extract the multiple-layer content features wherein the last-layer's feature can represent the abstract domain-shared content. In the second stage, we train a generator to integrate the content features with the style feature to generate a new image. During the generation process, dynamic skip connections and multiple-layer content features are used to build multiple-level content correspondences. Furthermore, we design a style reconstruction loss to make the style of the generated image consistent with that of the input style image. Numerous experimental results show that our SC-UNIT outperforms state-of-the-art methods in image quality, style diversity, and style consistency, even for domains with significant visual differences. The code is available at https://github.com/GZHU-DVL/SC-UNIT.

Keywords: Unsupervised Image-to-Image Translation · Prior Distillation · Generative Adversarial Network · Style Consistency

1 Introduction

The primary goal of Unsupervised Image-to-Image Translation (UNIT) is to translate an image from one domain to another and ensure the quality of the generated images. This technology can help create virtual images using existing real images. Many approaches have been proposed to enhance the quality of generated images, and make them more natural and realistic. CycleGAN [30]

© The Author(s), under exclusive license to Springer Nature Singapore Pte Ltd. 2024
Q. Liu et al. (Eds.): PRCV 2023, LNCS 14430, pp. 322–334, 2024.
https://doi.org/10.1007/978-981-99-8537-1_26

Fig. 1. Results of SC-UNIT. What we need to do is to translate the input (left of each pair of images) into output (right of each pair of images).

introduces the concept of cycle consistency to establish mappings between different domains. This approach has shown good performance in color and texture translation, such as transforming winter to summer, horses to zebras, and photos to Monet-like paintings. However, it does not perform well when it comes to translations that involve significant geometric transformations. Additionally, images generated by CycleGAN also lack diversity. With a single generator and one input image, it can only produce a single corresponding output image.

As the growing research in features decoupling, the majority of researchers have adopted the encoder-decoder framework and have applied a content encoder and a style encoder [7,13,14,22,23,27] to learn domain-shared features and domain-specific features respectively. Since Huang et al. [12] have proved the effectiveness of AdaIN layer in style transformation, almost all of the recent works in UNIT task have exploited the AdaIN layer to transfer the style of images.

StarGAN2 [7] is an excellent work that applies the encoder-decoder framework and the style encoder to deal with the problem of diversity. Furthermore, StarGAN2 provides a high-quality animal face dataset called AFHQ, which greatly assists in assessing the capability of models in translations across domains with significant variations in shape and appearance. However, it still does not work well in translations that involve large changes in appearance and shape, since such translations require models to build more abstract semantic correspondences.

GP-UNIT [27] proposes a two-stage framework that distills the generative prior from BigGAN [2] in the first stage to extract abstract content features which can represent common attributes across domains with dramatic differences in shape and appearance. This approach greatly improves the deficiencies of the aforementioned methods in building abstract semantic correspondences. However, we find that in GP-UNIT, the style consistency, such as species, between the style input and the generated image cannot be guaranteed very well.

In this paper, we propose a new framework called Unsupervised Image-to-Image Translation with Style Consistency (SC-UNIT) based on a style consistency module to improve the style consistency between the input style image and the generated image. Our framework contains two stages. In the first stage, we train a content encoder using the data with strong content correlation as input to extract domain-shared content features. In the second stage, we train a generator using a dynamic skip connection to build multi-level correspondences, and use a style consistency module to ensure the style similarity between the input style image and the generated image during the training. Experimental results demonstrate that our framework has addressed the aforementioned problems and achieved satisfactory performance, as depicted in Fig. 1.

2   Related Work

Unsupervised Image-to-Image Translation. UNIT aims to generate diverse and high-quality virtual images based on the data without label. Such a method can help reduce labor costs. Many works in this field have made outstanding contributions. For example, CycleGAN [30] exploits Cycle consistency constraints to build a mapping between two domains. Then, the introduction of discriminator [10] against the generator contributes to generating more realistic images. MUNIT [13] proposes a new framework where the latent space of images can be divided into a content space (domain-shared) and a style space (domain-specific). It employs a content encoder and a style encoder to extract content features and style features, respectively. COCO-FUNIT [26] utilizes a content-conditioned style encoder to alleviate the missing content problem. StarGAN2 [7] introduces FAN [3] to accurately localize the positions of facial features, resulting in good performance in face translation. Additionally, it utilizes diversity regularization to enhance the diversity of the generated images. GP-UNIT [27] proposes a two-stage framework which utilizes the prior distilled from BigGAN [2] to help learn abstract semantic content features.

Prior Distillation. Generally, prior can be helpful for models to accelerate convergence and reduce the learning difficulty. Another function of prior is to improve the stability of training. A large number of works [4,8,15,24,27,29] have demonstrated the effectiveness of prior distillation in improving models' performance. Those works generally distill prior from StyleGAN [19] or BigGAN [2] to improve image quality or better capture abstract content features.

Generative Adversarial Network. Generative Adversarial Network (GAN) [10] is the mainstream technique in UNIT. It can force a generator to generate more realistic images by introducing a discriminator to predict whether the image is fake. The discriminator and generator take turns making progress through the adversarial loss. As the discriminator becomes more accurate in distinguishing real and fake images, the generator learns to generate more authentic images, ultimately achieving a Nash equilibrium. Many works [1,2,6,7,13,17–19,21,27] with excellent performance in image generation have adopted GAN to improve the authenticity of generated images.

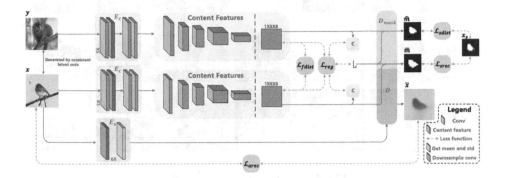

Fig. 2. First Stage: Training of Content Encoder.

3 Proposed Method

The training process is divided into two stages. In the first stage, we train a content encoder that distills the prior from the data generated by BigGAN to learn the domain-shared features better. In the second stage, we train a generator to build multiple-level correspondences between input and output using a dynamic skip connection. At the same time, we built the style consistency module to prevent the deviation of the learned style from the input style.

3.1 First Stage: Training of Content Encoder

The first stage is designed to train a content encoder that can extract domain-shared content features at an abstract semantic level. These features can represent the content of images, such as the orientation of the main object and rough shape correspondence. Figure 2 shows the training framework in the first stage.

During the training process, a pair of images (x, y) are used as input for each iteration. Every image pair (x, y), which is randomly selected from domain X and Y, is generated by a consistent latent code in BigGAN. Because the images generated by BigGAN using the same latent code show a strong correlation in content. Some examples are shown in Fig. 3. Then, we employ the content encoder E_c and style encoder E_s to extract content features $E_c(x)$ and style features $E_s(x)$, respectively. In the next step, we define a decoder D to reconstruct the original input x. The reconstructed image is denoted as $\bar{x}$. Meanwhile, $D(E_c(x), E_s(x), l_x)$ produces the mask $\bar{m}$ based on content feature $E_c(x)$ and domain label l_x to predict the mask of image x (l_x represents the domain of

Domain$_1$ Domain$_2$ Domain$_3$ Domain$_4$ Domain$_5$ Domain$_6$

Fig. 3. Generation of BigGAN [2].

the image x). The generated mask $\bar{m} = D_{mask}(E_c(x), l_x)$ (D_{mask} is part of D) should be as similar as x_s (x_s has been generated by applying HTC [5] in image x, which predicts the mask of x) for the purpose that the content feature extracted by E_c could include the information about the shape of input. So we use $\mathcal{L}_1$ loss as the shape reconstruction loss (denoted as $\mathcal{L}_{srec}$) to calculate the difference between $\bar{m}$ and x_s. Thus, $\mathcal{L}_{srec}$ can be defined as:

$$\mathcal{L}_{srec} = \lambda_{srec}\mathbb{E}_x[\|\bar{m} - x_s\|_1]. \tag{1}$$

To ensure that the content feature is domain-shared, we continue to execute the above operation with the single different point that we place the content features of x with the content of y when generating the mask $\hat{m} = D_{mask}(E_c(y), l_x)$ of the pair of images (x, y). Then, we calculate the difference between the generated mask $\hat{m}$ and x_s by the shape distance loss $\mathcal{L}_{sdist}$. $\mathcal{L}_{sdist}$ is defined as:

$$\mathcal{L}_{sdist} = \lambda_{sdist}\mathbb{E}_{x,y}[\|\hat{m} - x_s\|_1]. \tag{2}$$

Because the input images (x, y) are synthesized by the same latent code, which exhibits a strong relationship in content. The content features extracted by the content encoder respectively from images (x, y) should theoretically be the same as possible. Based on this opinion, the feature distant loss $\mathcal{L}_{fdist}$ is added, which could help the content encoder to find the domain-shared content features. $\mathcal{L}_{fdist}$ is defined as:

$$\mathcal{L}_{fdist} = \lambda_{fdist}\mathbb{E}_{x,y}[\|E_c(x) - E_c(y)\|_1]. \tag{3}$$

For the cycle consistency, the reconstructed image $\bar{x} = D(E_c(x), E_s(x), l_x)$ should be the same as the original input x. To this aim, the appearance

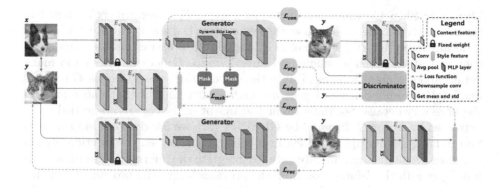

Fig. 4. Second Stage: Training of Generator.

reconstruction loss $\mathcal{L}_{arec}$ consisting of MSE loss and VGG perceptual loss [16] is added to ensure cycle consistency. $\mathcal{L}_{arec}$ is defined as:

$$\mathcal{L}_{arec} = \lambda_{arec}\mathbb{E}_x[\|\bar{x} - x\|_2 + VGG(\bar{x}, x)]. \tag{4}$$

In addition, a domain classifier C is introduced as part of the adversarial network. The goal is to guide C to classify the domain of the content feature more accurately. Confronting C could help train the content encoder to extract domain-shared content features. $\mathcal{L}_2$ normalization is used to raise the stability and generalization capability of the model. Both terms are included in a loss function $\mathcal{L}_{reg}$ (R indicates gradient reversal layer [9]):

$$\mathcal{L}_{reg} = \lambda_{reg}\mathbb{E}_x[\|E_c(x)\|_2] + \lambda_{cls}\mathbb{E}_x[-l_x \log C(R(E_c(x)))]. \tag{5}$$

Considering the above, the completed objective function of the first stage can be written as:

$$\min_{E_c, E_s, D, C} \mathcal{L}_{srec} + \mathcal{L}_{sdist} + \mathcal{L}_{fdist} + \mathcal{L}_{arec} + \mathcal{L}_{reg}. \tag{6}$$

3.2   Second Stage: Training of Generator

The second stage aims to train a generator G which could build multiple-level correspondences between the generated image $\hat{y}$ and the content input x. G can generate images at a more detailed level, incorporating content features $E_c(x)$ from content input x and style features $E_s(y)$ from style input y. Two images (x, y) will be inputs. At this stage, the content encoder E_c is trained in the first stage and fixed in weight in the second stage. Figure 4 illustrates the training process of the second stage.

Dynamic Skip Connection. In the first stage, E_c generates six content features, and the decoder D only uses the last content features in the last layer. This is because the primary purpose of D is to guide E_c to extract better coarse-level content features, such as shape and direction. So, D does not need too many details about content features. But in the second stage, the generator G requires fine-level correspondences to generate high-quality images that encompass all the necessary details. So, the content features of six layers are all needed to be extracted by E_c for G. Inspired by GP-UNIT [27], we use dynamic skip connection to integrate multiple-layer features better. Dynamic skip connection utilizes two gates (reset gate and update gate) to integrate the information from the current layer with the information from the previous layer. The process of dynamic skip connection is below: Every dynamic skip connection layer will use S^{i-1} (hidden state from previous layer), E_c^i (content feature from current layer), and G^{i-1} (generated feature from previous layer) as input (i indicates i-th layer of G). To match the scale of S^{i-1} with E_c^i, upsampling operation will be applied to S^{i-1}. For concise, upsampling will be replaced by the symbol $\uparrow$ hereinafter. For the same reason, σ and $\circ$ indicate the activation and convolutional layers, respectively. W_m^i, W_E^i, W_r^i, and W_S^i represent the convolutions weight of the update gate, new content, reset gate, and state, respectively. $Con(\cdot, \cdot)$ indicates concatenation. Content feature E_c^0 will be used as an initial hidden state S^0. Then, in layer i, the following operations will be performed:

$$\hat{S}^{i-1} = \sigma(W_S^i \circ \uparrow S^{i-1}), \quad r^i = \sigma(W_r^i \circ Con(\hat{S}^{i-1}, E_c^i)),$$

$$m^i = \sigma(W_m^i \circ Con(\hat{S}^{i-1}, E_c^i)), \quad S^i = r^i \hat{S}^{i-1}, \tag{7}$$

$$\hat{E}_c^i = \sigma(W_E^i \circ Con(S^i, E_c^i)), \quad Y = (1 - m^i)G^{i-1} + m^i \hat{E}_c^i.$$

It is worth mentioning that the mask loss function $\mathcal{L}_{msk}$ will be used in m^i to avoid the impact of useless elements and only retain the most important content. $\mathcal{L}_{msk}$ is defined as:

$$\mathcal{L}_{msk} = \lambda_{msk} \sum_i \mathbb{E}_x[\|m^i\|_1]. \tag{8}$$

Style Consistency Loss. Although high-quality image generation has been achieved in UNIT, the style of the generated image $\hat{y}$ fails to align with the style of the input image y. This problem also affects the diversity of styles in the generated images. There are two underlying reasons for this phenomenon. One reason is the absence of the style consistency loss function, resulting in a deviation of the learned style. The other is that the model prioritizes learning styles that appear more frequently in the style input. Based on these two points, we design a style reconstruction loss function $\mathcal{L}_{styr}$ which is calculated between the reconstructed image $\bar{y} = G(E_c(y), E_s(y))$ and style input y in style domain through MSE loss. $\mathcal{L}_{styr}$ can be defined as:

$$\mathcal{L}_{styr} = \lambda_{styr} \mathbb{E}_y[\|E_s(\bar{y}) - E_s(y)\|_2]. \tag{9}$$

Full Objection Function. Our completed loss function in the second stage is written below:

$$\min_{E_s,G} \max_D \mathcal{L}_{adv} + \mathcal{L}_{sty} + \mathcal{L}_{styr} + \mathcal{L}_{msk} + \mathcal{L}_{con} + \mathcal{L}_{rec}, \qquad (10)$$

where $\mathcal{L}_{adv}$ indicates a loss function of the discriminator P [10], $\mathcal{L}_{adv}$ can be written as:

$$\mathcal{L}_{adv} = \lambda_{adv}\mathbb{E}_{x,y}[\log(1 - P(\hat{y}))] + \mathbb{E}_y[\log P(y)], \qquad (11)$$

where $\hat{y} = G(E_c(x), E_s(y))$ is generated by integrating the content features $E_c(x)$ of input x and the style features $E_s(y)$ of input y. And because of this point, the generated image $\hat{y}$ should be the same as x in the content domain, and be the same as y in the style domain. Thus, $\mathcal{L}_{con}$ and $\mathcal{L}_{sty}$ are applied:

$$\mathcal{L}_{con} = \lambda_{con}\mathbb{E}_{x,y}[\|E_c(\hat{y}), E_c(x)\|_1], \qquad (12)$$

$$\mathcal{L}_{sty} = \lambda_{sty}\mathbb{E}_{x,y}[\|E_s(\hat{y}), E_s(y)\|_1]. \qquad (13)$$

The purpose of $\mathcal{L}_{rec}$ in the second stage is consistent with $\mathcal{L}_{arec}$ in the first stage. $\mathcal{L}_{rec}$ is defined below:

$$\mathcal{L}_{rec} = \lambda_{rec}\mathbb{E}_y[\|\bar{y} - y\|_1 + VGG(\bar{y}, y)], \qquad (14)$$

where $\bar{y} = G(E_c(y), E_s(y))$ denotes the reconstruction of the original y.

Fig. 5. Visual comparison. Our generated images are more realistic and have higher correspondences with inputs.

4   Experiment

4.1   Setting

Dataset. In the first stage, both real and synthetic data are used. ImageNet [25] with HTC is applied as data input for real data. HTC aims to find and tailor the object region. Each domain contains 600 images except the face domain, which includes 29K images from CelebA-HQ [17], for training. For synthetic data, 655 images of each domain are generated by BigGAN. Images with the same label in each domain correlate strongly because BigGAN uses the same latent code. This dataset is called symImageNet291, which contains 291 domains generated by BigGAN. Six hundred images per domain will be used for training. In the second stage, three datasets are prepared for five image translation tasks. For Dog↔Cat translation, the AFHQ dataset [7] is used, and each domain uses 4K images for training. As done in GP-UNIT, for Bird↔Dog translation, four classes of birds and four classes of dogs are selected from ImageNet291 for training (each class contains 600 images). For Male↔Female, CelebA-HQ is used for training. For Cat↔Face, 4K images of cats from AFHQ and 29K of faces from CelbeA-QH are used for training. For Bird↔Car, four classes of birds and four classes of dogs from ImageNet291 are chosen (each class contains 600 images) for training.

Setting of Hyper-Parameter. We use one NVIDIA GeForce RTX 3090 to train the model with Pytorch. The batch size is set to 10. The training time of the first stage is about one day. And the training time for the second stage is about two days. The number of iterations is set to 180K. The model occupies about 22GB. In the first stage, hyper-parameters are set for preferable training. We set $\lambda_{srec} = 5.0$, $\lambda_{sdist} = 5.0$, $\lambda_{fdist} = 1.0$, $\lambda_{arec} = 1.0$, $\lambda_{reg} = 0.001$, and $\lambda_{cls} = 1.0$. In the second stage, we set $\lambda_{adv} = 1.0$, $\lambda_{sty} = 50.0$, $\lambda_{styr} = 0.5$, $\lambda_{msk} = 1.0$, $\lambda_{con} = 1.0$. It is worth noting that dynamic skip connection is only used in the second and third layers of the generator for better preserving the most essential features.

4.2   Experimental Results

Qualitative Comparison. Visual comparisons in different translation tasks are shown in Fig. 5. We can see from Fig. 5 that GP-UNIT performs poorly in style consistency, outline recovery, and detail rebuilding. StarGAN2 can perform well in domains involving faces, but the performance is highly reduced in tasks involving objects without the five sense organs. And in terms of style consistency, StarGAN2 performs worse than GP-UNIT. Interestingly, our method could perform better than its competitors in style similarity or content correspondence.

Quantitative Comparison. For the evaluation of generated image quality and diversity, Frechet Inception Distance (FID) [11] and Learned Perceptual Image Patch Similarity (LPIPS) [28] are introduced. FID could estimate the quality of images by calculating the distance of feature vectors between the generated and original images. The lower FID is achieved, the higher the generated image

Table 1. Quantitative comparison. FID and LPIPS are used to assess the quality and diversity of generated images. Except for the diversity of generated images in StarGAN2, our model performs better than others.

Task	Male ↔ Female		Dog ↔ Cat		Human Face ↔ Cat	
Metric	FID	LPIPS	FID	LPIPS	FID	LPIPS
TraVeLGAN [1]	66.60	–	58.91	–	85.28	–
U-GAT-IT [20]	29.47	–	38.31	–	110.57	–
MUNIT [13]	22.64	0.37	80.93	0.47	56.89	0.53
COCO-FUNIT [26]	39.19	0.35	97.08	0.08	236.90	0.33
StarGAN2 [7]	14.61	**0.45**	22.08	0.45	11.35	0.51
GP-UNIT [27]	14.63	0.37	15.29	0.51	13.04	0.49
SC-UNIT	**11.45**	0.37	**10.38**	**0.54**	**9.04**	**0.53**
Task	Bird ↔ Dog		Bid ↔ Car		Average	
Metric	FID	LPIPS	FID	LPIPS	FID	LPIPS
TraVeLGAN [1]	169.98	–	164.28	–	109.01	–
U-GAT-IT [20]	178.23	–	194.05	–	110.12	–
MUNIT [13]	217.68	0.57	121.02	0.60	99.83	0.51
COCO-FUNIT [26]	30.27	0.51	207.92	0.12	122.27	0.28
StarGAN2 [7]	20.54	0.52	29.28	0.58	19.57	0.50
GP-UNIT [27]	11.29	0.60	13.93	0.61	13.64	0.52
SC-UNIT	**10.36**	**0.63**	**11.88**	**0.62**	**10.64**	**0.54**

content input   *style input*   *w/o SC*   *full model*

Fig. 6. Effect of style consistency module.

input   *output*   *input*   *output*

Fig. 7. Limitation. SC-UNIT fails to build direction correspondences in extremely distant domains.

quality. LPIPS is introduced for estimating the diversity of generated images by calculating perceptual loss between images to each other. The higher the LPIPS score is, the more diverse the generated images are. The comparison results are listed in Table 1. What can be found is that except for the diversity of Male↔Female translation, our method exceeds other methods in all tasks in terms of both FID and LPIPS.

Ablation Study. Figure 6 shows what will be generated if we do not use the style consistency (SC) module for our model. What we could find is that the generated result without a style consistency module cannot keep the style consistency with the input style. It seems that the subjects of these two images do not seem to belong to the same species. And because of the absence of the style consistency module, the generated image learns the left eye of the style input and the right eye of the content input. It is not a good performance in UNIT.

4.3   Limitations

Although our SC-UNIT outperforms state-of-the-art methods in image quality, style consistency, and style diversity, SC-UNIT still has some limitations. What we see in Fig. 7 is a standard limitation. SC-UNIT sometimes translates the tail of a bird into the front of a car because they are both the thinner parts of the objects. SC-UNIT pays more attention to shape reconstruction, ignoring direction.

5   Conclusion

In this paper, we design a novel two-stage Unsupervised Image-to-image Translation framework based on a style consistency module to make the style of generated images more similar to that of the style input, thereby reducing style deviation. It also improves feature fusion, making the resulting images more natural and diverse in style. If there is a work that can make a breakthrough in feature decoupling, it will be beneficial in UNIT because it can help the model learn domain-shared and domain-specific features more precisely. An interesting extension would be to apply the model to more domains with vast differences. Furthermore, the core ideas of the feature fusion method may apply to other research domains to generate data that would otherwise be unobtainable. We expect to see more exciting research on UNIT in tasks with greater challenges.

Acknowledgement. This work was partly supported by the National Natural Science Foundation of China under Grant 62272116. The authors acknowledge the Network Center of Guangzhou University for providing HPC computing resources.

References

1. Amodio, M., Krishnaswamy, S.: Travelgan: image-to-image translation by transformation vector learning. In: CVPR, pp. 8983–8992 (2019)

2. Brock, A., Donahue, J., Simonyan, K.: Large scale GAN training for high fidelity natural image synthesis. In: ICLR. vol. abs/1809.11096 (2019)
3. Bulat, A., Tzimiropoulos, G.: How far are we from solving the 2D & 3D face alignment problem? (and a dataset of 230,000 3d facial landmarks). In: ICCV, pp. 1021–1030 (2017)
4. Chan, K.C., Wang, X., Xu, X., Gu, J., Loy, C.C.: GLEAN: generative latent bank for large-factor image super-resolution. In: CVPR, pp. 14245–14254 (2021)
5. Chen, K., et al.: Hybrid task cascade for instance segmentation. In: CVPR, pp. 4974–4983 (2019)
6. Choi, Y., Choi, M., Kim, M., Ha, J.W., Kim, S., Choo, J.: Stargan: unified generative adversarial networks for multi-domain image-to-image translation. In: CVPR, pp. 8789–8797 (2018)
7. Choi, Y., Uh, Y., Yoo, J., Ha, J.W.: Stargan v2: diverse image synthesis for multiple domains. In: CVPR, pp. 8188–8197 (2020)
8. Collins, E., Bala, R., Price, B., Susstrunk, S.: Editing in style: uncovering the local semantics of GANs. In: CVPR, pp. 5771–5780 (2020)
9. Ganin, Y., Lempitsky, V.: Unsupervised domain adaptation by backpropagation. In: ICML, pp. 1180–1189 (2015)
10. Goodfellow, I., et al.: Generative adversarial nets. In: NeurIPS, pp. 2672–2680 (2014)
11. Heusel, M., Ramsauer, H., Unterthiner, T., Nessler, B., Hochreiter, S.: GANs trained by a two time-scale update rule converge to a local NASH equilibrium. In: NeurIPS, pp. 6629–6640 (2017)
12. Huang, X., Belongie, S.: Arbitrary style transfer in real-time with adaptive instance normalization. In: ICCV, pp. 1501–1510 (2017)
13. Huang, X., Liu, M.Y., Belongie, S., Kautz, J.: Multimodal unsupervised image-to-image translation. In: ECCV, pp. 172–189 (2018)
14. Jiang, L., Zhang, C., Huang, M., Liu, C., Shi, J., Loy, C.C.: TSIT: a simple and versatile framework for image-to-image translation. In: Vedaldi, A., Bischof, H., Brox, T., Frahm, J.-M. (eds.) ECCV 2020. LNCS, vol. 12348, pp. 206–222. Springer, Cham (2020). https://doi.org/10.1007/978-3-030-58580-8_13
15. Jiang, Y., Huang, Z., Pan, X., Loy, C.C., Liu, Z.: Talk-to-edit: fine-grained facial editing via dialog. In: ICCV, pp. 13799–13808 (2021)
16. Johnson, J., Alahi, A., Fei-Fei, L.: Perceptual losses for real-time style transfer and super-resolution. In: Leibe, B., Matas, J., Sebe, N., Welling, M. (eds.) ECCV 2016. LNCS, vol. 9906, pp. 694–711. Springer, Cham (2016). https://doi.org/10.1007/978-3-319-46475-6_43
17. Karras, T., Aila, T., Laine, S., Lehtinen, J.: Progressive growing of GANs for improved quality, stability, and variation. In: ICLR. vol. abs/1710.10196 (2018)
18. Karras, T., et al.: Alias-free generative adversarial networks. In: NeurIPS, pp. 852–863 (2021)
19. Karras, T., Laine, S., Aila, T.: A style-based generator architecture for generative adversarial networks. In: CVPR, pp. 4401–4410 (2019)
20. Kim, J., Kim, M., Kang, H., Lee, K.H.: U-gat-it: unsupervised generative attentional networks with adaptive layer-instance normalization for image-to-image translation. In: ICLR. vol. abs/1907.10830 (2019)
21. Liu, M., et al.: STGAN: a unified selective transfer network for arbitrary image attribute editing. In: CVPR, pp. 3673–3682 (2019)
22. Liu, M.Y., Breuel, T., Kautz, J.: Unsupervised image-to-image translation networks. In: NeurIPS, pp. 700–708 (2017)

23. Liu, M.Y., et al.: Few-shot unsupervised image-to-image translation. In: ICCV, pp. 10551–10560 (2019)
24. Patashnik, O., Wu, Z., Shechtman, E., Cohen-Or, D., Lischinski, D.: StyleCLIP: text-driven manipulation of stylegan imagery. In: ICCV, pp. 2085–2094 (2021)
25. Russakovsky, O., et al.: Imagenet large scale visual recognition challenge. IJCV **115**, 211–252 (2015)
26. Saito, K., Saenko, K., Liu, M.-Y.: COCO-FUNIT: few-shot unsupervised image translation with a content conditioned style encoder. In: Vedaldi, A., Bischof, H., Brox, T., Frahm, J.-M. (eds.) ECCV 2020. LNCS, vol. 12348, pp. 382–398. Springer, Cham (2020). https://doi.org/10.1007/978-3-030-58580-8_23
27. Yang, S., Jiang, L., Liu, Z., Loy, C.C.: Unsupervised image-to-image translation with generative prior. In: CVPR, pp. 18332–18341 (2022)
28. Zhang, R., Isola, P., Efros, A.A., Shechtman, E., Wang, O.: The unreasonable effectiveness of deep features as a perceptual metric. In: CVPR, pp. 586–595 (2018)
29. Zhu, J., Shen, Y., Zhao, D., Zhou, B.: In-domain GAN inversion for real image editing. In: Vedaldi, A., Bischof, H., Brox, T., Frahm, J.-M. (eds.) ECCV 2020. LNCS, vol. 12362, pp. 592–608. Springer, Cham (2020). https://doi.org/10.1007/978-3-030-58520-4_35
30. Zhu, J.Y., Park, T., Isola, P., Efros, A.A.: Unpaired image-to-image translation using cycle-consistent adversarial networks. In: ICCV, pp. 2223–2232 (2017)

SemanticCrop: Boosting Contrastive Learning via Semantic-Cropped Views

Ya Fang[1], Zipeng Chen[1], Weixuan Tang[2]([✉]), and Yuan-Gen Wang[1]([✉])

[1] School of Computer Science and Cyber Engineering, Guangzhou University,
Guangzhou, China
[2] Institute of Artificial Intelligence and Blockchain, Guangzhou University,
Guangzhou, China
{yafangna,czp}@e.gzhu.edu.cn, {tweix,wangyg}@gzhu.edu.cn

Abstract. Siamese-structure-based contrastive learning has shown excellent performance in learning visual representations due to its ability to minimize the distance between positive pairs and increase the distance between negative pairs. Existing works mostly employ RandomCrop or ContrastiveCrop to obtain positive pairs of an image. However, Random-Crop causes the cropped views to contain many useless backgrounds, while ContrastiveCrop produces positive pairs that are too similar. In this paper, we propose a novel SemanticCrop to yield cropped views containing as much semantic information as possible. Specifically, Semantic-Crop first computes a heatmap of an image. Then, an empirical threshold is tuned to box out a semantic region whose heatmap values are over this threshold. Finally, we design a center-suppressed probabilistic sampling to avoid excessive similarity between positive pairs, making the cropped view contain more parts of an object. As a plug-and-play module, the MoCo, SimCLR, SimSiam, and BYOL models equipped with our SemanticCrop module achieve an accuracy improvement from 0.5% to 2.34% on the CIFAR10, CIFAR100, IN-200, and IN-1K datasets. The code is available at https://github.com/GZHU-DVL/SemanticCrop.

Keywords: Contrastive learning · Self-supervised learning · Data agumentation

1 Introduction

With the rapid development of deep learning technology, many supervised methods have been proposed [12,17,22]. These methods utilized sufficient label information to train models and achieved reliable performance in downstream tasks. However, obtaining adequate labeled data in real scenarios is challenging, which limits further application of the supervised methods. In contrast, self-supervised learning (SSL) has been drawing increasing attention for its ability to train models on unlabeled data. As one of the most promising directions in SSL, contrastive learning stands out with its superior performance on downstream tasks [2,7,8,11,14,19,24,26].

© The Author(s), under exclusive license to Springer Nature Singapore Pte Ltd. 2024
Q. Liu et al. (Eds.): PRCV 2023, LNCS 14430, pp. 335–346, 2024.
https://doi.org/10.1007/978-981-99-8537-1_27

Contrastive learning first pairs two different augmented views of the same input image as positive samples and uses the augmented views of other input images as negative samples. Then, it works by reducing the distance between positive samples and increasing the distance between negative samples in the feature space. As such, the quality of positive sample pairs is crucial for contrastive learning. [23] aimed to enhance the quality of positive sample pairs, while [21] achieved good performance through strategies like image blending and label smoothing. In addition, [27] directly modified the values in the feature space to increase the distance between the positive and negative samples. Although these approaches employ different strategies to obtain positive samples, they all use RandomCrop to generate input samples.

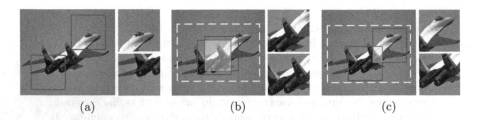

(a)   (b)   (c)

Fig. 1. The motivation of our proposed SemanticCrop. (a) show the cropped image obtained through RandomCrop, (b) show the cropped image obtained through ContrastiveCrop, and (c) show the cropped image obtained through SemanticCrop.

RandomCrop is widely used for data augmentation. However, it fails to utilize semantic information during the cropping process fully and may even damage the semantic information of the images. As shown in Fig. 1(a), the cropped images from RandomCrop may contain excessive background. ContrastiveCrop [16] designs semantic-aware localization and center-suppressed sampling to utilize the semantic information of the images during the cropping process. Semantic-aware localization provides an operational area for cropping based on semantic information, ensuring that each cropped image contains the object. According to β distribution, the center-suppressed sampling selects a point as the center point per argumentation to reduce the similarity of the cropped samples (see Fig. 1(b)). However, such center-suppressed sampling still picks out the center points of the cropped images that are very close in the distance, making the positive sample pairs highly similar.

To address this issue, this paper proposes a plug-and-play method called SemanticCrop. Specifically, we apply heatmap localization to determine the cropping operating area for an input image. We design a new center-suppressed probabilistic sampling approach within this area, where a small probability value is introduced as a constraint during the cropping process. By incorporating this additional constraint, we can further push the centers of the two consecutive crops apart, resulting in a stronger center suppression effect (as illustrated in

Fig. 1(c)). This leads to lower similarity between positive sample pairs, increasing the diversity of input samples to the model and improving its generalization ability. The main contributions of our work can be summarized as follows:

- To the best of our knowledge, this is the first study to investigate how to optimize center-suppressed sampling to make the cropping more effective.
- We propose a new method that adds a small constraint to sampling the center point during the cropping process. This constraint helps alleviate cropping instability while preserving valuable semantic information.
- Through extensive experiments, we have determined the empirically optimal probability values for different contrastive learning models. This strategy ensures that our method can perform best across various contrastive learning models.

2   Related Work

2.1   Contrastive Learning

Contrastive learning has attracted much attention in unsupervised learning because it exploits unlabeled data to train models. In contrastive learning, each input image is randomly augmented to generate two different augmented images. Then, the feature embedding of positive samples is narrowed and the feature embedding of negative examples is pushed further through contrastive loss.

In MoCo [10], a queue-based approach was proposed to store many negative samples, reducing the storage space required for the memory bank. It also utilized momentum updates to ensure the consistency of the features encoded by the encoder in the network architecture. In SimCLR [3], a fully connected layer was introduced to increase the diversity of sample features, and a sufficiently large batch size was used to ensure that the model could incorporate a significant number of negative samples. In BYOL [9], the traditional dictionary lookup task in contrastive learning was transformed into a prediction task, enabling the model to overcome the limitation of the number of negative samples, and beneficial techniques from previous works were also integrated into the network. In Sim-Siam [5], it was discovered through model simplification that siamese networks were natural and effective core tools for maintaining invariance in contrastive learning models. In addition to model architecture, researchers [1,2,18,25,28] have also explored the selection of training samples. Furthermore, theoretical analysis and empirical studies have been proposed to understand the essence of contrastive learning better.

2.2   Positive Sample Generation

The quality of positive samples is crucial for contrastive learning. Previous works have mostly focused on studying how to obtain high-quality positive sample pairs. In this regard, SimCLR [3] studied the impact of various data augmentation techniques on model performance and found that cropping and color variations are two effective methods for enhancing model performance. Additionally, due to the limitations of self-supervised learning on complex scene images with multiple objects, CAST [20] introduced unsupervised saliency maps to adjust the cropping process, making the models more robust and [15] took object-scene relation into account when making crops, but require additional object proposal algorithms. SimCLR [3], CAST, and [15] have improved the data augmentation methods, but fundamentally they still rely on the equiprobable selection of cropping center points in RandomCrop. ContrastiveCrop [16] addressed this issue by introducing semantic-aware localization and center-suppressed sampling to leverage semantic information while avoiding excessively high similarity in positive sample pairs. However, ContrastiveCrop fails to effectively suppress high similarity in positive sample pairs. In other words, even with the use of center-suppressed sampling, there can still be instances where the similarity in positive sample pairs is too high(see Fig. 1(b)). In our paper, we propose SemanticCrop. Specifically, we introduce probability value as a secondary constraint during center point sampling. When a sampled center point falls below this probability value, it is discarded and re-sampled. This further increases the distance between sampled center points during cropping, enhancing the suppression ability of center-suppressed sampling and improving control over the similarity of positive sample pairs.

3   Proposed Method

In this section, we introduce SemanticCrop. As shown in Fig. 2 and Algorithm 1, we first provide the parameters required for cropping and input an image into the contrastive learning model. Then, the heatmap location obtains a bounding box, as the cropping operational area. Next, the center-suppressed probabilistic sampling is used to obtain the cropping center point within the given operational area. Finally, the cropped images are obtained with the help of the sampled center point and the provided parameters. Regarding the aforementioned heatmap location and center-suppressed probabilistic sampling, we provide detailed explanations in Sect. 3.2 and Sect. 3.3, respectively.

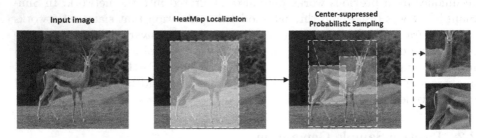

Fig. 2. Flowchart of the cropping process for *SemanticCrop*. The curve in the graph represents the function curve of the Beta distribution. The red line indicates the points that need to be discarded, and the blue line represents the points that can be chosen as center points. (Color figure online)

3.1 Preliminaries

RandomCrop is a widely used data augmentation technique in supervised and unsupervised learning. It increases input data diversity, enhancing the model's generalization ability. Specifically, given an input image I with a scale s and an aspect ratio r, the width and height of the cropped image are determined based on s and r. Then, we randomly choose a point inside the image I as the center to crop this image. The formula for RandomCrop can be described as:

$$(x, y, h, w) = \mathbb{R}_{crop}(s, r, I), \tag{1}$$

where (x, y) represents the center coordinates of the crop, (h, w) represents the height and width of the crop window. s represents the scale, r represents the aspect ratio, and I represents the input image.

While RandomCrop enhances the diversity of samples in the original dataset, the practice of equally selecting each point in the image as the center point for cropping fails to utilize the useful semantic information inherent in the images. As shown in Fig. 1(a), it may even harm the semantic information of the images due to excessive cropping into the background.

3.2 Heatmap Localization

The semantic information of an image is valuable for a contrastive learning model. Based on the research on ContrastiveCrop [16], we adopted heatmap localization to exclude background and locate targets. Specifically, in the early training stage, we use the RandomCrop method as a warm-up to give the model better feature extraction capabilities. Then, in the feature map output by the model based on the input image, we highlight the regions with feature values above t as the highlighted regions in the heatmap. Next, we introduce an indicator function to obtain the bounding box of an object from the heatmap, which serves as the cropping operational area for subsequent processes (as shown in Fig. 2). This process can be described simply as

$$H = L(\mathbb{1}[M > t]), \tag{2}$$

where M represents heatmap, $t \in [0, 1]$ is the threshold of activations, $\mathbb{1}$ is the indicator function, and L calculates the rectangular closure of activated positions. Then, we modify the formula of RandomCrop as

$$(\widetilde{x}, \widetilde{y}, \widetilde{h}, \widetilde{w}) = \mathbb{R}_{crop}(s, r, H), \tag{3}$$

where the definitions of $\widetilde{x}$, $\widetilde{y}$, $\widetilde{h}$, $\widetilde{w}$, s, r, and $\mathbb{R}_{crop}$ are similar to Eq. (1).

3.3 Center-Suppressed Probabilistic Sampling

After introducing heatmap localization, we better use the semantic information of the images. However, due to the reduced cropping operational area, the similarity between the cropped images is significantly increased. It makes the model receive more features with extremely high similarity and a decrease in generalization capability. To address this issue, we design center-suppressed probabilistic sampling to reduce the similarity of positive sample pairs. Specifically, when selecting the cropping center point within the given cropping operational area, we utilize a beta distribution $\beta(\alpha, \alpha)$, a U-shaped distribution, to adjust the probability of each point being selected as a cropping center point. This helps increase the likelihood of center points being closer to the boundaries of the cropping area and reduces the possibility of center points clustering together(see Fig. 2). Additionally, we introduce a probability value k as a secondary constraint. When the selected center point falls below k, it is discarded and re-sampled, further pushing the cropping center point toward the boundaries of the operational area. This method transforms the beta distribution into

$$\hat{\beta} = \beta(\alpha, \alpha) > k, \tag{4}$$

where k is a value $\in [0, 1]$. This allows us to discard points below the threshold value k. And by adjusting the value of k, we can easily fine-tune the technique for different datasets to achieve better results. Finally, we formulate our SemanticCrop as

$$(\hat{x}, \hat{y}, \hat{h}, \hat{w}) = \mathbb{S}_{crop}(s, r, H), \tag{5}$$

where $\mathbb{S}_{crop}$ refers to the sampling function incorporating center-suppressed probabilistic sampling, and H represents the bounding box of the earlier heatmap.

Algorithm 1. *SemanticCrop*

Input Image I, Crop scale s, Crop ratio r, Threshold of activations t, Probability value k, Parameter of β distribution α.

$h = \sqrt{s \cdot r}$ // Height of the crop
$w = \sqrt{s/r}$ // Width of the crop
$F = Foward(I)$ // Features of last layer
$N = Normalization(F)$ // Feature map after normalizing
$H = L(\mathbb{1}[M > t])$ // The cropping operational area from Eq. 2
while $u > k$ **and** $v > k$ **do**   // The crop center point (x, y) form $\hat{\beta}(\alpha, \alpha)$ by Eq. 4
$\quad x = H_{x0} + (H_{x1} - H_{x0}) \cdot v, u \sim \hat{\beta}(\alpha, \alpha)$
$\quad y = H_{y0} + (H_{y1} - H_{y0}) \cdot v, u \sim \hat{\beta}(\alpha, \alpha)$
end while
Output $\mathbb{S}_{crop} = (x, y, h, w)$

---

4 Experiments

4.1 Datasets

CIFAR10 [13] is a commonly used image classification dataset comprising 60,000 color images of 10 different categories. Each category contains 6,000 images. The images are of size 32×32 pixels and are divided into training and testing sets, with 50,000 images in the training set and 10,000 images in the testing set.

CIFAR100 [13] consists of 60,000 color images belonging to 100 different categories, each containing 600 images. The images are also 32×32 pixels and are divided into training and testing sets for training and evaluating the performance of models.

IN-200 [16] is a subset derived from the ImageNet dataset. Based on the source code provided by ContrastiveCrop [16], 200 classes were selected from ImageNet to create this new dataset for quick model training and evaluation. Each class in Imagenet-200 consists of several hundred images, resulting in a total of approximately 20,000 images.

ImageNet [6] is widely used for computer vision tasks such as image classification, object detection, and image recognition. It contains millions of images from various categories. The training set of ImageNet consists of approximately 1.2 million images from 1,000 different categories. These images cover a wide range of objects, scenes, and concepts. The validation set contains 50,000 images and is used for model validation and fine-tuning. The test set consists of 100,000 images and is used to evaluate the performance of models on unseen images.

4.2 Comparative Methods

To evaluate our method against the baseline models, we follow the same experimental protocol as ContrastiveCrop [16]. We employ four commonly used contrastive learning methods: MoCo [10], SimCLR [3], BYOL [9], and SimSiam [5]. We train and test these baseline methods using the source code provided by ContrastiveCrop. The accuracy values regarding ContrastiveCrop and RandomCrop in Table 1 are obtained by running the source code provided by ContrastiveCrop.

4.3 Implementation Details

According to the experimental protocol of ContrastiveCrop [16], we employ linear classification to validate the effectiveness of our method. We froze the weights and trained a supervised linear classifier on top of it. The top-1 classification accuracy on the validation set is reported as the result.

For CIFAR10 and CIFAR100, during the pre-training phase, we trained the ResNet-18 model for 500 epochs with a learning rate of 0.5. The linear classifier is initially trained with a learning rate of 10.0 for 100 epochs, and the learning rate is multiplied by 0.1 at the 60th and 80th epochs.

In the experiments on ImageNet, ResNet-50 is used as the base model. Pretrain settings of MoCo V1 [4], MoCo V2 [10] and SimSiam [5] exactly follow their original works. The parameter settings for SimCLR [3] remain consistent with ContrastiveCrop [16].

4.4   Experiment Resluts

Our method improves the accuracy of contrastive learning models trained using this approach on datasets with many samples and challenging training scenarios. In conclusion, SemanticCrop is effective for mainstream contrastive learning models, regardless of the size and diversity of the dataset.

Table 1. Linear classification results on CIFAR10 and CIFAR100.

Method	CIFAR10			CIFAR100		
	Rcrop	Ccrop	Scrop	Rcrop	Ccrop	Scrop
MoCo [10]	86.64	87.71	**88.39**	59.35	61.27	**61.69**
SimCLR [3]	89.55	89.74	**90.03**	60.71	61.85	**62.20**
BYOL [9]	91.53	91.80	**92.33**	63.86	64.90	**65.15**
SimSiam [5]	90.48	91.04	**91.17**	63.82	64.16	**64.74**

Results on CIFAR10 and CIFAR100. As shown in Table 1, under the same experimental settings as ContrastiveCrop, SemanticCrop trains MoCo and improves the classification accuracy by 0.4% and 0.68% on CIFAR10 and CIFAR100, respectively. SimCLR shows improvements of 0.3% and 0.35%, BYOL shows improvements of 0.53% and 0.25%, and SimSiam shows improvements of 0.13% and 0.58% on the same datasets. The experimental results demonstrate that SemanticCrop has a positive effect on the classification performance of different contrastive learning models on small datasets.

Table 2. Linear classification results on IN-200.

Method	Arch.	Epoch	IN-200 Top-1	
			Ccrop	*Ours*
MoCo V1	Restnet50	50	56.56	**56.71**
MoCo V2	Restnet50	50	52.83	**53.11**
SimCLR	Restnet50	50	52.65	**52.86**
SimSiam	Restnet50	50	51.51	**52.17**

Table 3. Linear classification results on IN-1K.

Method	Arch	Epoch	IN-1K Top-1
MoCo V1+*Ccrop*	Restnet50	100	57.22
MoCo V1+*Ours*	Restnet50	100	**57.42**
MoCo V2+*Ccrop*	Restnet50	100	63.21
MoCo V2+*Ours*	Restnet50	100	**63.49**

Results on ImageNet. As shown in Table 2 and Table 3, our results demonstrate that different contrastive learning models trained with SemanticCrop improve linear classification accuracy on ImageNet, which has larger sizes and more challenging classification tasks. On IN-200, MoCo V1 showed an improvement of 0.15%. MoCo V2 shows an improvement of around 0.5%. SimCLR showed an improvement of 0.21%. SimSiam shows an improvement of 0.62%. On IN-1K, MoCo V1 showed an improvement of 0.20%, and MoCo V2 showed an improvement of around 0.28%. The experimental results show that the proposed method is still effective on large datasets.

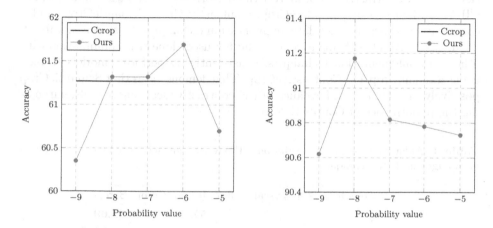

Fig. 3. Ablation results on CIFAR100 and CIFAR10. The line on the left represents the classification accuracy of MoCo on CIFAR100 under the influence of different thresholds, while the line on the right represents the classification accuracy of SimCLR on CIFAR10 under the influence of different probability values. The probability value are represented as 10 raised to the power of a negative exponent.

So far, according to Table 1, Table 2 and Table 3, it can be seen that there is a greater improvement in performance on the smaller datasets of CIFAR-10 and CIFAR-100, while the improvement on the larger datasets of IN-1K and IN-200 is relatively small.

4.5   Ablation Experiment

Probability Value Selection for Different Comparative Methods. We conduct ablation experiments to explore the influence of different probability values on various contrastive learning models and confirm that each model has its optimal probability value. As shown in Fig. 3, the performance of different contrastive learning models varies when trained with different thresholds on different datasets. However, at a specific probability value, the contrastive learning models consistently outperform the models trained using the ContrastiveCrop method. For instance, MoCo achieves the highest classification accuracy when the probability value is set to 10^{-6}, while SimCLR performs best when the probability value is set to 10^{-8}. This demonstrates that different contrastive learning models have different optimal probability values, and using too large or too small probability values may decrease model accuracy (see Fig. 3). Based on these observations, we conjecture that these disparities stem from the divergent focal points of different models within the feature space, leading to varied sensitivity towards the similarity of positive samples.

SemanticCrop with Other Transformations. To better compare SemanticCrop with ContrastiveCrop, we test the performance of models trained on the CIFAR10 dataset using RandomCrop, ContrastiveCrop, SemanticCrop, and a combination of various other data augmentation techniques. As shown in Table 4, it can be observed that SemanticCrop, whether used alone or in combination with other data augmentation techniques, consistently outperforms models trained using RandomCrop and ContrastiveCrop. This demonstrates that SemanticCrop positively impacts the performance of the contrastive learning models, surpassing the performance of ContrastiveCrop.

Table 4. Ablation experiment results on CIFAR-100 dataset with various combined data augmentation techniques.

Flip	ColorJitter + Grayscale	Blur	Rcrop	Ccrop	Scrop
√	√	√	59.35	61.27	**61.69**
√			47.83	48.37	**48.77**
	√		58.37	60.34	**60.57**
		√	23.27	24.54	**24.79**
			46.00	48.00	**48.26**

5   Conclusion

In this paper, we have presented a new cropping method (called SemanticCrop) for contrastive learning, which addresses the issue of high similarity in positive sample pairs when performing cropping operations within small regions. By incorporating heatmap localization, SemanticCrop can avoid including too much background information in the cropped images, thus reducing the highly dissimilar positive sample pairs. Specifically, we propose a center-suppressed probabilistic sampling to reduce the highly similar positive sample pairs. Furthermore, we empirically obtain the more suitable probability value for the leading contrastive learning models. Meanwhile, we find that too large or too small probability values harm the performance of contrastive learning models. Ablation study shows that SemanticCrop remains effective when combined with other data augmentation methods. Extensive experimental results demonstrate that our approach improves the performance of mainstream contrastive learning models regardless of the size and complexity of datasets.

Acknowledgement. This work was supported in part by the National Natural Science Foundation of China under Grants 62272116 and 62002075, in part by the Basic and Applied Basic Research Foundation of Guangdong Province under Grant 2023A1515011428, and in part by the Science and Technology Foundation of Guangzhou under Grant 2023A04J1723. The authors acknowledge the Network Center of Guangzhou University for providing HPC computing resources.

References

1. Arora, S., Khandeparkar, H., Khodak, M., Plevrakis, O., Saunshi, N.: A theoretical analysis of contrastive unsupervised representation learning. arXiv preprint arXiv:1902.09229 (2019)
2. Caron, M., et al.: Emerging properties in self-supervised vision transformers. In: ICCV, pp. 9650–9660 (2021)
3. Chen, T., Kornblith, S., Norouzi, M., Hinton, G.: A simple framework for contrastive learning of visual representations. In: ICML, pp. 1597–1607 (2020)
4. Chen, X., Fan, H., Girshick, R., He, K.: Improved baselines with momentum contrastive learning. arXiv preprint arXiv:2003.04297 (2020)
5. Chen, X., He, K.: Exploring simple Siamese representation learning. In: CVPR, pp. 15750–15758 (2021)
6. Deng, J., Dong, W., Socher, R., Li, L.J., Li, K., Fei-Fei, L.: Imagenet: a large-scale hierarchical image database. In: CVPR, pp. 248–255 (2009)
7. Everingham, M., Van Gool, L., Williams, C.K., Winn, J., Zisserman, A.: The pascal visual object classes (VOC) challenge. IJCV **88**, 303–338 (2010)
8. Faster, R.: Towards real-time object detection with region proposal networks. NeurIPS **9199**(10.5555), 2969239–2969250 (2015)
9. Grill, J.B., et al.: Bootstrap your own latent-a new approach to self-supervised learning. NeurIPS **33**, 21271–21284 (2020)
10. He, K., Gkioxari, G., Dollár, P., Girshick, R.: Mask R-CNN. In: ICCV, pp. 2961–2969 (2017)
11. Khosla, P., et al.: Supervised contrastive learning. NeurIPS **33**, 18661–18673 (2020)
12. Khosla, P., et al.: Supervised contrastive learning. NeurIPS **33**, 18661–18673 (2020)
13. Krizhevsky, A., Hinton, G., et al.: Learning multiple layers of features from tiny images (2009)
14. Lin, T.-Y., et al.: Microsoft COCO: common objects in context. In: Fleet, D., Pajdla, T., Schiele, B., Tuytelaars, T. (eds.) ECCV 2014. LNCS, vol. 8693, pp. 740–755. Springer, Cham (2014). https://doi.org/10.1007/978-3-319-10602-1_48
15. Mishra, S., et al.: Object-aware cropping for self-supervised learning. arXiv preprint arXiv:2112.00319 (2021)
16. Peng, X., Wang, K., Zhu, Z., Wang, M., You, Y.: Crafting better contrastive views for siamese representation learning. In: CVPR, pp. 16031–16040 (2022)
17. Peng, Y., He, X., Zhao, J.: Object-part attention model for fine-grained image classification. TIP **27**(3), 1487–1500 (2017)
18. Purushwalkam, S., Gupta, A.: Demystifying contrastive self-supervised learning: invariances, augmentations and dataset biases. NeurIPS **33**, 3407–3418 (2020)

19. Ren, S., He, K., Girshick, R., Sun, J.: Faster R-CNN: towards real-time object detection with region proposal networks. TPAMI **39**(6), 1137–1149 (2017)
20. Selvaraju, R.R., Desai, K., Johnson, J., Naik, N.: Casting your model: learning to localize improves self-supervised representations. In: CVPR, pp. 11058–11067 (2021)
21. Shen, Z., Liu, Z., Liu, Z., Savvides, M., Darrell, T.: Rethinking image mixture for unsupervised visual representation learning (2020). **3**(7), 8, arXiv preprint arXiv:2003.05438
22. Sun, M., Yuan, Y., Zhou, F., Ding, E.: Multi-attention multi-class constraint for fine-grained image recognition. In: Ferrari, V., Hebert, M., Sminchisescu, C., Weiss, Y. (eds.) ECCV 2018. LNCS, vol. 11220, pp. 834–850. Springer, Cham (2018). https://doi.org/10.1007/978-3-030-01270-0_49
23. Tian, Y., Krishnan, D., Isola, P.: Contrastive multiview coding. In: Vedaldi, A., Bischof, H., Brox, T., Frahm, J.-M. (eds.) ECCV 2020. LNCS, vol. 12356, pp. 776–794. Springer, Cham (2020). https://doi.org/10.1007/978-3-030-58621-8_45
24. Touvron, H., Vedaldi, A., Douze, M., Jégou, H.: Fixing the train-test resolution discrepancy. NeurIPS **32**, 8250–8260 (2019)
25. Xiao, T., Wang, X., Efros, A.A., Darrell, T.: What should not be contrastive in contrastive learning. arXiv preprint arXiv:2008.05659 (2020)
26. Zhang, J., Wang, Y., Zhou, Z., Luan, T., Wang, Z., Qiao, Y.: Learning dynamical human-joint affinity for 3d pose estimation in videos. TIP **30**, 7914–7925 (2021)
27. Zhu, R., Zhao, B., Liu, J., Sun, Z., Chen, C.W.: Improving contrastive learning by visualizing feature transformation. In: ICCV, pp. 10306–10315 (2021)
28. Zoph, B., et al.: Rethinking pre-training and self-training. NeurIPS **33**, 3833–3845 (2020)

Transformer-Based Multi-object Tracking in Unmanned Aerial Vehicles

Jiaxin Li and Hongjun Li(✉)

School of Information Science and Technology, Nantong University, Nantong 226019, Jiangsu,
People's Republic of China
lihongjun@ntu.edu.cn

Abstract. Unmanned aerial vehicles (UAVs) possess a wide field of view, maneuverability, and autonomy in the field of multi-object tracking. By combining the high-altitude perspective and multidimensional perception capabilities of UAVs with artificial intelligence image processing and object recognition technologies, efficient and accurate multi-object tracking can be achieved. To enhance the processing capacity of relevant information, a transformer-based multi-object tracking model is proposed in UAV perspectives. The backbone network of detector is based on the transformer architecture to extract target features. In this structure, the network is further deepened to optimize feature fusion and improve the ability to capture small targets. To address the tracking challenges arising from dynamic variations between different targets, the tracker implements multi-object tracking in UAV scenarios based on the motion characteristics of targets and optimizes the tracking process accordingly. The model was evaluated on the Visdrone dataset, achieving a detection accuracy of 90.1% and a MOTA of 51.9%. Furthermore, the model demonstrated a remarkable speed of 20 frames per second. In summary, the proposed algorithm demonstrates excellent performance in target tracking from UAV perspectives and exhibits certain advantages over other algorithms.

Keywords: Multi-object tracking · transformer · unmanned aerial vehicles

1 Introduction

With the rapid development of UAVs and artificial intelligence technologies, the integration of UAVs with artificial intelligence has shown tremendous potential in the field of multi-object tracking. The flexible flight capability and wide field of view of UAVs provide an ideal platform for capturing high-quality video information, while the fast algorithms and advanced visual processing capabilities of artificial intelligence offer accurate multi-object tracking capabilities [1]. The technology of multi-object tracking using UAVs plays an important and meaningful role in the current stage of social development [2]. It plays a critical role in public safety by enabling real-time monitoring and tracking of potential threats such as criminal activities and terrorist attacks, thereby enhancing societal security. Additionally, UAV-based multi-object tracking technology plays a significant role in traffic management and urban planning by accurately monitoring and analyzing vehicles, pedestrians, and traffic flow, providing data support for

Q. Liu et al. (Eds.): PRCV 2023, LNCS 14430, pp. 347–358, 2024.
https://doi.org/10.1007/978-981-99-8537-1_28

traffic optimization and urban design. It is also widely applied in environmental monitoring, disaster response, and agriculture sectors, enabling real-time monitoring of natural resources, disaster situations, and crop growth conditions, thereby providing decision support and guidance for rescue operations [3, 4]. In recent years, significant research achievements have emerged in the field of UAV-based multi-object tracking, providing effective solutions to the problem of target tracking from the UAV perspective [5]. For example, the JDE algorithm achieves accurate target localization and robust tracking performance by jointly training the object detection and feature embedding networks [6]. The CenterTrack algorithm achieves efficient and accurate multi-object tracking through center point estimation and motion prediction techniques [7].

However, UAV-based multi-object tracking still faces several challenges [8], such as the need for improved tracking performance in complex scenes and the issue of tracking speed. To address these challenges and better cope with the vast scenes from the UAV perspective, this paper proposes a deep network-based multi-object tracking model. The contributions of this model are as follows:

1. By improving the detection algorithm, the overall processing speed is enhanced while also strengthening the feature extraction capability. This improvement enables more accurate recognition and detection of targets in wide-angle scenarios. By effectively utilizing feature information, the algorithm can rapidly and accurately localize and identify multiple targets.
2. In the tracking algorithm, taking into account the characteristics of the targets, further enhancement is achieved in terms of tracking performance, efficiency, and capability in specific scenarios. By considering target motion patterns and context information, the algorithm can better predict target positions and trajectories, achieving stable and accurate multi-object tracking.
3. This model is designed for UAV perspective scenarios, enabling high-speed processing of UAV-captured videos, and it is also applicable to similar wide-angle scenarios, providing valuable references for related algorithms.

2   Relation Work

In the field of computer vision, Convolutional Neural Network (CNN) is a prominent model used for feature extraction. It utilizes multiple convolutional layers to extract features from deep networks, enabling analysis and understanding of visual data such as images and videos. CNN networks hold significant value in this field, and since the introduction of AlexNet [9], they have become the mainstream architecture and have evolved into various improved network structures, such as ResNet [10], DenseNet [11], and EfficientNet [12]. These enhanced network architectures exhibit superior feature extraction capabilities, thereby enhancing the accuracy and efficiency of the models.

Self-attention layers are a technique that enhances the feature extraction capability of CNN networks. They analyze and decode features by computing the dependencies between different positions in the input data. Self-attention layers have been widely applied in natural language processing and have recently been introduced to the field of computer vision. Transformer networks, based on the self-attention mechanism, have found extensive use in natural language processing [13] and have started to be applied to object detection tasks in computer vision in recent years. Visual Transformers and their improved networks [14, 15] have demonstrated high accuracy and speed in image classification. Visual Transformers directly apply the Transformer architecture to non-overlapping medium-sized image patches, striking a balance between speed and accuracy. This architecture has also been applied to object detection tasks with good results. Recently, researchers have applied the Visual Transformer model to dense visual tasks of object detection through upsampling or transposed convolution and have made improvements to optimize the related architectures for better performance in tasks such as classification and detection [16]. These techniques provide more accurate and efficient solutions for target detection tasks from the UAV perspective.

In the target tracking phase, factors such as localization, motion, and appearance are critical for multi-frame object matching. Recent target tracking algorithms [17–19] have introduced neural networks for learning motion to achieve more robust results in the presence of camera motion or low frame rates. These methods utilize position and motion information for short-term matching, improving tracking accuracy. For long-term matching, some algorithms employ appearance features. The Deepsort algorithm [20] extracts appearance features of detected targets using an independent re-identification model and combines them with position and motion information, utilizing a cascaded matching strategy for target tracking. Although embedding the re-identification model between detection and tracking is relatively convenient, the use of multiple models increases computational costs and slows down the overall processing speed. Currently, more tracking algorithms combine object detection with appearance feature extraction, using detection and vector feature models for target detection and feature extraction, omitting the separate re-identification module, and significantly improving the processing speed.

3 Multi-object Tracking Based on Transformer

To address the challenges of detecting and tracking targets in complex environments from the perspective of UAVs, this paper aims to enhance the practical application capabilities of tracking algorithms. The paper propose a transformer-based multi-object tracking algorithm that takes a balanced approach to detection and tracking, catering to a wide range of viewing angles. This algorithm employs a multi-stage feature extraction process to deepen the semantic understanding of features and enhance their extraction capabilities, thereby improving the ability to capture small targets. The extracted features

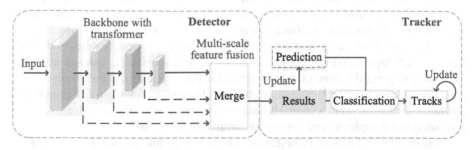

Fig. 1. Algorithm structure diagram

are then incorporated into the tracking algorithm, leveraging the relational characteristics among targets in wide-angle views to achieve higher tracking accuracy. By considering the comprehensive relationship among targets, our algorithm can accurately predict the positions and motion trajectories of targets, thus enhancing tracking accuracy and stability. The overall framework of the proposed algorithm is illustrated in Fig. 1.

3.1 Transformer-Based Detector

Backbone Network Section: Feature extraction plays a crucial role in object detection. Currently, the transformer module has achieved breakthroughs in the feature extraction module, making it essential for object detection tasks. To effectively detect small objects in large-scale visual images, the current solution involves using a multi-stage encoding-decoding structure for feature extraction and compressing image size. Through non-convolutional downsampling algorithms, it is possible to preserve detailed features as much as possible while leveraging the transformer module for more in-depth algorithmic convolution. This approach enables obtaining more accurate object detection results without sacrificing image quality.

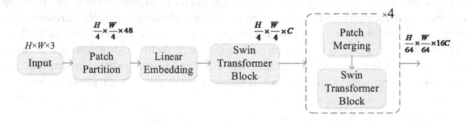

Fig. 2. Backbone network architecture

Figure 2 illustrates the structure of the backbone network. In this backbone network, an non-convolutional segmentation model is first used to divide the RGB image into non-overlapping patches, with each patch size being 4×4. Therefore, the feature dimension of each patch is $4 \times 4 \times 3 = 48$. Next, the original value features are projected using a linear embedding layer, and the transformer module with a mobile standard multi-head self-attention mechanism is employed for feature extraction, resulting in a multiple-fold increase in output dimension. Additionally, the non-convolutional downsampling module combines the transformer with a mobile multi-head self-attention mechanism, and through multiple iterations, it obtains an output size of H/64 $\times$ W/64 $\times$ 16C, achieving preliminary feature extraction for small objects. In this process, features of different scales at each stage are preserved and applied to the subsequent multi-task feature fusion. Compared to general feature extraction networks, the design of this backbone network has a deeper layer to better handle small objects in large scenes. Moreover, the network's downsampling avoids convolutional behavior in the neural network, effectively reducing computational complexity and improving computational speed.

Feature Fusion Section: In object detection, due to the varying sizes of objects, analyzing only one feature layer is often insufficient. Fusing features from different spatial scales can more effectively enhance detection performance. Lower-level features have higher resolution and focus more on fine details, but they lack semantic information due to their shallow network depth. On the other hand, higher-level features have richer semantic information but lower resolution, resulting in poorer perception of details due to their deep network structure. In real-world scenarios, wide-angle shooting leads to small scales for most objects. As the perspective changes, objects captured by the camera may become extremely small, making their recognition more challenging. To address this situation, a multi-scale fusion module with multiple detection layers is used to improve the detection capability for small objects.

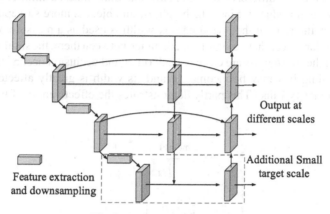

Fig. 3. Feature fusion structure

Figure 3 illustrates the feature fusion architecture. After the initial slicing convolution, the feature size becomes one-fourth of the input image. It then goes through four transformer modules, with each module reducing the feature size by half. The output of each module is retained, and different convolutions are applied to obtain features under different attention levels. Finally, multiple feature modules are fused together to generate four different-scale results. To enhance the anti-interference capability of the smallest detection layer and incorporate contextual information, the context information is added to the output of each layer. Each downsampling operation brings about a change in scale, and features of different scales are used as input to the corresponding output layer. This enriches the feature information of each layer and allows for further reference to the features of adjacent output layers, enabling the fusion of multiple input features for each scale output. By utilizing the feature maps during the feature extraction process for feature fusion, both feature loss and computational cost are reduced, thereby maintaining computational speed while improving the detection capability for small objects.

3.2 Tracking Algorithm Combining Spatial Attributes

With different shooting perspectives, the distribution of objects also varies. In the case of aerial views captured by UAVs, the high shooting angle results in minimal differences in size between objects and clear spatial distribution. Unlike the complexities of occlusion and other factors in low-angle views, tracking difficulty is relatively low.

In wide-angle views, most objects are generally dispersed and independent, although there may be some partial occlusions. To achieve more accurate tracking of objects in different scenes, the objects are inputted into a classifier for classification. When there is overlap between the bounding boxes of different objects and the overlapping region extends beyond the center point of one of the objects, the object is labeled as occluded. For occluded objects, improvements are made to the cosine distance in the tracking algorithm, taking into account spatial hierarchy to achieve object matching. On the other hand, non-occluded objects are labeled as dispersed and undergo simple bounding box matching.

To capture the size difference between two bounding boxes, a similarity parameter, denoted as 's', is introduced. Since the height of an object is more susceptible to local motion effects, the ratio of the bounding box width is used as a measure of similarity. When two similar objects have a significant distance between them, the similarity tends to be low. When the two objects are close to each other and occlusion occurs, the occluded object's bounding box may be incomplete and its width is greatly affected, resulting in a lower similarity value. This partly demonstrates the effectiveness of the similarity measure.

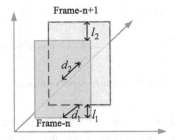

Fig. 4. Spatial depth relationships for different target frames

In the case of Fig. 4, considering the object's movement in the spatial depth direction, we focus on the object in the previous frame (frame-n) and the object detected in the current frame that intersects with the anchor box (frame-n + 1). These two objects may not necessarily be the same tracking target. The paper assume that the bottom edges of all bounding boxes lie on the ground, allowing us to place all bottom edges on the same plane and calculate the distance between the two objects more accurately. By analyzing the figure, we can see that the distance between the previous bounding box and the subsequent bounding box is denoted as d_1, while d_2 represents the distance between their center points. However, d_2 is not suitable for capturing the spatial depth differences between the two targets as it is influenced not only by the target's position but also by the target's height and the variation in perspective. Due to the different shooting directions, the direction of the extended line of d_1 varies. To unify the measurement, the paper introduce the vertical distance l_1 between the two bottom edges as the distance. To assess the target more reasonably, we also introduce the vertical distance l_2 between the two top edges and compare it with l_1 to calculate the contrast. Compared to l_1, this value includes the size difference between the two objects. Calculated as shown in Eq. (1).

$$r = \frac{l1}{l1 + l2} \tag{1}$$

where r is the contrast calculated based on the preliminary calculation of two vertical distances. When the target is in motion with a small range of movement, the difference between l_1 and l_2 is small, and the contrast approaches 0.5. When the target exhibits larger spatial depth movement, the contrast fluctuates around 0.5 as the target moves. Therefore, we use 0.5 as a basic threshold and introduce it into the calculation of contrast. This allows us to calculate the final formula for contrast, where a higher contrast value corresponds to a higher confidence level in the calculated intersection-over-union (IoU) results. Calculated as shown in Eq. (2).

$$c = 1 - |\frac{l1}{l1 + l2} - 0.5| \tag{2}$$

Finally, we incorporate the similarity 's' and contrast 'c' as corresponding parameters into the calculation of confidence. Since they are observed from different perspectives, we calculate them separately and use the root mean square to balance their values. Calculated as shown in Eq. (3).

$$out = \sqrt{(iou \times s)^2 + (iou \times c)^2} \tag{3}$$

4 Experiment and Analysis

The GPU model used for training and testing in this study is GeForce RTX 1080Ti with a VRAM size of 11 GB. The system has a total memory size of 31 GB. The software configuration includes Python 3.8, PyTorch 1.8.0, and CUDA 10.1.

4.1 Data Set and Evaluation Indicators

In this experiment, the VisDrone2019 dataset will be used for training and testing. VisDrone2019 is a large-scale dataset specifically designed for object detection and tracking, focusing on drone imagery. The dataset consists of thousands of high-resolution images and millions of annotated object instances, covering a wide range of scenes and environments such as urban areas, rural landscapes, and coastal regions. Due to the high complexity of drone-captured images, including variations in pose, occlusion, and multi-scale objects, the VisDrone2019 dataset is considered one of the most challenging datasets for drone imagery analysis.

In multi-object tracking algorithms, common evaluation metrics include Multiple Object Tracking Accuracy (MOTA), Multiple Object Tracking Precision (MOTP), False Positives (FP), False Negatives (FN), and ID F_1 Score (IDF_1). Accuracy will also be introduced as a performance indicator in terms of detection.

4.2 Experiments and Simulations

In order to comprehensively evaluate the performance of the proposed detection algorithm, we selected the VisDrone dataset as the experimental dataset and conducted separate detection experiments on it. To provide a more intuitive understanding of the algorithm's performance, we present the comparison results of different detectors and tracking algorithms in Table 1.

Table 1. Experiments with different modules

Detector	Tracker	IDF$_1$	MOTP	MOTA
Yolov3	Sort	44.2	23.4	41.9
	Deepsort	45.0	23.5	42.4
	Ours	**47.1**	**24.1**	**44.3**
Ours	Sort	48.9	25.7	45.6
	Deepsort	49.5	25.5	47.1
	Ours	**50.6**	**26.8**	**49.1**

The table reveals that when using YOLOv3 as the detector, different tracking algorithms yield minor differences in performance, but our proposed tracking algorithm still outperforms them. Furthermore, when using our proposed network, whether in conjunction with SORT or DeepSORT, the tracking performance is further improved. Particularly when compared to DeepSORT, our algorithm achieves a 4.5% increase in MOTA value. By considering multiple different tracking algorithms, it can be observed that our proposed detection algorithm consistently outperforms YOLOv3 in various tracking algorithm combinations. Overall, both the detector and tracker proposed in this paper demonstrate favorable performance, exhibiting overall satisfactory tracking results and providing an effective solution for research in the field of drone-based object tracking from aerial perspectives.

4.3 Ablation Experiments

In practical applications of tracking algorithms, it is crucial to consider both tracking performance and processing speed, as these two metrics directly impact the algorithm's practicality. In this regard, through ablation experiments, this paper conducted targeted improvements on each module of the JDE algorithm to evaluate the impact of each module on the algorithm's performance and the overall improvement of the algorithm.

The results from Table 2 indicate significant improvements of our algorithm compared to the JDE algorithm in terms of IDF1, MOTA, and MOTP metrics. These enhancements are attributed to ablative experiments conducted on three key modules of the JDE algorithm: the backbone network, the feature fusion Neck module, and optimization for drone perspective tracking. Regarding the backbone network, we optimized the network structure based on the transformer, resulting in a 2.5% improvement in MOTA and a 10

Table 2. Ablation experiments

Model	IDF$_1$	MOTA	MOTP	FPS
JDE	45.0	42.4	23.5	12
+Backbone	48.2	44.9	**24.3**	**22**
+Neck	49.5	47.2	25.6	20
+ Tracker	**50.6**	**49.1**	26.8	20

FPS increase in overall speed. In the Neck module, our algorithm utilized a multi-scale feature fusion approach, further improving MOTA by 2.3%. However, this approach led to a decrease in speed by 2 FPS. Finally, in the optimization for drone perspective tracking, our algorithm achieved a 1.9% improvement in MOTA while maintaining the overall processing speed, which saw a 66.7% increase. The experimental results demonstrate the significant performance enhancements of our proposed tracking algorithm in practical applications.

Fig. 5. Spatial depth relationships for different target frames

Fig. 6. Spatial depth relationships for different target frames

In order to better illustrate the performance of the algorithm, Figs. 5 and 6 show the actual tracking results comparing the Baseline and the proposed method in the context of unmanned aerial vehicle scenarios. The Baseline is a tracking model based on the JDE algorithm, which performs suboptimally in high-altitude UAV views. By observing Fig. 5, it is evident that the Baseline encounters difficulties in detecting small targets, leading to poorer tracking performance. In contrast, the proposed algorithm demonstrates more stability in target detection on the road. Furthermore, for targets in the middle of the road, both methods exhibit good tracking performance.

Additionally, Fig. 6 presents the performance of different methods across multiple frames. We can observe occlusions caused by vegetation obstructing vehicles and the overlapping of vehicles. Such complex scenarios often pose challenges to tracking. In the Baseline, occlusion by vegetation leads to target loss, and the overlapping of vehicles causes changes in target IDs, thereby affecting the tracking performance. However, the proposed algorithm maintains stable tracking performance in the face of such challenges. It can mitigate the tracking difficulties caused by occlusions to a certain extent, enabling continuous tracking of targets. In summary, the proposed tracking algorithm demonstrates superior performance in dealing with complex UAV scenarios, providing a reliable solution for UAV tracking tasks.

4.4 Comparison With Other Algorithms

Table 3. Algorithm comparison

Method	Precison	IDF$_1$	MOTA
JDE [6]	89.4	45.0	42.4
MOTDT [21]	86.1	47.3	45.8
Centertrack [7]	87.9	50.0	47.1
QuasiDense [22]	85.7	44.4	42.1
CorrTracker [23]	87.1	49.4	46.4
MOTR [24]	88.5	**51.2**	47.9
Ours	**90.1**	50.6	**49.1**

Table 3 presents a comparison of several state-of-the-art algorithms in recent years to evaluate their performance in multi-object tracking tasks. Among these algorithms, the proposed algorithm in this paper demonstrates better results in terms of IDF1 and MOTA metrics. This indicates that our algorithm has significant advantages in both the accuracy and overall performance of object tracking.For the detection module, the JDE algorithm performs best in detection accuracy but poorly in feature extraction for tracking, resulting in overall subpar tracking performance. In contrast, our algorithm not only maintains high detection accuracy but also further improves tracking performance. Compared to the worst-performing QuasiDense algorithm, our algorithm achieves a 7.0% increase in MOTA, and compared to the second-best performing MOTR algorithm, it achieves a 1.2% increase.

In conclusion, the proposed algorithm exhibits good tracking performance in addressing the challenges of object tracking from UAV perspectives, and it possesses certain advantages compared to other algorithms.

5 Conclusion

The use of multi-object tracking offers a broad perspective and enhanced maneuverability. This paper focuses on addressing the multi-object tracking problem and proposes a transformer-based model specifically designed for UAV perspective. Compared to recent algorithms, this model achieves a significant improvement in the MOTA metric, enabling more accurate object detection and tracking.

By combining the perspective of UAVs and artificial intelligence algorithms, we can achieve real-time tracking and analysis of multiple objects in complex scenarios. This has significant implications for applications such as domain monitoring, security, and disaster response using UAVs. In the future, we will continue to explore more advanced algorithms and technologies to further drive the development of multi-object tracking from the UAV perspective. The paper aim to apply these advancements to a wider range of fields, bringing greater convenience and benefits to people's lives and work.

References

1. Alavi, A.H., Jiao, P., Buttlar, W.G., et al.: Internet of Things enabled smart cities: state-of-the-art and future trends. Measurement **129**, 589–606 (2018)
2. Chanak, P., Banerjee, I.: Congestion free routing mechanism for IoT enabled wireless sensor networks for smart healthcare applications. IEEE Trans. Consum. Electron. **66**(3), 223–232 (2020)
3. Sharma, S.K., Wang, X.: Toward massive machine type communications in ultra-dense cellular IoT networks: current issues and machine learning-assisted solutions. IEEE Commun. Surv. Tutorials. **22**(1), 426–471 (2020)
4. Sreenu, G., Durai, S.: Intelligent video surveillance: a review through deep learning techniques for crowd analysis. J. Big Data **6**(1), 1–27 (2019)
5. Ren, W.H., Wang, X.C., Tian, J.D., et al.: Tracking-by-counting: using network flows on crowd density maps for tracking multiple targets. IEEE Trans. Image Process. **30**, 1439–1452 (2020)
6. Wang, Z.D., Zheng, L., Liu, Y.X., et al.: Towards real-time multi-object tracking. In: Computer Vision-ECCV 2020: 16th European Conference, pp. 107–122 (2020)
7. Zhou, X.Y., Koltun, V., Krähenbühl, P.: Tracking objects as points. In: Computer Vision-ECCV 2020: 16th European Conference, pp. 474–490 (2020)
8. Rezaee, K., Zadeh, H.G., Chakraborty, C., et al.: Smart visual sensing for overcrowding in COVID-19 infected cities using modified deep transfer learning. IEEE Trans. Industr. Inf. **19**(1), 813–820 (2022)
9. Krizhevsky, A., Sutskever, I., Hinton, G.E.: Imagenet classification with deep convolutional neural networks. Commun. ACM **60**(6), 84–90 (2017)
10. He, K.M., Zhang, X.Y., Ren, S.Q., et al.: Deep residual learning for image recognition. In: Proceedings of the IEEE Conference on Computer Vision and Pattern Recognition, pp. 770–778 (2016)

11. Huang, G., Liu, Z., Van Der Maaten, L., et al.: Densely connected convolutional networks. In: Proceedings of the IEEE Conference on Computer Vision and Pattern Recognition, pp. 4700–4708 (2017)
12. Tan, M.X., Le, Q.: EfficientNet: rethinking model scaling for convolutional neural networks. In: International Conference on Machine Learning, pp. 6105–6114 (2019)
13. DoVaswani, A., Shazeer, N., Parmar, N., et al.: Attention is all you need. In: Proceedings of the 31st International Conference on Neural Information Processing System, pp. 6000–6010 (2017)
14. Liu, Z., Lin, Y.T., Cao, Y., et al.: Swin transformer: hierarchical vision transformer using shifted windows. In: Proceedings of the IEEE/CVF International Conference on Computer Vision. pp. 10012–10022 (2021)
15. Han, K., Xiao, A., Wu, E.H., et al.: Transformer in transformer. Adv. Neural. Inf. Process. Syst. **2021**(34), 15908–15919 (2021)
16. Yuan, L., Chen, Y.P., Wang, T., et al.: Tokens-to-token VIT: training vision transformers from scratch on imagenet. In: Proceedings of the IEEE/CVF International Conference on Computer Vision, pp. 558–567 (2021)
17. Bewley, A., Ge, Z.Y., Ott, L., et al.: Simple online and realtime tracking. In: 2016 IEEE International Conference on Image Processing, pp. 3464–3468. (2016)
18. Chu, P., Wang, J., You, Q.Z., et al.: Transmot: Spatial-temporal graph transformer for multiple object tracking. In: Proceedings of the IEEE/CVF Winter Conference on Applications of Computer Vision, pp. 4870–4880 (2023)
19. Wu, J.L., Cao, J.L., Song, L.C., et al.: Track to detect and segment: an online multi-object tracker. In: Proceedings of the IEEE/CVF Conference on Computer Vision and Pattern Recognition, pp. 12352–12361 (2021)
20. Wojke, N., Bewley, A., Paulus, D.: Simple online and realtime tracking with a deep association metric. In: 2017 IEEE International Conference on Image Processing, pp. 3645–3649. IEEE (2017)
21. Chen, L., Ai, H.Z., Zhuang, Z.J., et al.: Real-time multiple people tracking with deeply learned candidate selection and person re-identification. In: 2018 IEEE International Conference on Multimedia and Expo, pp. 1–6. IEEE (2018)
22. Pang, J.M., Qiu, L.Y., Li, X., et al.: Quasi-dense similarity learning for multiple object tracking. In: Proceedings of the IEEE/CVF Conference on Computer Vision and Pattern Recognition, pp 164–173 (2021)
23. Wang, Q., Zheng, Y., Pan, P., et al.: Multiple object tracking with correlation learning. In: Proceedings of the IEEE/CVF Conference on Computer Vision and Pattern Recognition, pp. 3876–3886 (2021)
24. Zeng, F.G., Dong, B., Zhang, Y., et al.: Motr: End-to-end multiple-object tracking with transformer. In: Computer Vision-ECCV 2022: 17th European Conference, pp. 659–675 (2022)

HEI-GAN: A Human-Environment Interaction Based GAN for Multimodal Human Trajectory Prediction

Zihao Wang$^{(\boxtimes)}$ ⓘ, Xuguang Chen, Sichao Wen, and Yaonong Wang

Zhejiang Leapmotor Technology Co., Ltd., HangZhou 310051, China
`wang_zihao@leapmotor.com`

Abstract. Human trajectory prediction is an indispensable key component in autonomous driving systems and robot systems. The difficulty of human motion prediction lies in its inherent stochasticity and multimodality. Recently, some studies model the multimodality of human motion by predicting multiple possible future goals, which improves the accuracy of trajectory prediction. However, such methods fail to fully consider the influence of the environment on the target motion to be predicted. To solve the problem above, we propose HEI-GAN, a goal prediction model, with a human-environment interaction modeling method. The proposed method fuses the information from both the human motion and the environment to guide the model to learn the impact of human-environment interaction. We tested the performance of HEI-GAN on Stanford Drone Dataset and ETH/UCY dataset. The results show that HEI-GAN has a better prediction performance than the existing goal prediction models. At the same time, the prediction accuracy of the proposed method for the complete trajectory also reaches the advanced level.

Keywords: Environment Interaction · Goal Prediction · Human Trajectory Prediction · Multimodal Modeling

1 Introduction

With the development of technology, human trajectory prediction has been widely used in intelligent systems such as autonomous driving, robots and video surveillance [1].

The difficulty of human trajectory prediction lies in the inherent stochasticity of human motion. As shown in Fig. 1, different human behavior patterns and goals make the future trajectory multimodal. At the same time, the motion of surrounding pedestrians, obstacles and other environmental factors also makes this task much more complex.

Previous studies [2, 3] have established physical kinematics models based on human motion trajectories to model the relation between past motion and future trajectories. With the continuous evolvement of human trajectory prediction research, trajectory prediction methods based on social interaction [4–7], environment information [8, 9],

---

Z. Wang and X. Chen—Equal contribution.

Q. Liu et al. (Eds.): PRCV 2023, LNCS 14430, pp. 359–370, 2024.
https://doi.org/10.1007/978-981-99-8537-1_29

and final destination [10, 11] have been proposed successively. In addition, the focus of trajectory prediction has shifted from uni-modal prediction to multimodal prediction which gives multiple possible future trajectories [12–15]. Among all of the proposed methods, goal-based multimodal human trajectory prediction methods achieve superior results. However, the performance of these methods is strongly affected by the accuracy of the predicted goals. When making goal prediction, the existing models show some problems, such as incomplete utilization of scene information, lack of social interaction and inaccurate modeling, which reduce the accuracy of goal prediction.

Fig. 1. The future trajectory of a human is determined by the destination, surrounding agents and obstacles. In this scenario, the person in black is the pedestrian whose trajectory should be predicted, while all other gray characters are considered as obstacles.

Aiming at overcoming the difficulties in human trajectory prediction and the existing problems of goal-based methods stated above, we propose a multimodal human trajectory prediction method that makes full use of social interaction, environment knowledge and past motion information. The proposed method establishes the goal prediction model HEI-GAN based on human-environment interaction heatmap. We adopt a particular method to construct a heatmap to represent the social interaction relationship. At the same time, in order to better capture the stochasticity of human motion, a special GAN structure is innovatively adopted in the goal prediction model, which makes the predicted goal heatmap not constrained by the ground truth, and hence leads to better multimodality. We propose a trajectory prediction system that combines a deep learning model with a social force model (SFM) [2], defines two network modules to make efficient use of the predicted goals and sematic maps, and improves the accuracy of trajectory prediction.

The contributions of this work can be summarized as follows:

- We propose a novel interaction modeling method that integrates human-environment interaction information, and construct the interaction in the form of a heatmap. It solves the problem of incomplete use of environment information in the existing methods, and increase the effective information input to the network model.
- We construct HEI-GAN, a goal prediction network to solve the problem of inaccurate goal prediction in existing methods. HEI-GAN strengthens the feature learning based on adversarial process, and uses the discriminator to improve the accuracy of predicted goals from different receptive fields.

- We conduct complete tests on Stanford Drone and ETH/UCY, two popular public pedestrian datasets. Experimental results show that the proposed method reaches a top goal prediction level on both datasets, and, particularly, achieves better prediction results than the existing state-of-the-art methods on Stanford Drone datasets.

2 Related Works

With the continuous development of deep learning, the dominant models adopted in human trajectory prediction research have changed from early manual kinematic models [2, 3] to data-driven deep network models [4, 6]. Some studies formulate trajectory prediction as a sequence prediction problem and use recurrent neural networks (RNNs) to model human trajectory sequences [4]. Subsequently, more and more methods take social interaction as an important factor to construct trajectory prediction models.

2.1 Interaction Models

It has become one of the hottest topics to establish the interaction model between people and the environment. Gupta *et al.* [6] propose a pooling mechanism to encode the surrounding pedestrians into a global pooling vector. Mohamed *et al.* [16] improve the quality of aggregated interaction features [4, 6] by modeling the pedestrian interaction in the scene using a spatio-temporal graph structure. Salzmann *et al.* [12] also use a model based on graph structure to obtain the interaction relationship among pedestrians. With the attention mechanism [17] showing its remarkable effect in sequence regression, many studies propose more excellent prediction methods based on attention and Transformer [5, 18]. Sadeghian *et al.* [9] combine attention networks and GAN [19] and use social and physical attention mechanisms to extract social interaction features in scene context. Yu *et al.* [20] also use Transformer to improve the Graph Attention Network.

2.2 Multimodal Modeling

Due to the influence of the surrounding environment, the future trajectory of human motion becomes more uncertain. Thus, more and more recent trajectory prediction methods take the uncertainty of human motion into account. Some, for instance, use generative models to predict the distribution of future trajectories. Typical generative trajectory prediction models include CVAE [5, 12] and Gaussian mixture model [21]. Recently, goal-based methods [13, 22] have achieved higher prediction accuracy in the multimodal trajectory prediction task. Mangalam et al. [22] propose PECNet, a model that considers potential goals within human motion and uses the CVAE structure to predict multiple future goals. In another work of Mangalam *et al.* [10]'s, the past pedestrian trajectory is constructed into a heatmap. They use convolution neural networks to predict the probability distribution map, and sample multimodal goals from it. Xu *et al.* [15] store observed trajectories and their corresponding future endpoints as a set of empirical data in a repository. When performing complete trajectory prediction, multiple approximate goals are looked up in this repository. Chiara *et al.* [11] propose an attention-based trajectory prediction backbone on the basis of [10].

3 Method

3.1 Problem Definition

The ego-motion X of the pedestrian of interests constructed by the observed trajectory samples during past T_{obs} time steps, and represented in the form of a sequence $X = \{X^{t_0-T_{obs}}, X^{t_0-T_{obs}+1}, \ldots, X^{t_0}\}, X \in \mathbb{R}^{2 \times T_{obs}}$. $X^t = (x^t, y^t)$ is the position coordinate at time step t, t_0 is the last observed time step during the tracked period. The future trajectory during T_{fut} future time steps is modeled by a coordinate sequence $Y = \{Y^{t_0+1}, \ldots, Y^{t_0+T_{fut}-1}, Y^{t_0+T_{fut}}\}, Y \in \mathbb{R}^{2 \times T_{fut}}$.

Ego Motion Heatmap. Similar to the work conducted by Mangalam *et al.* [10], we transfer the observed trajectory X to image sequence $P = \{P^{t_0-T_{obs}}, P^{t_0-T_{obs}+1}, \ldots, P^{t_0}\}, P \in \mathbb{R}^{T_{obs} \times h \times w}$ with T_{obs} channels as a trajectory heatmap, where h, w are the height and the width of the scene RGB map, respectively. For each time step t, P^t is calculated using the distance $d_{i,j}^t$ between each pixel and X^t. The trajectory heatmap P^t and $d_{i,j}^t$ can be represented as:

$$P^t = \sum_{i=0}^{w} \sum_{j=0}^{h} \frac{2d_{i,j}^t}{\max d^t} \tag{1}$$

$$d_{i,j}^t = \|(p_x, p_y) - X^t\|_{2(x,y) \in P} \tag{2}$$

where (p_x, p_y) are the pixel coordinates.

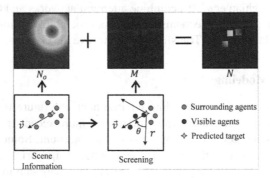

Fig. 2. The generation of social interaction heatmap N.

Scene Constraints. Scene Constraints include buildings, trees, sidewalks, roads and barriers. Our proposed method uses U-Net [23] to semantically separate the scene RGB map and divides the scene constraints into C categories. In this way, the scene sematic map $S \in \mathbb{R}^{C \times h \times w}$ is obtained, and its height and width are the same as those of the trajectory heatmap P.

Masked Social Interaction Heatmap. We model the social interaction based on three elements: time, location of agents and visual field of predicted target. In order to reduce the complexity of the model, the proposed method only considers pedestrians in the scene.

We construct a distance heatmap $N_o = \left\{ N_o^{t_0 - T_{obs}}, N_o^{t_0 - T_{obs} + 1}, \dots, N_o^{t_0} \right\}$, $N_o \in \mathbb{R}^{T_{obs} \times h \times w}$. The distance heatmap N_o^t at time step t can be represented as:

$$N_o^t = \sum_{i=0}^{s} \sum_{j=0}^{s} \left(\frac{s - d_{i,j}^t}{s} \right)^a \tag{3}$$

where $d_{i,j}^t$ is calculated by (2), α is a tunable hyper parameter and s is the length of N_o. N_o summarizes the influence of all surrounding area on the predicted pedestrian in the distance dimension during the observed period. By introducing α, we are able to control how much those agents that are far away from the pedestrian of interest should be accounted in this influence.

As shown in Fig. 2, We filter all surrounding agents in the scene through several key parameters such as the motion direction $\vec{v}_i$, field of view radius r and field of view angle θ of the pedestrian of interest. After that, we set the maximum number of surrounding agents as m, and set the influence range of agents as r_M. According to r_M and position coordinates, we define the field of view mask M^t, $M^t \in \{0, 1\}$ at time step t as:

$$M^t = \sum_{i=0}^{s} \sum_{j=0}^{s} \begin{cases} 1, & if\ i, j \in X_k \pm r_M \\ 0 & else \end{cases} \tag{4}$$

where X_k is the position of the kth agent.

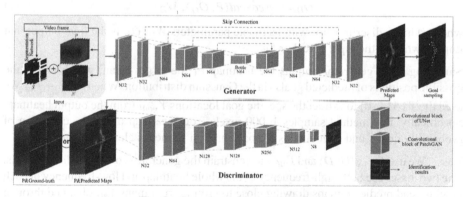

Fig. 3. Overall structure of HEI-GAN.

We generate the masked social interaction heatmap $N = \left\{ N^{t_0 - T_{obs}}, N^{t_0 - T_{obs} + 1}, \dots, N^{t_0} \right\}$, $N \in \mathbb{R}^{T_{obs} \times h \times w}$, by combining the distance heatmap N_o and the field of view mask M. This process can be represented as:

$$N^t = M^t \circ N_o^t \tag{5}$$

where $\circ$ stands for Hadamard product.

3.2 HEI-GAN

HEI-GAN uses heatmaps instead of trajectory sequences to complete the mapping from past information to future state. We creatively apply the adversariality of GAN for heatmaps processing, and get more reasonable predicted goals. We adjust the classic model Pix2Pix [24] in the field of image style transfer, and use U-Net generator and the discriminator named PatchGAN to achieve pedestrian trajectory prediction. The complete process is shown in Fig. 3.

Generator. In order to consider dynamic environment comprehensively, we concatenate the motion information P, interaction information N, and semantic information S as piror data H_{env} to represent the predicted scene. Specifically, the prior data $H_{env} = concat(S, N, P), H_{env} \in \mathbb{R}^{(C+2T_{obs}) \times s \times s}$, is the input of our generator. Then we use the codec with U-Net architecture to extract and transfer interlayer features. The skip connection layers inside U-Net can well preserve the spatial details of hidden features. The whole process is modeled as:

$$O_G = G(H_{env}; W_G) \tag{6}$$

where G is the generator with learnable parameters W_G. O_G represents the output heatmap, through which the final goal positions are sampled.

Discriminator. We choose PatchGAN, which is a fully convolutional structure, to judge the rationality of the prediction results from different receptive fields corresponding to each patch. As shown in (7), The convolutional layer is used to extract the features of input heatmaps and try to identify the authenticity of each s × s patch.

$$O_D = D(concat(P, O_G); W_D) \tag{7}$$

where D is the discriminator with learnable parameters W_D. O_D is the final output that implies whether the input is real or fake.

Sampling. The generator outputs the heatmap of the target agent, which marks the possible positions of predicted goals via the Gaussian distribution. When $t = t_0 + T_{fut}$. We used the TTST [10] to extract the specific goal locations $\hat{Y}_{goal}$ from the output heatmap. Specifically, this method samples 10000 pixels from heatmap under the condition of probability firstly, and then obtains K goals through K-means clustering algorithm.

Loss. We use $\mathcal{L}_{GAN}(G, D)$ and $\mathcal{L}_{BCE}$ to constrain the generation of the prediction goals. The two losses use both high-frequency of the whole heatmap and low-frequency of each pixel to lead predicted maps drawing close to ground truth maps Y_{gt}, obtained through the Gaussian distributions from destiation $Y^{t_0+T_{fut}}$. The total loss function $\mathcal{L}_{HEI-GAN}$ can be represented as:

$$\mathcal{L}_{GAN}(G, D) = \mathbb{E}\big[logD\big(concat\big(P, Y_{gt}\big)\big)\big]$$
$$+\mathbb{E}[\log(1 - O_D)] \tag{8}$$

$$\mathcal{L}_{BCE} = BCE(Y_{gt}, O_G) \tag{9}$$

$$\mathcal{L}_{HEI-GAN} = \mathcal{L}_{GAN}(G, D) + \lambda L_{BCE} \tag{10}$$

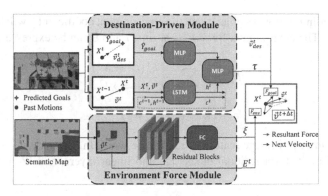

Fig. 4. Overall structure of the trajectory prediction system.

3.3 Trajectory Prediction System

As shown in Fig. 4, we propose a trajectory prediction system based on SFM [2], which includes destination-driven module and environment force module.

Destination-Driven Module. We introduce the modeling method of the influence of destination on pedestrian motion in SFM. When there is a certain destination $\widehat{Y}_{goal}$, the pedestrian movement to $\widehat{Y}_{goal}$ can be constructed as a polyline through destination driving force F_{goal}. In the future duration T_{fut}, F_{goal}^t of $\widehat{Y}_{goal}$ to the pedestrian's next velocity is determined by the expected direction $\overrightarrow{e}_{des}^{\,t} = \frac{\widehat{Y}_{goal} - X^t}{\|\widehat{Y}_{goal} - X^t\|_2}$, expected velocity $v_{des}^t = \frac{\|\widehat{Y}_{goal} - X^t\|_2}{T_{fut} - t}$ and the current velocity $\overrightarrow{v}^t = X^t - X^{t-1}$:

$$F_{goal}^t = \frac{1}{\tau}(v_{des}^t \overrightarrow{e}_{des}^{\,t} - \overrightarrow{v}^t) \tag{11}$$

$$\tau = \phi\left(\overrightarrow{v}^t, X^t, \widehat{Y}_{goal}, W_\tau\right) \tag{12}$$

where τ is the time interval describing that the pedestrian adjusts the current velocity to reach $\overrightarrow{v}_{des}^{\,t}$ again, which is fitted by neural network $\phi(\cdot)$. $\phi(\cdot)$ is composed of LSTM and MLPs, and W_τ is the network parameter. The input vector $[\overrightarrow{v}^t, X^t]$ of LSTM is concatenated by velocity and position.

Environment Force Module. We define the impact of scene constraints on the pedestrian trajectory as environment force F_{env}. For example, building entrances may be attractive to pedestrians, which makes pedestrians approach continuously during movement. However, greenbelts, obstacles and other structures will give pedestrians repulsive force, making them stay away from these structures. Therefore, the construction of F_{env} will be determined by the location, category and distance of stationary objects around pedestrians.

In semantic map S, we crop a $s_{env} \times s_{env}$ size image I_{env}, $I_{env} \in \mathbb{R}^{C \times s_{env} \times s_{env}}$, according to the velocity direction of the predicted pedestrian at time t. The position E

of the surrounding stationary objects and the environment coefficient ξ will be calculated based on I_{env}. The environment force F_{env}^t at time t and ξ can be expressed as:

$$F_{env}^t = \sum_{k=0}^{C} \frac{\xi_k^t (E^t - X^t)}{E^t - X_2^{t2}} \tag{13}$$

$$\xi = \psi(I_{env}, W_\xi) \tag{14}$$

where $\psi(\cdot)$ can be any image feature extraction network. In our work, we use a network composed of multi-layer residual blocks [2] and linear layers, and W_ξ is the network parameter.

Trajectory Prediction. Based on F_{goal} and F_{env}, we establish pedestrian dynamic model in discrete state. The process of the current velocity $\vec{v}^t$ changing into $\vec{v}^{t+\Delta t}$ after the time interval Δt can be calculated through the variation $\frac{d\vec{v}^t}{dt} = F_{goal} + F_{env}$:

$$\vec{v}^{t+\Delta t} = \vec{v}^t + \Delta t \frac{d\vec{v}^t}{dt} \tag{15}$$

Finally, we predict the position $\hat{X}^{t+\Delta t}$ of the pedestrian in next time step by following formula:

$$\hat{X}^{t+\Delta t} = \hat{X}^t + \Delta t \vec{v}^{t+\Delta t} \tag{16}$$

4 Experiment

4.1 Datasets and Metrics

We use two commonly-used public pedestrian datasets, Stanford Drone Dataset [25] and ETH/UCY [26, 27], to evaluate our method. To evaluate the accuracy of the predicted future trajectories given by our method, two common metrics, Average Displacement Error (ADE) and Final Displacement Error (FDE), are used to measure the quality of the predicted future trajectories in each experiment.

Table 1. Comparison with existing works on SDD. Bold and underlined numbers indicate the best and second-best.

Method	$Min_{20}ADE / Min_{20}FDE \downarrow$, $K = 20$ Samples						
	Social-GAN [6]	Sophie [9]	PECNet [22]	MemoNet [15]	Y-Net [10]	Goal-SAR [11]	Ours
	27.23/41.44	16.27/29.38	9.96/15.88	8.56/12.66	7.85/11.85	<u>7.75/11.83</u>	**7.72/11.59**

Table 2. Comparison with existing works on ETH/UCY dataset.

Method	$Min_{20}ADE/Min_{20}FDE \downarrow$, $K = 20$ Samples					
	ETH	HOTEL	UNIV	ZARA1	ZARA2	AVG
Social-GAN [6]	0.81/1.52	0.72/1.61	0.60/1.26	0.34/0.69	0.42/0.84	0.58/1.18
Sophie [9]	0.70/1.43	0.76/1.67	0.54/1.24	0.30/0.63	0.38/0.78	0.54/1.15
Transformer-TF [18]	0.61/1.12	0.18/0.36	0.35/0.65	0.22/0.38	0.17/0.32	0.31/0.55
STAR [20]	0.36/0.65	0.17/0.36	0.32/0.62	0.26/0.55	0.22/0.46	0.26/0.53
PECNet [22]	0.54/0.87	0.18/0.24	0.35/0.60	0.22/0.39	0.17/0.30	0.29/0.48
Trajectron ++ [12]	0.39/0.83	0.12/0.21	**0.20**/0.44	**0.15**/0.33	**0.11**/0.25	<u>0.19</u>/0.39
AgentFormer [5]	0.45/0.75	0.14/0.22	0.25/0.45	0.18/<u>0.30</u>	0.14/<u>0.24</u>	0.23/0.39
MemoNet [15]	0.40/0.61	<u>0.11</u>/0.17	<u>0.24</u>/<u>0.43</u>	0.18/0.32	0.14/<u>0.24</u>	0.21/0.35
Y-Net [10]	**0.28**/<u>0.33</u>	**0.10/0.14**	<u>0.24</u>/<u>0.41</u>	<u>0.17</u>/**0.27**	<u>0.13</u>/**0.22**	**0.18/0.27**
Ours	<u>0.34</u>/**0.32**	<u>0.11</u>/<u>0.15</u>	0.29/0.47	0.19/0.27	0.14/**0.22**	0.21/<u>0.29</u>

Table 3. Ablation study on model structures and inputs.

Networks	P	S	N	$Min_{20}FDE \downarrow$, K = 20 Samples					
				ETH	HOTEL	UNIV	ZARA1	ZARA2	SDD
CNN	√	√		0.43	0.16	0.50	0.33	0.26	11.88
	√	√	w/o sc	0.42	0.16	0.53	0.33	0.28	12.05
	√	√	√	0.41	**0.15**	**0.47**	0.31	0.25	11.79
GAN	√	√		0.50	0.16	0.49	0.30	0.24	11.71
	√	√	w/o sc	0.45	0.15	0.51	0.33	0.27	12.17
	√	√	√	**0.32**	**0.15**	**0.47**	**0.27**	**0.22**	**11.59**

(a) Coupa0   (b) Coupa1   (c)Hyang

(d) Little   (e) ZARA1   (f) HOTEL

Fig. 5. Visualization of predicted multimodal future trajectories and goal heatmaps.

4.2 Comparison with Existing Works

Stanford Drone Dataset: Table 1 list the results of bench-marking models on SDD. Note that we use the exact same test set as the other models in the table and achieve better results. Specifically, HEI-GAN achieves an $Min_{20}ADE$ of 7.72 and an $Min_{20}FDE$ of 11.59, which outperforms the existing state-of-the-art method, Goal-SAR.

ETH/UCY Dataset. We compare our approach with the existing top-ranking methods on ETH/UCY, and the results are shown in Table 2. In the ETH scene, HEI-GAN achieves an $Min_{20}FDE$ of 0.32, which is a better result compared to the current SOTA method Y-Net. Moreover, HEI-GAN reaches the same $Min_{20}FDE$ as Y-Net docs in ZARA1 and ZARA2. It can be seen that our model is able to provide accurate goals and guide the trajectory prediction system to predict the future trajectory closer to the ground truth in this way. However, there is still a performance gap between our model and Y-Net when it comes to the UNIV scene. The reason for this may be the prediction error is relatively high for some test samples that are not affected by the environment.

4.3 Ablation Study

In order to verify the effectiveness of each module in HEI-GAN to improve the accuracy of goal prediction, we conducted ablation study on submodels with different networks and input features under the $Min_{20}FDE$ metric. The first test is the model structure, to verify the effectiveness of GAN for goal prediction. We set up two models: CNN (only use U-Net), GAN (model composed of CNN generator and discriminator) for test. The second test is the model input, to verify the effect of social interactions and their different modeling methods on the model predictions.

As shown in Table 3, we show the $Min_{20}FDE$ of the submodels of each test. In the table, P is the heatmap of input trajectory, S is the semantic map of input scene, and N is the social interaction heatmap. "w/o sc" means that N is constructed using all the information from the surrounding agents without filtering them. The test results show that the $Min_{20}FDE$ of most of the submodels added with GAN is better than that only using CNN. In terms of model input, the social interaction heatmap was constructed after the nearby pedestrians were screened according to the visual field to further improve the accuracy of goal prediction.

4.4 Qualitative Results

In Fig. 5, we show the visualization of the predicted multimodal trajectories and goal heatmaps given by our proposed model. Among them, dots represent way points, starts represent endpoints, the ground truth future trajectory is represented in green, past trajectory is represented in blue, and multimodal predicted trajectories are represented in red. The first four pairs (a)–(d) in Fig. 5 are the four scenes in SDD, and the last two pairs (e)–(f) in Fig. 5 are scenes in ETH/UCY dataset. The qualitive results shows that HEI-GAN can well capture the multimodality of the future pedestrian motion, in addition to predicting a trajectory that is very close to the ground truth for different scenes.

5 Conclusion

This paper presents a goal prediction model HEI-GAN that performs well in pedestrian destination prediction task and a trajectory prediction system based on the predicted goals. In terms of input modeling, the proposed method fully considers information such as surrounding agents and obstacles, providing comprehensive interaction features to the model. In respect of the model structure, we propose a novel goal prediction network HEI-GAN. The proposed model creatively uses GAN for heatmap processing, and improves the accuracy of the predicted goals through the adversarial process.

The experiment results show that the proposed model reaches the top goal prediction level on the Stanford Drone Dataset and ETH/UCY dataset, and, in addition, outperforms the existing state-of-the-art method with regard to trajectory prediction on SDD.

References

1. Rudenko, A., et al.: Human motion trajectory prediction: a survey. Int. J. Robot. Res. **39**(8), 895–935 (2020)
2. Helbing, D., Molnar, P.: Social force model for pedestrian dynamics. Phys. Rev. **51**(5), 4282 (1995)
3. Schöller, C., Aravantinos, V., Lay, F., Knoll, A.: What the constant velocity model can teach us about pedestrian motion prediction. IEEE Robot. Autom. Lett. **5**(2), 1696–1703 (2020)
4. Alahi, A., et al.: Social LSTM: human trajectory prediction in crowded spaces. In: Proceedings of CVPR, pp. 961–971 (2016)
5. Yuan, Y., Weng, X., Ou, Y., Kitani, K.M.: Agentformer: agent-aware transformers for socio-temporal multi-agent forecasting. In: Proceedings of ICCV, pp. 9813–9823 (2021)
6. Gupta, A., Johnson, J., Fei-Fei, L., Savarese, S., Alahi, A.: Social gan: Socially acceptable trajectories with generative adversarial networks. In: Proceedings of CVPR, pp. 2255–2264 (2018)
7. Tang, H., Wei, P., Li, H., Li, J., Zheng, N.: Relation reasoning for video pedestrian trajectory prediction. In: 2022 IEEE International Conference on Multimedia and Expo (ICME), pp. 1–6 (2022)
8. Sadeghian, A., et al.: Car-net: Clairvoyant attentive recurrent network. In: Proceedings of ECCV, pp. 151–167 (2018)
9. Sadeghian, A., et al.: Sophie: an attentive gan for predicting paths compliant to social and physical constraints. In: Proceedings of CVPR, pp. 1349–1358 (2019)
10. Mangalam, K., An, Y., Girase, H., Malik, J.: From goals, waypoints & paths to long term human trajectory forecasting. In: Proceedings of ICCV, pp. 15233–15242 (2021)
11. Chiara, L.F., et al.: Goal-driven self-attentive recurrent networks for trajectory prediction. In: CVPR, pp. 2518–2527 (2022)
12. Salzmann, T., Ivanovic, B., Chakravarty, P., Pavone, M.: Trajectron++: dynamically-feasible trajectory forecasting with heterogeneous data. In: Vedaldi, A., Bischof, H., Brox, T., Frahm, J.-M. (eds.) ECCV 2020. LNCS, vol. 12363, pp. 683–700. Springer, Cham (2020). https://doi.org/10.1007/978-3-030-58523-5_40
13. Dendorfer, P., Osep, A., Leal-Taixé, L.: Goal-gan: Multimodal trajectory prediction based on goal position estimation. In: Proceedings of ACCV (2020)
14. Dendorfer, P., Elflein, S., Leal-Taixé, L.: Mg-GAN: a multi-generator model preventing out-of-distribution samples in pedestrian trajectory prediction. In: Proceedings of ICCV, pp. 13158–13167 (2021)

15. Xu, C., Mao, W., Zhang, W., Chen, S.: Remember intentions: retrospective-memory-based trajectory prediction. In: Proceedings of CVPR, pp. 6488–6497 (2022)
16. Mohamed, A., Qian, K., Elhoseiny, M., Claudel, C.: Social-STGCNN: a social spatio-temporal graph convolutional neural network for human trajectory prediction. In: Proceedings of CVPR, pp. 14424–14432. (2020)
17. Vaswani, A., et al.: Attention is all you need. Adv. Neural Inf. Process. Syst. **30** (2017)
18. Giuliari, F., Hasan, I., Cristani, M., Galasso, F: Transformer networks for trajectory forecasting. In: 2020 25th International Conference on Pattern Recognition (ICPR), pp. 10335–10342 (2021)
19. Goodfellow, I., et al.: Generative adversarial networks. Commun. ACM **63**(11), 139–144 (2020)
20. Yu, C., Ma, X., Ren, J., Zhao, H., Yi, S.: Spatio-temporal graph transformer networks for pedestrian trajectory prediction. In: Vedaldi, A., Bischof, H., Brox, T., Frahm, J.-M. (eds.) ECCV 2020. LNCS, vol. 12357, pp. 507–523. Springer, Cham (2020). https://doi.org/10.1007/978-3-030-58610-2_30
21. Ivanovic, B., Pavone, M.: The trajectron: probabilistic multi-agent trajectory modeling with dynamic spatiotemporal graphs. In: Proceedings of ICCV, pp. 2375–2384 (2019)
22. Mangalam, K., Girase, H., Agarwal, S., Lee, K.-H., Adeli, E., Malik, J., Gaidon, A.: It is not the journey but the destination: Endpoint conditioned trajectory prediction. In: Vedaldi, A., Bischof, H., Brox, T., Frahm, J.-M. (eds.) ECCV 2020. LNCS, vol. 12347, pp. 759–776. Springer, Cham (2020). https://doi.org/10.1007/978-3-030-58536-5_45
23. Ronneberger, O., Fischer, P., Brox, T: U-net: Convolutional networks for biomedical image segmentation. In: International Conference on Medical Image Computing and Computer-Assisted Intervention, pp. 234–241 (2015)
24. Isola, P., Zhu, J.Y., Zhou, T., Efros, A.A.: Image-to-image translation with conditional adversarial networks. In: Proceedings of CVPR, pp. 1125–1134 (2017)
25. Robicquet, A., Sadeghian, A., Alahi, A., Savarese, S.: Learning social etiquette: Human trajectory understanding in crowded scenes. In: Leibe, B., Matas, J., Sebe, N., Welling, M. (eds.) ECCV 2016. LNCS, vol. 9912, pp. 549–565. Springer, Cham (2016). https://doi.org/10.1007/978-3-319-46484-8_33
26. Pellegrini, S., Ess, A., Schindler, K., Van Gool, L.: You'll never walk alone: Modeling social behavior for multi-target tracking. In: 2009 IEEE 12th International Conference on Computer Vision, pp. 261–268 (2009)
27. Lerner, A., Chrysanthou, Y., Lischinski, D.: Crowds by example. Comput. Graph. Forum **26**(3), 655–664 (2007)

CenterMatch: A Center Matching Method for Semi-supervised Facial Expression Recognition

Linhuang Wang[1], Xin Kang[1(✉)], Satoshi Nakagawa[2], and Fuji Ren[3]

[1] Department of Information Science and Intelligent Systems, Tokushima University, Tokushima, Japan
kang-xin@is.tokushima-u.ac.jp
[2] Graduate School of Information Science and Technology, The University of Tokyo, Tokyo, Japan
[3] School of Computer Science and Engineering, University of Electronic Science and Technology of China, Chengdu, China

Abstract. The label uncertainty in large-scale qualitative facial expression datasets, caused by low-quality images and subjective annotations, combined with the high similarity between facial expression categories, makes facial expression recognition (FER) more challenging compared to traditional classification tasks. To address this problem, this paper proposes a new method called CenterMatch for semi-supervised facial expression recognition. Our approach sets class centers in a high-dimensional space and independent adaptive confidence thresholds for each category. As the positions of class centers are continuously updated by high-quality sample features, the confidence thresholds for each category adapt accordingly. Subsequently, high-confidence unlabeled samples are selected and assigned high-quality pseudo-labels, effectively suppressing label uncertainty. Additionally, we introduce a distance loss to constrain the updating direction of class centers, encouraging them to move away from each other and overcome the challenge of high similarity between expression categories. Experimental results demonstrate that our model outperforms other semi-supervised learning methods by adaptively adjusting confidence thresholds based on the learning status and difficulty of different categories, achieving a favorable balance between the quality and quantity of pseudo-labels. Furthermore, our approach exhibits excellent performance on multiple widely-used challenging in-the-world datasets, confirming its effectiveness and generalizability.

Keywords: Facial Expression Recognition · Semi-supervised Learning · Affective Computing

This research has been supported by the Project of Discretionary Budget of the Dean, Graduate School of Technology, Industrial and Social Sciences, Tokushima University.

Q. Liu et al. (Eds.): PRCV 2023, LNCS 14430, pp. 371–383, 2024.
https://doi.org/10.1007/978-981-99-8537-1_30

1   Introduction

Facial expression is a potent, natural, and universal signal for humans to communicate their emotional states and intentions [1,2]. Facial Expression Recognition (FER) has wide applications in the real world, such as driver fragile detection, affective computing, service robots, and human-computer interaction [3]. The advancements in FER heavily rely on large-scale annotated datasets [4]. However, collecting such datasets with a large number of labeled samples is challenging and expensive, as it requires the expertise of multiple human annotators [5,6]. Therefore, semi-supervised facial expression recognition (SS-FER) holds significant research value, as it enables training models on a large amount of unlabeled data.

Consistency regularization and pseudo-labeling are two powerful techniques that leverage unlabeled data and have been widely adopted in modern semi-supervised learning (SSL) algorithms [10]. Recently proposed [8,10–12] SSL methods have achieved competitive results by combining these techniques with both weak and strong data augmentations.

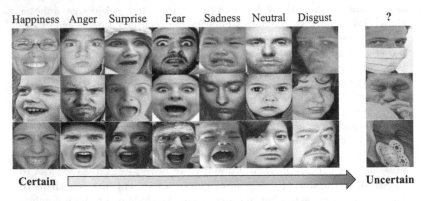

Fig. 1. Some contrasting examples from the RAFDB [5] training set. Certain facial expressions are highly similar, such as *Surprise* and *Fear*, *Neutral* and *Disgust*. Additionally, identifying the expression on the right is exceedingly challenging for both machines and even humans.

Despite the impressive performance of these methods on common classification tasks, they still face challenges when applied to SS-FER. There are two main reasons for this: (1) High similarity among FER categories [13]. Unlike common datasets, certain facial expressions exhibit extremely high similarity. As shown in Fig. 1, distinguishing between *fear* and *surprise* is challenging. (2) Uncertainty in labeling [7]. As shown in Fig. 1, collecting high-quality labels for in-the-wild facial expression datasets sourced from the internet is challenging due to factors such as annotators' subjectivity and image artifacts like occlusion and blurring.

In this approach, we construct class centers in a high-dimensional feature space and independent adaptive confidence thresholds for each category. During training, high-quality sample features are used to update the positions of the class centers, and the corresponding confidence thresholds adaptively change based on the distribution of the class centers. Categories that are far apart from other class centers indicate better learning states and lower learning difficulties, thereby maintaining higher confidence thresholds. Conversely, categories with closer distances have lower confidence thresholds. The adaptive thresholds enhance the model's attention towards categories with higher learning difficulties while ensuring high-quality pseudo-labels and improving the utilization of unlabeled data in the early stages of training. In this process, unlabeled data may be assigned more accurate pseudo-labels than the original labels, effectively suppressing label uncertainty in the datasets. Additionally, we introduce a distance loss to constrain the update of class centers. The distance loss encourages the class centers to move farther apart, thereby improving the recognition performance of highly similar facial expression categories. In summary, our main contributions can be summarized as follows:

- We propose a novel end-to-end SS-FER method called CenterMatch, which addresses the challenge of differentiating highly similar facial expressions categories through the constraint of distance loss. To the best of our knowledge, this is the first exploration in tackling the issue of high similarity between classes in SS-FER.
- Our adaptive confidence thresholds achieve an optimal balance between the quality and quantity of pseudo-labels, effectively suppressing label uncertainty through the utilization of high-quality pseudo-labels.
- Our CenterMatch outperforms other SSL algorithms on multiple widely used and challenging in-the-wild datasets, demonstrating superior performance.

2   Background

We first formulate the SSL framework for a C-class classification problem. Denote a batch of the labeled and unlabeled datasets as $\mathcal{D}_L = \left\{ \mathbf{x}_i^l, \mathbf{y}_i^l \right\}_{i=1}^{N_L}$ and $\mathcal{D}_U = \{\mathbf{x}_i^u\}_{i=1}^{N_U}$, respectively, where $\mathbf{x}_i^l, \mathbf{x}_i^u$ is the labeled and unlabeled training sample, and $\mathbf{y}_i^l$ is the one-hot ground-truth label for labeled data. We use N_L and N_U to represent the number of training samples in $\mathcal{D}_L$ and $\mathcal{D}_U$, respectively. Each unlabeled sample is individually subjected to weak augmentation and strong augmentation, resulting in augmented sample sets $\mathcal{W}_U = \{\mathbf{x}_i^w\}_{i=1}^{N_U}$ and $\mathcal{S}_U = \{\mathbf{x}_i^s\}_{i=1}^{N_U}$, respectively. This process can be represented as follows:

$$\mathbf{x}_i^w = \omega(\mathbf{x}_i^u) \tag{1}$$
$$\mathbf{x}_i^s = \Omega(\mathbf{x}_i^u)$$

where $\omega(\cdot)$ represents weak augmentation, and $\Omega(\cdot)$ represents strong augmentation.

All samples are fed into the model to obtain both the sample feature and the model prediction. It is worth noting that the term *sample feature* refers to the feature obtained from the last layer before making predictions. This process can be represented as follows:

$$f_i^l, \ p_i^l = Net(\mathbf{x}_i^l)$$
$$f_i^w, \ p_i^w = Net(\mathbf{x}_i^w) \tag{2}$$
$$f_i^s, \ p_i^s = Net(\mathbf{x}_i^s)$$

where $Net(\cdot)$ represents the model. $f \in \mathbb{R}^d$ and $p \in \mathbb{R}^C$ respectively denote the sample feature and the model's prediction.

We employ cross-entropy loss as the objective function for labeled data and strongly-augmented unlabeled data, where the pseudo-labels generated from predictions on weakly-augmented data serve as the labels for strongly-augmented data. This process can be described as follows:

$$\hat{y}_i^u = \mathrm{argmax}(p_i^w) \tag{3}$$

$$\mathcal{L} = \frac{1}{N_L} \sum_{i=1}^{N_L} CE(\mathbf{y}_i^l, p_i^l) + \frac{1}{N_U} \sum_{i=1}^{N_U} \mathbb{1}(\max(p_i^w) \geq \tau) \cdot CE(\hat{y}_i^u, p_i^s) \tag{4}$$

where $\hat{y}_i^u$ is the one-hot pseudo-label. $CE(\cdot)$ represents the cross-entropy loss, and τ denotes the confidence threshold.

From Eq. 4, it can be observed that in previous works [8–12], there has been a tendency to transform semi-supervised learning into supervised learning as much as possible. While this approach may be suitable for general classification tasks, it is not effective for FER due to the high similarity among certain facial expressions that are difficult to distinguish, and the uncertainty in labels may interfere with the effectiveness of supervised learning. Additionally, in some prior works, τ is set to a high fixed value (e.g., FixMatch [8] sets τ to 0.95). This not only leads to low sample utilization during the early stages of training but also results in poor learning performance for challenging class distinctions. Therefore, to address these issues specifically in FER tasks, we propose a novel semi-supervised algorithm called CenterMatch.

3   CenterMatch

The overall architecture of our proposed method, CenterMatch, is depicted in Fig. 2. Our model maps the samples into a high-dimensional feature space, where each category is associated with an updating class center. In addition to the commonly used cross-entropy loss, we introduce a distance loss to encourage samples of the same class to be closer and samples from different classes to be farther apart, thereby enhancing the discriminability of highly similar facial expressions. Moreover, we employ independent adaptive confidence thresholds for each category, which not only ensure the quality of pseudo-labels but also improve the model's learning effectiveness for challenging categories and the utilization of unlabeled data.

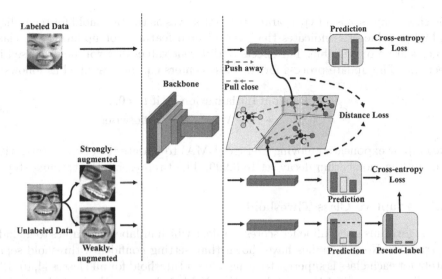

Fig. 2. The overall architecture of our CenterMatch. Where C_i representing the class center of the i-th class.

3.1 Class Center

We construct a high-dimensional feature space where samples of the same class are encouraged to cluster together, while samples from different classes are pushed apart as far as possible, aiming to increase the distinguishability of similar facial expressions. Each class is represented by a class center $c_j \in \mathbb{R}^d(j \in \{1, 2, ..., C\})$, which is iteratively updated during the training process.

During training, we update the class centers by selecting sample features with higher confidence from the labeled data and strongly-augmented data. The selection process can be described as follows:

$$\Delta f_j^l = \frac{\sum_{n=1}^{N_L} \mathbb{1}(\max(p_i^l) \geq \delta) \cdot \mathbb{1}(y_i^l = j) \cdot f_i^l}{\sum_{n=1}^{N_L} \mathbb{1}(\max(p_i^l) \geq \delta) \cdot \mathbb{1}(y_i^l = j)} \tag{5}$$

$$\Delta f_j^s = \frac{\sum_{n=1}^{N_U} \mathbb{1}(\max(p_i^s) \geq \delta) \cdot \mathbb{1}(\hat{y}_i^u = j) \cdot f_i^s}{\sum_{n=1}^{N_U} \mathbb{1}(\max(p_i^s) \geq \delta) \cdot \mathbb{1}(\hat{y}_i^u = j)} \tag{6}$$

$$\Delta c_j = \frac{\Delta f_j^l + \Delta f_j^s}{2} \tag{7}$$

where Δf_j^l and Δf_j^s represent the average sample features selected from labeled and strongly-augmented data, respectively. Δc_j represents the update value for the class centers. δ is a hyperparameter that serves as the threshold for reliability. A higher value of δ indicates that the selected features for updating the class centers are more reliable. The effects of different values of δ will be discussed in Sect. 4.3. The update process for the class centers can be described as follows:

$$c_j^t = \begin{cases} \text{Random Initialization,} & \text{if } t = 0, \\ \lambda c_j^{t-1} + (1 - \lambda)\Delta c_j^t, & \text{otherwise,} \end{cases} \tag{8}$$

We employ exponential moving average (EMA) to update the class centers, with λ being the momentum decay set to 0.999. The t represents the training step.

3.2 Adaptive Class Threshold

SSL algorithms use confidence scores to select which unlabeled samples to include in training. Recent studies have shown that setting confidence threshold separately for each class is superior to using a fixed threshold for all classes [4, 10–12]. Specifically, we adjust the confidence thresholds independently for each category based on their learning status and difficulty. Higher confidence thresholds are set for easily distinguishable categories, while lower thresholds are assigned to challenging categories (e.g., facial expressions that are prone to confusion), allowing the model to focus more on learning from samples of those categories. Moreover, lower confidence thresholds are beneficial for improving the utilization of unlabeled data in the early stages of training.

Therefore, we establish independent adaptive confidence thresholds for each category, which are dynamically adjusted according to the distribution of class centers in the feature space. Specifically, when a class center is further away from other centers, it indicates that the category is more easily distinguishable by the model, implying better learning status and lower learning difficulty. As a result, a higher confidence threshold is assigned to that category. Conversely, if the distance is smaller, suggesting a higher similarity to other categories, a lower threshold is set. To achieve this, we compute the distances between each class center and other class centers, and use these distances to update the confidence thresholds for each category. The specific process can be described as follows:

$$dis(c_j) = \sum_{k=1, k \neq j}^{C} ||c_j - c_k||_2 \tag{9}$$

$$\tau(c_j) = \text{MaxNorm}(dis(c_j)) \cdot \tau_{max} = \frac{dis(c_j)}{\max\{dis(c) : c \in [C]\}} \cdot \tau_{max} \tag{10}$$

where $dis(c_j)$ represents the sum of distances between the j-th class center and other class centers. $\tau(c_j)$ denotes the confidence threshold for the j-th class. We set τ_{max} to 0.95 as the upper limit for the confidence threshold. After scaling with maximum value normalization, more distant classes (corresponding to easily distinguishable expressions) maintain higher confidence thresholds, while closer classes (associated with difficult-to-distinguish expressions) have lower thresholds. Furthermore, during training, these confidence thresholds continuously change as the class centers are updated.

3.3 Loss Function

Following the previous SSL methods [4, 8–12], we compute the cross-entropy loss separately for labeled samples (denoted as $\mathcal{L}_{ce}^{l}$) and strongly-augmented samples (denoted as $\mathcal{L}_{ce}^{u}$). This process can be described as follows:

$$\mathcal{L}_{ce}^{l} = \frac{1}{N_L} \sum_{i=1}^{N_L} CE(\mathbf{y}_i^l, p_i^l) \tag{11}$$

$$\mathcal{L}_{ce}^{u} = \frac{1}{N_U} \sum_{i=1}^{N_U} \mathbb{1}(\max(p_i^w) \geq \tau(\hat{y}_i^u)) \cdot CE(\hat{y}_i^u, p_i^s) \tag{12}$$

In contrast to previous SSL methods [4, 8–12], inspired by CenterLoss [14], we introduce a distance loss that aims to bring similar samples closer together in the feature space while pushing different class samples further apart. For any given sample x, with its corresponding features f and label y, the distance loss can be computed as follows:

$$\mathcal{L}_{dis}(y, f) = \frac{||f - c_y||_2^2}{||f - c_y^{closest}||_2^2 + \epsilon} \tag{13}$$

where c_y represents the correct class center to which x belongs, while $c_y^{closest}$ represents the class center which is closest to c_y, indicating the most confusable class. To prevent a denominator of zero, we introduce a constant term ϵ set to 0.0001. The purpose of the $\mathcal{L}_{dis}(y, f)$ is to decrease the distance between x and c_y while increasing the distance between x and $c_y^{closest}$ thereby enhancing the distinguishability of c_y and $c_y^{closest}$.

We avoid using the sum of distances between c_y and all other class centers as the denominator because not all classes are equally confusable with c_y. If we consider the sum of distances, the model may overly emphasize increasing the distances from class centers that are already far away, leading to make the loss smaller. Consequently, it could neglect the closest class centers. Therefore, we only consider the distances between c_y and $c_y^{closest}$ allowing the model to focus on distinguishing the most confusable classes.

Specifically, we compute the distance loss separately for labeled samples (denoted as $\mathcal{L}_{dis}^{l}$) and strongly-augmented samples (denoted as $\mathcal{L}_{dis}^{u}$). This process can be described as follows:

$$\mathcal{L}_{dis}^{l} = \frac{1}{N_L} \sum_{i=1}^{N_L} \mathcal{L}_{dis}(y_i^l, f_i^l) \tag{14}$$

$$\mathcal{L}_{dis}^u = \frac{1}{N_U} \sum_{i=1}^{N_U} \mathbb{1}(\max(p_i^w) \geq \tau(\hat{y}_i^u)) \cdot \mathcal{L}_{dis}(\hat{y}_i^u, f_i^s) \tag{15}$$

Therefore the overall loss function of the network is defined as follows:

$$\mathcal{L} = \mathcal{L}_{ce}^l + \alpha_{dis}\mathcal{L}_{dis}^l + \alpha_u(\mathcal{L}_{ce}^u + \alpha_{dis}\mathcal{L}_{dis}^u) \tag{16}$$

where α_u and α_{dis} represent the loss weights for unlabeled samples and the distance loss respectively. Following previous works [8,10,11], we set α_u to 1.

4   Experiments

4.1   Datasets

We evaluate our method on three widely used in-the-wild datasets: RAF-DB [5], AffectNet [6] and FER-2013 [15]. The **FER-2013** [15] dataset is collected and labeled automatically using the Google Image Search API. It consists of over 35,000 images representing 7 basic expressions, with 28,000 images allocated for the training split. The **RAF-DB** [5] dataset comprises approximately 15,000 images, with 12,000 images allocated to the training split. The annotation process involved the efforts of 315 human annotators, with each label being carefully reviewed by approximately 40 individual annotators. This dataset encompasses 7 expression classes. **AffectNet** [6] is an immensely large dataset collected from the Internet. We utilized images depicting 7 fundamental facial expressions, with approximately 28K images in the training set and 3,500 images in the validation set.

4.2   Implementation Details

We conducted experiments using the PyTorch toolbox on an NVIDIA TITAN RTX GPU. Building upon previous studies [2,16], we adopted ResNet-18 [17] as the backbone network, with a parameter count of 11.18M and a computational complexity of 1.82G FLOPs. All images were resized to a dimension of 224 × 224 before being fed into the network. Consequently, the term f in Eq. 2 refers to a 512-dimensional feature vector obtained through global average pooling. We applied *RandomCrop* and *RandomHorizontalFlip* as weak augmentation methods and used RandAugment [18] as strong augmentation method. We set δ to 0.9. We used the SGD optimizer with a momentum of 0.9 and a weight decay of 0.0001. The initial learning rate was set to 0.1. For labeled data, a batch size of 64 was utilized, while for unlabeled data, the batch size was twice as large. All models were trained for 2^{20} iterations. To ensure fair comparison, we maintained these settings across all experiments. It is worth noting that class centers and the distance loss are exclusively computed during the training phase, thus not introducing any additional computational load during the inference stage.

In the semi-supervised experiments, we randomly selected 10, 25, 100, and 250 samples per class from the original training set as labeled data, while the remaining samples were treated as unlabeled data. Since all the datasets consisted of 7 categories, the resulting numbers of labeled samples were 70, 175, 700, and 1750, respectively. Additionally, for comparison purposes, we conducted experiments under the fully supervised setting using the same backbone network. To evaluate the models' performance, we utilized all samples from the original validation set of the dataset and employed accuracy as the evaluation metric. Following the SSL evaluation protocol, we conducted five experiments with different random seeds for semi-supervised learning and reported the average and standard deviation of the results across the five runs to ensure reliability and provide comprehensive insights.

4.3 Ablation Study

We evaluated the impact of different components of CenterMatch, and the specific results are presented in Table 1. We can observe that both the adaptive threshold and distance loss have improved the performance of the model, with the distance loss showing a greater improvement. This indicates that our proposed distance loss effectively enhances the discriminability between different categories.

Table 1. Evaluation of the Impact of CenterMatch Components

$\mathcal{L}_{dis}$	τ_c	RAF-DB	
		70 labels	1750 labels
✗	✗	58.45±3.12	75.89±2.14
✗	✓	60.41±2.60	78.45±1.45
✓	✗	71.35±0.54	81.35±1.37
✓	✓	74.32±1.02	83.12±1.21

Table 2. Evaluation of the Impact of δ

δ	RAF-DB	
	70 labels	1750 labels
0.14	34.12±2.45	39.46±2.14
0.25	40.15±4.21	45.17±3.45
0.9	74.32±1.02	83.12±1.21
0.95	72.45±0.13	81.32±0.24

Table 3. Evaluation of the Impact of α_{dis}

α_{dis}	RAF-DB	
	70 labels	1750 labels
0.01	64.34±1.01	76.91±1.62
0.1	67.63±1.14	78.75±1.52
1	74.32±1.02	83.12±1.21
10	70.84±1.25	80.64±1.07

Furthermore, we evaluated the impact of δ and α_{dis}. As shown in Table 2, when the value of δ is too small, it fails to ensure the reliability of the selected features, resulting in ineffective differentiation of class centers and thereby affecting the model performance. On the other hand, an excessively large δ value leads to overfitting, despite selecting highly reliable sample features, which in turn hampers the model's generalization performance. Hence, we set δ to 0.9 for all experiments. Based on the results presented in Table 3, we set α_{dis} to 1.

4.4 Quantitative and Qualitative Analysis

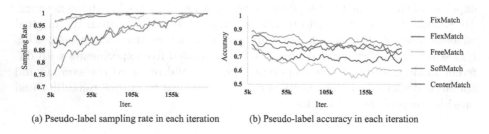

(a) Pseudo-label sampling rate in each iteration   (b) Pseudo-label accuracy in each iteration

Fig. 3. Comparison of the utilization of unlabeled data during training.

We conducted further analysis on the quantity and quality of pseudo-labels on RAF-DB [5] with 1750 labels. To quantitatively analyze the utilization of unlabeled data, we computed the ratio of the number of unlabeled samples used for training to the total number of unlabeled samples in a batch during each training iteration. This metric reflects the adoption rate of unlabeled data. Simultaneously, we calculated the proportion of correctly assigned pseudo-labels among the unlabeled samples used for training as a qualitative analysis metric. This metric measures the accuracy of the pseudo-labels.

As shown in Fig. 3(a), our CenterMatch significantly improves the utilization of unlabeled data in the early stages of training compared to fixed-threshold methods like FixMatch [8]. Additionally, Fig.3(b) illustrates that, except for FixMatch [8], our CenterMatch achieves the highest pseudo-label accuracy. As training progresses, the selected unlabeled samples inevitably contain increasingly challenging examples, leading to a decrease in the accuracy of pseudo-labels. However, our method still maintains a high level of performance. In conclusion, our CenterMatch strikes a favorable balance between the quantity and quality of pseudo-labels, ensuring both a high sampling rate and accuracy of pseudo-labels.

Table 4. The performance of different semi-supervised methods. Bold represents the best accuracy for each setting.

Labels	FER-2013				RAF-DB				AffcetNet			
	70	175	700	1750	70	175	700	1750	70	175	700	1750
MixMatch [9]	$45.69_{\pm1.34}$	$46.41_{\pm1.60}$	$55.73_{\pm0.68}$	$58.27_{\pm0.45}$	$36.34_{\pm0.74}$	$43.12_{\pm0.86}$	$64.14_{\pm0.67}$	$73.66_{\pm0.89}$	$30.80_{\pm1.21}$	$32.40_{\pm1.42}$	$39.77_{\pm1.30}$	$48.31_{\pm1.51}$
FixMatch [8]	$47.88_{\pm0.45}$	$49.9_{\pm0.32}$	$59.46_{\pm0.12}$	$62.20_{\pm1.20}$	$43.25_{\pm0.32}$	$52.44_{\pm0.16}$	$64.34_{\pm0.65}$	$75.51_{\pm1.31}$	$30.08_{\pm0.51}$	$38.31_{\pm0.32}$	$46.37_{\pm0.47}$	$51.25_{\pm0.61}$
FlexMatch [10]	$48.21_{\pm0.33}$	$50.12_{\pm1.30}$	$60.42_{\pm0.65}$	$63.45_{\pm1.21}$	$59.45_{\pm0.64}$	$63.46_{\pm0.49}$	$66.14_{\pm1.63}$	$77.13_{\pm0.72}$	$31.03_{\pm0.24}$	$39.45_{\pm0.21}$	$47.35_{\pm0.81}$	$52.46_{\pm0.13}$
FreeMatch [11]	$49.31_{\pm0.45}$	$51.32_{\pm1.47}$	$61.21_{\pm0.34}$	$64.32_{\pm1.45}$	$69.45_{\pm0.78}$	$71.65_{\pm0.64}$	$72.45_{\pm2.41}$	$79.64_{\pm0.45}$	$33.45_{\pm0.26}$	$42.65_{\pm0.32}$	$49.65_{\pm1.01}$	$54.61_{\pm0.33}$
SoftMatch [12]	$47.34_{\pm1.27}$	$50.47_{\pm1.03}$	$62.78_{\pm0.45}$	$62.32_{\pm1.32}$	$71.35_{\pm1.56}$	$70.14_{\pm0.23}$	$73.45_{\pm1.80}$	$80.32_{\pm0.14}$	$34.65_{\pm0.31}$	$43.44_{\pm0.65}$	$50.32_{\pm1.24}$	$54.14_{\pm0.43}$
CenterMatch (ours)	$\mathbf{50.31}_{\pm1.64}$	$\mathbf{52.32}_{\pm0.77}$	$\mathbf{63.61}_{\pm0.89}$	$\mathbf{65.45}_{\pm1.03}$	$\mathbf{74.32}_{\pm1.02}$	$\mathbf{75.45}_{\pm1.45}$	$\mathbf{78.45}_{\pm1.17}$	$\mathbf{83.12}_{\pm1.21}$	$\mathbf{38.45}_{\pm0.43}$	$\mathbf{45.64}_{\pm0.54}$	$\mathbf{52.14}_{\pm0.35}$	$\mathbf{56.26}_{\pm1.12}$
Same Backbone in Fully Supervised	66.78				81.21				57.13			

4.5 Comparison with State-of-the-Art Methods

Table 4 presents the comparison results between our method and other approaches. Our method outperforms the others in all experimental settings, even surpassing the performance of supervised learning on the RAF-DB dataset. We attribute this improvement to the significant label uncertainty present in the RAF-DB dataset. In fully supervised learning, the inclusion of noisy labels can cause the model to learn erroneous features, ultimately impacting its performance. However, in the context of semi-supervised learning, when a noisy label corresponds to an unlabeled sample, our method assigns a pseudo-label that better aligns with the sample's distinctive features. Essentially, it reallocates a label that is more accurate than the original one. This approach enhances the model's resilience to label uncertainty, contributing to its improved performance.

Fig. 4. t-SNE [19] visualization of facial expression features obtained by different methods.

4.6 Visualization

We visualized facial expression features obtained from different methods using t-SNE [19]. All models were trained on RAF-DB [5] with 1750 labels. As shown in Fig. 4, our CenterMatch significantly clusters features of the same category together and exhibits clear boundaries between different categories. Compared to other methods, our approach enables easier discrimination between different expressions, even those with high similarity. Furthermore, we calculated the confusion matrices for the different methods. As depicted in Fig. 5, CenterMatch notably improves the recognition accuracy of highly similar expressions, such as *Surprise* and *Fear*, *disgust* and *neutral*, when compared to other methods.

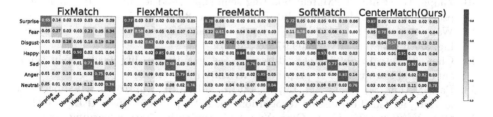

Fig. 5. Comparison of the confusion matrices among different methods.

5   Conclusion

In this paper, we propose a novel method called Centermatch for semi-supervised facial expression recognition. Our approach involves constructing a class center and an independent adaptive confidence threshold for each category. The confidence threshold dynamically adapts as the class center is continuously updated, thereby mitigating the label uncertainty caused by low-quality images and annotator subjectivity. Additionally, we introduce a distance loss to encourage the class centers to move farther apart, facilitating better discrimination between highly similar facial expression categories. Our method achieves state-of-the-art results on multiple widely used challenging int-the-wild datasets, demonstrating its effectiveness and generalization ability.

References

1. Tian, Y.I., Kanade, T., Cohn, J.F.: Recognizing action units for facial expression analysis. IEEE Trans. Pattern Anal. Mach. Intell. **23**(2), 97–115 (2001)
2. Darwin, C., Prodger, P.: The Expression of the Emotions in Man and Animals. Oxford University Press, USA (1998)
3. She, J., Hu, Y., Shi, H., Wang, J., Shen, Q., Mei, T.: Dive into ambiguity: latent distribution mining and pairwise uncertainty estimation for facial expression recognition. In: CVPR, pp. 6248–6257 (2021)
4. Li, H., Wang, N., Yang, X., Wang, X., Gao, X.: Towards semi-supervised deep facial expression recognition with an adaptive confidence margin. In: CVPR, pp. 4166–4175 (2022)
5. Li, S., Deng, W., Du, J.: Reliable crowdsourcing and deep locality-preserving learning for expression recognition in the wild. In: CVPR, pp. 2852–2861 (2017)
6. Mollahosseini, A., Hasani, B., Mahoor, M.H.: AffectNet: a database for facial expression, valence, and arousal computing in the wild. IEEE Trans. Affect. Comput. **10**(1), 18–31 (2017)
7. Wang, K., Peng, X., Yang, J., Lu, S., Qiao, Y.: Suppressing uncertainties for large-scale facial expression recognition. In: CVPR, pp. 6897–6906 (2020)
8. Sohn, K., et al.: FixMatch: simplifying semi-supervised learning with consistency and confidence. Adv. Neural. Inf. Process. Syst. **33**, 596–608 (2020)
9. Berthelot, D., Carlini, N., Goodfellow, I., Papernot, N., Oliver, A., Raffel, C.A.: MixMatch: a holistic approach to semi-supervised learning. Adv. Neural. Inf. Process. Syst. **32** (2019)
10. Zhang, B., et al.: FlexMatch: boosting semi-supervised learning with curriculum pseudo labeling. Adv. Neural. Inf. Process. Syst. **34**, 18408–18419 (2021)
11. Wang, Y., et al.: FreeMatch: self-adaptive thresholding for semi-supervised learning. arXiv preprint arXiv:2205.07246 (2022)
12. Chen, H., et al.: SoftMatch: addressing the quantity-quality trade-off in semi-supervised learning. arXiv preprint arXiv:2301.10921 (2023)
13. Pan, X., Liu, W., Wang, Y., Lu, X., Liu, B.: MSL-FER: mirrored self-supervised learning for facial expression recognition. In: ICIP, pp. 1601–1605. IEEE (2022)
14. Wen, Y., Zhang, K., Li, Z., Qiao, Yu.: A discriminative feature learning approach for deep face recognition. In: Leibe, B., Matas, J., Sebe, N., Welling, M. (eds.) ECCV 2016. LNCS, vol. 9911, pp. 499–515. Springer, Cham (2016). https://doi.org/10.1007/978-3-319-46478-7_31

15. Goodfellow, I.J., et al.: Challenges in representation learning: a report on three machine learning contests. In: Lee, M., Hirose, A., Hou, Z.-G., Kil, R.M. (eds.) ICONIP 2013. LNCS, vol. 8228, pp. 117–124. Springer, Heidelberg (2013). https://doi.org/10.1007/978-3-642-42051-1_16

16. Roy, S., Etemad, A.: Analysis of semi-supervised methods for facial expression recognition. In: ACII, pp. 1–8. IEEE (2022)

17. He, K., Zhang, X., Ren, S., Sun, J.: Deep residual learning for image recognition. In: CVPR, pp. 770–778 (2016)

18. Cubuk, E.D., Zoph, B., Shlens, J., Le, Q.V.: RandAugment: practical automated data augmentation with a reduced search space. In: CVPR, pp. 702–703 (2020)

19. Van der Maaten, L., Hinton, G.: Visualizing data using t-SNE. J. Mach. Learn. Res. 9(11), 2579–2605 (2008)

Cross-Dataset Distillation with Multi-tokens for Image Quality Assessment

Timin Gao[1,2], Weixuan Jin[2], Bokai Lai[2], Zhen Chen[2], Runze Hu[3],
Yan Zhang[1,4(✉)], and Pingyang Dai[1,2]

[1] Key Laboratory of Multimedia Trusted Perception and Efficient Computing,
Ministry of Education of China, Xiamen University,
Xiamen 361005, People's Republic of China
bzhy986@gmail.com

[2] Department of Artificial Intelligence, School of Informatics, Xiamen University,
Xiamen 361005, People's Republic of China

[3] School of Information and Electronics, Beijing Institute of Technology,
Beijing 100081, People's Republic of China

[4] Institute of Artificial Intelligence, Xiamen University,
Xiamen 361005, People's Republic of China

Abstract. No Reference Image Quality Assessment (NR-IQA) aims to accurately evaluate image distortion by simulating human assessment. However, this task is challenging due to the diversity of distortion types and the scarcity of labeled data. To address these issues, we propose a novel attention distillation-based method for NR-IQA. Our approach effectively integrates knowledge from different datasets to enhance the representation of image quality and improve the accuracy of predictions. Specifically, we introduce a distillation token in the Transformer encoder, enabling the student model to learn from the teacher across different datasets. By leveraging knowledge from diverse sources, our model captures essential features related to image distortion and enhances the generalization ability of the model. Furthermore, to refine perceptual information from various perspectives, we introduce multiple class tokens that simulate multiple reviewers. This not only improves the interpretability of the model but also reduces prediction uncertainty. Additionally, we introduce a mechanism called Attention Scoring, which combines the attention-scoring matrix from the encoder with the MLP header behind the decoder to refine the final quality score. Through extensive evaluations of six standard NR-IQA datasets, our method achieves performance comparable to the state-of-the-art NR-IQA approaches. Notably, it achieves SRCC values of **0.932** (compared to 0.892 in TID2013) and **0.964** (compared to 0.946 in CSIQ).

Keywords: Image quality assessment · Distillation · Transformer

1 Introduction

No Reference Image Quality Assessment (NR-IQA) is a fundamental research domain that aims to simulate the subjective evaluation of distorted images by

Q. Liu et al. (Eds.): PRCV 2023, LNCS 14430, pp. 384–395, 2024.
https://doi.org/10.1007/978-981-99-8537-1_31

human perceptual systems, without reference images. The state-of-the-art methods [1–4] in this field currently utilize pre-trained upstream backbones to extract semantic features, which are subsequently fine-tuned on NR-IQA datasets. However, the scarcity of IQA datasets poses a challenge, as simple fine-tuning often yields unsatisfactory results. Consequently, numerous researchers are dedicated to fully leveraging the limited information available in IQA datasets. For example, Zheng et al. [5] extracted reference information from degraded images by distilling knowledge from pristine-quality images, enabling the capture of deep image priors that prove useful for quality assessment. Yin et al. [6] incorporated non-IQA datasets as reference images to expand the dataset and utilized knowledge distillation to transfer distribution differences across various distorted images.

We believe that distillation solely from the original dataset or other task datasets is inadequate for acquiring crucial quality feature representations, and it also hampers the fusion of knowledge between students and teachers. To overcome this challenge, we propose a novel approach called CDMT-IQA, which leverages knowledge distillation across datasets to acquire more essential quality representations. Specifically, we augment the Vision Transformer architecture with a distillation token, which serves a similar purpose as the class token but focuses on reproducing the teacher's pseudo labels. It is noteworthy that the student and teacher models are associated with different datasets. Through distillation, the student can acquire cross-dataset knowledge from the teacher, and through cross-dataset knowledge fusion, further enhance the model's analytical capabilities. Furthermore, we introduce multiple class tokens and an Attention Scoring mechanism based on self-attention to mitigate scoring uncertainties. Our contributions are the following:

- To the best of our knowledge, we are the first to introduce a distillation token for cross-dataset knowledge distillation in the NR-IQA task. By incorporating knowledge from diverse datasets, our model is capable of capturing crucial information related to image quality perception. This integration further enhances the generalization capability of the model.
- We introduce multiple class tokens to simulate different image quality judges and introduce a new Attention Scoring mechanism to generate quality predictions. Multiple class tokens reduce the randomness of prediction, and Attention Scoring thoroughly explores the relationship between class tokens and the image, as well as the underlying distortion information.
- Our method demonstrated state-of-the-art performance across all four synthetic datasets and remained competitive on two authentic datasets.

2 Related Work

2.1 NR-IQA Based on CNN

Recently, deep learning-based NR-IQA methods have gained significant popularity due to their ability to effectively learn intricate perceptual features from images. CNN-based NR-IQA methods commonly employ image blocks as input or extract learned features from different layers of the CNN to create multi-scale representations, with shallow layers capturing local details and deep layers capturing high-level semantics, which are then combined and mapped to quality scores. Several relevant works [7–9] have been acknowledged and advocated. Although these strategies have been proven effective, they inevitably introduce a substantial computational burden during training and inference. Moreover, CNN-based methods are frequently limited by the inadequate representation of non-local features, resulting from the inherent local bias of CNN. This limitation negatively impacts the NR-IQA task.

2.2 NR-IQA Based on Vision Transformer

Vision Transformer (ViT) [10] has demonstrated impressive performance in various visual-related applications. The NR-IQA model based on ViT can be divided into two categories: hybrid transformers and pure ViT transformers. The method based on the hybrid Transformer [2,11] employs CNN to extract perceptual features, which are then used as input to the Transformer encoder. On the other hand, pure ViT transformers [4,12] directly utilize image patches as input to the Transformer encoder. The Transformer-based methods have shown promising performance. However, their capability to accurately characterize image quality remains limited due to their exclusive reliance on Transformer encoders. Furthermore, existing ViT approaches commonly rely on the class token for determining image quality. It should be noted that class tokens were initially designed for describing image content in tasks such as object recognition. Consequently, directly utilizing a single class token to represent the perceived image quality features may not be the most suitable approach.

3 Methodology

3.1 Overview

We present CDMT-IQA, a novel Transformer-based framework for NR-IQA. The overall architecture of CDMT-IQA is illustrated in Fig. 1. Given an input image, we partition it into a sequence of patches. We introduce multiple learnable class tokens as well as a distillation token and incorporate them into Vision Transformer. By employing self-attention operations, these tokens capture both local and non-local dependencies from the patch embeddings, thereby preserving comprehensive information regarding image quality. Subsequently, we feed these tokens back into the Transformer decoder. Following multi-head self-attention, the tokens engage in cross-attention with the image features obtained by the decoder. Finally, we feed the quality embeddings obtained from class tokens into a score refinement module named Attention Scoring to obtain the quality score and feed the quality embeddings obtained from distillation tokens into an MLP head to obtain the distillation score. During training, we calculate $\mathcal{L}_1$ loss and

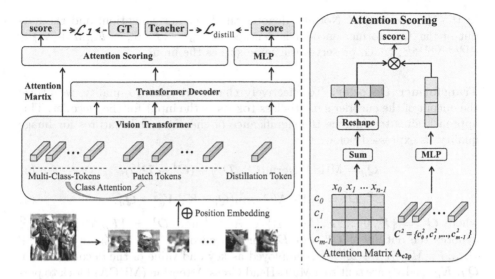

Fig. 1. The overview of our CDMT-IQA.

distillation loss for the two scores, respectively. During inference, we weigh the sum of two scores to obtain the final score. This method ensures the accuracy and generalization of predicting image quality.

3.2 Multi-tokens Image Quality Transformer

Multi-tokens Transformer Encoder. We introduce a Multi-Tokens Transformer Encoder based on the Vision Transformer [10]. Vision Transformer encoder combines both local and non-local information from a sequence of input patches in a way that minimizes any inherent biases, enabling it to effectively capture and analyze the perceptual features of an image in a comprehensive manner. Compared to it, we employ multiple class tokens and a distillation token instead of a single class token. More specifically, we first divide each image into N patches, each patch is transformed into a D-dimensional embedding using a linear projection layer. By incorporating position embeddings into each patch embedding, we generate a sequence of patch tokens $\boldsymbol{X} = \{\boldsymbol{x_0}, \boldsymbol{x_1}, ..., \boldsymbol{x_n}\} \in \mathbb{R}^{N \times D}$. Furthermore, we introduce M additional learnable class tokens $\boldsymbol{C^0} = \{c_0^0, c_1^0, ..., c_m^0\} \in \mathbb{R}^{M \times D}$ and a distillation token $\boldsymbol{D^0} = \{\boldsymbol{d^0}\} \in \mathbb{R}^{1 \times D}$. The final input token $\boldsymbol{T} = \{\boldsymbol{C^0}, \boldsymbol{D^0} \ \boldsymbol{X}\} \in \mathbb{R}^{(M+N+1) \times D}$ comprises the patch tokens, the class tokens, and the distillation token. Three linear projection layers are employed to convert $\boldsymbol{T}$ into matrices $\boldsymbol{Q}, \boldsymbol{K}, \boldsymbol{V} \in \mathbb{R}^{(M+N+1) \times D}$, representing the query, key, and value, respectively. The Multi-Head Self-Attention (MHSA) operation is utilized to generate the output feature $\boldsymbol{F_o}$, which is obtained as follows:

$$\boldsymbol{F_M} = \text{MHSA}(\boldsymbol{Q}, \boldsymbol{K}, \boldsymbol{V}) + \boldsymbol{T}, \tag{1}$$

$$\boldsymbol{F_o} = \text{MLP}(\text{Norm}(\boldsymbol{F_M})) + \boldsymbol{F_M}, \tag{2}$$

In the above equation, Norm($\cdot$) denotes the layer normalization, and the output of the Transformer encoder is denoted as $\boldsymbol{F}_o = \{\boldsymbol{F}_o[0]; ...; \boldsymbol{F}_o[M + N]\} \in \mathbb{R}^{(M+N+1)\times D}$, which preserves the same size as the input.

Transformer Decoder. To effectively characterize image quality, we utilize the output of the encoder's multi-class-tokens as the input for the Decoder. This approach aims to enhance the significance of the extracted features for image quality, as expressed below:

$$Q_d = \text{MHSA}(\text{Norm}(\boldsymbol{C^1}, \boldsymbol{D^1}) + (\boldsymbol{C^1} + \boldsymbol{D^1})), \tag{3}$$

$$\boldsymbol{C^2}, \boldsymbol{D^2} = \text{MHCA}(\text{ Norm}(\boldsymbol{Q}_d), \boldsymbol{K}_d, \boldsymbol{V}_d) + \boldsymbol{Q}_d, \tag{4}$$

where $\boldsymbol{C^1} = \{\boldsymbol{F}_o[0]; ...; \boldsymbol{F}_o[M - 1]\} \in \mathbb{R}^{M \times D}$ and $\boldsymbol{D^1} = \{\boldsymbol{F}_o[M]\} \in \mathbb{R}^{1 \times D}$ denotes the output of $\boldsymbol{C^0}$ and $\boldsymbol{D^0}$ for the encoder. $\boldsymbol{K}_d = \boldsymbol{V}_d = \{\boldsymbol{F}_o[M + 1]; ...; \boldsymbol{F}_o[M + N]\} \in \mathbb{R}^{N \times D}$ is employed as key and value of the decoder. Then, $\boldsymbol{Q}_d$, $\boldsymbol{K}_d$, and $\boldsymbol{V}_d$ are sent to a Multi-Head Cross-Attention (MHCA) block to perform the cross-attention. This interaction ensures that the attentional features become more significant for image quality.

3.3   Image Quality Prediction

The output of the Decoder is commonly regarded as a consolidation of perceptual information pertaining to image quality. This output is then forwarded to the MLP header to facilitate the regression task for quality prediction. By using multiple class tokens, the output of each token can be interpreted as the quality preferences of judges with different subjective opinions. Furthermore, we introduce an **Attention Scoring** mechanism that combines the MLP information with the attention information derived from the encoder to improve the prediction accuracy of the class tokens.

Specifically, within the Transformer encoder, we can establish a token-to-token attention map. By utilizing the global pairwise attention matrix $\mathbf{A}_{t2t} \in \mathbb{R}^{(M+N+1)\times(M+N+1)}$ and $\mathbf{A}_{t2t} = \text{softmax}(\mathbf{QK}^\top/\sqrt{D})$. We can extract the class attention to patches $\mathbf{A}_{c2p} \in \mathbb{R}^{M \times N}$, where $\mathbf{A}_{c2p} = \mathbf{A}_{t2t}[0 : M-1, M+1 : M+N]$, which signifies the relationship between each judge and each patch. Each row of the attention matrix $\mathbf{A}_{c2p}$ corresponds to the attention scores of a judgment across all patches. We can extract the attention map from each layer of the Transformer encoder. Considering that higher layers acquire more sophisticated discriminative representations, whereas earlier layers capture general and low-level visual information. We suggest merging the class-to-patch connections from the last K Transformer encoding layers to establish a more precise correspondence between judgments and images. The formula for this process can be expressed as:

$$\hat{\mathbf{A}}_{c2p} = \frac{1}{K} \sum_{l}^{l+K} \mathbf{A}_{c2p}^l, \tag{5}$$

$$\mathbf{A}_{c2i} = \sum_{j=0}^{N} \hat{\mathbf{A}}_{c2p}[:,j]. \tag{6}$$

Subsequently, the obtained scores from $\mathbf{A}_{c2i} \in \mathbb{R}^{M \times 1}$ and the MLP will be subjected to dot-product, yielding the final predicted quality score $Y_{\text{class\_token}}$:

$$Y_{\text{class\_token}} = \mathbf{A}_{c2i}^{T} \cdot \text{MLP}(\boldsymbol{C^2}). \tag{7}$$

For the final predicted quality score of the distillation token $Y_{\text{distillation\_token}}$, we simply use an MLP layer for the calculation:

$$Y_{\text{distill\_token}} = \text{MLP}(\boldsymbol{D^2}). \tag{8}$$

When inference, we use β to balance $Y_{\text{class\_token}}$ and $Y_{\text{distill\_token}}$ to obtain the final score of image quality Y_{final}.

$$Y_{\text{final}} = \beta \, Y_{\text{class\_token}} + (1 - \beta) \, Y_{\text{distill\_token}}. \tag{9}$$

3.4 Attention Distillation Cross Datasets

We develop an attention distillation across datasets to enhance model accuracy and generalization. As depicted in Fig. 1, we introduce a distillation token into the initial embedding. As discussed in Sect. 3.3, our student model generates two scores corresponding to class tokens and distillation tokens. These scores are utilized to compute the hard loss for the labels and the distillation loss for the teacher. The global loss function is defined as follows:

$$\mathcal{L}_{\text{global}} = \lambda \| Y_{\text{class\_token}} - Y \|_1 + (1 - \lambda) \| Y_{\text{distill\_token}} - Y_T \|_1, \tag{10}$$

where $Y_{\text{class\_token}}$ and $Y_{\text{distill\_token}}$ denote the outputs of multiple class tokens and the distillation token, Y represents the ground truth, Y_T represents the predictions of the teacher model, and $\| \cdot \|_1$ represents the $\mathcal{L}_1$ norm. It is important to note that we employ a cross-dataset teacher model, implying that the dataset for training the teacher differs from the one currently used for student training. We incorporate these additional perturbations into the training of student models, aiming to acquire more fundamental representations of image quality by leveraging knowledge across different datasets.

4 Experiments

4.1 Benchmark Datasets and Evaluation Protocols

We evaluate the effectiveness of the proposed CDMT-IQA model by conducting an assessment using six standard NR-IQA datasets. These datasets consist of four synthetic datasets, namely LIVE [13], CSIQ [14], TID2013 [15], and KADID [16], as well as two authentic datasets, namely LIVEC [17] and SPAQ [18]. The synthetic datasets are generated by applying different types of

Table 1. Performance comparison measured by averages of SRCC and PLCC on four synthetic datasets and two authentic datasets, where bold entries indicate the best results, underlines indicate the second-best.

Method	LIVE		CSIQ		TID2013		KADID		LIVEC		SPAQ	
	PLCC	SRCC	PLCC	SRCC	PLCC	SRCC	PLCC	SRCC	PLCC	SRCC	PLCC	SRCC
DIIVINE [19]	0.908	0.892	0.776	0.804	0.567	0.643	0.435	0.413	0.591	0.588	0.600	0.599
BRISQUE [20]	0.944	0.929	0.748	0.812	0.571	0.626	0.567	0.528	0.629	0.629	0.817	0.809
ILNIQE [21]	0.906	0.902	0.865	0.822	0.648	0.521	0.558	0.534	0.508	0.508	0.712	0.713
WaDIQaM [8]	0.955	0.960	0.844	0.852	0.855	0.835	0.752	0.739	0.671	0.682	–	–
DBCNN [22]	0.971	0.968	0.959	0.946	0.865	0.816	0.856	0.851	0.869	0.851	0.915	0.911
TIQA [11]	0.965	0.949	0.838	0.825	0.858	0.846	0.855	0.850	0.861	0.845	–	–
P2P-BM [23]	0.958	0.959	0.902	0.899	0.856	0.862	0.849	0.840	0.842	0.844	–	–
HyperIQA [1]	0.966	0.962	0.942	0.923	0.858	0.840	0.845	0.852	0.882	0.859	0.915	0.911
TReS [2]	0.968	0.969	0.942	0.922	0.883	0.863	0.858	0.859	0.877	0.846	–	–
MUSIQ [3]	0.911	0.940	0.893	0.871	0.815	0.773	0.872	0.875	0.746	0.702	0.921	0.918
DACNN [24]	0.980	0.978	0.957	0.943	0.889	0.871	0.905	0.905	0.884	0.866	0.921	0.915
DEIQT [4]	0.982	0.980	0.963	0.946	0.908	0.892	0.887	0.889	**0.894**	**0.875**	0.923	0.919
CDMT-IQA (ours)	**0.985**	**0.983**	**0.974**	**0.964**	**0.942**	**0.932**	**0.907**	**0.906**	**0.894**	0.868	**0.925**	**0.920**

distortions, such as JPEG compression and random noise, to a limited set of original images. Specifically, the LIVE dataset contains 799 synthetically distorted images, while CSIQ contains 866, with five and six distortion types, respectively. In contrast, the TID2013 dataset comprises 3,000 images with 24 distortion types, and the KADID dataset consists of 10,125 images with 25 distortion types. On the other hand, the authentic datasets consist of images taken by diverse photographers using a variety of mobile devices. LIVEC comprises 1162 images, and SPAQ contains 11,000 images obtained from public multimedia resources.

In order to evaluate the predictive performance of the CDMT-IQA model in relation to accuracy and monotonicity, we utilize two commonly employed criteria: Spearman's Rank Correlation Coefficient (SRCC) and Pearson's Linear Correlation Coefficient (PLCC). These metrics are widely recognized and provide a measurement scale ranging from 0 to 1. A value closer to 1 indicates a higher level of performance for both SRCC and PLCC, thereby reflecting superior predictive capabilities.

4.2 Implementation Details

To train the CDMT-IQA, we follow a standard procedure that involves randomly cropping the input image into ten patches, each with a resolution of 224×224 pixels. These patches are subsequently reshaped into a sequence of patches with a patch size of 16, and an input token dimension is 384. For the model architecture, we adopt the ViT-S Transformer encoder proposed in DeiT III [25]. The attention score is computed as the average of the encoder's last three layers, and the number of class tokens is set to 6. As for the decoder, we employ a depth of 1. The model underwent training for 9 epochs, with a learning rate of 2×10^{-4}.

Table 2. SRCC on the cross datasets validation. The best performances are highlighted in boldface. Underlines indicate the second-best.

| Training | LIVEC | KonIQ | LIVE | CSIQ |
Testing	KonIQ	LIVEC	CSIQ	LIVE
DBCNN	0.754	0.755	0.758	0.877
P2P-BM	0.74	0.77	0.712	–
HyperIQA	0.772	0.785	0.744	0.926
TReS	0.733	0.786	0.761	–
DEIQT	0.744	0.794	0.781	0.932
CDMT-IQA	**0.805**	**0.810**	**0.806**	**0.960**

Additionally, a decay factor of 10 is applied every 3 epochs. The β in Eq. 9 is set to 0.99, and the λ in Eq. 10 is set to 10^{-6}. To ensure a comprehensive evaluation, we randomly partitioned the dataset into 80% of the training set and 20% of the testing set. To reduce performance bias, we repeated this process ten times and reported the average values of PLCC and SRCC.

4.3 Overall Prediction Performance Comparison

The results of the comparison between CDMT-IQA and 12 classical or state-of-the-art NR-IQA methods, which include hand-crafted feature-based NR-IQA methods like ILNIQE [21] and BRISQUE [20], as well as deep learning-based methods such as HyperIQA [1] and TRes [2]. Table 1 shows the performance on four synthetic datasets, revealing that CDMT-IQA has outperformed all previous methods, particularly in CSIQ and TID2013, where significant improvements have been achieved. In terms of SRCC, CDMT-IQA has shown enhancements of 0.018 and 0.04 compared to the second-best performing method. The right half of Table 1 presents our performance on authentic datasets, demonstrating that we achieve the highest accuracy on SPAQ and performed competitively with other methods on LIVEC. Due to the diversity of image content and distortion types, achieving excellent performance on so many datasets simultaneously is a daunting challenge. Therefore, these results provide strong evidence to support the effectiveness and superiority of CDMT-IQA in accurately characterizing image quality.

4.4 Generalization Capability Validation

To further assess the generalization ability of CDMT-IQA, we conduct cross-dataset validation experiments. Specifically, our model is trained on one dataset and subsequently tested on another dataset without any fine-tuning or parameter adaptation. To ensure simplicity and universality, we perform four sets of experiments, comprising two synthetic datasets and two authentic datasets. The experimental results of SRCC averages on the four datasets are presented in

Table 3. Ablation experiments on TID2013 and CSIQ datasets. Bold entries indicate the best performance.

Cross-Dataset Distillation	Multi-tokens Attention Scoring	TID2013		CSIQ	
		PLCC	SRCC	PLCC	SRCC
✗	✗	0.896	0.873	0.960	0.945
		±0.050	±0.063	±0.010	±0.014
✔	✗	0.922	0.906	0.969	0.960
		±0.028	±0.033	±0.006	±0.008
✗	✔	0.923	0.906	0.971	0.960
		±0.015	±0.016	±0.005	±0.007
✔	✔	**0.942**	**0.932**	**0.974**	**0.964**
		±0.012	±0.014	±0.005	±0.007

Table 2. It is evident that CDMT-IQA demonstrates the best performance on all four cross datasets. Notably, CDMT-IQA's cross-dataset performance not only significantly surpasses previous SOTA methods but also rivals some supervised methods on LIVE. These results strongly indicate the robust generalization ability of CDMT-IQA.

4.5   Ablation Study

CDMT-IQA is an innovative NR-IQA model based on Transformer architecture, comprising two key components: Cross-Dataset Distillation and Multi-tokens Attention Scoring. Both components play pivotal roles in accurately characterizing image quality and enhancing the overall model performance. To gain deeper insights into the significance of each component, we conduct ablation experiments and analyze their impact on the TID2013 and CSIQ. The results of the ablation study are presented in Table 3, revealing substantial improvements in the baseline when utilizing either Cross-Dataset Distillation or Multi-tokens Attention Scoring alone across the two datasets. The combination of Cross-Dataset Distillation and Multi-tokens Attention Scoring yields the maximum enhancement over the baseline. Notably, on the TID2013 dataset, PLCC and SRCC increase by 0.046 and 0.059, respectively. In conclusion, the effectiveness of the ablation experiment underscores the significant contributions of all components in our method toward accurately characterizing image quality.

4.6   Visualization of Class Activation Map

We utilize GradCAM to visualize the feature attention map. Figure 2 reveals that our model comprehensively and accurately directs its attention to the target area, while the baseline [4] often focuses on non-target areas or fails to identify the focal region. Furthermore, the quality score prediction results demonstrate

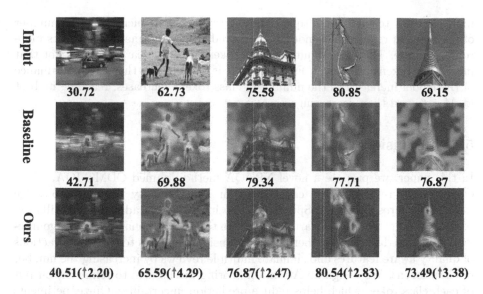

Fig. 2. Activation maps of baseline [4] and CDMT-IQA using Grad-CAM. Rows 1–3 represent input images, CAMs generated by baseline [4], and generated by CDMT-IQA, respectively. The numbers below each row represent the Ground Truth, and the predicted scores of the baseline model and our model, respectively.

Table 4. Analysis of the number of class tokens on synthetic datasets. Bold entries indicate the best performance.

Number of tokens	TID2013		CSIQ	
	PLCC	SRCC	PLCC	SRCC
M = 1	0.922	0.906	0.969	0.960
M = 2	0.928	0.913	0.970	0.960
M = 4	0.932	0.920	0.966	0.955
M = 6	**0.942**	**0.932**	**0.974**	**0.964**
M = 8	0.927	0.910	0.960	0.951
M = 10	0.928	0.914	0.962	0.947

the superior image quality evaluation capabilities of our model compared to the baseline, with the predicted scores exhibiting closer alignment with the Ground Truth. In short, these phenomena serve as compelling evidence of the effectiveness of our proposed method.

4.7 Analysis of the Number of Class Tokens

We conduct experiments on the impact of different numbers of class tokens on TID2013 and CSIQ, and the results are shown in Table 4. As intuition suggests, using only one class token is not enough. This is easy to explain. We consider multiple class tokens as multiple different reviews, and increasing the number of reviews can reduce the randomness of predictions and make the results more representative. Therefore, multiple class tokens are necessary. But it's not that more class tokens are better. From Table 4, it can be seen that the performance of the model increases as the number of class tokens increases, reaching its best at $M = 6$, and then decreasing.

5 Conclusion

In this paper, we propose a novel NR-IQA method named CDMT-IQA, which enhances the representation capabilities for image quality through knowledge distillation across datasets. Specifically, we introduce an additional distillation token to facilitate student learning from the teacher and enable knowledge integration across datasets. Furthermore, we consider the class token as abstractions of quality-aware features and simulate multiple reviews by increasing the number of class tokens. We design an Attention Scoring mechanism to refine the output of each class token which helps reduce prediction uncertainty. Our experiments conducted on six IQA datasets have demonstrated the superiority of the proposed method.

Acknowledgement. This work was supported by National Key R&D Program of China (No. 2022ZD0118202), the National Science Fund for Distinguished Young Scholars (No. 62025603), the National Natural Science Foundation of China (No. U21B2037, No. U22B2051, No. 62176222, No. 62176223, No. 62176226, No. 62072386, No. 62072387, No. 62072389, No. 62002305 and No. 62272401), and the Natural Science Foundation of Fujian Province of China (No. 2021J01002, No. 2022J06001).

References

1. Su, S., et al.: Blindly assess image quality in the wild guided by a self-adaptive hyper network. In: CVPR, pp. 3667–3676 (2020)
2. Golestaneh, S.A., Dadsetan, S., Kitani, K.M.: No-reference image quality assessment via transformers, relative ranking, and self-consistency. In: Proceedings of the IEEE/CVF Winter Conference on Applications of Computer Vision, pp. 1220–1230 (2022)
3. Ke, J., Wang, Q., Wang, Y., Milanfar, P., Yang, F.: MUSIQ: multi-scale image quality transformer. In: CVPR, pp. 5148–5157 (2021)
4. Qin, G., et al.: Data-efficient image quality assessment with attention-panel decoder. In: Proceedings of the Thirty-Seventh AAAI Conference on Artificial Intelligence (2023)
5. Zheng, H., Yang, H., Fu, J., Zha, Z.-J., Luo, J.: Learning conditional knowledge distillation for degraded-reference image quality assessment. In: CVPR, pp. 10242–10251 (2021)

6. Yin, G., Wang, W., Yuan, Z., Han, C., Ji, W., Sun, S., Wang, C.: Content-variant reference image quality assessment via knowledge distillation. In: Proceedings of the AAAI Conference on Artificial Intelligence, vol. 36, pp. 3134–3142 (2022)
7. Kang, L., Ye, P., Li, Y., Doermann, D.: Convolutional neural networks for no-reference image quality assessment. In: CVPR, pp. 1733–1740 (2014)
8. Bosse, S., Maniry, D., Müller, K.-R., Wiegand, T., Samek, W.: Deep neural networks for no-reference and full-reference image quality assessment. IEEE Trans. Image Process. **27**(1), 206–219 (2017)
9. Liu, X., van de Weijer, J., Bagdanov, A.D.: RankIQA: learning from rankings for no-reference image quality assessment. In: ICCV (2017)
10. Dosovitskiy, A., et al.: An image is worth 16×16 words: transformers for image recognition at scale. In: ICLR (2021)
11. You, J., Korhonen, J.: Transformer for image quality assessment. In: ICIP, pp. 1389–1393. IEEE (2021)
12. Ke, J., Wang, Q., Wang, Y., Milanfar, P., Yang, F.: MUSIQ: multi-scale image quality transformer. In: Proceedings of the IEEE/CVF International Conference on Computer Vision, pp. 5148–5157 (2021)
13. Sheikh, H.R., Sabir, M.F., Bovik, A.C.: A statistical evaluation of recent full reference image quality assessment algorithms. IEEE Trans. Image Process. **15**(11), 3440–3451 (2006)
14. Larson, E.C., Chandler, D.M.: Most apparent distortion: full-reference image quality assessment and the role of strategy. J. Electron. Imaging **19**(1), 011006 (2010)
15. Ponomarenko, N., et al.: Image database TID2013: peculiarities, results and perspectives. Signal Process.: Image Commun. **30**, 57–77 (2015)
16. Lin, H., Hosu, V., Saupe, D.: KADID-10k: a large-scale artificially distorted IQA database. In: 2019 Eleventh International Conference on Quality of Multimedia Experience (QoMEX), pp. 1–3. IEEE (2019)
17. Ghadiyaram, D., Bovik, A.C.: Massive online crowdsourced study of subjective and objective picture quality. IEEE Trans. Image Process. **25**(1), 372–387 (2015)
18. Fang, Y., Zhu, H., Zeng, Y., Ma, K., Wang, Z.: Perceptual quality assessment of smartphone photography. In: Proceedings of the IEEE/CVF Conference on Computer Vision and Pattern Recognition, pp. 3677–3686 (2020)
19. Saad, M.A., Bovik, A.C., Charrier, C.: Blind image quality assessment: a natural scene statistics approach in the DCT domain. IEEE Trans. Image Process. **21**(8), 3339–3352 (2012)
20. Mittal, A., Moorthy, A.K., Bovik, A.C.: No-reference image quality assessment in the spatial domain. IEEE Trans. Image Process. **21**(12), 4695–4708 (2012)
21. Zhang, L., Zhang, L., Bovik, A.C.: A feature-enriched completely blind image quality evaluator. IEEE Trans. Image Process. **24**(8), 2579–2591 (2015)
22. Zhang, W., Ma, K., Yan, J., Deng, D., Wang, Z.: Blind image quality assessment using a deep bilinear convolutional neural network. IEEE Trans. Circuits Syst. Video Technol. **30**(1), 36–47 (2018)
23. Ying, Z., Niu, H., Gupta, P., Mahajan, D., Ghadiyaram, D., Bovik, A.: From patches to pictures (PaQ-2-PiQ): mapping the perceptual space of picture quality. In: CVPR, pp. 3575–3585 (2020)
24. Pan, Z., et al.: DACNN: blind image quality assessment via a distortion-aware convolutional neural network. IEEE Trans. Circuits Syst. Video Technol. **32**(11), 7518–7531 (2022)
25. Touvron, H., Cord, M., Jégou, H.: Deit III: revenge of the VIT. arXiv preprint arXiv:2204.07118 (2022)

Quality-Aware CLIP for Blind Image Quality Assessment

Wensheng Pan[1,2], Zhifu Yang[2], DingMing Liu[2], Chenxin Fang[2],
Yan Zhang[1,3(✉)], and Pingyang Dai[1,2]

[1] Key Laboratory of Multimedia Trusted Perception and Efficient Computing,
Ministry of Education of China, Xiamen University,
Xiamen 361005, People's Republic of China
bzhy986@gmail.com
[2] School of Informatics, Xiamen University,
Xiamen 361005, People's Republic of China
[3] Institute of Artificial Intelligence, Xiamen University,
Xiamen 361005, People's Republic of China

Abstract. Blind Image Quality Assessment (BIQA) aims to simulate
human perception of image quality without reference images. Pretrained
visual-linguistic models, like CLIP, have shown excellent performance in
various visual tasks and have been successfully applied in BIQA. However, existing CLIP-based approaches typically employ a coarse classification method, dividing images into two or five quality levels based on
CLIP's text-image comparison ability. In this work, we propose a novel
approach for BIQA that introduces a fine-grained quality-level stratification strategy. This strategy enables a more precise assessment of
image quality across a wider range of levels. Additionally, we present
a two-stage training model called Quality-Aware CLIP (QA-CLIP). In
the first stage, we leverage a set of learnable text tokens to optimize
the text description and fully utilize the representation capabilities of
CLIP's text encoder. In the second stage, we further optimize the image
encoder and quality-aware block to capture features that are highly relevant to perceived quality. Experimental results demonstrate that QA-
CLIP achieves comparable performance with state-of-the-art methods
on various synthetic and real datasets. Notably, in CSIQ, TID2013, and
KADID datasets, QA-CLIP outperforms the state-of-the-art by 1.2%,
4.7%, and 4.8% respectively in terms of Spearman Rank Correlation
Coefficient (SRCC).

Keywords: Blind Image Quality Assessment · CLIP ·
Quality-Aware · Learnable Text Tokens

1 Introduction

The importance of digital images has been increasing in fields such as medical
imaging, multimedia communication, and computer vision. Therefore, accurate

© The Author(s), under exclusive license to Springer Nature Singapore Pte Ltd. 2024
Q. Liu et al. (Eds.): PRCV 2023, LNCS 14430, pp. 396–408, 2024.
https://doi.org/10.1007/978-981-99-8537-1_32

evaluation of image quality has become crucial. Image Quality Assessment (IQA) aims to objectively measure image quality by simulating human visual perception. Recently, research has focused on Blind Image Quality Assessment (BIQA) due to the unavailability of reference images.

Conventional BIQA approaches rely on hand-engineered features, such as natural scene statistics (NSS), or shallow feature learning using codebooks. In recent years, deep learning-based methods have gained popularity for their ability to extract intricate image perceptual features. These methods employ a feature extractor to extract image features and then predict the image quality score.

OpenAI introduced the CLIP model [14], a self-supervised pretraining framework that encodes images and text into a shared vector space without requiring annotated data. CLIP demonstrates strong zero-shot transfer learning capabilities across various tasks. Wang et al. [18] was the first to explore CLIP's potential in challenging image perception evaluation tasks. They proposed a prompt pairing strategy using antonym prompts (e.g., "Good photo." and "Bad photo.") to reduce prompt ambiguity. Building on this work, Zhang et al. [23] proposed multi-task learning to fine-tune CLIP by combining image quality assessment with distortion-type classification and scene classification tasks. They employed a Likert-scale with five quality levels, and the model outputs scores as logit-weighted sums of the quality levels. However, this multi-task training approach incurs additional annotation costs, and the coarse division of quality levels presents performance bottlenecks.

To address this issue, we propose a fine-grained quality-level stratification strategy that expands the quality levels to 100. Each quality level is mapped to a specific range of score values. For example, images with scores in the range of [49,50) are categorized as level [50], corresponding to class [50]. By leveraging the CLIP model and training on text-image pairs, our aim is to develop a feature representation model that is highly correlated with image quality.

In existing methods based on Vision Transformer (ViT) for BIQA, the CLS token is typically considered to contain aggregated perceptual information for image quality. However, directly utilizing the CLS token for optimal representation is challenging. Qin et al. [13] proposed a Quality-Aware Decoder to further explain the CLS token and extract features that closely align with image quality. Building on the work of Qin, we propose a Quality-Aware Block (QAB) that consists of a Quality-Aware Decoder and a Feature Fusion Module. The Feature Fusion Module combines features from multiple perspectives to reduce uncertainty derived from single perspectives.

Additionally, we present a two-stage training model called Quality-Aware CLIP (QA-CLIP). In the first stage, inspired by Li et al. [7], we introduce a set of learnable text tokens to describe quality levels. Only the learnable tokens are optimized, while other module parameters are frozen. This allows us to fully leverage the representation ability of the CLIP text encoder and learn the optimal text description. In the second stage, we freeze the learnable tokens and the text encoder and update the parameters of the image encoder and Quality-

Aware Block. This enables us to learn features that are highly relevant to image perception and quality.

In summary, our contributions are as follows: 1) We propose a fine-grained quality-level stratification strategy for BIQA, enabling the learning of features that are more closely related to image quality. 2) We present a two-stage training model called QA-CLIP. In our model, a set of learnable text tokens is incorporated to leverage the representation ability of the text encoder. Inspired by Qin et al. [13], we propose the QAB module to evaluate image quality from multiple perspectives and extract deep features purely related to quality levels. 3) Experimental results demonstrate that QA-CLIP outperforms state-of-the-art methods on both synthetic and natural datasets, showcasing the significant potential of CLIP-based models in the field of BIQA.

2   Related Work

2.1   BIQA

Conventional BIQA methods traditionally relied on hand-engineered features like natural scene statistics (NSS) or shallow feature learning through codebooks. However, in recent years, there has been a surge in the popularity of deep learning-based approaches that excel at extracting complex image perceptual features. These methods typically employ a feature extractor to capture image features, which are then used to predict the image quality score. For instance, Su et al. [16] developed a CNN-based IQA model that mimics the top-down perception mechanism of the Human Visual System (HVS) by splitting the IQA task into two sequential subtasks based on the characterization of image semantics. Deep learning-based approaches have demonstrated their effectiveness in capturing and leveraging intricate image features, surpassing the performance of traditional methods in BIQA tasks.

2.2   CLIP Applications

CLIP is a pre-training method that leverages contrastive learning of text-image pairs and has shown remarkable transfer learning capabilities across various visual tasks. Initially, Radford et al. [14] collected 400 million text-image pairs from the internet to pre-train the CLIP model using contrastive learning. Building upon CLIP's success, Zhou et al. [24] introduced the CoOp model, which further improved transfer learning by incorporating prompt tuning, and replacing fixed prompts with dynamically adjustable ones. Shortly after its inception, CLIP was also applied to semantic segmentation and object detection tasks.

The work closest to ours is Wang et al. [18] and Zhang et al. [23]. Wang et al. [18]demonstrated that CLIP can be directly applied to visual perception assessment without task-specific fine-tuning. They employed a prompt pairing strategy, utilizing antonym prompts in pairs (e.g., "Good photo." and "Bad photo.") to guide the evaluation. Zhang et al. [23] proposed fine-tuning the CLIP

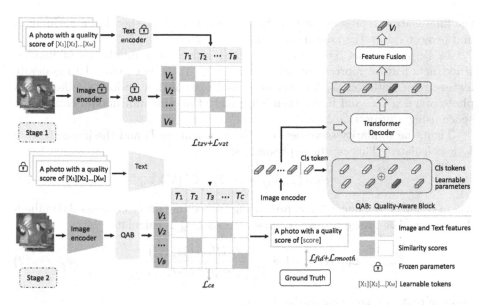

Fig. 1. The overview of QA-CLIP. In the first stage, only a set of learnable tokens are optimized, while other module parameters are frozen. In the second stage, we optimize the image encoder and Quality-Aware Block.

model using multi-task learning, incorporating the image quality assessment task with distortion-type classification and scene classification tasks. They employed a Likert-scale with five quality levels, and the model outputs scores as logit-weighted sums of the quality levels. In contrast to the aforementioned works, we propose a fine-grained quality-level stratification strategy that expands the quality levels to 100. Moreover, we employ a two-stage training strategy that leverages the representation ability of the text encoder.

3   Method

3.1   Overview of CLIP

We provide a brief introduction to vision-language pre-training, focusing on the CLIP model. The CLIP model comprises a text encoder $\mathcal{T}: \mathbb{T} \mapsto \mathbb{R}^K$ and an image encoder $\mathcal{V}: \mathbb{R}^N \mapsto \mathbb{R}^K$. The text encoder $\mathcal{T}$ generates feature representations of natural language text descriptions. On the other hand, the image encoder $\mathcal{V}$ extracts deep semantic features from images using architectures like ResNet-50 or ViT. To encode text descriptions, each word in a given description (e.g., "A photo of a [class]") is first converted into a lowercase byte pair encoding (BPE) representation. This representation assigns a unique numeric ID from a vocabulary of length 49152. To ensure computational efficiency, each text sequence has a fixed length of 77 and includes start [SOS] and end [EOS] tokens. After passing through the transformer layers, the [EOS] token serves as the final feature representation of the encoded text. It is then layer normalized and projected into the multimodal embedding space.

To be specific, let $D = \{x_1, x_2, ..., x_B\}$ represent a minibatch of images. $\mathcal{V}$ extracts feature representations V_i for image x_i. $\mathcal{T}$ is employed to generate feature representations T_i for corresponding textual descriptions similar to "A photo of a dog". T_i and V_i are then mapped to the same multimodal embedding space using a projection layer.

Then, the similarity score between the text feature T_i and the image feature V_i is calculated. This can be expressed as:

$$Sim\,(T_i, V_i) = G\,(T_i) \cdot G'\,(V_i) \tag{1}$$

where G and G' are linear layers projecting embedding into a cross-modal embedding space. The image-to-text contrastive loss $\mathcal{L}_{v2t}$ is calculated as:

$$\mathcal{L}_{v2t}\,(i) = -\log \frac{exp\,(Sim\,(T_i, V_i))}{\sum_{j=1}^{B} exp\,(Sim\,(T_j, V_i))} \tag{2}$$

and the text-to-image contrastive loss $\mathcal{L}_{t2v}$:

$$\mathcal{L}_{t2v}\,(i) = -\log \frac{exp\,(Sim\,(T_i, V_i))}{\sum_{j=1}^{B} exp\,(Sim\,(T_i, V_j))} \tag{3}$$

3.2   QA-CLIP

Preliminaries. Given an image $x_i \in \mathbb{R}^N$ that may undergo several stages of degradations, the goal of a BIQA model $\hat{q}: \mathbb{R}^N \mapsto \mathbb{R}$ is to predict the perceptual quality of x_i, close to its Mean Opinion Score (MOS), denoted as $q_i \in \mathbb{R}$.

Stage 1. QA-CLIP is illustrated in Fig. 1. LIQE [23] employed a Likert-scale with five quality levels, and the model outputs scores as logit-weighted sums of the quality levels. However, for BIQA, dividing the quality of images into five levels corresponding to the scores lacks the desired precision for more nuanced applications. Differently, we propose a fine-grained quality-level stratification strategy that expands the quality levels to 100. Each quality level is mapped to a specific range of score values. For example, images with scores in the range of [49,50) are categorized as level [50], corresponding to class [50]. As a result, each image x_i is assigned a quality level label I_i and a MOS label q_i. The final predicted score of the model for an image can be represented as:

$$\hat{q}\,(x) = \sum_{c=1}^{C} P\,(c|x) \times c \tag{4}$$

where $P\,(c|x)$ represents the class probabilities after applying softmax, and C represents the total number of classes. However, it is difficult to exploit CLIP in our approach where the classes are numbers instead of specific text.

To address this issue, we introduce a set of learnable text tokens for each quality level. Specifically, we design the textual input for the text encoder as follows: "A photo with a quality score of $[X]_1[X]_2[X]_3...[X]_M$". M represents the number of learnable tokens. In the first stage of training, we only train the learnable text tokens while freezing the parameters of the remaining modules. The loss function used in the training process is similar to the CLIP model, but there may be multiple images belonging to the same score interval in a minibatch, which means T_i may have multiple positive samples. Therefore, the loss is modified as follows:

$$\mathcal{L}_{t2v}(i) = -\frac{1}{|A|}\sum_{a \in A} \log \frac{exp\left(Sim\left(T_i, V_a\right)\right)}{\sum_{j=1}^{B} exp\left(Sim\left(T_i, V_j\right)\right)} \tag{5}$$

where A represents the set of all images within a minibatch that share the same class as image x_i. The overall loss function is represented as:

$$\mathcal{L}_{Stage1} = \mathcal{L}_{t2v} + \mathcal{L}_{v2t} \tag{6}$$

Stage 2. In the previous training stage, QA-CLIP learned text descriptions that correspond to quality levels. In this subsequent stage, we keep the text descriptions input to the text encoder and the parameters of the text encoder fixed and focus on optimizing the image encoder and Quality-Aware Block. To enhance the overall performance, we adopt a pairwise learning-to-rank model estimation approach for BIQA, following the methodology introduced in LIQE. Specifically, we transform the training set, which includes MOS q_x, into another set where the input consists of pairs of images (x_1, x_2), and the target $p(x_1, x_2)$ is a binary label.

$$p(x_1, x_2) = \begin{cases} 1 & \text{if } q_{x_1} \geq q_{x_2} \\ 0 & \text{otherwise} \end{cases} \tag{7}$$

Following Thurstone's Case V model, we make the assumption that the true perceptual quality of the image follows a Gaussian distribution with mean estimated by $\hat{q}$. Based on this assumption, we can express the probability that the quality of x_1 is higher than x_2 as:

$$\hat{p}(x_1, x_2) = \Phi\left(\frac{\hat{q}(x_1) - \hat{q}(x_2)}{\sqrt{2}}\right) \tag{8}$$

Φ represents the standard normal cumulative distribution function with a variance of 1. In order to minimize the difference between two discrete probability distributions, we employ the fidelity loss [17].

$$\mathcal{L}_{fid}(p, \hat{p}) = 1 - \sqrt{p\hat{p}} - \sqrt{(1-p)(1-\hat{p})} \tag{9}$$

Additionally, we utilize the Smooth L1 Loss $\mathcal{L}_{smooth}$ and Cross-entropy loss with label smoothing $\mathcal{L}_{ce}$ for the purpose of optimization. $\mathcal{L}_{ce}$ is denoted as:

$$\mathcal{L}_{ce} = \sum_{k=1}^{C} -I_k' \log P_k \tag{10}$$

where $I'_k = (1 - \varepsilon) I_k + \varepsilon/C$ denotes value in the quality level target distribution, and P_k represents prediction logits of class k. In summary, the loss used in our second training stage is summarized as:

$$\mathcal{L}_{Stage2} = \mathcal{L}_{fid} + \alpha\mathcal{L}_{smooth} + \beta\mathcal{L}_{ce} \tag{11}$$

where α and β are the coefficients balancing $\mathcal{L}_{smooth}$ and $\mathcal{L}_{ce}$.

Quality-Aware Block. Inspired by the Attention-Panel Mechanism proposed in [13], we propose a Quality-Aware Block to generate multi-view subjective opinions for images, as shown in Fig. 1. Differently, we utilize a feature fusion module to obtain an average feature embedding from multiple feature embeddings output by the transformer decoder. Specifically, the image encoder generates feature representations for image x_i, denoted as $Z_i = \{F_{cls}; F_1, F_2, F_3, ..., F_P\} \in \mathbb{R}^{(P+1)\times d}$, where P is the number of patches. Subsequently, we then slice Z_i to obtain $Z' = \{F_1, F_2, F_3, ..., F_P\} \in \mathbb{R}^{P\times d}$. Let L be the number of views, and we create attention-panel embeddings $J = \{J_1, ..., J_L\} \in \mathbb{R}^{L\times d}$. Additionally, we expand the CLS token L times to obtain $T = \{F_{cls}, ..., F_{cls}\} \in \mathbb{R}^{L\times d}$.

We use $(J + T)$ and Z' as inputs to the transformer decoder to obtain the multiple feature embeddings output $S = \{S_1, S_2, S_3, ..., S_L\} \in \mathbb{R}^{L\times d}$. S is then fed into the feature fusion module to obtain the final integrated multi-view image feature V_i. Specifically, we perform averaging on L features denoted as:

$$V_i = \frac{1}{L}\sum_{j=1}^{L} S_j \tag{12}$$

4 Experiments

4.1 Datasets and Evaluation Protocols

We conduct a comprehensive evaluation of our proposed model using six widely recognized image quality evaluation datasets. These datasets consist of four authentic datasets and two synthetic datasets. The authentic datasets we use are LIVE [15], CSIQ [6], TID2013 [12], and KADID [8]. The synthetic datasets are LIVEC [2] and KonIQ [4]. To assess the performance of our model, we employ Pearson's Linear Correlation Coefficient (PLCC) and Spearman's Rank order Correlation Coefficient (SRCC) as evaluation metrics. PLCC measures the accuracy of our model's predictions, while SRCC evaluates the monotonicity of the BIQA algorithm predictions. Both metrics range from 0 to 1, where higher values indicate better performance in terms of prediction accuracy and monotonicity.

4.2 Implementation Details

Table 1. Performance comparison is measured by averages of PLCC and SRCC, where bold entries indicate the best results.

Dataset	LIVE	CSIQ	TID2013	KADID	LIVEC	KonIQ
Criterion	PLCC					
BRISQUE [10]	0.944	0.748	0.571	0.567	0.629	0.685
ILNIQE [21]	0.906	0.865	0.648	0.558	0.508	0.537
MEON [9]	0.955	0.864	0.824	0.691	0.710	0.628
DBCNN [22]	0.971	0.959	0.865	0.856	0.869	0.884
TIQA [20]	0.965	0.838	0.858	0.855	0.861	0.903
MetaIQA [25]	0.959	0.908	0.868	0.775	0.802	0.856
P2P-BM [19]	0.958	0.902	0.856	0.849	0.842	0.885
HyperIQA [16]	0.966	0.942	0.858	0.845	0.882	0.917
TReS [3]	0.968	0.942	0.883	0.858	0.877	0.928
MUSIQ [5]	0.911	0.893	0.815	0.872	0.746	0.928
DACNN [11]	0.980	0.957	0.889	0.905	0.884	0.912
DEIQT [13]	**0.982**	0.963	0.908	0.887	0.894	**0.934**
LIQE [23]	0.951	0.939	–	0.931	**0.910**	0.908
ours	0.978	**0.970**	**0.944**	**0.981**	0.898	0.873
Criterion	SRCC					
BRISQUE [10]	0.929	0.812	0.626	0.528	0.629	0.681
ILNIQE [21]	0.902	0.822	0.521	0.534	0.508	0.523
MEON [9]	0.951	0.852	0.808	0.604	0.697	0.611
DBCNN [22]	0.968	0.946	0.816	0.851	0.851	0.875
TIQA [20]	0.949	0.825	0.846	0.850	0.845	0.892
MetaIQA [25]	0.960	0.899	0.856	0.762	0.835	0.887
P2P-BM [19]	0.959	0.899	0.862	0.840	0.844	0.872
HyperIQA [16]	0.962	0.923	0.840	0.852	0.859	0.906
TReS [3]	0.969	0.922	0.863	0.859	0.846	0.915
MUSIQ [5]	0.940	0.871	0.773	0.875	0.702	0.916
DACNN [11]	0.978	0.943	0.871	0.905	0.866	0.901
DEIQT [13]	**0.980**	0.946	0.892	0.889	0.875	**0.921**
LIQE [23]	0.970	0.936	–	0.930	**0.904**	0.919
ours	0.978	**0.958**	**0.939**	**0.978**	0.869	0.873

Model. The backbone of our image and text feature extractor are the image encoder $\mathcal{V}$ and the text encoder $\mathcal{T}$ from CLIP, respectively. We specifically employ the ViT-B/16 architecture for the image encoder, which comprises 12 transformer layers with a hidden size of 768 dimensions. To align with the output of $\mathcal{T}$, the dimension of the image feature vector is reduced from 768 to 512 using a linear layer. In our Quality-Aware Block, we employ a single-layer transformer decoder.

Training Details. In the first training stage, we employ the Adam optimizer with an initial learning rate of 3×10^{-5}, which is decayed using a cosine schedule. The training process includes a warm-up period of 5 epochs, during which random cropping is applied as data augmentation. Each original image is randomly cropped into 8 images of size 224×224. In this stage, only the learnable tokens $[X]_1[X]_2[X]_3...[X]_M$ are optimized. In the second training stage, we also use the Adam optimizer, but with a warm-up period of 10 epochs. The learning rate is linearly increased from 9.5×10^{-7} to 5×10^{-6}. At the 30th and 50th epochs, the learning rate is decreased by a factor of 0.1. Random horizontal flipping and random cropping are applied as data augmentation during this stage. And we set the coefficient α to 0.001 and β to 0.1 for $\mathcal{L}_{Stage2}$. Both training stages consist of 60 epochs.

For the LIVE and CSIQ datasets, the batch size is set to 32, and 64 for other datasets. We split the data into 80% for training and 20% for testing. To mitigate performance bias, we repeat each experiment 10 times and calculate the average PLCC and SRCC.

4.3   Overall Prediction Performance Comparison

Table 1 reports the comparison results between the proposed QA-CLIP and 13 state-of-the-art BIQA methods, including both hand-crafted methods such as ILNIQE [21] and BRISQUE [10], as well as deep learning-based methods such as MUSIQ [5], MetaIQA [25], DEIQT [13], and LIQE [23] (based on CLIP). Across all datasets, QA-CLIP demonstrates competitive performance compared to the state-of-the-art methods. Notably, on the CSIQ, TID2013, and KADID datasets, QA-CLIP achieves state-of-the-art performance, surpassing existing methods by improving the PLCC metric by 0.7%, 3.6%, and 5.0% respectively, and the SRCC metric by 1.2%, 4.7%, and 4.8% respectively. These results highlight the effectiveness and leading-edge performance of QA-CLIP in image quality assessment.

4.4   Generalization Capability Validation

We assess the generalization performance of QA-CLIP using cross-dataset validation. Specifically, we train a BIQA model on one dataset and directly evaluate it on another dataset without fine-tuning or adjusting parameters. The results of the comparisons with DBCNN, P2P-BM, HyperIQA, TRes, and DEIQT are shown in Table 2. We utilize four datasets and report the median experimental results. QA-CLIP exhibits superior performance compared to other methods on the KonIQ and CSIQ datasets and remains competitive on the remaining datasets. These experimental findings highlight the robust generalization performance of QA-CLIP.

Table 2. SRCC on the cross datasets validation. The best performances are highlighted with boldface.

Training	LIVEC	KonIQ	LIVE	CSIQ
Testing	KonIQ	LIVEC	CSIQ	LIVE
DBCNN	0.754	0.755	0.758	0.877
P2P-BM	0.740	0.770	0.712	–
HyperIQA	0.772	0.785	0.744	0.926
TReS	0.733	0.786	0.761	–
DEIQT	0.744	**0.794**	0.781	**0.932**
ours	**0.800**	0.772	**0.789**	0.925

4.5   Ablation Study

Necessity of Learnable Tokens. Our textual description is "A photo with a quality score represented by $[X]_1[X]_2[X]_3...[X]_M$". Without learnable tokens, the text encoder's representation capabilities could not be fully utilized. To validate the necessity of learnable tokens, we attempt to remove them. Table 5 demonstrates a significant performance decline compared to QA-CLIP.

Necessity of Quality-Aware Block. To validate the effectiveness of the Quality-Aware Block (QAB), we attempt to replace the QAB with other structures: (1) nothing (no additional structure), and (2) Adaptor in [1], which includes two linear layers followed by ReLU activation layers. The experimental results in Table 3 demonstrate that incorporating QAB yields significantly better results compared to the other experimental configurations.

Table 3. Ablation study on each component of QA-CLIP

Variants	TID2013		KADID	
	PLCC	SRCC	PLCC	SRCC
w/o Learnable Tokens	0.743	0.787	0.712	0.759
Nothing Instead QAB	0.916	0.905	0.974	0.972
Adaptor Instead QAB	0.936	0.926	0.976	0.975
QA-CLIP	**0.944**	**0.939**	**0.981**	**0.978**

Effect of Each Loss. Table 4 illustrates the impact of each loss function on the performance of QA-CLIP on the TID2013 dataset. Without utilizing fidelity loss, the performance experiences a respective decrease of 1.1% and 1.4%. Similarly, the absence of smooth L1 loss results in performance reductions of 0.9% and

1.6%, respectively. Furthermore, omitting cross-entropy loss with label smoothing leads to performance decreases of 0.5% and 0.9%, respectively.

Table 4. Ablation study on each loss on the TID2013 dataset

$\mathcal{L}_{fid}$	$\mathcal{L}_{smooth}$	$\mathcal{L}_{ce}$	PLCC	SRCC
✓	✓		0.939	0.930
✓		✓	0.935	0.923
	✓	✓	0.933	0.925
✓	✓	✓	**0.944**	**0.939**

Number of Learnable Tokens M. We conduct an analysis on the parameter M in Table 5. Interestingly, increasing the number of tokens to 6 or 8 does not result in significant performance improvements. Instead, there is only a slight increase in performance on LIVE. Ultimately, we choose M = 4 as it yields the largest performance improvement across different settings.

Table 5. Ablation study on the number of learnable tokens

Number	LIVE		CSIQ	
	PLCC	SRCC	PLCC	SRCC
2	0.974	0.972	0.959	0.945
4	**0.978**	**0.978**	**0.970**	0.958
6	0.976	0.976	0.962	0.950
8	0.972	0.973	**0.970**	**0.962**

5   Conclusion

To better leverage CLIP's cross-modal representation abilities, we propose Quality-Aware CLIP with two training stages. In contrast to previous approaches that coarsely divide the quality levels of BIQA into two or five categories, we introduce a fine-grained quality-level stratification strategy that expands the number of quality levels to 100. In the first stage, we utilize learnable text tokens to optimize the text descriptions and fully exploit the representation capabilities of CLIP's text encoder. This stage focuses on refining the textual representations associated with quality levels. In the second stage, we optimize the image encoder and quality-aware block to capture image features that are highly relevant to perceived quality. Experiments on various synthetic and real-world datasets demonstrate the effectiveness and superiority of our proposed model.

Acknowledgement. This work was supported by National Key R&D Program of China (No. 2022ZD0118202), the National Science Fund for Distinguished Young Scholars (No. 62025603), the National Natural Science Foundation of China (No. U21B2037, No. U22B2051, No. 62176222, No. 62176223, No. 62176226, No. 62072386, No. 62072387, No. 62072389, No. 62002305 and No. 62272401), and the Natural Science Foundation of Fujian Province of China (No. 2021J01002, No. 2022J06001).

References

1. Gao, P., et al.: CLIP-adapter: better vision-language models with feature adapters. arXiv preprint arXiv:2110.04544 (2021)
2. Ghadiyaram, D., Bovik, A.C.: Massive online crowdsourced study of subjective and objective picture quality. IEEE Trans. Image Process. **25**(1), 372–387 (2015)
3. Golestaneh, S.A., Dadsetan, S., Kitani, K.M.: No-reference image quality assessment via transformers, relative ranking, and self-consistency. In: Proceedings of the IEEE/CVF Winter Conference on Applications of Computer Vision, pp. 1220–1230 (2022)
4. Hosu, V., Lin, H., Sziranyi, T., Saupe, D.: KonIQ-10k: an ecologically valid database for deep learning of blind image quality assessment. IEEE Trans. Image Process. **29**, 4041–4056 (2020)
5. Ke, J., Wang, Q., Wang, Y., Milanfar, P., Yang, F.: MUSIQ: multi-scale image quality transformer. In: Proceedings of the IEEE/CVF International Conference on Computer Vision, pp. 5148–5157 (2021)
6. Larson, E.C., Chandler, D.M.: Most apparent distortion: full-reference image quality assessment and the role of strategy. J. Electron. Imaging **19**(1), 011006 (2010)
7. Li, S., Sun, L., Li, Q.: CLIP-ReID: exploiting vision-language model for image re-identification without concrete text labels. In: Proceedings of the AAAI Conference on Artificial Intelligence, vol. 37, pp. 1405–1413 (2023)
8. Lin, H., Hosu, V., Saupe, D.: KADID-10k: a large-scale artificially distorted IQA database. In: 2019 Eleventh International Conference on Quality of Multimedia Experience (QoMEX), pp. 1–3. IEEE (2019)
9. Ma, K., Liu, W., Zhang, K., Duanmu, Z., Wang, Z., Zuo, W.: End-to-end blind image quality assessment using deep neural networks. IEEE Trans. Image Process. **27**(3), 1202–1213 (2017)
10. Mittal, A., Moorthy, A.K., Bovik, A.C.: No-reference image quality assessment in the spatial domain. IEEE Trans. Image Process. **21**(12), 4695–4708 (2012)
11. Pan, Z., et al.: DACNN: blind image quality assessment via a distortion-aware convolutional neural network. IEEE Trans. Circuits Syst. Video Technol. **32**(11), 7518–7531 (2022)
12. Ponomarenko, N., et al.: Image database TID2013: peculiarities, results and perspectives. Sig. Process. Image Commun. **30**, 57–77 (2015)
13. Qin, G., et al.: Data-efficient image quality assessment with attention-panel decoder. arXiv preprint arXiv:2304.04952 (2023)
14. Radford, A., et al.: Learning transferable visual models from natural language supervision. In: International Conference on Machine Learning, pp. 8748–8763. PMLR (2021)
15. Sheikh, H.R., Sabir, M.F., Bovik, A.C.: A statistical evaluation of recent full reference image quality assessment algorithms. IEEE Trans. Image Process. **15**(11), 3440–3451 (2006)

16. Su, S., et al.: Blindly assess image quality in the wild guided by a self-adaptive hyper network. In: Proceedings of the IEEE/CVF Conference on Computer Vision and Pattern Recognition, pp. 3667–3676 (2020)

17. Tsai, M.F., Liu, T.Y., Qin, T., Chen, H.H., Ma, W.Y.: FRank: a ranking method with fidelity loss. In: Proceedings of the 30th Annual International ACM SIGIR Conference on Research and Development in Information Retrieval, pp. 383–390 (2007)

18. Wang, J., Chan, K.C., Loy, C.C.: Exploring clip for assessing the look and feel of images. arXiv preprint arXiv:2207.12396 (2022)

19. Ying, Z., Niu, H., Gupta, P., Mahajan, D., Ghadiyaram, D., Bovik, A.: From patches to pictures (PaQ-2-PiQ): mapping the perceptual space of picture quality. In: Proceedings of the IEEE/CVF Conference on Computer Vision and Pattern Recognition, pp. 3575–3585 (2020)

20. You, J., Korhonen, J.: Transformer for image quality assessment. In: 2021 IEEE International Conference on Image Processing (ICIP), pp. 1389–1393. IEEE (2021)

21. Zhang, L., Zhang, L., Bovik, A.C.: A feature-enriched completely blind image quality evaluator. IEEE Trans. Image Process. **24**(8), 2579–2591 (2015)

22. Zhang, W., Ma, K., Yan, J., Deng, D., Wang, Z.: Blind image quality assessment using a deep bilinear convolutional neural network. IEEE Trans. Circuits Syst. Video Technol. **30**(1), 36–47 (2018)

23. Zhang, W., Zhai, G., Wei, Y., Yang, X., Ma, K.: Blind image quality assessment via vision-language correspondence: A multitask learning perspective. In: Proceedings of the IEEE/CVF Conference on Computer Vision and Pattern Recognition, pp. 14071–14081 (2023)

24. Zhou, K., Yang, J., Loy, C.C., Liu, Z.: Learning to prompt for vision-language models. Int. J. Comput. Vision **130**(9), 2337–2348 (2022)

25. Zhu, H., Li, L., Wu, J., Dong, W., Shi, G.: MetaIQA: deep meta-learning for no-reference image quality assessment. In: Proceedings of the IEEE/CVF Conference on Computer Vision and Pattern Recognition, pp. 14143–14152 (2020)

Multi-agent Perception via Co-attentive Communication Mechanism

Ning Gong[1], Zhi Li[1], Shaohui Li[1], Yuxin Ke[1], Zhizhuo Jiang[1], Yaowen Li[1], and Yu Liu[2]($\boxtimes$)

[1] Tsinghua Shenzhen International Graduate School, Tsinghua University, Shenzhen, China
{gn21,keyx22}@mails.tsinghua.edu.cn,
{zhilizl,lishaohui,jiangzhizhuo,liyw23}@sz.tsinghua.edu.cn
[2] Department of Electronics, Tsinghua University, Beijing, China
liuyu77360132@126.com

Abstract. Multi-agent collaborative perception has the potential to significantly enhance perception performance by facilitating the exchange of complementary information among agents through communication. Effective communication plays a crucial role in enabling agents to collaborate and exchange valuable information. However, traditional methods face challenges when it comes to communication scheduling. To address these issues, we propose the Co-Attentive Multi-agent Perception (CAMP) model, a learning-based framework for multi-agent collaborative perception. In CAMP, we propose a novel co-attentive scheduler to construct communication among agents from both pixel-level and feature-level perspectives. We conduct extensive experiments on a public dataset and the experimental results clearly illustrate the effectiveness of our proposed CAMP in the multi-agent perception task while using the least amount of bandwidth resources.

Keywords: Multi-Agent System · Collaborative Perception · Communication Mechanism · Transformer

1 Introduction

In the past decade, there has been significant success in single agent perception tasks using deep learning methods [1]. However, when it comes to real-world scenarios like automatic driving and unmanned aerial vehicles, systems are often comprised of multiple autonomous and interactive agents [3]. Given the complexity of these real-world environments, the traditional perception approaches designed for single agents are limited in capturing a comprehensive view of the surroundings. This limitation arises from the fact that each agent typically has limited sensing capabilities and a restricted local perspective. Recognizing this constraint, the multi-agent collaborative perception [11] recently emerges as an intriguing problem to be addressed.

© The Author(s), under exclusive license to Springer Nature Singapore Pte Ltd. 2024
Q. Liu et al. (Eds.): PRCV 2023, LNCS 14430, pp. 409–421, 2024.
https://doi.org/10.1007/978-981-99-8537-1_33

Indeed, multi-agent collaborative perception aims to overcome the limitations of single-agent approaches by facilitating agents to cooperate and share information. Then, the communication between different agents plays a pivotal role. Previous studies have predominantly focused on modeling multi-agent communication through fully-connected graphs, where agents share features or messages via broadcast communication [2,4]. However, when it comes to real-world applications, the availability of limited bandwidth renders fully-connected communication impractical. Moreover, an increase in the number of agents would result in an unacceptable surge in computing resources. To address these challenges, recent studies begin to learn dynamic communication connections between agents [8,11,12]. These efforts primarily utilize feature maps derived from multi-agent perceptions to construct a communication graph. Although they have made significant progress in addressing bandwidth limitations in multi-agent perception, the feature and pixel semantics within communication scheduling of multi-agent perceptions remains largely unexplored.

Unfortunately, achieving effective communication scheduling in multi-agent perceptions is a challenging task. On the one hand, learning an effective communication mechanism under bandwidth constraints remains an open issue. It is important but difficult to make choices in effectively selecting which agents should communicate with each other and construct a communication graph to optimize the exchange of information within limited resources. On the other hand, once the communication targets have been determined, the fusion of perceptual information from different agents presents another significant challenge. Specifically, incorporating the feature-level and pixel-level semantics of multi-agent perceptions and achieving a comprehensive understanding of environments is particularly difficult.

To tackle the aforementioned challenges, in this paper, we propose a Co-Attentive Multi-agent Perception (CAMP) model to learn a communication scheduler from both feature-level and pixel-level semantics. Specifically, CAMP consists of four modules, *i.e.*, a feature encoder, a co-attentive scheduler, a perceptive fusion module, and a downstream task decoder. To begin with, we develop a feature encoder to extract multi-agent perceptive features from the observations. Then, we propose a novel co-attentive scheduler to determine how the agents construct communication from both pixel-level and feature-level perspectives. Next, with a perceptive fusion module, we can generate fused perceptive representations after communication. Finally, we utilize a downstream task decoder to output the prediction for the given task. We conduct extensive experiments on a public dataset to evaluate the effectiveness of our proposed CAMP. Experimental results illustrate that our proposed CAMP outperforms all compared methods while using the least amount of bandwidth resources.

2 Related Works

2.1 Multi-agent Communication

Multi-agent communication refers to the process in which agents exchange information to coordinate their actions and share knowledge within a multi-agent system. Over the past few years, researchers have been dedicated to developing various methods and protocols for multi-agent communication to enhance the performance and efficiency of multi-agent systems. Previous studies [10,15,17] mainly focus on establishing communication rules by employing predefined communication protocols. For example, CommNet [16] utilizes continuous communication to accomplish fully cooperative tasks and learns communication protocols among a dynamically changing number of agents, VAIN [7] introduces attention mechanisms to determine which agents can communicate to share information. DIAL [4] uses a reinforcement learning method to derive the individual Q-value [20] of each agent based on its observation and the messages from other agents. The majority of these works rely on rule-based methods or reinforcement learning techniques, employing fully connected communication approaches. However, these approaches exhibit significant drawbacks, especially in scenarios where intelligent agents operate within dynamic environments and face bandwidth limitations. As a response to these challenges, subsequent research has progressively introduced distributed communication methods and established dynamic connections among intelligent agents to address the challenges posed by complex environments. For example, Who2com [12] employs cross-attention module to identify specific communication objects, thereby reducing bandwidth usage. When2com [11] develops a self-attention mechanism to determine the necessity of communication with other agents. Airpooling [13] adopts a pseudo-centralized approach where an intelligent agent serves as a server in the system, effectively reducing communication bandwidth. CRCN [14] incorporates channel attention to extract matching relationships between different agents, facilitating the establishment of communication connections. In this work, we focus on the critical aspect of establishing communication relationships, specifically the task of generating confidence scores between different agents. Moreover, we propose a novel co-attentive scheduler to determine how the agents communicate from both feature-level and pixel-level perspectives.

2.2 Information Fusion in Multi-agent Perception

Information fusion refers to the integration of data and knowledge from multiple information sources to provide more comprehensive and accurate information [19]. In multi-agent perception tasks, the agent needs to obtain a comprehensive understanding of the surroundings by integrating information from other agents. To that end, how to fuse the information from diverse agents is another pivotal concern in multi-agent perception. The currently effective approach for information fusion is weighted fusion using attention mechanisms. For instance, ATOC [9] applies an attention unit to determine what to broadcast

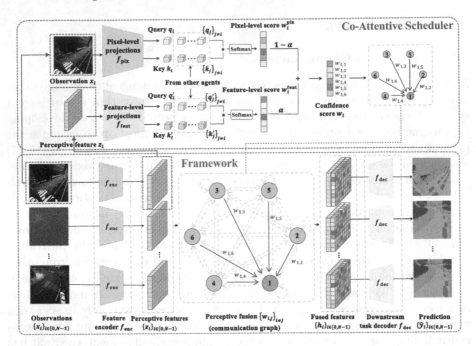

Fig. 1. Illustration of the proposed Co-Attentive Multi-agent Perception (CAMP) framework. CAMP employs a co-attentive scheduler to achieve efficient and robust communication among the agents. All agents share identical feature encoders, pixel-level projections, feature-level projections, and downstream task decoders.

and how to fuse the feature. There are also works that utilize graph neural networks (GNN [21]) for information fusion. V2VNet [5] introduces a multi-round message-passing approach based on graph neural networks, aiming to enhance perception and prediction performance. Where2comm [8] uses spatial confidence to extract spatial features from point cloud data for information fusion. In this paper, we utilize an attention-based perceptive fusion module to use the received perceptive features from other communicated agents.

3   Methods

This section introduces the Co-Attentive Multi-agent Perception (CAMP), a multi-agent collaborative perception framework that leverages co-attention for improved communication of noisy observations. As shown in Fig. 1, each agent in the proposed framework adopts several modules, *i.e.*, a feature encoder, a co-attentive scheduler, a perceptive fusion module, and a downstream task decoder. Among these modules, the co-attentive scheduler determines how the agents communicate with each other. Specifically, we employ pixel-level and feature-level attention to predict the confidence in transmitting other agents' messages precisely. In the remaining parts of this section, we sequentially formulate the communication problem, overview the proposed framework, and elaborate on the co-attentive scheduler.

3.1 Problem Formulation

We denote the number of agents in a multi-agent perception system with N, and their observations with $\mathcal{X} = \{x_i\}_{i=1,...,N}$. Besides, each agent is expected to finish a task with the perceptive supervision $\mathcal{Y} = \{y_i\}_{i=1,...,N}$. However, the observations of several agents may be noisy due to the unstable conditions for environments and perceptive devices. These noisy observations may degrade the performance of the downstream tasks. Therefore, we wish to schedule the transmissions of the observations and achieve optimal performance of the multi-agent system. The optimization of our framework corresponds to Eq. (1).

$$\{\theta, \mathcal{P}\}^* = \arg\max_{\theta, \mathcal{P}} \sum_{i=1}^{N} g\left(\phi_\theta\left(x_i, \{\mathcal{P}_{i\rightarrow j}\}_{j=1}^{N}\right), y_i\right), \tag{1}$$

where $g(\cdot, \cdot)$ is the metric evaluating performance, $\phi_\theta(\cdot)$ is the perception network (*i.e.*, the feature encoder, perceptive fusion, and downstream task decoder in this work) with trainable parameters θ, and $\mathcal{P}_{i\rightarrow j}$ denotes the message transmitted from the ith agent to the jth agent. In this paper, we specify $\mathcal{P}_{i\rightarrow j}$ with confidence scores $\{w_{i,j}\}_{i\neq j}$ and utilize the co-attentive scheduler to obtain it.

3.2 Overview of the Proposed Framework

Existing collaborative perception approaches can be broadly categorized into three main types: raw-measurement-based early collaboration, output-based late collaboration, and feature-based intermediate collaboration. We focus on feature-based intermediate collaboration due to the balance between performance and bandwidth utilization. The framework of the proposed method is presented in the bottom half of Fig. 1, and the components of the proposed framework are introduced as follows.

Feature Encoder. The feature encoder extracts perceptive features from the observations. Note that CAMP could be incorporated with diverse inputs, such as RGB images and point clouds, to tackle various tasks. This work adopts a pre-trained ResNet-18 [6] model as the backbone of the feature encoder. The model is pre-trained on the ImageNet dataset and is truncated to fit semantic segmentation. We denote the input of the ith agent as x_i, then

$$z_i = f_{\text{enc}}(x_i; \theta_e), \tag{2}$$

where z_i represents the perceptive feature, $f_{\text{enc}}(\cdot; \theta_e)$ is the encoder parameterized by θ_e. The extracted perceptive features are further transmitted according to the communication graph obtained via the co-attentive scheduler.

Co-attentive Scheduler. The co-attentive scheduler determines how the agents communicate. We establish the communication topology with a graph

$G = (V, E)$, where the vertices V represent agents and edges E denote the communication between two agents. Besides, we assign the estimated confidence score $w_{i,j}$ as the weight of each edge. Here, we adopt the original observation x_i and the perceptive feature z_i as the inputs. We extract two groups of queries and keys (q_i, k_i) and (q_i', k_i') from x_i and z_i with a pixel-level attention $f_{\text{pix}}(\cdot; \theta_t)$ and a latent-level attention $f_{\text{feat}}(\cdot; \theta_c)$, respectively. The queries and keys are further weighted to get the confidence scores $w_i = \{w_{i,j}\}$, which are utilized in the further perceptive fusion. The details of the co-attentive scheduler are elaborated in Sect. 3.3.

Perceptive Fusion. We employ a weighted sum to fuse the received perceptive features $\{z_j\}$ ($j \neq i$). Without loss of generality, we adopt a zero-tensor $z_j = 0$ for the dropped feature to keep a consistent formulation of the fusion module. Then the fused feature h_i for the ith agent is obtained as follows.

$$h_i = \text{concat} \left(z_i, \sum_{j \neq i} w_{i,j} z_j \right), \tag{3}$$

where $\text{concat}(a, b)$ concatenate the two tensors a and b on the channel-wise dimension.

Downstream Task Decoder. The downstream task decoder outputs the prediction for the given task. For example, in the context of semantic segmentation, the downstream task decoder outputs the predicted pixel-level segmentation. Given the fused feature h_i, the decoder $f_{\text{dec}}(\cdot)$ outputs the result $\hat{y}_i$ as

$$\hat{y}_i = f_{\text{dec}}(h_i; \theta_d), \tag{4}$$

where θ_d denotes the learned parameters for the decoder. Note that $\{\hat{y}_i\}$ is the final result produced by the agents in the collaborative perception system, which could be regarded as $\phi_\theta \left(x_i, \{\mathcal{P}_{i \rightarrow j}\}_{j=1}^N \right)$ term in the problem formulation in Eq. (1).

Loss Function and Training Strategy. To optimize the whole framework, we use the label for downstream tasks as supervision and take the loss function in Eq. (5) for back-propagation and weight updates.

$$L = \underset{\theta_e, \theta_f, \theta_p, \theta_d}{\arg \max} \sum_{i=1}^N g(\hat{y}_i, y_i), \tag{5}$$

where θ_f and θ_p are parameters for co-attentive scheduler, which are elaborated in Sect. 3.3. We subsequently update the weights of our model using the aforementioned loss in an end-to-end fashion. For semantic segmentation, $g(\cdot, \cdot)$ is the mean intersection of union (MIoU). Please note that we do not use the communication labels included in the dataset during training, which were only used as an indicator to evaluate communication accuracy during the test.

3.3   Co-attentive Scheduler

We illustrate the proposed co-attentive scheduler on the top half of Fig. 1. The scheduler determines the communications between the agents, that is, whether two agents are expected to transmit messages. Here, we specify the perceptive features $\{z_i\}_{i=1,\cdots,N}$ as the transmitted messages. To achieve efficient communication of noisy observations, we adopt a pixel-level attention module $f_{\text{pix}}(\cdot;\boldsymbol{\theta}_p)$ and a latent-level attention module $f_{\text{feat}}(\cdot;\boldsymbol{\theta}_f)$. The co-attentions further contribute to the confidence score $\{w_{i,j}\}_{i\neq j}$ that measures the importance of transmitting the perceptive feature of the jth agent to the ith agent. These components are introduced in detail as follows.

Feature-Level Attention. The feature attention module takes the perceptive feature z_i as the input and outputs feature-level scores $\boldsymbol{w}_i^{\text{feat}}$. This module projects z_i into a query $\boldsymbol{q}_i \in \mathbb{R}^{1\times M}$ and a key $\boldsymbol{k}_i \in \mathbb{R}^{1\times M}$:

$$\boldsymbol{q}_i,\ \boldsymbol{k}_i = f_{\text{feat}}(\boldsymbol{z}_i;\boldsymbol{\theta}_f), \tag{6}$$

where $f_{\text{feat}}(\cdot;\boldsymbol{\theta}_f)$ is a nonlinear projection parameterized by $\boldsymbol{\theta}_f$. The projection utilizes multiple convolutional layers to downsample the feature map. Subsequently, the downsampled features are flattened and passed through a fully connected layer to obtain the query and key.

Each scheduler gathers the queries $\{\boldsymbol{q}_j\}_{j\neq i}$ and keys $\{\boldsymbol{k}_j\}_{j\neq i}$ from other agents. Then the feature-level score $\boldsymbol{w}_i^{\text{feat}}$ is obtained by

$$\boldsymbol{w}_i^{\text{feat}} = [w_{i,0}^{\text{feat}},\cdots,w_{i,N-1}^{\text{feat}}] = \text{softmax}\left(\mathbf{Q}\boldsymbol{k}_i^{\text{T}}\right), \tag{7}$$

where $\mathbf{Q} = \text{concat}(\boldsymbol{q}_0,\cdots,\boldsymbol{q}_{N-1})$ is the concatenation of N queries on the first dimension. Note that $\boldsymbol{w}_i^{\text{feat}}$ contains N elements, and we use $N-1$ of them for estimating the final scores.

Pixel-Level Attention. The query generated by feature-level attention may have a perceptual bias due to pre-training and can not precisely model the full spatial correlation between two agents. Therefore, we additionally employ pixel-level attention to capture the comprehensive correlation and ensure effective communication between agents. Inspired by Transformer [18], we decouple Q, K and V of attention module and apply the Transformer-based encoder module to obtain pixel-level query $\boldsymbol{q}_i' \in \mathbb{R}^{1\times M}$ and pixel-level key $\boldsymbol{k}_i' \in \mathbb{R}^{1\times M}$. The projection is

$$\boldsymbol{q}_i',\ \boldsymbol{k}_i' = f_{\text{pix}}(\boldsymbol{x}_i;\boldsymbol{\theta}_p), \tag{8}$$

where $f_{\text{pix}}(\cdot;\boldsymbol{\theta}_p)$ is a projection parameterized by $\boldsymbol{\theta}_p$. The projection first applys several convolutional layers to extract features from the original input image. These features are then subject to downsampling, flattening, and full connected layer to obtain a compact vector representation. Subsequently, the transformer-based encoder module is employed to generate the pixel-level query and key.

Then the pixel-level score w_i^{pix} is obtained as follows by gathering the queries $\{q_j'\}_{j\neq i}$ and keys $\{k_j'\}_{j\neq i}$ from other agents.

$$w_i^{\text{pix}} = \left[w_{i,0}^{\text{pix}}, \cdots, w_{i,N-1}^{\text{pix}} \right] = \text{softmax}\left(\mathbf{Q}' k_i'^{\text{T}} \right), \tag{9}$$

where $\mathbf{Q}' = \text{concat}(q_0', \cdots, q_{N-1}')$ is the concatenation of N queries on the first dimension.

Confidence Score Estimation. Finally, the confidence scores generated by the two modules are weighted in the following way:

$$w_{i,j} = \alpha \cdot w_{i,j}^{\text{pix}} + (1-\alpha) \cdot w_{i,j}^{\text{feat}}, \quad \forall j \neq i, \tag{10}$$

where $w_{i,j}$ is the ultimate confidence score for transmitting the perceptive feature from the jth agent to the ith agent. In addition, we further reduce the bandwidth for necessary communication by removing the connections with a confidence score $w_{i,j}$ less than threshold τ. Thus, the communication graph could be sparse and efficient.

4   Experiment

4.1   Dataset and Experiment Settings

Dataset. We use the publicly available dataset **AirSim-MAP** [11] to evaluate the effectiveness of our model. Airsim-MAP [11] contains multi-view images captured by five unmanned aerial vehicles (UAVs) flying simultaneously in a simulated city environment. In the noisy version of the dataset, Airsim-MAP applies Gaussian blurs with a random kernel size ranging from 1 to 100 [12]. Each view has a probability ($\geq 50\%$) of being corrupted, and each frame contains at least one noisy image in the five views. The noisy version of the dataset comprises 4.2K aerial RGB images of size 512×512 and their corresponding 3D bounding boxes.

Experiment Settings. We follow the experimental scenario of multi-agent collaborative perception in When2com [11] and set up a Multiple Request Multiple Support (MRMS) case as the main experiment of this work. In this case, we aim to generate precise 2D semantic segmentation masks for each agent based on their observations. The agent responsible for performing the segmentation is referred to as the requesting agent, while the other agents providing supplementary information are referred to as supporting agents. By leveraging the collaboration of all agents, the objective is to achieve accurate and comprehensive 2D semantic segmentation results for all the agents. In the multi-agent system, we introduce noise to half of the frames. We randomly replace one of the remaining agents with the non-degraded frame from the original agent. It is ensured that each degraded requesting agent has a corresponding non-degraded image among its supporting agents.

Baselines and Evaluation Metrics. We evaluate several baseline models, including fully-connected (FC) and distributed communication (DistCom) models. The FC models fuse all agents' observations, either using weighted or unweighted approaches. On the other hand, DistCom models selectively fuse a subset of the available observations, focusing on specific agents or information sources.

- CatAll (FC) is a naive FC model baseline that concatenates the encoded image features of all agents before subsequent network stages.
- RandCom (DistCom) serves as a naive distributed baseline method that randomly selects one of the other agents as a supporting agent.
- Who2com [12] (DistCom) is a distributed communication (DistCom) model that excludes the self-attention mechanism.
- OccDeg presents the performance when agents operate independently with degraded data. AllNorm corresponds to the performance when agents operate independently with clean, non-degraded data.

We also consider communication modules of CommNet [16], and TarMac [2], MRGNN [22], When2com [11], CRCN [14] as our baseline methods for comparison. For a fair comparison, we use ResNet18 [6] as the feature backbone for our and all mentioned baseline models. We evaluate the performance of all the models with mean intersection over union (MIoU), overall accuracy (PA), and frequency weighted accuracy (FA). In addition, we report the bandwidth of all FC and DistCom models in Megabyte per frame (MBpf). The bandwidth is obtained by summing up the sizes of transmitted queries, keys, and perceptive features.

Table 1. Experimental results on MRMS. Note that we evaluate these models with the metric of the mean intersection of union (MIoU), overall accuracy (PA), frequency weighted accuracy (FA), and use MBytes per frame (Mbpf) and the averaged number of links per agent for measuring bandwidth.

Models	Bandwidth	Noisy			Normal			Avg		
		PA	FA	MIoU	PA	FA	MIoU	PA	FA	MIoU
AllNorm	–	–	–	–	–	–	–	89.10	**80.72**	57.39
OccDeg	–	65.74	48.31	30.01	88.39	79.43	56.29	82.75	70.61	49.37
CatAll	2.5/5	71.85	55.72	34.91	88.18	78.99	55.07	84.16	72.65	49.87
VAIN	2.5/5	69.72	53.85	34.14	88.85	80.27	57.47	84.14	72.94	51.18
CommNet	2.5/5	70.09	53.72	33.95	88.67	79.93	56.60	84.09	72.70	50.70
TarMac	2.5/5	86.63	76.53	52.90	87.43	77.78	54.31	87.23	77.47	53.82
MRGNN	2.5/5	88.66	79.91	56.95	88.98	80.44	57.49	88.90	80.31	57.36
RandCom	0.5/1	56.15	34.85	17.12	84.61	73.21	47.84	77.60	62.67	40.19
Airp	0.5/1.67	62.77	43.31	25.59	83.93	72.27	47.40	78.72	64.35	41.92
Who2com	0.5/1	69.72	53.85	34.14	88.85	80.27	57.47	84.14	72.94	51.18
When2com	**0.38/0.77**	87.84	78.15	56.15	88.76	79.99	57.35	88.53	79.62	57.05
CRCN	1.5/3	88.54	79.67	56.95	88.93	80.30	57.72	88.84	80.15	57.53
CAMP (Ours)	**0.38/0.76**	**89.04**	**80.58**	**57.43**	**89.21**	**80.79**	**58.02**	**89.11**	80.62	**57.88**

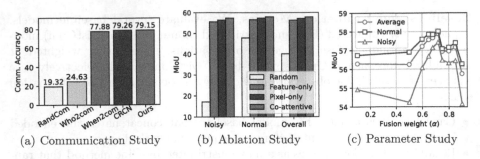

(a) Communication Study   (b) Ablation Study   (c) Parameter Study

Fig. 2. Results of comparative experiments on communication accuracy, ablation study, and parameter study.

4.2   Quantitative Evaluation

We list the experiment results in the MRMS case in Table 1. Although CatAll and RandCom are with communication, they perform even worse than Occdeg (without communication), which means that the noisy observations can reduce the performance on the downstream task. By weighting the received features via attention, VAIN [7] and Who2com [12] get more stable fused features and thus improve the performance with respect to NoCom, CatAll, and RandCom.

In addition to the above baselines, we also evaluated several full-connected communication methods (*i.e.*, CommNet [16], MRGNN [22], and TarMac [2]) in the MRMS scenario. Even though CommNet [16] fuses the information from other agents by pooling, which is inefficient in processing degraded observations, as shown in the comparison to CatAll. TarMac [2] significantly improve the performance compared to CatAll. However, the adopted one-way communication results in large bandwidth usage, which is not feasible to real-world applications. MRGNN [22] employs graph neural networks to promote the communication topology. However, the fully connected communication adopted in MRGNN [22] also renders inefficient bandwidth usage. Since fully connected communication suffers from inefficient bandwidth usage, we design the proposed framework based on distributed communication.

We also compare the proposed method with other distributed communication models (*i.e.*, Airpooling [13], When2com [11], and CRCN [14]). Airpooling [13] exploits the waveform superposition property of a multiaccess channel to implement fast over-the-air averaging of pooled features. However, the performance on the downstream task is sacrificed. When2com [11] uses a self-attention to establish an adaptive communication topology. Compared with other methods, it has significantly improved bandwidth utilization and downstream task accuracy. CRCN [14] adopts channel attention to reduce the redundancy of the agents' observations. Compared to these distributed communication models, the proposed method achieves state-of-the-art performance on both bandwidth usage and downstream task performance. In other words, the proposed co-attentive scheduler is proved to be more efficient than self-attention used in When2com [11] and channel attention used in CRCN [14]. Moreover, the proposed method works stable on noisy observations, which benefits real-world applications.

4.3 Studies on Co-attentive Scheduler

Communication Accuracy Analysis. We compute communication accuracy metrics on the Single Request Multiple Support (SRMS) case of the AirSim-MAP dataset [11] to investigate the source of our model's improvement. SRMS is a scenario in a Multi-agent system where a single requester sends a request, and multiple supporters support it. Communication accuracy measures the effectiveness of communication between requesters and supporters. The SRMS case is offered with the ground-truth communication labels to evaluate the efficiency of collaborative perception algorithms. The communication labels are the rankings of communication confidence scores from a single agent's perspective, indicating the necessity of adopting supporters' information. As demonstrated in Fig. 2(a), we provide the result of RandCom, Who2com [12], When2com [11], CRCN [14] and the proposed method. The attention modules significantly improve communication accuracy compared to RandCom. Besides, the proposed method surpasses most methods on communication accuracy, which demonstrates that proposed co-attention is effective. The proposed method achieves superior performance on the downstream task compared to CRCN [14], as presented in Table 1.

Ablation Study on Co-attentive Scheduler. We also conduct ablation studies on the co-attentive scheduler. We first evaluate the effectiveness of the proposed co-attention by training four models with different strategies for predicting the confidence scores, *i.e.*, the proposed co-attention, pixel-level only, feature-level only, and random scores. The results for these four models are presented in Fig. 2(b). The proposed co-attention achieves the best MIoU on semantic segmentation, indicating the proposed method's efficiency.

Parameter Study on Fusion Weight α. We traverse the fusion weight $\alpha \in [0, 1]$ to balance the pixel-level attention and feature-level attention. The results are presented in Fig. 2(c), showing that $\alpha = 0.7$ is the optimal fusion weight. The results suggest that both pixel-level and feature-level attentions are important for constructing efficient communication.

5 Conclusion

In this paper, we proposed CAMP, a learning-based multi-agent collaborative perception framework. Our framework employs a novel co-attentive scheduler, leading to efficient agent communication and stable performance on downstream tasks. Experimental results show the proposed CAMP achieves state-of-the-art, and the proposed co-attentive scheduler is effective.

Acknoledgement. This work was partially supported by grants from the National Key Research and Development Program of China (No. 2021YFA0715202), Natural Science Foundation of China (No. 62293544 and 62022092) and China Postdoctoral Science Foundation (No. 2022M720077).

References

1. Chen, S., Liu, B., Feng, C., Vallespi-Gonzalez, C., Wellington, C.: 3D point cloud processing and learning for autonomous driving: impacting map creation, localization, and perception. IEEE Sig. Process. Mag. **38**(1), 68–86 (2020)
2. Das, A., et al.: TarMAC: targeted multi-agent communication. In: International Conference on Machine Learning, pp. 1538–1546. PMLR (2019)
3. Ferber, J.: Multi-Agent Systems: An Introduction to Distributed Artificial Intelligence, 1st edn. Addison-Wesley Longman Publishing Co., Inc. (1999)
4. Foerster, J., Assael, I.A., De Freitas, N., Whiteson, S.: Learning to communicate with deep multi-agent reinforcement learning. In: Advances in Neural Information Processing Systems, vol. 29 (2016)
5. Harding, J., et al.: Vehicle-to-vehicle communications: readiness of V2V technology for application. Technical report, United States, National Highway Traffic Safety Administration (2014)
6. He, K., Zhang, X., Ren, S., Sun, J.: Deep residual learning for image recognition. In: Proceedings of the IEEE Conference on Computer Vision and Pattern Recognition, pp. 770–778 (2016)
7. Hoshen, Y.: VAIN: attentional multi-agent predictive modeling. In: Advances in Neural Information Processing Systems, vol. 30 (2017)
8. Hu, Y., Fang, S., Lei, Z., Zhong, Y., Chen, S.: Where2comm: communication-efficient collaborative perception via spatial confidence maps. arXiv preprint arXiv:2209.12836 (2022)
9. Jiang, J., Lu, Z.: Learning attentional communication for multi-agent cooperation. In: Advances in Neural Information Processing Systems, vol. 31 (2018)
10. Li, Y., Bhanu, B., Lin, W.: Auction protocol for camera active control. In: 2010 IEEE International Conference on Image Processing, pp. 4325–4328. IEEE (2010)
11. Liu, Y.C., Tian, J., Glaser, N., Kira, Z.: When2com: multi-agent perception via communication graph grouping. In: Proceedings of the IEEE/CVF Conference on Computer Vision and Pattern Recognition, pp. 4106–4115 (2020)
12. Liu, Y.C., Tian, J., Ma, C.Y., Glaser, N., Kuo, C.W., Kira, Z.: Who2com: collaborative perception via learnable handshake communication. In: 2020 IEEE International Conference on Robotics and Automation (ICRA), pp. 6876–6883. IEEE (2020)
13. Liu, Z., Lan, Q., Kalør, A.E., Popovski, P., Huang, K.: Over-the-air multi-view pooling for distributed sensing. arXiv preprint arXiv:2302.09771 (2023)
14. Luo, G., Zhang, H., Yuan, Q., Li, J.: Complementarity-enhanced and redundancy-minimized collaboration network for multi-agent perception. In: Proceedings of the 30th ACM International Conference on Multimedia, pp. 3578–3586 (2022)
15. Melo, F.S., Spaan, M.T.J., Witwicki, S.J.: QUERYPOMDP: POMDP-based communication in multiagent systems. In: Cossentino, M., Kaisers, M., Tuyls, K., Weiss, G. (eds.) EUMAS 2011. LNCS (LNAI), vol. 7541, pp. 189–204. Springer, Heidelberg (2012). https://doi.org/10.1007/978-3-642-34799-3_13

16. Sukhbaatar, S., Fergus, R., et al.: Learning multiagent communication with back-propagation. In: Advances in Neural Information Processing Systems, vol. 29 (2016)
17. Tan, M.: Multi-agent reinforcement learning: independent vs. cooperative agents. In: Proceedings of the Tenth International Conference on Machine Learning, pp. 330–337 (1993)
18. Vaswani, A., et al.: Attention is all you need. In: Advances in Neural Information Processing Systems, vol. 30 (2017)
19. Wang, Z., Ziou, D., Armenakis, C., Li, D., Li, Q.: A comparative analysis of image fusion methods. IEEE Trans. Geosci. Remote Sens. **43**(6), 1391–1402 (2005)
20. Watkins, C.J., Dayan, P.: Q-learning. Mach. Learn. **8**, 279–292 (1992). https://doi.org/10.1007/BF00992698
21. Wu, Z., Pan, S., Chen, F., Long, G., Zhang, C., Philip, S.Y.: A comprehensive survey on graph neural networks. IEEE Trans. Neural Netw. Learn. Syst. **32**(1), 4–24 (2020)
22. Zhou, Y., Xiao, J., Zhou, Y., Loianno, G.: Multi-robot collaborative perception with graph neural networks. IEEE Robot. Autom. Lett. **7**(2), 2289–2296 (2022)

DBRNet: Dual-Branch Real-Time Segmentation NetWork for Metal Defect Detection

Tianpeng Zhang[1,2,3], Xiumei Wei[1,2,3], Xiaoming Wu[1,2,3], and Xuesong Jiang[1,2,3](✉)

[1] Key Laboratory of Computing Power Network and Information Security, Ministry of Education, Shandong Computer Science Center, Qilu University of Technology (Shandong Academy of Sciences), Jinan, China
jxs@qlu.edu.cn
[2] Shandong Engineering Research Center of Big Data Applied Technology, Faculty of Computer Science and Technology, Qilu University of Technology (Shandong Academy of Sciences), Jinan, China
[3] Shandong Provincial Key Laboratory of Computer Networks, Shandong Fundamental Research Center for Computer Science, Jinan, China

Abstract. Metal surface defect detection is an important task for quality control in industrial production processes, and the requirements for accuracy, and running speed are becoming increasingly high. However, maintaining the realization of real-time surface defect segmentation remains a challenge due to the complex edge details of metal defects, inter-class similarity, and intra-class differences. For this reason, we propose Dual-branch Real-time Segmentation NetWork (DBRNet) for pixel-level defect classification on metal surfaces. First, we propose the Low-params Feature Enhancement Module (LFEM), which improves the feature extraction capability of the model with fewer parameters and does not significantly reduce the inference speed. Then, to solve the problem of inter-class similarity, we design the Attention Flow-semantic Fusion Module (AFFM) to effectively integrate the high-dimensional semantic information into the low-dimensional detail feature map by generating flow-semantic offset positions and using global attention. Finally, the Deep Connection Pyramid Pooling Module (DCPPM) is proposed to aggregate multi-scale context information to realize the overall perception of the defect. Experiments on NEU-Seg, MT, and Severstal Steel Defect Dataset show that the DBRNet outperforms the other state-of-the-art approaches in balance accuracy, speed, and params. The code is publicly available at https://github.com/fffcompu/DBRNet-Defect.

Keywords: Metal surface defect detection · real-time · semantic segmentation

This work was supported in part by Major innovation projects of the pilot project of science, education and industry integration (2022JBZ01-01), and in part by Taishan Scholars Program (tsqn202211203).

Q. Liu et al. (Eds.): PRCV 2023, LNCS 14430, pp. 422–434, 2024.
https://doi.org/10.1007/978-981-99-8537-1_34

1   Introduction

In the industrial production process, many types of defects inevitably appear on the surface of metal materials, such as spots, scratches, and surface inclusions, which will have a negative impact on the performance of the product or even pose a safety hazard, so surface metal defect detection methods become critical. However, metal defect detection is still a challenge due to the complex edge details, large differences in the same defects class (Intra-class differences), and the local similarity problem of different defects class (Inter-class similarity).

With the development of deep learning. Semantic segmentation is proposed, which is different from Faster RCNN [15] and YOLO [2,9] series of target detection methods, which can achieve pixel-level classification and more accurate prediction of boundaries. Therefore, generic segmentation methods [1,3,17] are widely used in metal defect segmentation. Zhang et al. [13], used a dual parallel attention module added to the encoder of DeepLabV3+ to improve local representation and enrich context dependence achieved 89.95 mIoU on the steel defect dataset. Zhang et al. [21] proposed MCNet, using a dense block pyramid pool to make full use of context information to accomplish surface defect segmentation of tracks. Dong et al. [5] proposed PGA-Net with pyramidal feature fusion blocks and global context attention blocks, which effectively connect various feature maps extracted from the backbone network, establishe a deep supervision mechanism and obtains precise bounds for segmenting various defects. Damacharla [4] proposes the TL-UNet approach, which uses a migration learning approach to accomplish automatic surface segmentation. However, although the above network structure can segment the defects well, it does not achieve real-time results due to the complexity of its structure.

In recent years, many scholars are researching fast segmentation methods for road scenes. BiseNetV2 [20] proposes a bilateral structure with two branches to extract semantic features and detailed features respectively, which ensures real-time performance while extracting spatial information efficiently. ShuffleNetV2 [12] uses channel shuffle and group convolution to reduce computational cost. STDCNet [6] designs a short-term dense cascade module to extract multi-scale features and proposes a detail aggregation module to learn the decoder, which can preserve the underlying spatial details more accurately without increasing the inference time. DDRNet [7] introduces bilateral connections to enhance the information exchange between context and detail branches for precise segmentation. However, metal defects differ significantly from road scenes, thus affecting the accuracy and robustness of these algorithms when processing images containing metal defects. To achieve high-precision defect segmentation while maintaining real-time speed, this paper proposes DBRNet with pixel-level metal defect prediction, and the contributions of this paper are summarized as follows:

(1) We propose a Low-params Feature Enhancement Module (LFEM), which uses depth-wise convolution with different kernel sizes to reduce the number of parameters and enhance feature extraction without significantly slowing down the model's inference speed.

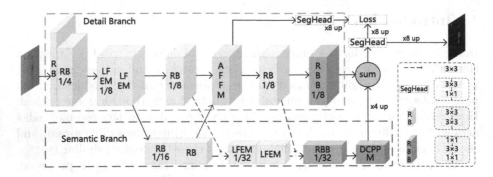

Fig. 1. The overall architecture diagram of the network. 3×3, 1×1 represents the standard convolution using the corresponding kernel size. $\oplus$ and "sum" indicated element-wise addition. "up" represents upsampling operation.

(2) To deal with the problem of inter-class similarity, Attention Flow-semantic Fusion Module(AFFM) is proposed, using global attention and semantic flow makes it effectively fuse high-level semantic features while preserving detailed features.

(3) The Deep Connection Pyramid Pooling Module (DCPPM) is designed to be added to the semantic branch to improve the overall perception of defects by using dense connections to fuse different pooling features.

(4) The experimental results on three publicly available defect datasets demonstrate the effectiveness of the proposed DBRNet for metal surface defect segmentation.

2   Method

The overall structure of the network is shown in Fig. 1. The model is based on dual-branch architecture. Where the semantic branch is concerned with the category information of defects and the detail branch is concerned with the location details of defects. The whole network consists of the LFEM, AFFM, DCPPM, and ResNet's Residual Bottleneck Block (RBB) and Residual Basic Block (RB) modules. Firstly, two LFEM are used to replace the original RB blocks in the detail branch and semantic branch to generate feature maps of the 1/8, 1/32 image resolution, respectively, to enhance the feature extraction capability of the model. Afterwards, AFFM is added before the fourth RB of the detail branch to efficiently fuse semantic features into the high-resolution image. The feature maps of the detail branch are fused into the semantic branch by 3×3 convolution downsampling before LFEM and RBB of the semantic branch to improve edge recognition ability. Finally, the "sum" element-wise addition combines the detail branch and semantic branch information. And the output is obtained by SegHead operation. It is important to note that the processing of SegHead and Auxiliary loss is kept consistent with that of DDRNet [7].

2.1 Low-Params Feature Enhancement Module

The number of parameters and the model inference speed are not a single linear relationship, as most devices are optimized for 3×3 standard convolution, giving RB blocks composed of 3×3 convolution an advantage in running speed. However, simply stacking the RB block makes the number of parameters of the model rise dramatically and the feature extraction method is homogeneous. To balance params, speed, and accuracy, We propose Low-params Feature Enhancement Module (LFEM) inspired by ShuffleNetV2 [12] and LAANet [22] for replacing the original RB block which generates feature maps of 1/8 image resolution size at the detail branch and 1/32 image resolution size at the semantic branch.

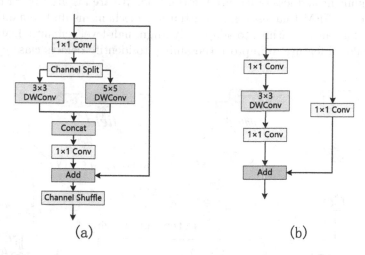

(a)   (b)

Fig. 2. The details of LFEM. (a) is utilized when the stride = 1. (b) is based the mobile inverted bottleneck, which is used when the stride = 2. Conv represents standard convolution. DWConv is the depth-wise convolution.

LFEM is shown in Fig. 2. It can be divided into two situations. When the stride is equal to 1, at the beginning of LFEM, the channel number of input is expanded 2× to obtain an upgrade feature map through 1×1 convolution to enhance the ability of feature information extraction. To maintain computing efficiency, we will then divide the upgraded feature maps into two branches. Each branch obtains a half-channel upgrade feature map by channel split, and then uses 3×3 depth-wise convolution on the first branch to extract local information, use 5×5 depth-wise convolution on the second branch to obtain more complex feature information. The outputs of two branches are connected through channel connection. Afterward, 1×1 convolution is used to increase information sharing between branches, followed by performing residual add connections to prevent gradient disappearance and explosion. Finally, channel shuffle is used to promote information fusion between channels, thereby achieving higher segmentation accuracy. When the stride is equal to 2, it is similar to the MobileNet structure. However, in order to improve the information retention ability during the downsampling process, 1×1 standard convolution is used for downsampling the residual part.

2.2   Attention Flow-Semantic Fusion Module

In defect segmentation, the local similarity of different defects makes it difficult to identify the class information of defects. Integrating semantic information into detail feature maps can enable segmentation results to retain detailed information while containing rich category information. However, directly using element-wise operations to fuse semantic features into detail feature maps leads to misalignment and loss of feature information. We are inspired by SFNet [11]. We propose AFFM that uses global attention mechanisms and semantic offset position of adjacent features to adaptively align high-level semantic feature into low-level detail feature to improve the ability to identify defects class.

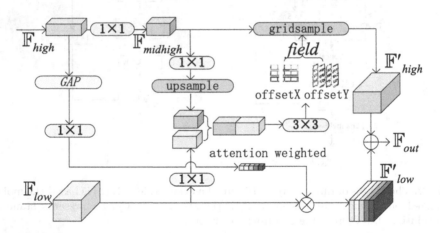

Fig. 3.   The architecture of AFFM. $\otimes$ indicated element-wise multiplication operation. GAP indicated global average pooling operation. Gridsample represents the grid_sample function.

As Shown in Fig. 3. The input of AFFM is the $\mathbb{F}_{high} \in \mathbb{R}^{H/2 \times W/2 \times 2C}$ represents the high-level feature map from the semantic branch and the $\mathbb{F}_{low} \in \mathbb{R}^{H \times W \times C}$ represents the low-level feature map from the detail branch, with the output being $\mathbb{F}_{out} \in \mathbb{R}^{H \times W \times C}$. We first use global average pooling for the high-level feature to quickly extract global context information, followed immediately by using 1×1 convolution to adjust to the same number of channels as low-level feature and multiplying the obtained attention weighted $\in \mathbb{R}^{1 \times 1 \times C}$ with low-level feature to achieve a simple way to weight low-level feature to highlight detailed information. To cope with the inefficient feature fusion between high-level and low-level features caused by the difference in resolution and information contained, we resample the high-level feature to achieve feature alignment by setting a learnable semantic offset position $field \in \mathbb{R}^{H \times W \times 2}$. Specifically, we use 1×1 convolution to perform channel compression on the high-level feature to get the $\mathbb{F}_{midhigh} \in \mathbb{R}^{H/2 \times W/2 \times C}$, and then use 1×1 convolution and upsampling to generate a contrast feature map. The low-level feature map is only processed with 1×1 convolution operation to generate a contrast feature map with the same shape $H \times W \times C/2$. These two comparison feature maps are fused in a channel-connected manner and fed into 3×3 convolution to explore the semantic offset positions between features at different levels. The semantic offset position $field$ consists of offsetX and offsetY, which represent the offset position on the horizontal axis and the vertical axis, respectively, of the sampling coordinates of each pixel point during the sampling process. The operation of gridsample is used to resample the $\mathbb{F}_{midhigh}$ to generate the $\mathbb{F}'_{high}$ embedded in the correct position in the low-level feature. Finally, the element-wise addition operation combines $\mathbb{F}'_{low}$ and $\mathbb{F}'_{high}$ to obtain the final output $\mathbb{F}_{out}$. The entire process can be formulated as:

$$\mathbb{F}'_{low} = \text{Conv}_{1 \times 1}\left(GAP\left(\mathbb{F}_{high}\right)\right) \otimes \mathbb{F}_{low}$$

$$\mathbb{F}_{midhigh} = \text{Conv}_{1 \times 1}\left(\mathbb{F}_{high}\right)$$

$$field = \text{Conv}_{3 \times 3}\left(\text{Concat}\left(\text{Conv}_{1 \times 1}\left(\mathbb{F}_{low}\right), \text{up}\left(\text{Conv}_{1 \times 1}\left(\mathbb{F}_{midhigh}\right)\right)\right)\right)$$

$$\mathbb{F}'_{high} = \text{gridsample}\left(\mathbb{F}_{midhigh}, field\right)$$

$$\mathbb{F}_{out} = \mathbb{F}'_{low} \oplus \mathbb{F}'_{high} \tag{1}$$

where $\text{Conv}_{i \times i}$ indicated standard convolution with $i \times i$ kernel size. Concat indicated channel connection operation.

2.3   Deep Connection Pyramid Pooling Module

For defect segmentation, some defects are intra-class differences, which require context information to achieve the overall perception of metal defects. Pooling operations can be used to get multi-scale context information quickly. However, traditional pooling modules lack information sharing after the pooling operation. For this purpose, we designed DCPPM, which uses pooling operations with different kernel sizes, and dense join to obtain more fine-grained multi-scale context information in the semantic branch.

As shown in Fig. 4. The input $x \in \mathbb{R}^{H \times W \times C}$ is fed to five parallel branches to perform multi-scale feature fusion operation. The 5th branch uses the global average pooling operation $P_g(.)$. The 2nd branch to 4th branch utilize P_i, $i \in \{2, 3, 4\}$ with kernel size $\{5, 9, 17\}$ pooling operations. Then Except for the first branch, the other four branches were subjected to pooling operation, after which $\mathbb{F}^i_{mid} \in \mathbb{R}^{H \times W \times C/4}$ was generated by 1×1 standard convolution and upsampling operation for subsequent fusion.

The branches of larger pooling kernel are fused with deeper feature information as a way to obtain multi-scale information at different fine-grained levels. Specifically, to get the output $\mathbb{F}^i_{out}$ of each branch, the 1st branch does not have

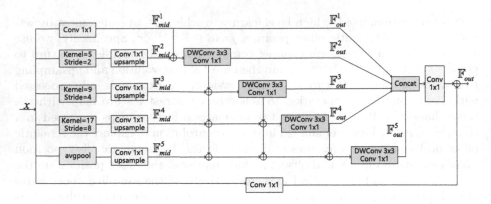

Fig. 4. The details of DCPPM.

any operation, $\mathbb{F}_{mid}^0$ is directly used as output $\mathbb{F}_{out}^0$. The output of 2nd branch and 3rd branch is the result of summing $\mathbb{F}_{mid}^i$ and $\mathbb{F}_{out}^{i-1}$ and then using the 3×3 depth-wise convolution and 1×1 standard convolution operations. The 4th and 5th branches are summed by $\mathbb{F}_{mid}^i$ and $\mathbb{F}_{out}^{i-1}$, $\mathbb{Q}_{i-1}$, immediately also followed by depth-wise convolution and standard convolution operation to obtain the output. $\mathbb{Q}_i$ represents the output of the intermediate addition operation of branch i. The $\mathbb{F}_{out}^i$ of all paths are stacked together and fed into 1×1 convolution, and 1×1 convolution is also used for the input x. Then, the two outputs are summed to obtain $\mathbb{F}_{out}$.

3   Experiments

3.1   Datasets and Evaluation Metrics

Datasets. The NEU-Seg [5] Dataset consists of 3600 hot-rolled steel strip images with a resolution of 200×200. It contains three types of defects, Inclusion, Patches and Scratches. Each type of defect contains 1200 samples. To evaluate the FPS, we resize the image's resolution to 512×512. Each image was annotated at the semantic level.

MT Dataset [8], it contains 5 types of magnetic-tile defects, porosity, cracks, wear, fracture, unevenness. It contains 392 defect images and 952 images without defects. All images are not the same resolution, so the image size resize to 512×512, and each image is made defective segmentation of the label.

The Severstal Steel Defect Dataset [18] contains 12568 steel images with a resolution of 1600×256 and pixel-level annotations of the four defect categories, and the sample distribution in the dataset contains 5902 defect-free samples and 6666 defect samples.

Mertics. In this paper, we use the evaluation metrics widely used in the field of Real-Time Segmentation. mean Intersection over Union (mIoU), Frames Per

Second (FPS), Params, and Floating Point operations (FLOPs) to measure the quality and efficiency of the model.

3.2 Implementation Details

Using the SGD optimizer with momentum and linear learning rate strategy. The SGD momentum value was set to 0.9, the initial learning rate was set to 1e−2, the weight decay factor was set to 5e−4. For data augmentation, the NEU-Seg and MT datasets, we used random augmentation to 0.5 to 2.5 followed by random cropping to 512 × 512, and the Severstal Steel Defect Dataset was randomly cropped to 512 × 256. The batch size during training was set to 8, all datasets were divided into train:val:test = 6:2:2, and the loss function was OHEM. All experiments were performed on NVIDIA RTX3080 with PyTorch 1.11.0 and cuda11.3 version.

3.3 Comparison Experiments

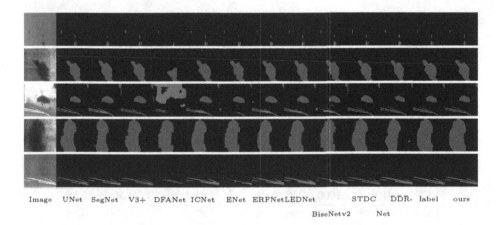

Fig. 5. Visualisation of results on NEU-Seg dataset.

We compared our DBRNet with 11 representative current segmentation methods on three publicly available typical defect segmentation datasets. Table 1 and Table 2 show the experimental results.

Results on NEU-Seg Dataset. On this dataset, DBRNet achieves 123.40 FPS at 512 × 512 of the image's resolution and achieves 83.14% mIoU. DBRNet's mIoU and FPS is the best compared to other methods. Specifically, compared with DDRNet, our method's Params, and FLOPs decreased by 2.35 MB and 1.45G, respectively, with a small increase in FPS. In terms of accuracy, DBRNet's mIoU is 1.89% higher than DDRNet. DBRNet is also competitive among DeepLabV3+, U-Net methods in accuracy. The comparison visualization on this

Table 1. Comparative experiments on NEU-Seg and MT datasets

Method	Resolution	Params(M)	FLOPs(G)	FPS	mIoU	
					NEU-Seg	MT
U-Net [17]	512 × 512	28.95	361.23	26.24	82.65	31.52
SegNet [1]	512 × 512	29.44	160.02	53.67	76.70	15.30
DeepLabV3+ [3]	512 × 512	54.70	83.20	47.24	82.79	30.50
DFANet [10]	512 × 512	2.15	**1.79**	39.26	45.02	20.07
ICNet [23]	512 × 512	26.70	5.58	56.00	79.13	61.35
ENet [14]	512 × 512	**0.35**	2.42	83.48	78.79	38.18
ERFNet [16]	512 × 512	2.06	12.89	84.87	79.09	35.13
LEDNet [19]	512 × 512	0.91	5.72	69.80	78.99	48.09
BiseNetV2 [20]	512 × 512	5.19	17.75	117.90	81.38	62.05
STDCNet [6]	512 × 512	9.34	11.37	121.81	82.09	62.58
DDRNet [7]	512 × 512	5.69	4.89	121.70	81.25	67.18
ours	512 × 512	3.34	3.44	**123.40**	**83.14**	**70.51**

Table 2. Comparative experiments on Severstal Steel Defect Dataset

Method	Resolution	Params(M)	FLOPs(G)	FPS	mIoU
U-Net [17]	1600 × 256	28.95	564.42	17.05	54.65
SegNet [1]	1600 × 256	29.44	251.53	33.43	44.20
DeepLabV3+ [3]	1600 × 256	54.70	130.00	36.40	60.07
DFANet [10]	1600 × 256	2.15	**2.79**	38.00	35.83
ICNet [23]	1600 × 256	26.70	8.73	55.00	60.02
ENet [14]	1600 × 256	**0.35**	7.04	72.31	47.89
ERFNet [16]	1600 × 256	2.06	20.14	83.87	55.30
LEDNet [19]	1600 × 256	0.91	8.94	70.50	45.79
BiseNetV2 [20]	1600 × 256	5.19	27.75	112.00	59.08
STDCNet [6]	1600 × 256	9.34	7.27	116.12	57.64
DDRNet [7]	1600 × 256	5.69	7.63	**122.30**	59.54
ours	1600 × 256	3.34	5.41	120.06	**60.61**

dataset is shown in Fig. 5. It can be seen that the results predicted by DBRNet are close to the ground truth (label) in the detail part, which demonstrates its effectiveness in metal surface defect segmentation.

Results on MT Dataset. On this dataset, the FPS, Params, and FLOPs of DBRNet are the same as in NEU-Seg, because the resolution of the images is the same. The difference is that the accuracy of our method on this dataset is significantly higher than the other comparison methods, with mIoU of 70.51%, which proves that DBRNet is equally effective in segmenting metal defects with a more number of types.

Results on Severstal Steel Defect Dataset. DBRNet's mIoU is also the best on this dataset. Although the FPS metric is lower than DDRNet 2.24, the mIoU of DBRNet is 1.07% higher than DDRNet. The mIoU of ICNet is similar to that of DBRNet, but our method is faster and has smaller number of parameters. Compared to other methods, our method achieves an excellent balance between speed, params, and accuracy on this dataset.

3.4   Ablation Study

To verify the effectiveness of the proposed method, ablation experiments were performed on three datasets. It should be noted that our baseline is modified DDRNet and ablation method is a replacement of the original module. Table 3 and Table 4 show the ablation experimental results.

Table 3. Ablation in NEU-Seg and MT datasets

Baseline	LFEM	AFFM	DCPPM	Params(M)	FLOPs(G)	FPS	mIoU	
							NEU-Seg	MT
✓				5.53	4.27	**137.01**	79.55	66.38
✓	✓			3.84	3.58	130.71	80.57	68.26
✓		✓		5.54	4.29	129.64	80.46	67.88
✓			✓	5.01	4.14	133.33	79.87	67.09
✓	✓		✓	**3.32**	**3.45**	126.59	81.40	68.87
✓	✓	✓		3.85	3.59	124.29	82.38	69.70
✓	✓	✓	✓	3.34	3.74	123.40	**83.14**	**70.51**

Table 4. Ablation in Severstal Steel Defect Dataset

Baseline	LFEM	AFFM	DCPPM	Params(M)	FLOPs(G)	FPS	mIoU
✓				5.53	6.68	**133.60**	58.64
✓	✓			3.84	5.59	125.31	60.02
✓		✓		5.54	6.70	123.40	59.83
✓			✓	5.01	6.47	132.00	59.31
✓	✓		✓	**3.32**	**5.39**	123.30	60.37
✓	✓	✓		3.85	5.62	122.74	60.54
✓	✓	✓	✓	3.34	5.41	120.06	**60.61**

Effectiveness of Single Module. Compared to the baseline. After adding LFEM to replace corresponding RB blocks, FPS only decreased by approximately 7 but the number of parameters decreased by 1.69 MB and mIoU increased by 1.02%, 1.88%, 1.38% in NEU-Seg, MT, Severstal Steel Defect Dataset respectively. This suggests that the addition of LFEM is more effective in extracting feature capability than using RB blocks alone. The number of parameters increased by just 0.01M after adding AFFM, and mIoU increased by 0.91%, 1.5%, and 1.19% on the three datasets. This demonstrates that AFFM can effectively fuse semantic information and solve the similarity problem of different defects without adding significant parameters. The addition of DCPPM increased miou by 0.32%, 0.71%, and 0.67% on the three datasets, demonstrating that it can aggregate multi-scale information to achieve an overall perception of defects.

Effectiveness of Different Module Combinations. In addition to the single analysis of the modules, some experiments were set up to evaluate the effect of the combination of the different modules. As seen in rows 5 to 7 of Table 3 and Table 4. Each combination improved the accuracy of the model. The best results were achieved when all three modules were added, specifically, the amount of parameters of the model were reduced by 2.19 MB and the mIoU reached 83.14%, 70.51%, and 60.61% on the three datasets. The results show that the addition of the three modules can effectively improve the performance of the model.

4   Conclusion

In this paper, we propose DBRNet that implements real-time metal defect segmentation. For the model to have high accuracy and efficiency. First, we design the LFEM module with fewer params to enhance the ability of feature extraction. Secondly, to solve the inter-class similarity problem, we use the AFFM module, which can more effectively fuse high-dimensional semantic information into low-dimensional detail information. In addition, we added DCPPM to the semantic branch to increase the multi-scale information of the model and thus realize the overall perception of defects. Based on the experimental results on three datasets, the algorithm shows excellent performance in terms of model efficiency and segmentation results. This indicates that the model can be deployed on metal defect detectors to achieve real-time segmentation of metal defects.

References

1. Badrinarayanan, V., Kendall, A., Cipolla, R.: SegNet: a deep convolutional encoder-decoder architecture for image segmentation. IEEE Trans. Pattern Anal. Mach. Intell. **39**(12), 2481–2495 (2017)
2. Bochkovskiy, A., Wang, C.Y., Liao, H.Y.M.: YOLOv4: optimal speed and accuracy of object detection. arXiv preprint arXiv:2004.10934 (2020)

3. Chen, L.-C., Zhu, Y., Papandreou, G., Schroff, F., Adam, H.: Encoder-decoder with atrous separable convolution for semantic image segmentation. In: Ferrari, V., Hebert, M., Sminchisescu, C., Weiss, Y. (eds.) ECCV 2018. LNCS, vol. 11211, pp. 833–851. Springer, Cham (2018). https://doi.org/10.1007/978-3-030-01234-2_49

4. Damacharla, P., Rao, A., Ringenberg, J., Javaid, A.Y.: TLU-Net: a deep learning approach for automatic steel surface defect detection. In: 2021 International Conference on Applied Artificial Intelligence (ICAPAI), pp. 1–6. IEEE (2021)

5. Dong, H., Song, K., He, Y., Xu, J., Yan, Y., Meng, Q.: PGA-Net: pyramid feature fusion and global context attention network for automated surface defect detection. IEEE Trans. Ind. Inf. **16**(12), 7448–7458 (2019)

6. Fan, M., et al.: Rethinking BiSeNet for real-time semantic segmentation. In: Proceedings of the IEEE/CVF Conference on CVPR, pp. 9716–9725 (2021)

7. Hong, Y., Pan, H., Sun, W., Jia, Y.: Deep dual-resolution networks for real-time and accurate semantic segmentation of road scenes. arXiv preprint arXiv:2101.06085 (2021)

8. Huang, Y., Qiu, C., Yuan, K.: Surface defect saliency of magnetic tile. Vis. Comput. **36**, 85–96 (2020). https://doi.org/10.1007/s00371-018-1588-5

9. Jocher, G.: Yolov5 (2021). https://github.com/ultralytics/yolov5

10. Li, H., Xiong, P., Fan, H., Sun, J.: DFANet: deep feature aggregation for real-time semantic segmentation. In: Proceedings of the IEEE/CVF Conference on CVPR, pp. 9522–9531 (2019)

11. Li, X., et al.: Semantic flow for fast and accurate scene parsing. In: Vedaldi, A., Bischof, H., Brox, T., Frahm, J.-M. (eds.) ECCV 2020. LNCS, vol. 12346, pp. 775–793. Springer, Cham (2020). https://doi.org/10.1007/978-3-030-58452-8_45

12. Ma, N., Zhang, X., Zheng, H.-T., Sun, J.: ShuffleNet V2: practical guidelines for efficient CNN architecture design. In: Ferrari, V., Hebert, M., Sminchisescu, C., Weiss, Y. (eds.) Computer Vision – ECCV 2018. LNCS, vol. 11218, pp. 122–138. Springer, Cham (2018). https://doi.org/10.1007/978-3-030-01264-9_8

13. Pan, Y., Zhang, L.: Dual attention deep learning network for automatic steel surface defect segmentation. Comput.-Aided Civ. Infrastruct. Eng. **37**(11), 1468–1487 (2022)

14. Paszke, A., Chaurasia, A., Kim, S., Culurciello, E.: ENet: a deep neural network architecture for real-time semantic segmentation. arXiv preprint arXiv:1606.02147 (2016)

15. Ren, S., He, K., Girshick, R., Sun, J.: Faster R-CNN: towards real-time object detection with region proposal networks. In: Advances in Neural Information Processing Systems, vol. 28 (2015)

16. Romera, E., Alvarez, J.M., Bergasa, L.M., Arroyo, R.: ERFNet: efficient residual factorized convnet for real-time semantic segmentation. IEEE Trans. Intell. Transp. Syst. **19**(1), 263–272 (2017)

17. Ronneberger, O., Fischer, P., Brox, T.: U-Net: convolutional networks for biomedical image segmentation. In: Navab, N., Hornegger, J., Wells, W.M., Frangi, A.F. (eds.) MICCAI 2015. LNCS, vol. 9351, pp. 234–241. Springer, Cham (2015). https://doi.org/10.1007/978-3-319-24574-4_28

18. Severstal: Steel defect detection, kaggle challange 2019 (2019). https://www.kaggle.com/c/severstal-steel-defect-detection

19. Wang, Y., et al.: LEDNet: a lightweight encoder-decoder network for real-time semantic segmentation. In: 2019 IEEE International Conference on Image Processing (ICIP), pp. 1860–1864. IEEE (2019)

20. Yu, C., Gao, C., Wang, J., Yu, G., Shen, C., Sang, N.: BiSeNet V2: bilateral network with guided aggregation for real-time semantic segmentation. Int. J. Comput. Vis. **129**, 3051–3068 (2021). https://doi.org/10.1007/s11263-021-01515-2
21. Zhang, D., Song, K., Xu, J., He, Y., Niu, M., Yan, Y.: MCnet: multiple context information segmentation network of no-service rail surface defects. IEEE Trans. Instrum. Meas. **70**, 1–9 (2020)
22. Zhang, X., Du, B., Wu, Z., Wan, T.: LAANet: lightweight attention-guided asymmetric network for real-time semantic segmentation. Neural Comput. Appl. **34**(5), 3573–3587 (2022). https://doi.org/10.1007/s00521-022-06932-z
23. Zhao, H., Qi, X., Shen, X., Shi, J., Jia, J.: ICNet for real-time semantic segmentation on high-resolution images. In: Ferrari, V., Hebert, M., Sminchisescu, C., Weiss, Y. (eds.) ECCV 2018. LNCS, vol. 11207, pp. 418–434. Springer, Cham (2018). https://doi.org/10.1007/978-3-030-01219-9_25

MaskDiffuse: Text-Guided Face Mask Removal Based on Diffusion Models

Jingxia Lu[1], Xianxu Hou[4], Hao Li[1], Zhibin Peng[1], Linlin Shen[1,2,3(✉)], and Lixin Fan[5]

[1] Computer Vision Institute, College of Computer Science and Software Engineering, Shenzhen University, Shenzhen, China
{lujingxia2021,lihao2021,pengzhibin2021}@email.szu.edu.cn
[2] WeBank Institute of Financial Technology, Shenzhen University, Shenzhen, China
llshen@szu.edu.cn
[3] National Engineering Laboratory for Big Data System Computing Technology, Shenzhen University, Shenzhen, China
[4] School of AI and Advanced Computing, Xi'an Jiaotong-Liverpool University, Suzhou, China
[5] WeBank Co., Ltd., Shenzhen, China

Abstract. As masked face images can significantly degrade the performance of face-related tasks, face mask removal remains an important and challenging task. In this paper, we propose a novel learning framework, called MaskDiffuse, to remove face masks based on Denoising Diffusion Probabilistic Model (DDPM). In particular, we leverage CLIP to fill the missing parts by guiding the reverse process of pretrained diffusion model with text prompts. Furthermore, we propose a multi-stage blending strategy to preserve the unmasked areas and a conditional resampling approach to make the generated contents consistent with the unmasked regions. Thus, our method achieves interactive user-controllable and identity-preserving masking removal with high quality. Both qualitative and quantitative experimental results demonstrate the superiority of our method for mask removal over alternative methods.

Keywords: Diffusion models · Mask removal · Text-to-image

1 Introduction

A common challenge for face perception is occlusion. Due to the large area of occlusion, face images with masks heavily degrade the performance of many face-related tasks, such as face detection [26], face identification [22], face recognition [8] and age estimation [12]. Removing masks is an effective way to improve the performance of these tasks.

Early works [7] mainly regard the mask removal problem as an inpainting task. They usually employ the generative adversarial network (GAN) to restore the masked facial areas while preserving the background of the original image. Although these approaches have provided appealing results, there are several

© The Author(s), under exclusive license to Springer Nature Singapore Pte Ltd. 2024
Q. Liu et al. (Eds.): PRCV 2023, LNCS 14430, pp. 435–446, 2024.
https://doi.org/10.1007/978-981-99-8537-1_35

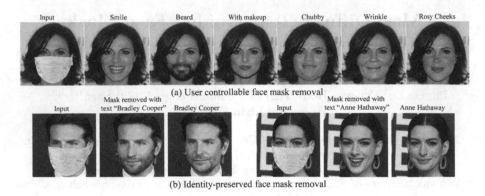

| Input | Smile | Beard | With makeup | Chubby | Wrinkle | Rosy Cheeks |

(a) User controllable face mask removal

| Input | Mask removed with text "Bradley Cooper" | Bradley Cooper | Input | Mask removed with text "Anne Hathaway" | Anne Hathaway |

(b) Identity-preserved face mask removal

Fig. 1. Our method enables (a) user controllable face mask removal guided by facial attribute descriptions and (b) identity-preserved mask removal guided by the name of a celebrity.

limitations. First, it is difficult to ensure that the filled content is visually consistent with the unmasked regions. Second, due to the deterministic process of GANs, the images generated by these approaches often lack diversity. In addition, existing methods mainly use predefined annotations such as landmarks and sketches as conditions. However, it's difficult for them to achieve highly intuitive and interactive image inpainting. Inspired by recent advances in vision-language pretraining [18] and diffusion models [10,24], we propose MaskDiffuse: an intuitive text-guided face mask removal method. The diffusion model is an emerging alternative paradigm for generative modeling and it can generate samples through an iterative denoising process. Our key idea is to fill in the missing parts by guiding the reverse process of pretrained diffusion model with text prompts. In particular, the guidance is achieved by employing CLIP [18], which has been recently shown to be effective for text-driven edits. In addition, we propose a multi-stage blending strategy to better preserve the unmasked areas during the reverse diffusion process. Furthermore, we adopt a conditional resampling approach to maintain the global visual consistency of the image. As a result, our proposed approach enables a simple and intuitive user interface for text-guided face mask removal. As shown in Fig. 1, we can achieve user-controllable face mask removal by using various facial attribute descriptions, and identity-preserved face mask removal by using the names of celebrities.

Concretely, the main contributions of this work are summarized as follows:

- We propose a diffusion model-based framework to achieve text-guided face mask removal. Our method makes full use of the semantics of images and textual descriptions in a joint framework.
- We further incorporate a multi-stage blending strategy and a conditional resampling approach during the reverse diffusion process, which can effectively preserve the unmasked facial areas and maintain the global visual consistency of the output images.

- We validate the effectiveness of our approach on various mask removal tasks, demonstrating the superiority of our method over other state-of-the-art techniques.

2 Related Work

2.1 Image Inpainting with GANs

GAN-based methods are commonly used for image inpainting with or without conditions. Unconditional image inpainting approaches directly predict the missing parts with the corrupted images and the corresponding masks. Early works achieve inpainting by using global and local discriminators [11], attention-based architecture [31] and multi-column convolutional neural networks [27]. By contrast, conditional image inpainting methods require additional information such as landmarks, sketches, or texts. For example, Yang et al. [30] achieve face inpainting conditioned on facial landmarks. Nazeri et al. [16] use the hallucinated edges as a prior to guide the inpainting process. Xiong et al. [28] propose to first predict the foreground contour and then restore the missing region using the predicted contour as guidance. Other works [13,32] achieve image inpainting based on text descriptions. However, these GAN-based methods usually require paired data and struggle to generate diverse images.

2.2 Image Inpainting with Diffusion Models

Another successful line of framework for image inpainting is based on diffusion models. Diffusion models [15,24,25] replace the unknown region of the image with a noised version of input image after each sampling step. Other methods such as GLIDE [17] and Palette [21] fine-tune the diffusion models to achieve inpainting. During fine-tuning, random regions of training examples are erased, and the remaining portions are then fed into the model along with a mask channel as additional information. However, in the context of face images, these methods usually generate artifacts.

2.3 Text-to-Image Synthesis

There are also works that generate images from text descriptions, known as text-to-image synthesis. Typically, a conditional model is trained to generate images from paired texts [29]. More recently, impressive results have been achieved by leveraging large scale auto-regressive [19] or diffusion models [20]. Rather than training condition models, several approaches [3,5] employ test-time optimization to explore the latent spaces of a pre-trained generator. These models typically guide the optimization to minimize a text-to-image similarity score derived from an auxiliary model such as CLIP [18]. However, existing methods are mainly designed for image generation tasks while we focus on inpainting the masked region of the facial image under the guidance of text descriptions.

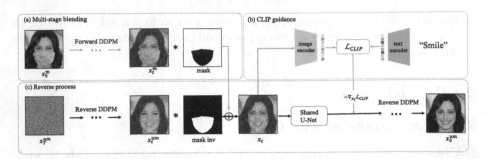

Fig. 2. Overview of our approach: (a) multi-stage blending: diffuse the clean image at step t and multiply it with a binary mask; (b) CLIP guidance: calculate the CLIP objective; (c) reverse process: denoise the image using the condition provided by (a) and (b).

3   Preliminaries: Denoising Diffusion Probabilistic Models

Denoising diffusion probabilistic models (DDPMs) [10,24] show promising performance in image generation, which consists of a forward diffusion process and a reverse diffusion process. Given a sample $x_0 \sim q(x_0)$, a Markov chain of latent variables $x_1, ..., x_T$ are produced in the forward process by progressively adding a small amount of Gaussian noise to the sample:

$$q\left(x_t \mid x_{t-1}\right) := \mathcal{N}\left(x_t; \sqrt{1 - \beta_t} x_{t-1}, \beta_t \mathbf{I}\right), \tag{1}$$

where $\beta_t \in (0, 1)$ is a noise schedule controlling the step size of adding noise. Since the Gaussian distributions are *i.i.d.*, we can directly sample from x_0 to any step x_t without generating the intermediate steps by denoting $\alpha_t := 1 - \beta_t$ and $\bar{\alpha}_t := \prod_{s=1}^{t} \alpha_s$:

$$q\left(x_t \mid x_0\right) := \mathcal{N}\left(x_t; \sqrt{\bar{\alpha}_t} x_0, (1 - \bar{\alpha}_t)\, \mathbf{I}\right), \tag{2}$$

furthermore, x_t can be expressed as a linear combination of x_0 and ϵ by reparameterize trick:

$$x_t = \sqrt{\bar{\alpha}_t} x_0 + \sqrt{1 - \bar{\alpha}_t}\, \epsilon, \tag{3}$$

where $\epsilon \sim \mathcal{N}(0, \mathbf{I})$ has the same dimension as x_0.

Since the reverse of the forward process $q\left(x_{t-1} \mid x_t\right)$ is intractable, DDPM learns parameterized Gaussian transitions $p_\theta\left(x_{t-1} \mid x_t\right)$. The generative (or reverse) process has the same functional form [24] as the forward process, and it is expressed as a Gaussian transition with learned mean and fixed variance [10]:

$$p_\theta\left(x_{t-1} \mid x_t\right) = \mathcal{N}\left(x_{t-1}; \mu_\theta\left(x_t, t\right), \sigma_t^2 \mathbf{I}\right). \tag{4}$$

Further, by decomposing μ_θ into a linear combination of x_t and the noise approximate ϵ_θ, the generative process is expressed as:

$$x_{t-1} = \frac{1}{\sqrt{\alpha_t}} \left(x_t - \frac{1 - \alpha_t}{\sqrt{1 - \bar{\alpha}_t}} \epsilon_\theta\left(x_t, t\right)\right) + \sigma_t \mathbf{z}, \tag{5}$$

where $\mathbf{z} \sim \mathcal{N}(0, \mathbf{I})$, and ϵ_θ represents a neural network used for noise prediction for each denoising process.

4 Method

Overview. We aim to remove the face mask and simultaneously generate the corresponding area consistent with the provided text descriptions. To achieve this goal, we seek to control the generation process of DDPM to fill the missing areas guided by a text-to-image similarity based on CLIP model. Furthermore, we propose a multi-stage blending strategy to preserve the unmasked areas and a conditional resampling approach to preserve the global visual consistency of the image.

4.1 Text-Guided Mask Removal

Problem Statement. Given a masked face image x^{m}, a text prompt d, and a binary mask m, our goal is to produce an unmasked image x^{um}, which should satisfy the following conditions: 1) the masked region $x^{\mathrm{um}} \odot m$ should be consistent with the text description d; 2) the unmasked areas $x^{\mathrm{um}} \odot (1 - m)$ are well preserved; 3) natural transition across the boundaries of inpainted regions.

Multi-stage Blending. First, we consider how to preserve the visual contents of unmasked areas. One straightforward way is to blend the inpainted image and the original image in pixel space. However, this naive blending could result in artifacts around the boundaries. As shown in Fig. 2(a), we propose a multi-stage blending strategy during the reverse diffusion process of DDPM. In particular, at each stage t of the reverse process, we first obtain the noised version of the masked image x_t^{m} according to Eq. 3 and the corresponding unmasked image x_t^{um} according to Eq. 5. Then the blended image is obtained as

$$x_{t-1} = m \odot x_{t-1}^{\mathrm{um}} + (1 - m) \odot x_{t-1}^{\mathrm{m}} \tag{6}$$

In our experiment, we perform the above blending for each reverse diffusion stage. As the structure of the generated content is mainly determined in the early steps of the diffusion process [9], natural transition on the mask boundary could be guaranteed with a multi-stage blending strategy.

CLIP Guidance. Inspired by guided diffusion [6] that uses the gradient information of a pretrained classifier to control the image generation process, we incorporate text information into the reverse diffusion process by leveraging CLIP model. Since the off-the-shelf CLIP models are trained on clean images, we estimate the clean image x_0 for current x_t according to the forward diffusion process as follows:

$$\widehat{x}_0 = \frac{x_t}{\sqrt{\bar{\alpha}_t}} - \frac{\sqrt{1 - \bar{\alpha}_t}\epsilon_\theta(x_t, t)}{\sqrt{\bar{\alpha}_t}} \tag{7}$$

Then, a CLIP-based loss $\mathcal{L}_{CLIP}(\widehat{x}_0, d)$ is used to measure the mismatch between the given text and image. Specifically, the cosine distance between the CLIP embedding of the text description and the embedding of the estimated clean image $\widehat{x}_0$ is calculated as:

$$\mathcal{L}_{CLIP}(\widehat{x}_0, d) = 1 - f(\widehat{x}_0) \cdot g(d) \tag{8}$$

where f and g denote the CLIP image encoder and text encoder, respectively. Finally, we use gradients of $\mathcal{L}_{CLIP}$ with respect to the x_t to guide the diffusion sampling process as:

$$x_{t-1} \sim \mathcal{N}(\mu_\theta(x_t, t) - s\Sigma\nabla_{x_t}\mathcal{L}_{CLIP}(\widehat{x}_0, d), \Sigma) \tag{9}$$

where s represents the scale of guidance.

4.2 Conditional Resampling

In order to strengthen the guidance of the CLIP, we propose a conditional resampling procedure during the reverse diffusion process. In particular, resampling is performed by adding Gaussian noise to a less noisy image $x_{t-\lambda}$ to generate a more noisy image x_t, which is then denoised back to image $x_{t-\lambda}$ again. λ is the step size at which the resampling is performed. This process would be repeated several times. Note that, the CLIP guidance is also incorporated during the denoising process. Thus, the CLIP guidance can be reinforced for each resampling process and more text-consistent content can be generated. Moreover, by resampling from the blended image, we can effectively preserve the global visual consistency of the generated image. The approach we mentioned above is described in Algorithm 1.

Algorithm 1 . Text guided mask removal: given a diffusion model $(\mu_\theta(x_t), \Sigma_\theta(x_t))$, and CLIP model

---

Input: masked facial image x^m, target text description d, input mask m
Output: edited image x with restored area according to the text description d
1:   $x_T \sim \mathcal{N}(0, I)$
2:   **for** $t = T, ..., 1$ **do**
3:     **for** $u = 1, ..., U$ **do**
4:       $\epsilon \sim N(0, \mathbf{I})$ if $t > 1$, else $\epsilon = 0$
5:       $\mu, \Sigma \leftarrow \mu_\theta(x_t), \Sigma_\theta(x_t)$
6:       $\widehat{x}_0 \leftarrow \frac{x_t}{\sqrt{\alpha_t}} - \frac{\sqrt{1-\bar{\alpha}_t}\,\epsilon_\theta(x_t, t)}{\sqrt{\alpha_t}}$
7:       $x_{t-1}^{um} \sim \mathcal{N}(\mu - \nabla_{x_t}\mathcal{L}_{CLIP}(\widehat{x}_0, d), \Sigma)$
8:       $x_{t-1}^m \sim \mathcal{N}(\sqrt{\bar{\alpha}_t}x^m, (1 - \bar{\alpha}_t)\mathbf{I})$
9:       $x_{t-1} \leftarrow x_{t-1}^{um} \odot m + x_{t-1}^m \odot (1 - m)$
10:      **if** $u < U$ and $t > 1$ **then**
11:        $x_t \sim \mathcal{N}\left(\sqrt{1 - \beta_{t-1}}x_{t-1}, \beta_{t-1}\mathbf{I}\right)$
12:      **end if**
13:     **end for**
14: **end for**
15: **return** x_0

---

5 Experiments and Applications

5.1 Experiment Settings

Pretrained Models. In our implementation, we choose CLIP model with pretrained ViT-B/32 weights. DDPM is pretrained on CelebA-HQ dataset provided by [14]. We use facial attribute descriptions to achieve user-controllable facial mask removal and use the names of celebrities for identity-preserved facial mask removal.

Training Details. We use $T = 250$ diffusion timesteps in our experiment and the resampling frequency λ is set to 10. We use Adam solver for optimization with a constant learning rate $3e^{-4}$. Our method is implemented using PyTorch on a single NVIDIA Tesla V100 GPU.

Fig. 3. User controllable mask removal: A comparison with PaintByWord++ [2, 4], Local CLIP-guided diffusion [3], Blended diffusion [1]. All baselines fail to well preserve the unmasked region, in contrast to the results of MaskDiffuse.

5.2 Comparison and Evaluation

Qualitative Results. We conduct qualitative and quantitative experiments to validate the effectiveness of the proposed method on the task of text-guided face mask removal. For comparison, we choose recently proposed CLIP-based methods, including PaintByWord++ [2,4], local CLIP guided diffusion [3] and blended diffusion [1]. The results of user-controllable face mask removal are shown in Fig. 3. we can see that although the local CLIP diffusion model can generate high-quality images, the unmasked region is modified. Though blended diffusion can preserve the unmasked region, it produces artifacts across the edge of the mask. In addition, the content generated by the two methods does not well match the text descriptions. By contrast, our MaskDiffuse is able to generate face inpainting consistent with the provided descriptions. Besides, we conduct a comparison between these methods on the task of identity-preserved face mask removal. As shown in Fig. 4, MaskDiffuse can better preserve the identity.

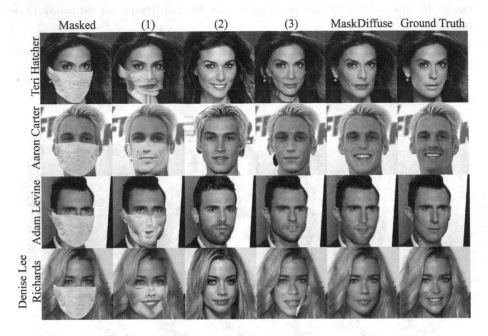

Fig. 4. Identity-preservation: A comparison with (1) PaintByWord++ [2,4], (2) Local CLIP-guided diffusion [3], (3) Blended diffusion [1]. Our method can better preserve the identity.

User Study. In addition, we conduct a user study for a quantitative comparison. Participants are asked to rate (scale of 1–5) the results shown in Fig. 3 and Fig. 4, in terms of realism, identity preservation, and agreement with the text description. Table 1 shows that MaskDiffuse outperforms the three baselines in all these aspects.

Table 1. User study results. ↑ indicates that higher is better.

Method	ID-preserve↑	Text match↑	Realism↑
PaintByWord++	1.38	1.82	1.28
Local CLIP GD	2.85	3.54	3.48
Blended diffusion	2.73	2.55	2.43
MaskDiffuse	**4.51**	**3.85**	**4.12**

5.3   Mask Removal for Face Recognition

Removing the mask has important applications in face recognition as the mask obscures important facial features such as the eyes, nose, and mouth. Our method can further be used to improve the accuracy of face recognition for masked faces by removing the masks. To this end, we conduct a face verification experiment following [23], which extracts the features of the two face images and calculates their Euclidean distance. Following the standard 1 : 1 verification setting, while the positive pair consists of the masked and the clean faces of the same identity, the negative pair contains the masked and clear faces from different identities. We collect both masked and clean faces of 100 celebrities from the Internet and randomly generate 100 positive and 100 negative pairs. The features are extracted by FaceNet and the similarity of two faces is compared with a threshold to decide whether the pair of faces is positive or negative. Figure 5 visualizes the ROC curve by setting the threshold from 0.5 to 1.5. We also compare our method with Local CLIP Guided Diffusion [3], PaintByWord++ [2,4], and Blended Diffusion [1]. As shown in Fig. 5, though Blended Diffusion, Paint-ByWord++, and Local CLIP Guided Diffusion are able to remove the mask and generate a clean face, the identity is not well preserved, i.e. the verification performances of the three approaches are even worse than that by using original masked faces. By contrast, MaskDiffuse can well preserve the identity when removing the mask, thus achieving the best verification accuracy. Furthermore, we conduct the comparison in terms of the Peak Signal Noise Ratio (PSNR) and Structural Similarity Index (SSIM) in Table 2. It can be seen that our approach outperforms others by a large margin.

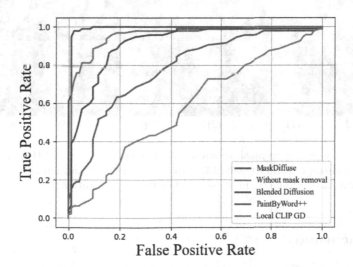

Fig. 5. ROC curves for PaintByWord++ [2,4], Local CLIP Guided Diffusion [3], Blended Diffusion [1], without mask removal and MaskDiffuse.

Table 2. Quantitative comparison for identity-preserved face mask removal. ↑ indicates that higher is better.

Method	PSNR↑	SSIM↑
PaintByWord++	9.76	0.43
Local CLIP GD	8.97	0.31
Blended diffusion	15.67	0.71
MaskDiffuse	**18.56**	**0.75**

5.4   Ablation Study

In this section, we conduct ablation studies to validate the effectiveness of individual components in our method: multi-stage blending, CLIP guidance, and conditional resampling. The random seeds are fixed throughout the experiments. Figure 6 shows the visual comparison by adding each component. We observe that the visual quality of faces generated by our full model is much better than other ablated versions. For example, when only using the CLIP guidance, the unmasked regions are significantly changed, and the identity is not well preserved when only using multi-stage blending. In addition, without conditional resampling, unwanted artifacts across the boundaries could be generated.

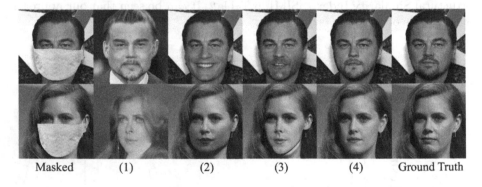

Masked   (1)   (2)   (3)   (4)   Ground Truth

Fig. 6. Identity-preserved mask removal for our model (1) with CLIP guidance only, (2) with multi-stage blending only, (3) with CLIP guidance and multi-stage blending, and (4) with CLIP guidance, multi-stage blending and conditional resampling.

6   Conclusion

In this paper, we present a novel denoising diffusion probabilistic model solution for text-guided face mask removal. By leveraging the knowledge learned from CLIP to guide the image generation process of diffusion models, we achieve user-controllable and identity-preserved face mask removal with high quality. Finally, we demonstrate the application of our method to the tasks of interactive face editing and face recognition, and its superiority over baselines.

Acknowledgement. This work was supported by the National Natural Science Foundation of China under Grant 62206180, 82261138629; Guangdong Basic and Applied Basic Research Foundation under Grant 2023A1515010688, 2020A1515111199 and 2022 A1515011018; Shenzhen Municipal Science and Technology Innovation Council under Grant JCYJ20220531101412030; and Swift Fund Fintech Funding.

References

1. Avrahami, O., Lischinski, D., Fried, O.: Blended diffusion for text-driven editing of natural images. In: Proceedings of the IEEE/CVF Conference on Computer Vision and Pattern Recognition, pp. 18208–18218 (2022)
2. Bau, D., et al.: Paint by word. arXiv preprint arXiv:2103.10951 (2021)
3. Crowson, K.: CLIP guided diffusion HQ 256x256 (2021). https://colab.research. google.com/drive/12a_Wrfi2_gwwAuN3VvMTwVMz9TfqctNj
4. Crowson, K.: VQGAN+CLIP (2021). https://colab.research.google.com/drive/ 1L8oL-vLJXVcRzCFbPwOoMkPKJ8-aYdPN
5. Crowson, K., et al.: VQGAN-CLIP: open domain image generation and editing with natural language guidance. In: Avidan, S., Brostow, G., Cissé, M., Farinella, G.M., Hassner, T. (eds.) ECCV 2022, Part XXXVII. LNCS, vol. 13697, pp. 88–105. Springer, Cham (2022). https://doi.org/10.1007/978-3-031-19836-6_6
6. Dhariwal, P., Nichol, A.: Diffusion models beat GANs on image synthesis. Adv. Neural. Inf. Process. Syst. **34**, 8780–8794 (2021)
7. Din, N.U., Javed, K., et al.: A novel GAN-based network for unmasking of masked face. IEEE Access **8**, 44276–44287 (2020)
8. He, X., Yan, S., Hu, Y., Niyogi, P., Zhang, H.J.: Face recognition using Laplacian faces. IEEE Trans. Pattern Anal. Mach. Intell. **27**(3), 328–340 (2005)
9. Hertz, A., Mokady, R., et al.: Prompt-to-prompt image editing with cross attention control. arXiv preprint arXiv:2208.01626 (2022)
10. Ho, J., Jain, A., Abbeel, P.: Denoising diffusion probabilistic models. Adv. Neural. Inf. Process. Syst. **33**, 6840–6851 (2020)
11. Iizuka, S., Simo-Serra, E., Ishikawa, H.: Globally and locally consistent image completion. ACM Trans. Graph. (ToG) **36**(4), 1–14 (2017)
12. Li, K., Xing, J., Hu, W., Maybank, S.J.: D2C: deep cumulatively and comparatively learning for human age estimation. Pattern Recogn. **66**, 95–105 (2017)
13. Lin, Q., Yan, B., Li, J., Tan, W.: MMFL: multimodal fusion learning for text-guided image inpainting. In: Proceedings of the 28th ACM International Conference on Multimedia, pp. 1094–1102 (2020)
14. Lugmayr, A., Danelljan, M., Romero, A., Yu, F., Timofte, R., Van Gool, L.: Repaint: inpainting using denoising diffusion probabilistic models. In: Proceedings of the IEEE/CVF Conference on Computer Vision and Pattern Recognition, pp. 11461–11471 (2022)
15. Meng, C., Song, Y., et al.: SDEdit: image synthesis and editing with stochastic differential equations. arXiv preprint arXiv:2108.01073 (2021)
16. Nazeri, K., Ng, E., Joseph, T., Qureshi, F.Z., Ebrahimi, M.: Edgeconnect: Generative image inpainting with adversarial edge learning. arXiv preprint arXiv:1901.00212 (2019)

17. Nichol, A., Dhariwal, P., et al.: Glide: towards photorealistic image generation and editing with text-guided diffusion models. arXiv preprint arXiv:2112.10741 (2021)

18. Radford, A., et al.: Learning transferable visual models from natural language supervision. In: International Conference on Machine Learning, pp. 8748–8763. PMLR (2021)

19. Ramesh, A., et al.: Zero-shot text-to-image generation. In: International Conference on Machine Learning, pp. 8821–8831. PMLR (2021)

20. Saharia, C., Chan, W., et al.: Photorealistic text-to-image diffusion models with deep language understanding. arXiv preprint arXiv:2205.11487 (2022)

21. Saharia, C., et al.: Palette: image-to-image diffusion models. In: ACM SIGGRAPH 2022 Conference Proceedings, pp. 1–10 (2022)

22. Samaria, F.S., Harter, A.C.: Parameterisation of a stochastic model for human face identification. In: Proceedings of 1994 IEEE Workshop on Applications of Computer Vision, pp. 138–142. IEEE (1994)

23. Schroff, F., Kalenichenko, D., Philbin, J.: FaceNet: a unified embedding for face recognition and clustering. In: Proceedings of the IEEE Conference on Computer Vision and Pattern Recognition, pp. 815–823 (2015)

24. Sohl-Dickstein, J., Weiss, E., Maheswaranathan, N., Ganguli, S.: Deep unsupervised learning using nonequilibrium thermodynamics. In: International Conference on Machine Learning, pp. 2256–2265. PMLR (2015)

25. Song, Y., Sohl-Dickstein, J., et al.: Score-based generative modeling through stochastic differential equations. arXiv preprint arXiv:2011.13456 (2020)

26. Viola, P., Jones, M.J.: Robust real-time face detection. Int. J. Comput. Vision **57**, 137–154 (2004)

27. Wang, Y., Tao, X., Qi, X., Shen, X., Jia, J.: Image inpainting via generative multi-column convolutional neural networks. In: Advances in Neural Information Processing Systems, vol. 31 (2018)

28. Xiong, W., et al.: Foreground-aware image inpainting. In: Proceedings of the IEEE/CVF Conference on Computer Vision and Pattern Recognition, pp. 5840–5848 (2019)

29. Xu, T., et al.: AttnGAN: fine-grained text to image generation with attentional generative adversarial networks. In: Proceedings of the IEEE Conference on Computer Vision and Pattern Recognition, pp. 1316–1324 (2018)

30. Yang, Y., Guo, X., Ma, J., Ma, L., Ling, H.: LaFIn: generative landmark guided face inpainting. arXiv preprint arXiv:1911.11394 (2019)

31. Yu, J., et al.: Generative image inpainting with contextual attention. In: Proceedings of the IEEE Conference on Computer Vision and Pattern Recognition, pp. 5505–5514 (2018)

32. Zhang, L., Chen, Q., Hu, B., Jiang, S.: Text-guided neural image inpainting. In: Proceedings of the 28th ACM International Conference on Multimedia, pp. 1302–1310 (2020)

Image Generation Based Intra-class Variance Smoothing for Fine-Grained Visual Classification

Zihan Yan[1], Ruoyi Du[1], Kongming Liang[1], Tao Wei[2], Wei Chen[2], and Zhanyu Ma[1(✉)]

[1] School of Artificial Intelligence, Beijing University of Posts and Telecommunications, Beijing 100876, China
mazhanyu@bupt.edu.cn
[2] Li Auto, Beijing, China

Abstract. Fine-grained visual classification (FGVC) is challenging because of the unsmooth intra-class data distribution caused by the combination of relatively significant intra-class variation and scarce training data. To this end, most works in FGVC focused on explicitly/implicitly enhancing the model representation ability. In this paper, however, we take a different stance – alleviating the unsmooth intra-class data distribution in FGVC datasets via data generation. In particular, we propose the following components for data augmentation: (i) SmoothGAN: an information-theoretic extension to the Generative Adversarial Network (GAN) that can generate high-quality fine-grained images with continuously varying intra-class differences. (ii) Dual-threshold-filtering: the generated data are selected according to both their reality and discriminability via SmoothGAN's discriminator and a basic FGVC model. Experiments on popular FGVC datasets demonstrate that training with augmented data can significantly boost model performance in the FGVC task. The code is available at https://github.com/PRIS-CV/SmoothGAN.

Keywords: Fine-Grained Visual Classification · Image Generation · Data Augmentation

1 Introduction

Fine-grained visual classification (FGVC) has made significant strides in various industrial and research domains, demonstrating its success in applications such

This work was supported in part by Beijing Natural Science Foundation Project No. Z200002, in part by National Natural Science Foundation of China (NSFC) No. U19B2036, U22B2038, 62106022, 62225601, in part by Youth Innovative Research Team of BUPT No. 2023QNTD02, in part by scholarships from China Scholarship Council (CSC) under Grant CSC No. 202206470055 and in part by BUPT Excellent Ph.D. Students Foundation No. CX2022152.

Q. Liu et al. (Eds.): PRCV 2023, LNCS 14430, pp. 447–459, 2024.
https://doi.org/10.1007/978-981-99-8537-1_36

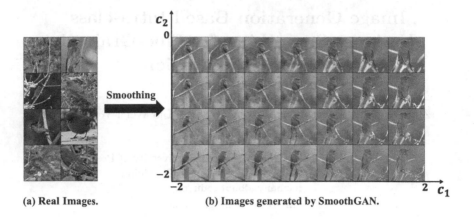

(a) Real Images.   (b) Images generated by SmoothGAN.

Fig. 1. Real images vs. Images generated by SmoothGAN. (a) Real images have large intra-class variation. (b) SmoothGAN disentangles representations on a category of CUB-200-2011. Varying only two latent codes c_1 from $[-2, 2]$ and c_2 from $[-2, 0]$, other latent codes and noise remain unchanged.

as automatic biodiversity monitoring [32], intelligent retail [19], intelligent transportation [37], *etc.* FGVC involves the classification of fine-grained sub-categories within an object category [13]. In comparison to conventional classification tasks, FGVC exhibits a greater degree of intra-class variation and inter-class similarity [9]. Additionally, fine-grained datasets tend to have a small-scale nature due to the presence of rare categories and the high cost associated with annotation. Consequently, this results in an uneven intra-class distribution within FGVC datasets, posing challenges for extracting discriminative features even for state-of-the-art deep models.

The majority of previous research in FGVC has primarily concentrated on improving the model's representational capacity by employing techniques such as localization or segmentation to explicitly extract discriminative features [14,34,38,39]. Despite significant progress, limited or no attempts have been made to address the root cause of the problem, namely, the mitigation of the non-uniform intra-class distribution in FGVC datasets. In this regard, an intuitive solution involves interpolating the data points and smoothing the data distribution through data augmentation techniques. This paper aims to investigate FGVC from this particular perspective.

Given the consistent advancements in generation models in recent years, such as Generative Adversarial Networks (GANs) [15] and Diffusion Models [18],*etc*, that excels in capturing the underlying distribution of real data and generating realistic samples. Hence, this paper specifically extends the framework of ProjectedGAN to achieve data interpolation. Nonetheless, generating samples directly using ProjectedGAN gives rise to two challenges. The first challenge is generating smooth data. As illustrated in Fig. 1 we aim to generate samples with continuous variations in intra-class attributes to facilitate data augmentation. The second challenge pertains to the potential unreliability of synthetic samples, characterized by their lack of naturalness or low distinguishability. These unreliable samples can adversely impact the performance of the FGVC model [29].

To overcome these challenges, our approach comprises two components: (i) **SmoothGAN**, an information-theoretic extension of ProjectedGAN capable of generating fine-grained images with high-quality and disentangled class-independent features, and (ii) **Dual-threshold-filtering**, a data selection strategy that filters out unreliable synthetic samples during the data augmentation process performed by the generation model.

Our main contribution lies in stating the optimization challenges caused by the non-smoothness of data and, for the first time, introducing data generation to achieve intra-class distribution smoothing for FGVC. Technically, building upon our motivation, we propose SmoothGAN – we employ a Conditional ProjectedGAN as the base model and incorporate the concept of mutual information constraint from InfoGAN to generate high-quality intra-class smooth-varying images. Additionally, we introduce the Dual-threshold-filtering mechanism to select high-quality images. Experiments on four widely used datasets demonstrate the effectiveness of the proposed method – the generated and filtered data can significantly boost model performance in the FGVC task.

2 Related Work

2.1 Fine-Grained Visual Classification

The success of deep learning has led to the development of numerous methods aimed at distinguishing between different fine-grained categories. Based on the level of annotation information utilized during training, FGVC methods relying on deep neural networks can be categorized into those based on strongly supervised information and those based on weakly supervised information.

Strongly supervised information-based methods [2,7,34,38] necessitate not only the inclusion of category annotations in the dataset but also annotated frame or part annotation point information during the model training process. Owing to the exorbitant labeling costs associated with strongly supervised information, recognition methods that solely rely on weakly supervised labeling information using category annotations have garnered increased attention [3,4,12,16,36]. In FGVC, the identification and acquisition of discriminative local information play a crucial role. Therefore, [6,14,24,35,39] enhance the accuracy of fine-grained image classification by precisely localizing and learning discriminative feature regions. While these ideas have worked well, there is less or no effort invested in getting to the root of the problem - smoothing out the intra-class distribution in the FGVC dataset.

2.2 Image Generation

Generative Adversarial Networks (GANs) [15] have emerged as one of the most successful techniques for generating highly realistic images. Traditional generative adversarial network models are trained through an adversarial game played between two entities: the discriminator $D(\cdot)$ and the generator $G(\cdot)$. The discriminator is trained to discern between generated and real samples, while the generator is trained to deceive the discriminator.

Image generation techniques have found extensive applications in both industrial and academic domains. These applications include image-to-image translation [41], face generation [20], image restoration [8], super-resolution image reconstruction [10], data augmentation [1], *etc.* As image generation techniques have evolved, it has become feasible to generate fine-grained images that contain distinctive features. Hence, this paper explores the generation of data for data augmentation using a generation model capable of producing high-quality fine-grained images that exhibit discernible features.

Fig. 2. Generated images of SmoothGAN on CUB-200-2011, Stanford Cars, Stanford Dogs, and FGVC-Aircraft datasets.

2.3   Data Augmentation

Data augmentation techniques enhance the training dataset by generating additional samples through transformations applied to real samples. Conventional data augmentation methods encompass geometric transformations such as rotation, scaling, white balancing, and image sharpening, which introduce geometric variations or distortions to the image [27]. The recent advancements in GAN have led to their application in data augmentation [1,40]. The majority of these studies employing generative models for data augmentation primarily aim to mitigate classifier overfitting. In our work, our objective is twofold: to alleviate overfitting by augmenting the sample size and to enhance the smoothness of the intra-class distribution within the Fine-grained Visual Classification (FGVC) dataset.

3   Methods

3.1   SmoothGAN

To address the issue of unsmooth intra-class distribution in the FGVC dataset through data augmentation, the generated images must meet the dual requirements of high quality and continuous intra-class variation. Considering the impressive performance of ProjectedGAN in generating high-quality fine-grained images, we selected it as the base model. Figure 1, 2 demonstrate that our generation model, SmoothGAN, built upon ProjectedGAN, is capable of generating high-quality images within the FGVC dataset. Furthermore, to enhance the smoothness of the intra-class data distribution and generate samples with continuously varying intra-class differences, we drew inspiration from [5] and developed SmoothGAN, which disentangles the class-independent features of each image class.

In contrast to traditional GANs, ProjectedGAN incorporates a collection of feature projectors $K_n(\cdot)$ that map real and generated images to the discriminator's input space, along with a set of independent discriminators $D_n(\cdot)$ that operate on distinct feature projections. The feature projectors $K_n(\cdot)$ are implemented using a feature extractor that produces multi-scale outputs. ProjectedGAN training objective $V(\cdot, \cdot)$ can thus be formulated as follows

$$\min_{G} \max_{\{D_n\}} V(\{D_n\}, G) = \sum_{n \in \mathcal{N}} \mathbb{E}_x[log D_n(K_n(x))] + \mathbb{E}_z[log(1 - D_n(K_n(G(z))))],$$

(1)

where x is real sample from real distribution $\mathbb{P}_x$ and z is latent vector sampled from a simple distribution $\mathbb{P}_z$.

To control the category of the generated images, we need to introduce the image label y as the condition. Hence, Conditional-ProjectedGAN training can be formulated as follows

$$\min_{G} \max_{\{D_n\}} V(\{D_n\}, G) = \sum_{n \in \mathcal{N}} \mathbb{E}_x[log D_n(K_n(x|y))] + \mathbb{E}_z[log(1 - D_n(K_n(G(z|y))))].$$

(2)

In SmoothGAN, we introduce an auxiliary distribution Q for each multi-scale discriminator of ProjectedGAN. Hence, SmoothGAN training can thus be formulated as follows

$$\min_{G, \{Q_n\}} \max_{\{D_n\}} V(\{D_n\}, G, \{Q_n\}) = V(\{D_n\}, G) - \sum_{n \in \mathcal{N}} \lambda L_I(G, Q_n),$$ (3)

where $L_I(G, Q_n)$ is a variational lower bound on the mutual information $I(c; G(z, c))$ of the latent code c and the generated distribution $G(z, c)$. With Monte Carlo simulation, $L_I(G, Q)$ is easy to approximate. In particular, L_I can be maximized with respect to Q directly and with respect to $G(\cdot)$ via the reparametrization trick [5]. Equation 3 is an extension of the formula in Info-GAN [5], primarily introduced to be compatible with ProjectedGAN by incorporating multiple auxiliary distributions Q.

3.2 Dual-Threshold-Filtering Strategy

As GAN sampling involves randomness, some generated samples may be unreliable and negatively impact the classifier. To address this issue, we propose a

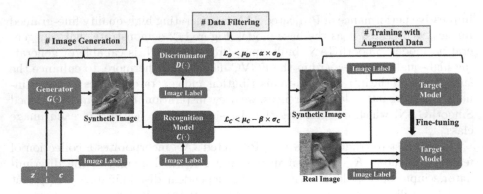

Fig. 3. Illustration of the proposed method. The input to the generator consists of random noise z', latent code c, and image label.

Dual-threshold-filtering strategy. In data augmentation, an ideal distribution of generated data should meet two criteria: proximity to the real data distribution and the presence of sufficient class-discriminative features. To align the generated data for data augmentation with the real data distribution, we employ the discriminator of SmoothGAN to filter the generated data. Furthermore, we utilize a fine-tuned classifier to select generated images with ample discriminative features for data augmentation.

In our approach, we evaluate the generated samples using both the discriminator $D(\cdot)$ of SmoothGAN and a fine-tuned classifier $C(\cdot)$, and retain only the samples with low discriminator and classifier losses. As for the classifier, we utilized a pre-trained model on ImageNet-1K and fine-tuned it on the training set. In practice, during the generation of samples for each class, k samples of that class are initially generated to compute the mean μ and variance σ of the discriminator loss $\mathcal{L}_D$ and classifier loss $\mathcal{L}_C$.

We filter the samples that will be used in the final training based on the conditions

$$\mathcal{L}_D < \mu_D - \alpha * \sigma_D, \tag{4}$$

and

$$\mathcal{L}_C < \mu_C - \beta * \sigma_C, \tag{5}$$

where α and β are adjustable hyperparameters.

3.3 Training Classifier with Augmented Data

Following the training of SmoothGAN and the generation of filtered generative data for data augmentation, we proceed to train the classification model using a combination of real data and the generated data until convergence is achieved. Figure 3 provides an illustration of the proposed method. Formally, the combined dataset contains N_1 real data and N_2 generated data. For a batch of real data $x_{real}, y_{real} \in \{(x_i, y_i)\}_{i=1}^{N_1}$ and a batch of generated data $x_{gen}, y_{gen} \in \{(x_i, y_i)\}_{i=1}^{N_2}$, where x_i is the image and y_i represents the corresponding one-hot category label, the classifier loss $\mathcal{L}_C$ can be formulated as follows

$$\mathcal{L}_C = \mathcal{H}(C(x_{real}), y_{real}) + \mathcal{H}(C(x_{gen}), y_{gen}), \tag{6}$$

where $C(\cdot)$ denotes the prediction of the classifier for the image and $\mathcal{H}(\cdot, \cdot)$ represents the cross-entropy function. In practice, we set $N_2 = 10 \times N_1$ to ensure the diversity of the generated samples, and the batch size for both real data and generated data is kept the same.

In order to mitigate overfitting to the generated data, we further fine-tune the model exclusively using real data until convergence is achieved. In this phase, the classifier loss $\mathcal{L}_C$ can be formulated as follows

$$\mathcal{L}_C = \mathcal{H}(C(x_{real}), y_{real}). \tag{7}$$

4 Experimental Results and Discussions

4.1 Datasets

We evaluate the proposed method on four FGVC datasets, including CUB-200-2011 [33], Stanford Cars [22], Stanford Dogs [21], and FGVC-Aircraft [26]. Detailed statistics about these datasets are listed in Table 1.

Table 1. Details of the FGVC datasets used in our experiments.

Dataset	Categories	Training	Test
CUB-200-2011	200	5994	5794
Stanford Cars	196	8144	8041
Stanford Dogs	120	12000	8559
FGVC-Aircreft	100	6667	3333

4.2 Implementation Details

For the training of SmoothGAN, we keep in line with ProjectedGAN. We use FastGAN_lite [25] as the generator and EfficientNet-Lite1 [31] as the feature network. Using the hinge loss [23], we train with a batch size of 64 until $8000k$ real images were presented to the discriminator, an amount sufficient to allow the generator to reach values close to convergence. The resolution of both the training and generated images is set to 256×256. The training of SmoothGAN is performed on the training set only. When training SmoothGAN and generating images, for the input latent vector $z = (z', c)$ of the generation model, we combine a 246-dimensional random Gaussian noise vector $z' \sim \mathcal{N}(0, 1)$ with a

Table 2. Performance of our data augmentation method on FGVC datasets, including CUB-200-2011, Stanford Cars, Stanford Dogs, and FGVC-Aircraft datasets. We take Accuracy (%) as the evaluation criterion.

Method	Backbone	Input size	CUB (%)	Cars (%)	Dogs (%)	Aircraft (%)
Baseline	VGG16	224 × 224	79.8	90.6	**84.3**	88.5
Baseline+Ours	VGG16	224 × 224	**80.4**	**90.9**	83.6	**88.9**
Baseline	ResNet50	224 × 224	81.6	92.6	87.7	88.5
Baseline+Ours	ResNet50	224 × 224	**82.8**	**93.0**	**87.9**	**90.2**
PMG [11]	ResNet50	224 × 224	83.3	92.5	86.8	89.2
PMG+Ours	ResNet50	224 × 224	**83.8**	**93.1**	**87.1**	**90.3**

10-dimensional latent code $c \sim U(-2, 2)$, resulting in a sequence of 10 consecutive latent codes. The coefficient λ in Eq. 3 is set to 0.1. For data filtering, we specify k as 1000, α as 1, and β as 0.3 in Sect. 3.2. We generate 300 filtered images for each class at a resolution of 256×256.

The classification model is trained using a pre-trained model on ImageNet-1K. Specifically, we employed ResNet50 [17] and VGG16 [30] as the pre-trained models for our experiments. During the training phase, the input images are resized to 256×256 and then randomly cropped to 224×224, only the class labels are introduced for training. Also, horizontal flipping is randomly applied to the data augmentation. We utilized the stochastic gradient descent (SGD) optimizer with a momentum of 0.9 and a weight decay of 0.0005. The learning rate of the fully connected (FC) layer was initialized to 0.01, while the learning rate of the pre-trained backbone was set to 1/10 of the FC layer. Initially, the model was trained for 100 epochs with a batch size of 32, maintaining the same ratio of real data to real data. Subsequently, we performed fine-tuning for 100 epochs exclusively on real data with a batch size of 16.

As a comparison, we replaced the generated data with real data for the initial 100 epochs of the aforementioned training strategy, while the final 100 epochs continued to be trained using only real data.

4.3   Main Results

We use ResNet50 and VGG16 as the backbone of the baseline method respectively to demonstrate the generalizability of the proposed method for different model structures. In addition, we test the proposed method on PMG [11], demonstrating that although the state-of-the-art FGVC model works well enough, our method can achieve further improvements on the existing state-of-the-art FGVC model. The experimental performance on the FGVC datasets is shown in Table 2.

As shown in Table 2, our proposed data augmentation method has significant improvement on the CUB-200-2011, Stanford Cars, and FGVC-Aircraft datasets for different backbone models as well as on the baseline and SOTA methods. For the Stanford Dogs dataset, the improvements are more pronounced with ResNet50 as the backbone method. For instance, ResNet50 equipped with our data augmentation approach achieves 82.8%/83.8% on Baseline/PMG method in the CUB-200-2011 dataset, which is +1.2%/+0.5% higher than the naive methods.

4.4 Ablation Study

In this section, we perform an ablation study on the CUB-200-2011 dataset, utilizing a ResNet50 backbone in the baseline method. The objective is to evaluate the efficacy of our proposed SmoothGAN structure, Dual-threshold-filtering strategy, and sampling strategy. This study allows us to assess the individual contributions of these components to the overall performance of the method.

SmoothGAN and Dual-Threshold-Filtering Strategy. The results of the ablation study conducted on the CUB-200-2011 dataset are presented in Table 3. The findings demonstrate the effectiveness of our proposed approaches. In particular, utilizing SmoothGAN as the generation model for data augmentation outperforms ProjectedGAN. This improvement can be attributed to ProjectedGAN's inability to generate images with continuous intra-class variation, leading to difficulties in smoothing the intra-class variation within the FGVC dataset. Moreover, the application of the Dual-threshold-filtering strategy yields benefits in terms of data augmentation from two distinct perspectives: ensuring that generated samples align closely with the true data distribution and selecting samples with discriminative features. Notably, the combination of both filtering strategies yields superior results compared to individual usage.

Table 3. Ablation study on CUB-200-2011, which demonstrates the effectiveness of the Dual-threshold-filtering Strategy and showcases the superior performance of Smooth-GAN compared to ProjectedGAN under the same filtering strategy.

Generation Model	Discriminator Filter	Classifier Filter	Accuracy (%)
Ours	✗	✗	81.4
	✓	✗	81.6
	✗	✓	81.7
	✓	✓	**82.8**
ProjectedGAN [28]	✗	✗	79.8
	✓	✓	81.4

Table 4. Study on sampling strategy, which demonstrates that SmoothGAN is capable of generating images with continuous variation by utilizing uniform distribution sampling on the latent code, thereby smoothing the intra-class distribution.

Generation Model	Sampling strategy	Accuracy (%)
Ours	Gaussian random+Uniform random	**82.8**
	Gaussian random	82.0
ProjectedGAN [28]	Gaussian random+Uniform random	81.4
	Gaussian random	81.4

SmoothGAN and Sampling Strategy. When generating images for data augmentation, two different sample strategies can be employed: (1) For the input latent vector z for the generation model, combining a 246-dimensional random Gaussian noise vector z' with a 10-dimensional latent code c as mentioned in Sect. 4.2. (2) Directly using a 256-dimensional random Gaussian noise vector as the input for the generation model. To ensure fairness, we applied the same data filtering strategy for both sampling strategies. The results are presented in Table 4, where the first sampling strategy utilized uniform sampling for the latent code, enabling the generation of images with continuous intra-class variation, as depicted in Fig. 1, thereby smoothing the intra-class data distribution.

5   Conclusion

In this paper, we approached the fine-grained visual classification problem from the data augmentation perspective – generating images to smooth intra-class variance within the FGVC dataset. Our primary contribution is identifying the optimization challenges arising from the non-smoothness of data and, for the first time, introducing data generation as a means to achieve intra-class distribution smoothing in the FGVC domain. Technically, building upon our motivation, we propose SmoothGAN – we employ a Conditional ProjectedGAN as the base model and incorporate the concept of mutual information constraint from Info-GAN to generate high-quality intra-class smooth-varying images. Additionally, we introduce the Dual-threshold-filtering mechanism to select images from both the perspective of reality and discriminability. We evaluate our method on four commonly used FGVC datasets and demonstrate its effectiveness.

References

1. Antoniou, A., Storkey, A., Edwards, H.: Data augmentation generative adversarial networks. arXiv preprint arXiv:1711.04340 (2017)
2. Branson, S., Van Horn, G., Belongie, S., Perona, P.: Bird species categorization using pose normalized deep convolutional nets. arXiv preprint arXiv:1406.2952 (2014)

3. Chang, D., et al.: Complex scenario-oriented fine-grained visual classification platform. In: 2022 IEEE 24th International Workshop on Multimedia Signal Processing (MMSP), p. 1. IEEE (2022)
4. Chang, D., Pang, K., Du, R., Ma, Z., Song, Y.Z., Guo, J.: Making a bird AI expert work for you and me. arXiv preprint arXiv:2112.02747 (2021)
5. Chen, X., Duan, Y., Houthooft, R., Schulman, J., Sutskever, I., Abbeel, P.: Infogan: interpretable representation learning by information maximizing generative adversarial nets. In: Advances in Neural Information Processing Systems, vol. 29 (2016)
6. Chen, Y., Bai, Y., Zhang, W., Mei, T.: Destruction and construction learning for fine-grained image recognition. In: Proceedings of the IEEE/CVF Conference on Computer Vision and Pattern Recognition, pp. 5157–5166 (2019)
7. Cui, Y., Zhou, F., Lin, Y., Belongie, S.: Fine-grained categorization and dataset bootstrapping using deep metric learning with humans in the loop. In: Proceedings of the IEEE Conference on Computer Vision and Pattern Recognition, pp. 1153–1162 (2016)
8. Denton, E., Gross, S., Fergus, R.: Semi-supervised learning with context-conditional generative adversarial networks. arXiv preprint arXiv:1611.06430 (2016)
9. Ding, Y., Zhou, Y., Zhu, Y., Ye, Q., Jiao, J.: Selective sparse sampling for fine-grained image recognition. In: Proceedings of the IEEE/CVF International Conference on Computer Vision, pp. 6599–6608 (2019)
10. Dong, C., Loy, C.C., He, K., Tang, X.: Image super-resolution using deep convolutional networks. IEEE Trans. Pattern Anal. Mach. Intell. **38**(2), 295–307 (2015)
11. Du, R., et al.: Fine-grained visual classification via progressive multi-granularity training of jigsaw patches. In: Vedaldi, A., Bischof, H., Brox, T., Frahm, J.-M. (eds.) ECCV 2020, Part XX. LNCS, vol. 12365, pp. 153–168. Springer, Cham (2020). https://doi.org/10.1007/978-3-030-58565-5_10
12. Du, R., Xie, J., Ma, Z., Chang, D., Song, Y.Z., Guo, J.: Progressive learning of category-consistent multi-granularity features for fine-grained visual classification. IEEE Trans. Pattern Anal. Mach. Intell. **44**(12), 9521–9535 (2021)
13. Dubey, A., Gupta, O., Guo, P., Raskar, R., Farrell, R., Naik, N.: Pairwise confusion for fine-grained visual classification. In: Proceedings of the European Conference on Computer Vision (ECCV), pp. 70–86 (2018)
14. Fu, J., Zheng, H., Mei, T.: Look closer to see better: recurrent attention convolutional neural network for fine-grained image recognition. In: Proceedings of the IEEE Conference on Computer Vision and Pattern Recognition, pp. 4438–4446 (2017)
15. Goodfellow, I., et al.: Generative adversarial networks. Commun. ACM **63**(11), 139–144 (2020)
16. Guo, Y., Du, R., Li, X., Xie, J., Ma, Z., Dong, Y.: Learning calibrated class centers for few-shot classification by pair-wise similarity. IEEE Trans. Image Process. **31**, 4543–4555 (2022)
17. He, K., Zhang, X., Ren, S., Sun, J.: Deep residual learning for image recognition. In: Proceedings of the IEEE Conference on Computer Vision and Pattern Recognition, pp. 770–778 (2016)
18. Ho, J., Jain, A., Abbeel, P.: Denoising diffusion probabilistic models. Adv. Neural. Inf. Process. Syst. **33**, 6840–6851 (2020)
19. Karlinsky, L., Shtok, J., Tzur, Y., Tzadok, A.: Fine-grained recognition of thousands of object categories with single-example training. In: Proceedings of the

IEEE Conference on Computer Vision and Pattern Recognition, pp. 4113–4122 (2017)

20. Karras, T., Laine, S., Aittala, M., Hellsten, J., Lehtinen, J., Aila, T.: Analyzing and improving the image quality of stylegan. In: Proceedings of the IEEE/CVF Conference on Computer Vision and Pattern Recognition, pp. 8110–8119 (2020)

21. Khosla, A., Jayadevaprakash, N., Yao, B., Li, F.F.: Novel dataset for fine-grained image categorization: Stanford dogs. In: Proceedings of CVPR Workshop on Fine-Grained Visual Categorization (FGVC), vol. 2. Citeseer (2011)

22. Krause, J., Stark, M., Deng, J., Fei-Fei, L.: 3D object representations for fine-grained categorization. In: Proceedings of the IEEE International Conference on Computer Vision Workshops, pp. 554–561 (2013)

23. Lim, J.H., Ye, J.C.: Geometric GAN. arXiv preprint arXiv:1705.02894 (2017)

24. Lin, T.Y., RoyChowdhury, A., Maji, S.: Bilinear CNN models for fine-grained visual recognition. In: Proceedings of the IEEE International Conference on Computer Vision, pp. 1449–1457 (2015)

25. Liu, B., Zhu, Y., Song, K., Elgammal, A.: Towards faster and stabilized GAN training for high-fidelity few-shot image synthesis. In: International Conference on Learning Representations (2020)

26. Maji, S., Rahtu, E., Kannala, J., Blaschko, M., Vedaldi, A.: Fine-grained visual classification of aircraft. arXiv preprint arXiv:1306.5151 (2013)

27. Perez, L., Wang, J.: The effectiveness of data augmentation in image classification using deep learning. arXiv preprint arXiv:1712.04621 (2017)

28. Sauer, A., Chitta, K., Müller, J., Geiger, A.: Projected GANs converge faster. Adv. Neural. Inf. Process. Syst. **34**, 17480–17492 (2021)

29. Shi, H., Wang, L., Ding, G., Yang, F., Li, X.: Data augmentation with improved generative adversarial networks. In: 2018 24th International Conference on Pattern Recognition (ICPR), pp. 73–78. IEEE (2018)

30. Simonyan, K., Zisserman, A.: Very deep convolutional networks for large-scale image recognition. arXiv preprint arXiv:1409.1556 (2014)

31. Tan, M., Le, Q.: Efficientnet: rethinking model scaling for convolutional neural networks. In: International Conference on Machine Learning, pp. 6105–6114. PMLR (2019)

32. Van Horn, G., et al.: Building a bird recognition app and large scale dataset with citizen scientists: the fine print in fine-grained dataset collection. In: Proceedings of the IEEE Conference on Computer Vision and Pattern Recognition, pp. 595–604 (2015)

33. Wah, C., Branson, S., Welinder, P., Perona, P., Belongie, S.: The Caltech-UCSD birds-200-2011 dataset (2011)

34. Wei, X.S., Xie, C.W., Wu, J., Shen, C.: Mask-CNN: localizing parts and selecting descriptors for fine-grained bird species categorization. Pattern Recogn. **76**, 704–714 (2018)

35. Xiao, T., Xu, Y., Yang, K., Zhang, J., Peng, Y., Zhang, Z.: The application of two-level attention models in deep convolutional neural network for fine-grained image classification. In: Proceedings of the IEEE Conference on Computer Vision and Pattern Recognition, pp. 842–850 (2015)

36. Yang, Z., Luo, T., Wang, D., Hu, Z., Gao, J., Wang, L.: Learning to navigate for fine-grained classification. In: Proceedings of the European Conference on Computer Vision (ECCV), pp. 420–435 (2018)

37. Yin, J., Wu, A., Zheng, W.S.: Fine-grained person re-identification. Int. J. Comput. Vision **128**(6), 1654–1672 (2020)

38. Zhang, N., Donahue, J., Girshick, R., Darrell, T.: Part-based R-CNNs for fine-grained category detection. In: Fleet, D., Pajdla, T., Schiele, B., Tuytelaars, T. (eds.) ECCV 2014. LNCS, vol. 8689, pp. 834–849. Springer, Cham (2014). https://doi.org/10.1007/978-3-319-10590-1_54

39. Zheng, H., Fu, J., Mei, T., Luo, J.: Learning multi-attention convolutional neural network for fine-grained image recognition. In: Proceedings of the IEEE International Conference on Computer Vision, pp. 5209–5217 (2017)

40. Zheng, Z., Zheng, L., Yang, Y.: Unlabeled samples generated by GAN improve the person re-identification baseline in vitro. In: Proceedings of the IEEE International Conference on Computer Vision, pp. 3754–3762 (2017)

41. Zhu, J.Y., Park, T., Isola, P., Efros, A.A.: Unpaired image-to-image translation using cycle-consistent adversarial networks. In: Proceedings of the IEEE International Conference on Computer Vision, pp. 2223–2232 (2017)

Cross-Domain Soft Adaptive Teacher for Syn2Real Object Detection

Weijie Guo[1], Boyong He[1], Yaoyuan Wu[1], Xianjiang Li[2], and Liaoni Wu[1,2](✉)

[1] Institute of Artificial Intelligence, Xiamen University, Xiamen 361005, China
wuliaoni@xmu.edu.cn
[2] School of Aerospace Engineering, Xiamen University, Xiamen 361102, China

Abstract. Current state-of-the-art object detectors are constructed using supervised deep-learning approaches. These approaches require a large amount of annotated training data. Although synthetic image-generation methods can provide a large amount of annotated data, unsupervised transfer of object-recognition models from synthetic to real domains is a complicated problem given the large gap between the domains. To mitigate this problem, in this paper, we propose a general synthetic-to-real cross-domain object-detection framework. In this framework, we establish a simple mean teacher model for most detectors and propose a teacher–student framework named soft adaptive teacher (SAT). This leverages domain adversarial learning and domain-adaption augmentation to address the domain gap. Specifically, we alleviate bias by augmenting training samples with image-level adaptations for the student model. Moreover, we employ feature-level adversarial training in the student model, allowing features derived from the source and target domains to share similar distributions. Finally, we introduce the soft teacher mechanism to select reliable pseudo-labels for the teacher model. By tackling the model-bias issue using these strategies, our SAT model was found to achieve average precision values of 57.2% (55.7%) on the Sim10k to Cityscape (Sim10k to BDD100k) benchmarks, 3.1 (10.4) percentage points higher than the previous state-of-the-art methods. Furthermore, we achieved an average precision of 66.2% on the dataset for object detection in aerial images (DOTA), and this is 31.2% points higher than the results from the Faster RCNN model without domain adaptation trained only with labeled source domain images.

Keywords: Unsupervised domain adaption · Synthetic to real · Object detection

1 Introduction

Undoubtedly, the success of machine-learning methods in visual-recognition tasks relies heavily on obtaining large annotated datasets. Current powerful com-

Supplementary Information The online version contains supplementary material available at https://doi.org/10.1007/978-981-99-8537-1_37.

puting capabilities have driven significant advances in modern object-detection methods based on convolutional neural networks (CNNs) [1–6]. However, the problem of data scarcity has become a major obstacle to the development and application of deep learning in certain fields. While collecting more training data can alleviate this issue, the expensive and time-consuming annotation process can make it impractical. In some scenarios, e.g., biomedical images, it may even be impossible to obtain extensive and precise annotations.

With the ongoing breakthroughs in related fields, using synthetic data to train models has become a leading way to address data scarcity. Data can be collected from artificially constructed scenarios [7], and generative models can be used to produce data with instance-level labels [8]. Although synthetic data offers the potential for low annotation costs and high labeling accuracy, if a model is trained using only synthetic data and then applied to real-world scenarios, its performance will be significantly degraded. The research focus of domain-adaptive object detection (DAOD) [9–15] is to address the domain shift between the source and target datasets. Typically, DAOD uses labeled data from a source domain and unlabeled data from a target domain to train a stable and robust detection model. To overcome the significant differences between synthetic and real images, in this work, we developed a universal domain-adaptation algorithm library. This algorithm library integrates current mainstream domain-adaptation methods and is compatible with multiple object-detection algorithms; it is not just limited to Faster RCNN [1].

The research achievements of this work can be summarized as follows.

- We established a universal domain-adaptation algorithm library, which includes commonly used domain-adaptation methods such as domain adversarial training, pseudo-labeling, image style transfer, and mean teacher (MT). We have adapted most of the current object-detection methods, including single-stage detectors, two-stage detectors, anchor-free series, and detection transformer (DETR) series, thus constructing a complete domain-adaptation algorithm benchmark.
- We introduced the domain-adaptation augmentation method to perform online image transformations. This method can align the source domain dataset with the target domain dataset in terms of image style, thereby further reducing the gap between the source and target domains.
- We established a simple MT model for most object detectors. To improve the quality of pseudo-labels, we introduced a soft teacher (ST) mechanism to handle pseudo-labels and proposed the soft adaptive teacher (SAT) framework. We further extended this pseudo-label handling method to the Cascade RCNN detectors.

2   Related Works

Object Detection. Modern object-detection methods can be grouped into two categories. The first type is the two-stage architectures, such as Faster RCNN [1] and Mask R-CNN [5], which first extract the regions of interest and then perform bounding-box classification and regression. The second type is the single-stage detectors, such as SSD [3], feature pyramid networks (FPN) [6], fully convolutional one-stage (FCOS) detection [2], adaptive training sample selection (ATSS) [4], and you only look once (YOLO) [16,17] series, which directly output bounding boxes and classes from predefined anchors. Although the former class of object-detection method has slightly higher accuracy, the latter is faster and more compact, making these approaches more suitable for time-sensitive applications and for computationally limited edge devices. Recently, the Transformer architecture [18] has been successfully applied in the field of image processing and has been embedded in various end-to-end DETR approaches [19,20]. Compared to current mainstream CNN methods, Transformer is more robust [21]. However, despite achieving high detection accuracy, DETRs still suffer from the sophistication of their architectures and issues with slow convergence.

Cross-Domain Object Detection. The early cross-domain object-detection methods mainly focused on using the two-stage detector Faster RCNN [9–11, 22,23]. The pioneering DA-Faster algorithm [9] incorporated a gradient-reversal layer [24], and this was the first approach to include image-level and instance-level alignment to improve cross-domain performance. Strong–weak distribution alignment (SWDA) [10] uses a similar method of strong local and weak global feature alignment to further improve performance. The stacked complementary losses (SCL) method [11] represents a gradient-detach-based approach that also uses Faster RCNN as its detector. The unbiased mean teacher (UMT) [25], cycle-consistent [26], and hierarchical transferability calibration network (HTCN) [23] approaches use CycleGAN [27] to generate simulated training images to alleviate domain bias.

In contrast, some recent research has attempted to solve the DAOD problem using one-stage detectors. For example, EPMDA [28] adapts FCOS [2] to explicitly extract object masks, I^3Net [29] introduces complementary modules specifically designed for the SSD [3] architecture. Inspired by semi-supervised learning methods, some works have adopted the MT framework [30] for unsupervised domain adaptation in object detection. Mean teacher with object relations (MTOR) [31] is based on MT but trains its teacher network by implementing region-level, image-level, and intra-image consistency. UMT [25] approaches cross-domain object detection from the perspective that detection models are prone to bias toward source images, and it uses three effective strategies to mitigate model bias. Finally, the adaptive teacher algorithm [15] applies weak–strong augmentation and adversarial training to address domain-discrepancy problems.

3 Proposed Method

3.1 Definition of Terms

For the cross-domain object-detection task, we have a set of source images $\mathbf{I}^s$ annotated with a total of N object bounding boxes $\mathcal{B} = \{B_j|_{j=1}^N, B_j =$

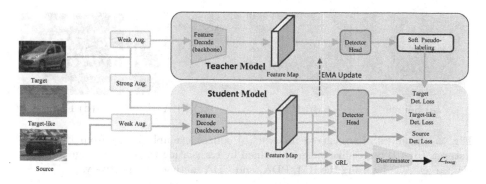

Fig. 1. Overview of our proposed SAT. In each mini-batch, three types of image are used: target images, target-like images, and source images. The source images and target-like images, which have annotations, are used to optimize the object-detection loss of the student model. The teacher model generates soft pseudo-labels [32] to train the student, while the student updates the teacher model with an exponential moving average (EMA). A discriminator with a gradient-reversal layer is employed to align the distributions across the two domains in the student model.

$(x_j, y_j, w_j, h_j)\}$ and corresponding class labels $\mathcal{C} = \{C_j|_{j=1}^N, \ C_j \in (0, 1, \ldots, c)\}$ from c object classes, and a set of unlabeled target images $\mathbf{I}^t$. With N_s source images $\mathbf{I}^s$, the source bounding-box coordinates set $\mathcal{B}^s$, and the class-labels set $\mathcal{C}^s$, we represent the source domain as $\mathcal{D}_s = \{(\mathbf{I}_i^s, \mathcal{B}_i^s, \mathcal{C}_i^s)|_{i=1}^N\}$. Similarly, we define the target domain with N_t label-invisible images as $\mathcal{D}_t = \{I_t|_{i=1}^N\}$.

Following these definitions, the loss for training a supervised model with the labeled source dataset $\mathcal{D}_s$ can be written as:

$$\mathcal{L}_{\text{det}}(\mathbf{I}^s, \mathcal{B}^s, \mathcal{C}^s) = \mathcal{L}_{\text{box}}(\mathcal{B}^s; \mathbf{I}^s) + \mathcal{L}_{\text{cls}}(\mathcal{C}^s; \mathbf{I}^s), \tag{1}$$

where $\mathcal{L}_{\text{box}}$ is the loss for predicted bounding boxes, and $\mathcal{L}_{\text{cls}}$ is the loss of classification probability. In the next sections, we introduce the unsupervised domain adaptation related to the unlabeled target domain images $\mathcal{D}_t$.

3.2 Domain-Adaption Augmentation

As noted in a previous report relating to the UMT [25], although the MT model can improve the robustness of predictions for the target domain, the learned model often has a bias toward the source domain because supervised training relies mainly on labeled samples from the source domain. As shown in Fig. 2, to solve this problem, we use three domain-adaptation augmentation methods to complete image style transfer. These are histogram matching (HM), Fourier-domain adaptation (FDA), and pixel-distribution adaptation (PDA). These methods are performed online and do not require pre-training like Cycle-GAN [27]. In short, this method can make the model more favorable to target

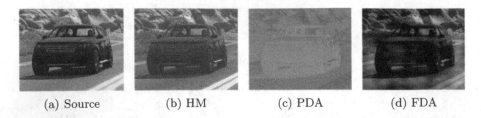

| (a) Source | (b) HM | (c) PDA | (d) FDA |

Fig. 2. Examples of translated images: (a) source example image from the Sim10k dataset; (b)–(d) target-like images obtained by translating the source image in (a) into the target style by using the HM, PDA, and FDA methods, respectively.

domain data, thereby reducing bias toward source domain data. The loss for target-like images can be written as:

$$\mathcal{L}_{\mathrm{det}}(\mathbf{P}^s, \mathcal{B}^s, \mathcal{C}^s) = \mathcal{L}_{\mathrm{box}}(\mathcal{B}^s; \mathbf{P}^s) + \mathcal{L}_{\mathrm{cls}}(\mathcal{C}^s; \mathbf{P}^s) \tag{2}$$

which has the same form as the loss in Eq. (1), but with $\mathbf{I}^s$ replaced by $\mathbf{P}^s$.

3.3 Feature Adversarial Learning

The primary objective of feature adversarial learning is to reduce the large domain distribution gap between the source domain and the target domain. As shown in Fig. 1, to better adapt to different detectors, we chose image-level feature adaptation [9] as the method for adversarial feature learning. Specifically, we extract image features from the backbone or neck feature layer of the model as the domain representation.

We minimize the distance between the two domains to align features by attaching domain discriminators to feature extractors at various feature levels. Then, we train the feature-extractor backbone and domain discriminators through adversarial training. We use the binary cross-entropy loss for the domain discriminator as follows:

$$\mathcal{L}_{\mathrm{img}} = -\sum_{i=1}^{N} [G_i \log P_i + (1 - G_i) \log (1 - P_i)], \tag{3}$$

where $i \in \{1, \ldots, N\}$ represents the N training images, $G_i \in \{1, 0\}$ is the ground truth of the domain label in the i-th training image (1 and 0 indicate the source and target domains, respectively), and P_i is the prediction of the domain classifier.

3.4 Mean Teacher and Soft Adaptive Teacher

As shown in Fig. 1, the MT model consists of two models with identical architectures: the student model and the teacher model. Inspired by the UMT [25], we first use available source data $\mathcal{D}_s$ to train the model, the loss $\mathcal{L}_{\mathrm{sup}}$ is the same as $\mathcal{L}_{\text{det}}$. The teacher model is used to generate pseudo-labels on the target domain images for student training. To filter out noisy pseudo-labels, we set a confidence threshold σ on the predicted bounding boxes of the teacher model to remove false positives. Therefore, after obtaining pseudo-labels on the target domain images from the teacher model, $\mathcal{L}_{\text{unsup}}$ is used to update the weights of the student model again. This is expressed:

$$\mathcal{L}_{\text{unsup}}(\mathbf{I}^s, \mathcal{B}^t, \mathcal{C}^t) = \mathcal{L}_{\text{box}}(\mathcal{B}^t; \mathbf{I}^s) + \mathcal{L}_{\text{cls}}(\mathcal{C}^t; \mathbf{I}^s), \tag{4}$$

where $\mathcal{B}_t$ and $\mathcal{C}_t$ are pseudo-labels generated by a teacher model. Here, we use both classification and bounding-box pseudo-labels. Following the basic MT model, we use the EMA formula to update the weights of the teacher model so that it generates more accurate pseudo-labels. The EMA formula is:

$$\theta_t \leftarrow \alpha\theta_t + (1 - \alpha)\theta_s, \tag{5}$$

where θ_t and θ_s represent the weight parameters of the teacher and student models, respectively. As pointed out for the ST model in a previous report [32], the performance of an MT model depends on the quality of the pseudo-labels; simply using a higher threshold can cause significant damage to the performance of the detector. To improve the quality of the pseudo-labels, we draw inspiration from the ST mechanism to handle the pseudo-labels and establish the SAT. Compared to the basic MT model, the SAT model greatly improves the synthetic-to-real transfer effect.

3.5 Full Objective

The overall loss of the MT and SAT models is:

$$\mathcal{L} = \mathcal{L}_{\text{det}}(\mathbf{I}^s, \mathcal{B}^s, \mathcal{C}^s) + \mathcal{L}_{\text{det}}(\mathbf{P}^s, \mathcal{B}^s, \mathcal{C}^s) + \lambda_{\text{unsup}}\mathcal{L}_{\text{det}}(\mathbf{I}^t, \mathcal{B}^t, \mathcal{C}^t) + \lambda_{\text{img}}\mathcal{L}_{\text{img}}, \tag{6}$$

where λ_{unsup} and λ_{img} are trade-off parameters that are used to control the corresponding loss weights.

4   Experiments

To validate the effectiveness of our approach, we compared it with state-of-the-art methods for cross-domain object detection on benchmark datasets in three sets of synthetic-to-real experiments.

Implementation Details. To compare the present approach with previous methods, we used the pre-trained ResNet-101 model on ImageNet as the backbone of the Faster RCNN [1] model. Furthermore, we also conducted extensive experiments on other detectors, such as ATSS [4] and DETR with improved denoising anchor boxes (DINO) [33]. For the MT and SAT models, unless otherwise stated, we set the trade-off parameters $\lambda_{unsup} = 1.0$ and $\lambda_{img} = 0.1$ for all the experiments. We set the confidence threshold $\sigma = 0.8$ in all our experiments, and the weight-smoothing coefficient parameter α of the exponential moving average for the teacher model was set to 0.99.

To aid understanding of the experimental results (for example, in Table 1), we provide the following definitions: (1) FAL is the normal model with our domain-adaption augmentation and a feature adversarial learning strategy, as described in Sects. 3.2 and 3.3; (2) MT is the mean teacher model with our domain-adaption augmentation strategy, as described in Sects. 3.2 and 3.3. (3) SAT is the mean teacher model with both of our strategies, as described in Sects. 3.2, 3.3, and 3.4.

Table 1. Average precision (AP, in %) of different methods for cross-domain object detection on the validation set of Cityscapes for Sim10k → Cityscapes adaptation.

Model	Method	AP on car
Faster RCNN	Source Only	43.1
	SCL [11]	42.6
	DA-Faster [9]	38.9
	SCDA [22]	43.0
	SWDA [10]	40.1
	HTCN [23]	42.5
	UMT [25]	43.1
	D-adapt [34]	51.9
	AWADA [35]	54.1
	FAL (ours)	51.1
	MT (ours)	**55.4**
	SAT (ours)	**57.2**
	Oracle	70.4

Model	Method	AP on cars
ATSS	Source Only	48.5
	FAL (ours)	**54.0**
	MT (ours)	**54.8**
	Oracle	69.8
DINO	Source Only	43.6
	FAL (ours)	**47.7**
	MT (ours)	**57.5**
	Oracle	74.4

4.1 Sim10k to Cityscapes

Datasets. Sim10k [36] includes 10,000 images rendered by the Grand Theft Auto game engine with bounding-box annotations for 58,701 cars. All of these images can be used for training. The Cityscapes [37] dataset is a city-scene dataset used in the field of autonomous driving; this includes 2,975 images in the training set and 500 images in the validation set.

In addition to the baselines compared previously, we further included SCL [11], DA-Faster [9], selective cross-domain alignment (SCDA) [22], SWDA [10], HTCN [23], D-adapt [34], UMT [25], and attention-weighted adversarial domain adaptation (AWADA) [35] for comparison. As mentioned above, we not only

Table 2. Average precision (AP, in %) of different methods for cross-domain object detection on the validation set of BDD100K for Sim10k → BDD100k adaptation.

Model	Method	AP on car
Faster RCNN	Source Only	32.3
	SWDA [10]	42.9
	CDN [38]	45.3
	FAL (ours)	42.4
	MT (ours)	**52.5**
	SAT (ours)	**55.7**
	Oracle	76.7
ATSS	Source Only	36.8
	FAL (ours)	**42.6**
	MT (ours)	**54.4**
	Oracle	74.7
DINO	Source Only	34.5
	FAL (ours)	**39.6**
	MT(ours)	**55.8**
	Oracle	77.7

Table 3. The average precision (AP, in %) of different methods for cross-domain object detection on the validation set of DOTA for RarePlanes → DOTA adaptation.

Method	Model	AP on airplane
Faster RCNN	Source Only	35.0
	FAL (ours)	36.3
	MT (ours)	**62.9**
	SAT (ours)	**66.2**
	Oracle	96.8
ATSS	Source Only	37.9
	FAL (ours)	**42.6**
	MT (ours)	**61.3**
	Oracle	96.1
DINO	Source Only	29.8
	FAL (ours)	**32.6**
	MT (ours)	**50.4**
	Oracle	94.5

report the experimental results for the Faster RCNN [1] model but also for the ATSS [4] and DINO [33] models.

Results. The experimental results of Sim10k [36] to Cityscapes [37] are shown in Table 1. We take the Sim10k [36] dataset and Faster RCNN [1] as an example to explain the experimental results. In particular, the normal FAL model obtains an average precision (AP) of 51.1%, which outperforms the result of 18% from the source-only baseline. This demonstrates that domain-adaption augmentation and feature adversarial learning can help to considerably improve the robustness of the object-detection model against data variance. We then replaced the normal model with the MT model, which caused a further improvement to 55.4%. Then, by introducing the ST mechanism, we finally improved the results to 57.2%, and this surpasses the state-of-the-art AWADA [35] results, which means that the present method represents the new state-of-the-art performance for cross-domain object detection on the Cityscape dataset. Additionally, we report the experimental results of FAL and MT on the ATSS [4] and DINO [33] models.

4.2 Sim10k to BDD100K

Datasets. To fully evaluate the proposed method, we replaced Cityscapes [37] with BDD100K [39] as the target dataset. This contains 100,000 high-definition video sequences and samples of key frames for each video at the 10th second, resulting in 100,000 images that have been instance annotated. The category-labeling space of BDD100k [39] is not completely consistent with Cityscapes

Table 4. Ablation study for AP on the Sim10k [36] → Cityscapes [37] experiment. In the first row, Img and Aug indicate the image-level adaptation module and domain-adaption augmentation, respectively, MT indicates the simple MT model, and Soft denotes the ST [32] mechanism. Source Only indicates the Faster RCNN model without domain adaptation trained only with labeled source domain images.

	Aug	Img	Soft	MT	AP
Source Only					43.1
Aug	✓				49.9
Img		✓			45.2
MT (Baseline)				✓	52.5
MT+Aug	✓			✓	54.9
MT+Img		✓		✓	54.3
MT+Soft			✓	✓	55.3
MT+Img+Aug+Soft (SAT)	✓	✓	✓	✓	**57.2**

[37], so we converted it to an eight-class labeling space to be consistent with Cityscapes [37].

Results. The experimental results are shown in Table 2. Similar to the Sim10k to Cityscapes experiment, we report the experimental results of Faster RCNN [1], ATSS [4], and DINO [33]. The SAT result improved by 31.2% compared to the source and surpassed that of conditional domain normalization (CDN) [38], achieving an AP of 57.9%. This once again demonstrates the effectiveness of our proposed method.

4.3   RarePlanes to DOTA

Datasets. RarePlanes [40] is a dataset that combines real and synthetic satellite images. The real part of the dataset includes 253 Maxar WorldView-3 satellite scenes spanning 112 locations, with 14,700 airplanes manually labeled. The accompanying synthetic dataset was generated using a novel simulation platform and contains 50,000 synthetic satellite images with approximately 630,000 airplane annotations. The dataset for object detection in aerial images (DOTA) [41] is a large-scale remote sensing dataset consisting of 2,806 aerial images. The targets cover various scales, positions, and shapes and are annotated by experts in the field of remote sensing into 15 categories. The fully annotated dataset consists of a total of 188,282 instances. In our experiments, we split the DOTA [41] dataset into a series of images of 800 × 800 pixels each and aligned both datasets using airplanes as one category.

Results. The experimental results are shown in Table 3. As in the previous experiments, we observe that our MT and SAT approach gradually improves the MT model by addressing its model bias with different strategies. This again demonstrates the effectiveness of our proposed approach.

4.4 Ablation Study on Components

The effect of each individual proposed component for the domain-adaptation detection method was investigated. All experiments were conducted with the same Faster RCNN and RestNet-101 backbone on the Sim10k to Cityscapes experiment. The results are presented in Table 4. The results of this ablation study clearly demonstrate the positive effect of each proposed component of the domain-adaptive object detection.

5 Conclusion

In this paper, we propose a framework for Syn2Real object detection including commonly used domain-adaptation methods such as domain adversarial training, pseudo-labeling, image transformation, and semi-supervised learning. This can be adapted to the majority of object-detection algorithms. Our training pipeline design, combined with online image transformation and adversarial learning, solves the problem of source domain bias in teacher–student models. By introducing the ST mechanism, we further improve the accuracy of pseudo-labels. We conducted transfer learning experiments from Sim10k to Cityscapes, from Sim10k to BDD100K, and from RarePlanes to DOTA. The experimental results show that the proposed method is quite effective.

References

1. Ren, S., He, K., Girshick, R., Sun, J.: Faster R-CNN: towards real-time object detection with region proposal networks. In: Advances in Neural Information Processing Systems, vol. 28 (2015)
2. Tian, Z., Shen, C., Chen, H., He, T.: FCOS: fully convolutional one-stage object detection. In: Proceedings of the IEEE/CVF International Conference on Computer Vision, pp. 9627–9636 (2019)
3. Liu, W., et al.: SSD: single shot MultiBox detector. In: Leibe, B., Matas, J., Sebe, N., Welling, M. (eds.) ECCV 2016. LNCS, vol. 9905, pp. 21–37. Springer, Cham (2016). https://doi.org/10.1007/978-3-319-46448-0_2
4. Zhang, S., Chi, C., Yao, Y., Lei, Z., Li, S.Z.: Bridging the gap between anchor-based and anchor-free detection via adaptive training sample selection. In: Proceedings of the IEEE/CVF Conference on Computer Vision and Pattern Recognition, pp. 9759–9768 (2020)
5. He, K., Gkioxari, G., Dollár, P., Girshick, R.: Mask R-CNN. In: Proceedings of the IEEE International Conference on Computer Vision, pp. 2961–2969 (2017)
6. Lin, T.Y., Dollár, P., Girshick, R., He, K., Hariharan, B., Belongie, S.: Feature pyramid networks for object detection. In: Proceedings of the IEEE Conference on Computer Vision and Pattern Recognition, pp. 2117–2125 (2017)

7. He, B., Li, X., Huang, B., Gu, E., Guo, W., Wu, L.: Unityship: a large-scale synthetic dataset for ship recognition in aerial images. Remote Sens. **13**(24), 4999 (2021)
8. Ge, Y., Xu, J., Zhao, B.N., Itti, L., Vineet, V.: DALL-E for detection: language-driven context image synthesis for object detection. arXiv preprint arXiv:2206.09592 (2022)
9. Chen, Y., Li, W., Sakaridis, C., Dai, D., Van Gool, L.: Domain adaptive faster R-CNN for object detection in the wild. In: Proceedings of the IEEE Conference on Computer Vision and Pattern Recognition, pp. 3339–3348 (2018)
10. Saito, K., Ushiku, Y., Harada, T., Saenko, K.: Strong-weak distribution alignment for adaptive object detection. In: Proceedings of the IEEE/CVF Conference on Computer Vision and Pattern Recognition, pp. 6956–6965 (2019)
11. Shen, Z., Maheshwari, H., Yao, W., Savvides, M.: SCL: towards accurate domain adaptive object detection via gradient detach based stacked complementary losses. arXiv preprint arXiv:1911.02559 (2019)
12. Khodabandeh, M., Vahdat, A., Ranjbar, M., Macready, W.G.: A robust learning approach to domain adaptive object detection. In: Proceedings of the IEEE/CVF International Conference on Computer Vision, pp. 480–490 (2019)
13. Yao, X., Zhao, S., Xu, P., Yang, J.: Multi-source domain adaptation for object detection. In: Proceedings of the IEEE/CVF International Conference on Computer Vision, pp. 3273–3282 (2021)
14. Li, W., Liu, X., Yuan, Y.: Sigma: semantic-complete graph matching for domain adaptive object detection. In: Proceedings of the IEEE/CVF Conference on Computer Vision and Pattern Recognition, pp. 5291–5300 (2022)
15. Li, Y.J., et al.: Cross-domain adaptive teacher for object detection. In: Proceedings of the IEEE/CVF Conference on Computer Vision and Pattern Recognition, pp. 7581–7590 (2022)
16. Redmon, J., Farhadi, A.: Yolov3: an incremental improvement. arXiv preprint arXiv:1804.02767 (2018)
17. Bochkovskiy, A., Wang, C.Y., Liao, H.Y.M.: Yolov4: optimal speed and accuracy of object detection. arXiv preprint arXiv:2004.10934 (2020)
18. Dosovitskiy, A., et al.: An image is worth 16x16 words: transformers for image recognition at scale. arXiv preprint arXiv:2010.11929 (2020)
19. Carion, N., Massa, F., Synnaeve, G., Usunier, N., Kirillov, A., Zagoruyko, S.: End-to-end object detection with transformers. In: Vedaldi, A., Bischof, H., Brox, T., Frahm, J.-M. (eds.) ECCV 2020. LNCS, vol. 12346, pp. 213–229. Springer, Cham (2020). https://doi.org/10.1007/978-3-030-58452-8_13
20. Zhu, X., Su, W., Lu, L., Li, B., Wang, X., Dai, J.: Deformable DETR: deformable transformers for end-to-end object detection. arXiv preprint arXiv:2010.04159 (2020)
21. Bhojanapalli, S., Chakrabarti, A., Glasner, D., Li, D., Unterthiner, T., Veit, A.: Understanding robustness of transformers for image classification. In: Proceedings of the IEEE/CVF International Conference on Computer Vision, pp. 10231–10241 (2021)
22. Zhu, X., Pang, J., Yang, C., Shi, J., Lin, D.: Adapting object detectors via selective cross-domain alignment. In: Proceedings of the IEEE/CVF Conference on Computer Vision and Pattern Recognition, pp. 687–696 (2019)
23. Chen, C., Zheng, Z., Ding, X., Huang, Y., Dou, Q.: Harmonizing transferability and discriminability for adapting object detectors. In: Proceedings of the IEEE/CVF Conference on Computer Vision and Pattern Recognition, pp. 8869–8878 (2020)

24. He, Z., Zhang, L.: Domain adaptive object detection via asymmetric tri-way faster-RCNN. In: Vedaldi, A., Bischof, H., Brox, T., Frahm, J.-M. (eds.) ECCV 2020. LNCS, vol. 12369, pp. 309–324. Springer, Cham (2020). https://doi.org/10.1007/978-3-030-58586-0_19

25. Deng, J., Li, W., Chen, Y., Duan, L.: Unbiased mean teacher for cross-domain object detection. In: Proceedings of the IEEE/CVF Conference on Computer Vision and Pattern Recognition, pp. 4091–4101 (2021)

26. Zhang, D., Li, J., Xiong, L., Lin, L., Ye, M., Yang, S.: Cycle-consistent domain adaptive faster RCNN. IEEE Access **7**, 123903–123911 (2019)

27. Zhu, J.Y., Park, T., Isola, P., Efros, A.A.: Unpaired image-to-image translation using cycle-consistent adversarial networks. In: Proceedings of the IEEE International Conference on Computer Vision, pp. 2223–2232 (2017)

28. Hsu, C.-C., Tsai, Y.-H., Lin, Y.-Y., Yang, M.-H.: Every pixel matters: center-aware feature alignment for domain adaptive object detector. In: Vedaldi, A., Bischof, H., Brox, T., Frahm, J.-M. (eds.) ECCV 2020. LNCS, vol. 12354, pp. 733–748. Springer, Cham (2020). https://doi.org/10.1007/978-3-030-58545-7_42

29. Chen, C., Zheng, Z., Huang, Y., Ding, X., Yu, Y.: I3Net: implicit instance-invariant network for adapting one-stage object detectors. In: Proceedings of the IEEE/CVF Conference on Computer Vision and Pattern Recognition, pp. 12576–12585 (2021)

30. Tarvainen, A., Valpola, H.: Mean teachers are better role models: weight-averaged consistency targets improve semi-supervised deep learning results. In: Advances in Neural Information Processing Systems, vol. 30 (2017)

31. Cai, Q., Pan, Y., Ngo, C.W., Tian, X., Duan, L., Yao, T.: Exploring object relation in mean teacher for cross-domain detection. In: Proceedings of the IEEE/CVF Conference on Computer Vision and Pattern Recognition, pp. 11457–11466 (2019)

32. Xu, M., et al.: End-to-end semi-supervised object detection with soft teacher. In: Proceedings of the IEEE/CVF International Conference on Computer Vision, pp. 3060–3069 (2021)

33. Caron, M., et al.: Emerging properties in self-supervised vision transformers. In: Proceedings of the IEEE/CVF International Conference on Computer Vision, pp. 9650–9660 (2021)

34. Jiang, J., Chen, B., Wang, J., Long, M.: Decoupled adaptation for cross-domain object detection. arXiv preprint arXiv:2110.02578 (2021)

35. Menke, M., Wenzel, T., Schwung, A.: Awada: attention-weighted adversarial domain adaptation for object detection. arXiv preprint arXiv:2208.14662 (2022)

36. Johnson-Roberson, M., Barto, C., Mehta, R., Sridhar, S.N., Rosaen, K., Vasudevan, R.: Driving in the matrix: can virtual worlds replace human-generated annotations for real world tasks? arXiv preprint arXiv:1610.01983 (2016)

37. Cordts, M., et al.: The cityscapes dataset for semantic urban scene understanding. In: Proceedings of the IEEE Conference on Computer Vision and Pattern Recognition, pp. 3213–3223 (2016)

38. Su, P., et al.: Adapting object detectors with conditional domain normalization. In: Vedaldi, A., Bischof, H., Brox, T., Frahm, J.-M. (eds.) ECCV 2020. LNCS, vol. 12356, pp. 403–419. Springer, Cham (2020). https://doi.org/10.1007/978-3-030-58621-8_24

39. Yu, F., et al.: BDD100K: a diverse driving dataset for heterogeneous multitask learning. In: Proceedings of the IEEE/CVF Conference on Computer Vision and Pattern Recognition, pp. 2636–2645 (2020)

40. Shermeyer, J., Hossler, T., Van Etten, A., Hogan, D., Lewis, R., Kim, D.: Rareplanes: synthetic data takes flight. In: Proceedings of the IEEE/CVF Winter Conference on Applications of Computer Vision, pp. 207–217 (2021)
41. Xia, G.S., et al.: DOTA: a large-scale dataset for object detection in aerial images. In: Proceedings of the IEEE Conference on Computer Vision and Pattern Recognition, pp. 3974–3983 (2018)

Dynamic Graph-Driven Heat Diffusion: Enhancing Industrial Semantic Segmentation

Jiaquan Li[1], Min Jiang[2(✉)], and Minghui Shi[2]

[1] Institute of Artificial Intelligence, Xiamen University, Xiamen 361005, China
jiaquanli@stu.xmu.edu.cn
[2] School of Informatics, Xiamen University, Xiamen 361005, China
{minjiang,smh}@xmu.edu.cn

Abstract. Dust significantly impacts construction progress and worker health, necessitating the use of machine learning for dust area identification and pollution mitigation. Existing dust semantic segmentation methods face limitations due to data quality, leading to suboptimal performance in real-world applications. These limitations stem from complex backgrounds and the model's inadequate ability to extract dust features, resulting in conflicting training processes. To overcome these challenges, we propose a thermal energy propagation training approach. Our method incorporates a thermal propagation mechanism into the self-training framework to dynamically adjust the weights of the training loss. This dynamic weight assignment is achieved through a graph-based approach. By utilizing ground truth, we generate an initial weight distribution that gradually weakens from the center to the surrounding areas, expanding as training progresses. This dynamic weighting scheme enables the model to gain insights into pixel clusters in the early stages and focus on learning edge areas in later stages. Throughout training iterations, the teacher model guides the dynamic adjustment of weight distribution, continuously updating it to capture essential regional information. This dynamic graph-driven weight assignment enhances the model's ability to accurately extract dust features. Experimental evaluations conducted on diverse datasets demonstrate the competitive performance of our approach compared to previous methods. The proposed dynamic graph-driven heat diffusion technique addresses the limitations posed by complex backgrounds and data quality, making it a valuable tool for dust area identification. It enables improved construction progress monitoring and worker health protection in industrial settings.

Keywords: Semantic Segmentation · Dynamic Graph · Self-Training

---

Supported in part by the National Natural Science Foundation of China under Grant 6227622.

Q. Liu et al. (Eds.): PRCV 2023, LNCS 14430, pp. 473–484, 2024.
https://doi.org/10.1007/978-981-99-8537-1_38

1   Introduction

Dust is a type of particulate matter that resembles smoke produced by flames and exhibits characteristics such as wide dispersion and high sensitivity to brightness. It is frequently encountered in various construction environments, posing detrimental effects not only on the overall progress of construction projects but also on the health of the workers involved. To mitigate the impact of dust, it is crucial to promptly identify and control its generation, often through measures like watering and dust suppression. However, conventional methods relying on physical sensors such as infrared and laser devices prove to be cost-intensive when applied to larger areas. Additionally, the limited receptive field of these sensors significantly hampers the detection of dust activities on a large scale.

The application of artificial intelligence has significantly permeated various industries and fields, offering numerous deep learning solutions aimed at addressing industrial challenges. For instance, employing a semantic segmentation model, which analyzes RGB images captured by cameras, can accurately identify the spatial location of dust and facilitate prompt responses. Leveraging deep learning solutions can effectively prevent extensive dust dispersion and mitigate substantial costs [16]. Nevertheless, these methods encounter several notable limitations. Firstly, their performance heavily relies on the quality of the collected data for modeling purposes. Recognizing this issue, our proposed approach treats the training process as a semi-supervised learning task, enabling the model to not only depend on data labels but also leverage the inductive characteristics of data features. Secondly, the presence of complex backgrounds often confuses the model, leading to incorrect classifications and yielding limited performance outcomes. Analyzing the model's performance results, we observed an increased number of mistakes caused by similar background colors. This highlights the model's failure to learn semantic and shape features, contributing to the aforementioned challenges. Inspired by this observation, we introduce the concept of heat propagation and propose a dynamic graph-driven weighting method to enhance the training process. Different from graph-based data mining methods to explore data feature preferences [2, 13]. The objective of this method is to provide more instructive guidance during training, enabling the model to acquire more effective semantic and shape features.

Our method draws inspiration from the principles of heat propagation and the law of energy conservation, equating unit heat with pixel weights. During the training process, we employ a self-training teacher-student model that assigns pseudo-labels and adjusts pixel weights to guide the student model's training. This serves two purposes: firstly, it prevents excessive reliance of the student model on the training data, acting as a form of regularization. Secondly, it encourages the student model to learn meaningful features from the data.

Our method was evaluated on the DSS dataset created by Yuan et al. [19], and the landing dust project dataset (DustProj) respectively. After adopting our proposed method, the accuracy improvement of 4.4% on the DSS dataset and 7.4% on the DustProj dataset were respectively obtained.

2   Related Work

Deep semantic segmentation is a contemporary approach that utilizes deep learning models to analyze image data and predict segmentation labels. Typically, deep semantic segmentation methods employ convolutional encoding-decoding structures to generate desired outcomes. The encoder's objective is to generate hierarchical, low-resolution image features in a progressive manner, while the decoder samples these features and maps them to category scores for each pixel. Mainstream semantic segmentation models can be categorized into two types. The first type includes models that leverage attention mechanisms, such as Segformer [15]. These methods exhibit strong fitting and memory capabilities, making them highly beneficial for pre-training. Many state-of-the-art semantic segmentation results are achieved using this type of model. However, they suffer from high computational complexity and struggle to quickly achieve performance advantages with limited data. The second type mainly relies on multi-scale feature convolution methods, such as Deeplabv3 [1], and DDRNet [4]. These methods, when combined with effective data augmentation techniques, achieve high accuracy requirements and demonstrate broader applicability across different devices.

2.1   Self-training Framework

The self-training framework is a widely used semi-supervised learning method in computer vision tasks. Over the years, numerous studies have employed this framework to enhance model performance in vision tasks effectively. The foundation of this framework lies in the teacher-student model concept. The teacher model predicts unlabeled data, assigns pseudo-labels based on a defined threshold, and then utilizes these pseudo-labels to guide the training of the student model. Pseudo-labels are used as a knowledge transfer medium to distill teacher model knowledge to student models. Many existing transfer methods have proved the effectiveness of knowledge transfer [6–8]. The self-training framework has demonstrated promising results in both semi-supervised and unsupervised domains. Notably, DAFormer [5] is a representative unsupervised method, while CReST [20] stands out as a semi-supervised method. In this study, we leverage the self-training framework as an extension of the effective guidance model training. We employ the pseudo-labels generated by the teacher model to guide dynamic weight assignment and update the gradient direction. This approach accelerates the fitting accuracy and efficiency of the student model.

3   Method

3.1   Industrial Semantic Segmentation

In semantic segmentation, our objective is to leverage a labeled dataset $D_l = \{x_i^l, y_i^l\}_{i=1}^{N^l}$ to train a semantic segmentation model that exhibits consistent and

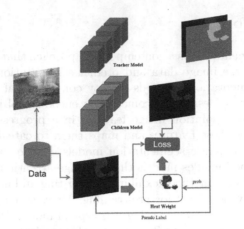

Fig. 1. Overview of training process, Teacher model generated pseudo label and probability to update Heat Weight.

reliable performance. However, in the context of data-poor environments, acquiring suitable labeled data poses significant challenges. Not only is the quantity of labeled data scarce, but the quality of data labeling is also compromised due to the influence of adverse conditions (Fig. 1).

To strike a balance between data acquisition costs and model robustness, semi-supervised semantic segmentation employs various methods to utilize unlabeled data for the same task. These methods aim to maximize the benefits derived from unlabeled data. Typically, semi-supervised approaches employ techniques such as adversarial training [10], self-training, and others to introduce an unlabeled loss component that enhances the model's generalization in the unlabeled domain. The overall loss function can be expressed as follows:

$$L = L_s + \lambda_u L_u \tag{1}$$

where L_s represents the supervised loss computed on the labeled dataset D_l, while L_u represents the unsupervised loss computed on the unlabeled dataset D_u. The weight coefficient λ_u controls the relative importance of the unsupervised loss component. This formulation allows for the integration of both labeled and unlabeled data into the training process, enabling the model to benefit from a broader range of information sources.

In contrast, industrial data is characterized by numerous adverse environmental factors. Simply introducing external unlabeled data can lead to challenges such as failure to converge and negative transfer problems. Recognizing this, we revisited the methodology and identified the importance of mitigating conflicting properties. To address this, we propose a novel approach utilizing the teacher-student model, which employs a labeled dynamic weight assignment coefficient instead of incorporating an unlabeled loss. The revised total loss function is defined as follows:

$$L = \frac{1}{N^l} \sum_{i=1}^{N^l} -\tau_i(y_i^l \log(p_i) + (1 - y_i^l) \log(1 - p_i)) \tag{2}$$

Here, τ_i represents the dynamic weight value (see Sect. 3.2 for details), while p_i corresponds to the predicted probability for pixel i. By incorporating this dynamic weight assignment scheme, we can effectively manage conflicting properties and enhance the convergence and performance of the model in industrial scenarios.

Pseudo-labels. Pseudo-labels bridge the gap between unlabeled data and the semantic segmentation model, enabling the model to leverage the informative aspects of unlabeled data and enhance its overall performance. In the context of binary semantic segmentation in our method, we utilize the labeled dataset D_l to generate pseudo-labels, which assist in heap propagation (see Sect. 3.2 for more details). The feature extraction module f and the segmentation head h collaborate to predict the sigmoid probability for the j-th pixel in the i-th image:

$$p_{i,j}^l = \text{sigmoid}(h(f(x_{i,j}^l))) \tag{3}$$

where $p_{i,j}^l$ represents the confidence level associated with the predicted classification label for the pixel $x_{i,j}^l$. A higher confidence value indicates a greater certainty in the model's classification. To ensure the reliability of online-generated pseudo-labels and minimize potential misguidance from incorrect labels, we apply a threshold to filter out pseudo-labels with low confidence. The pseudo-label generation process is as follows:

$$y_{i,j}^l = \begin{cases} 1, & \text{if } p_{i,j}^l > \gamma_t, \\ \text{ignore}, & \text{otherwise} \end{cases} \tag{4}$$

In this equation, γ_t represents the confidence threshold at the t-th iteration. The choice of setting γ_t as a constant or allowing it to vary during training depends on the specific task. In industrial environments, where labeling quality tends to be unreliable, online predicted labels often cluster within a floating interval. Consequently, employing a variable confidence threshold during training typically improves accuracy, as it can adapt to the data's specific characteristics and handle the prevalent labeling uncertainties found in industrial settings.

3.2   Dynamic Weight Graph by Heat Propagation

Conflicts in data labels often emerge due to the subjectivity of the data and the interference of background elements, such as varying observation perspectives. The aforementioned pseudo-labels offer a means to assess the training progress of the model. To mitigate unnecessary data conflicts, we propose a dynamic weight graph method that combines the pseudo-labels generated by the teacher model.

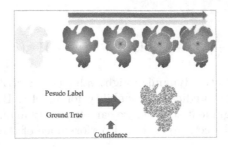

Fig. 2. The process of dynamic weight graph with heat propagation. The brightness indicates the weight value and transitions from the initial center high edge low to the edge high center low with training epochs.

To effectively guide the model's learning process through dynamic weight graph, it is essential to achieve a stable and accurate distribution of weights across the data. In practice, heat propagation demonstrates global stability, where the heat source generates high levels of heat, and as the heat diffuses, its energy gradually diminishes towards the edges. Ensuring the desired standard of overall heat distribution requires the heat source to be responsive to low-heat areas. Building upon this phenomenon, our work focuses on the guidance provided by heat propagation. The subsequent explanation will delve into the process, starting from heat initialization and progressing to heat propagation.

Heat Initialization. The heat emanates from the center of the heat source and progressively extends to encompass the entire region, undergoing gradual attenuation along the way. In our specific scenario, the heat source originates within the central region of the label and gradually diminishes as it expands outward from the label's center, ultimately forming a dynamic graph.

As depicted in Fig. 2, initially, the center of the label possesses the highest heat, which then evenly transfers heat to the surrounding regions. Adhering to the law of energy conservation, the sum of all heat values should remain constant. Assuming the heat per pixel in the segmentation map is denoted as τ, the total heat can be calculated as:

$$\Upsilon_s = \sum_{i,j} \tau \bigotimes I\{Seg_{i,j}! = ignore\}_s \quad \tau < 1 \tag{5}$$

Here, Υ_s represents the total heat of the s-th area, while $Seg_{i,j}$ corresponds to the binary segmentation map. The difference between N_s and Υ_s represents the heat that can be distributed to each pixel, and we temporarily store this excess heat in the center pixel.

Boundary areas often pose challenges for segmentation models, as they tend to exhibit prediction errors. Consequently, When dealing with dynamic graphs, where models possess coarse-grained segmentation capabilities, it becomes crucial to focus on the boundary regions. In order to address this requirement, we introduce a diffusion process that allows heat to propagate outward layer by layer from the graph center, employing a diffusivity factor denoted as $\alpha 1$. This dynamic diffusion process can be expressed as:

$$\tau_{s,l_{t+1}} = \Upsilon_{s,t}/N_{s,l_{t+1}} + \tau_{s,t+1} \qquad (6)$$

In the above equation, l_t represents the most recently updated layer, and $N_{s,l_{t+1}}$ denotes the number of nodes belonging to layer l_{t+1} within region s. By incorporating this diffusion mechanism, we can effectively consider the unique characteristics of each layer and achieve a more precise attention to the boundary areas within the dynamic graph.

Algorithm 1. Heat Spread With Pseudo Label

Input: $diffLocs, prob, heatInfo, layerNum$
Output: $heatMap$
//positional differences,sigmoid probabilities, initial heat, label layers
1: **function** UPDATEHEATWITHDISCREPANCY($diffLocs, prob, heatInfo, layerNum$)
2:      $hl \leftarrow 0.001$ ▷ loss rate of propagation
3:      Initialization heat with $prob$ and $diffLocs$ as $heatMap$
4:      $li \leftarrow layerNum$
5:      **while** $li > 0$ **do**
6:          $lossRateByLayer \leftarrow hl * (1 - (1 - hl)^{li+1})$
7:          **while** loc in $DiffLocs[li]$ **do**
8:              $cNeedHeat \leftarrow heatMap[loc][needHeat]$
9:              **if** $heatInfo[loc]$ has $heatMap[parents]$ **then**
10:                 $pNeedHeat \leftarrow heatMap[parents][needHeat]$
11:                 $pTempHeat \leftarrow heatMap[parents][tempHeat]$
12:                 $pTempHeat \leftarrow cNeedHeat * (1 - cNeedHeat * lossRateByLayer) +$
    $pTempHeat$
13:                 $pNeedHcat \leftarrow pNeedHeat + pTempHeat * hl$
14:             **else**
15:                 $heatMap[parents][needHeat] \leftarrow cNeedHeat * hl$
16:                 $heatMap[parents][tempHeat] \leftarrow cNeedHeat * (1 - cNeedHeat *$
    $lossRateByLayer)$
17:             **end if**
18:         **end while**
19:     **end while**
20:     **return** $heatMap$
21: **end function**

Heat Propagation. Heat initialization takes place after reading and clipping the data to obtain the initial dynamic graph distribution. The initial distribution is assigned values based on specific rules, such as higher values in the center, lower values at the edges, or fixed constant values. However, relying solely on the initial distribution is insufficient to effectively guide the model, as it merely introduces prior knowledge that requires the expertise of senior engineers.

After initializing the heat, we further introduce a heat propagation mechanism as illustrated in Algorithm 1. During the training process, we obtain pseudo-labels generated by the teacher model as well as the predicted sigmoid probabilities. The pseudo-labels are used to calculate the positional differences with the ground truth, thereby obtaining the necessary conditions for updates, namely, positional differences, sigmoid probabilities, initial heat, and effective label layers.

Inspired by the concept of energy loss in heat transfer, we propagate the heat from the center of the labels with a loss rate of hl. To optimize computational resources, we impose a constraint that lower-level nodes only receive heat from a single corresponding upper-level node. The heat propagation process consists of two steps: heat demand aggregation and heat allocation.

Aggregation: As outlined in lines 5–19 of Algorithm 1, heat demand aggregation gradually propagates the demand from the lower-level nodes to the upper-level nodes. The upper-level nodes calculate the discounted heat loss caused by the heat transfer to their respective layers.

Allocation: Based on the results of heat demand aggregation, the heat is allocated from the center of the labels to each positional difference. It is worth noting that the loss transferred to the lower-level nodes is absorbed by the upper-level nodes along the way, ensuring that all nodes along the path receive attention from the model, thereby improving the smoothness of the labels.

4   Experiments

4.1   Dataset Description

We evaluate our method using two datasets. The first dataset is DustProj, which is an industrial dust dataset comprising 1343 images for training, 199 images for validation, and 261 images for testing. Given the harsh conditions of industrial applications and the non-professional characteristics of the annotators, the training and validation set data were annotated by 11 non-professionals. However, for the test set, professionals annotated the data after the initial 11 personnel completed their annotations.

The second dataset we employ is the DSS Smoke Segmentation dataset, a synthetic dataset focusing on smoke. It contains a total of 73632 pictures, with 70632 images for training and 3000 images for test, which is further divided into three groups: DS01, DS02, and DS03.

4.2   Implementation Details

Our method employs the Deep Dual-resolution networks (DDRNet) [4] as the primary test model. To evaluate its performance, we compare it with other well-known semantic segmentation network models, namely UNet++ [22], PSP-Net, Segformer, and Deeplabv3. Among these models, we incorporate dynamic loss weighted learning of heat propagation using DDRNet and Segformer, which exhibit exceptional performance in our results.

During the training process, we apply various data enhancements such as color gamut adjustment, clipping, and blurring, along with image standardization. For DDRNet, UNet++, and PSPNet, we utilize the SGD optimizer with an initial learning rate of 0.01, gradually decreasing it over time. As for the Segformer model, we employ the Adam optimizer with an initial learning rate of 0.003, which decreases to its original value of 0.01 over the course of training.

4.3   Experimental Results

To evaluate the performance of our model, we utilize the mean Intersection over Union (mIoU) as the general semantic segmentation accuracy measure. Furthermore, we also compare the model's parameters, model size, and Frames Per Second (FPS) to assess its efficiency. These indicators serve as valuable references for selecting a model suitable for industrial environments. In the following sections, we provide a detailed comparison between the two datasets, highlighting the achievements obtained through our approach.

Evaluation on DustProj. The DustProj dataset is specifically designed for dust analysis in industrial environments. As a field-applied dataset, it exhibits certain limitations, including the challenges posed by harsh industrial conditions and the presence of flawed data. Nevertheless, these limitations provide valuable insights into the real-world application of machine learning algorithms in industrial settings. In Table 1, we present quantitative results comparing our approach with several popular models such as UNet++, Deeplabv3, and others. The experimental results reveal significant variations in data accuracy among the different models, with PSPNet and DDRNet exhibiting contrasting performance. Leveraging the feature extraction capabilities of pre-trained ResNet, both PSPNet and Deeplabv3 achieve notable prediction accuracy on this dataset. However, DDRNet, lacking pre-training and dual-branch guidance, struggles to adapt to the complexities of this industrial environment, resulting in an mIoU of only 53%.

Table 1. Comparison of all the methods in terms of MIoU [%] in DustProj

Method	mIOU(%) Dustproj	Parameters (M)	Model size (MB)	Image size (pixel)	FPS
pspnet [21]	59.1	35.7	144	(512,512)	37.4
unet++ [22]	58.0	8.7	35.8	(512,512)	32.2
deeplabv3 [1]	58.1	57.2	228	(512,512)	28.2
segformer [15]	59.2	42.5	171	(512,512)	41.4
ddrnet [4]	53.0	30.2	122	(512,512)	71.8
ddrnet+hp	60.4	30.2	122	(512,512)	71.8

In contrast, our method incorporates a teacher model and a heat propagation guidance model without altering the DDRNet architecture. By leveraging attention learning techniques, we successfully guide and unlock the fitting ability of the DDRNet model, leading to the highest mIoU accuracy.

Table 2. Comparison of all the methods in terms of MIoU [%] in DSS

Method	mIOU(%)			Parameters	Model size	FPS
	DS01	DS02	DS03	(M)	(MB)	
ERFNet [11]	69.9	67.9	68.7	2.06	15.8	60.5
LEDNet [12]	69.0	67.8	68.5	0.91	7.18	58.9
DFANet [9]	63.2	59.4	61.8	2.18	16.9	32.4
CGNet [14]	68.6	65.5	67.2	0.49	3.94	53.0
DSS [19]	71.0	71.0	69.8	29.9	-	32.5
Frizzi [3]	70.4	70.0	70.7	57.0	-	60.4
W-Net [18]	73.1	74.0	73.4	31.1	127	-
Light-weighted Net [17]	74.2	72.5	72.8	0.88	6.88	68.8
segformer [15]	74.8	75.4	75.1	42.5	171	43.7
ddrnet [4]	75.0	74.2	73.1	30.2	122	76.2
segformer+hp	78.1	78.0	77.7	42.5	171	43.7
ddrnet+hp	79.4	79.6	79.4	30.2	122	76.2

4.4   Evaluation on DSS

The DSS dataset is a publicly available smoke synthesis dataset that offers a large quantity of high-quality data. It combines smoke with various background images using synthesis technology, providing a comprehensive evaluation of air particle segmentation accuracy.

Table 2 presents a comparison of our proposed method's mIoU, FPS, and other indicators on the DSS dataset. DDRNet demonstrates good performance on this dataset, but we don't stop there. By introducing our dynamic heat spread weighting scheme, we can further enhance its performance, achieving the best mIoU. It's worth noting that we also apply the heat spread weighting scheme to the Segformer model, effectively improving its accuracy on this dataset. This further validates the effectiveness of our proposed method in guiding model learning.

In summary, our method effectively guides the model during training, avoiding overfitting caused by data fitting preferences and enhancing the model's focus on feature extraction from the data.

5   Conclusion

This paper presents a novel training framework that utilizes dynamic weight assignment to enhance model performance. The main contribution of this method is its ability to leverage a teacher model for comprehensive guidance during student model training, leading to improved fitting and generalization capabilities. However, it's important to note that the introduction of pixel-level dynamic weight allocation may increase CPU computing time, potentially affecting GPU utilization. Additionally, our current approach primarily focuses on weight allocation for instance targets and doesn't fully address sample proportion differences between targets. Future work will aim to address the challenge of dynamic weight distribution among targets more effectively. The goal is to unlock the potential of underestimated models and revitalize their performance.

References

1. Chen, L.C., Papandreou, G., Schroff, F., Adam, H.: Rethinking Atrous Convolution for Semantic Image Segmentation (2017). arXiv:1706.05587
2. Du, G., et al.: Graph-based class-imbalance learning with label enhancement. IEEE Trans. Neural Netw. Learn. Syst. 1–15 (2021). https://doi.org/10.1109/TNNLS.2021.3133262
3. Frizzi, S., Bouchouicha, M., Ginoux, J.M., Moreau, E., Sayadi, M.: Convolutional neural network for smoke and fire semantic segmentation. IET Image Proc. 15(3), 634–647 (2021). https://doi.org/10.1049/ipr2.12046
4. Hong, Y., Pan, H., Sun, W., Jia, Y.: Deep Dual-resolution Networks for Real-time and Accurate Semantic Segmentation of Road Scenes (2021). arXiv:2101.06085
5. Hoyer, L., Dai, D., Van Gool, L.: DAFormer: improving network architectures and training strategies for domain-adaptive semantic segmentation. In: 2022 IEEE/CVF Conference on Computer Vision and Pattern Recognition (CVPR), New Orleans, LA, USA, pp. 9914–9925. IEEE (2022). https://doi.org/10.1109/CVPR52688.2022.00969
6. Jiang, M., Huang, W., Huang, Z., Yen, G.G.: Integration of global and local metrics for domain adaptation learning via dimensionality reduction. IEEE Trans. Cybern. 47(1), 38–51 (2017). https://doi.org/10.1109/TCYB.2015.2502483
7. Jiang, M., Wang, Z., Guo, S., Gao, X., Tan, K.C.: Individual-based transfer learning for dynamic multiobjective optimization. IEEE Trans. Cybern. 51(10), 4968–4981 (2021). https://doi.org/10.1109/TCYB.2020.3017049
8. Jiang, M., Wang, Z., Hong, H., Yen, G.G.: Knee point-based imbalanced transfer learning for dynamic multiobjective optimization. IEEE Trans. Evol. Comput. 25(1), 117–129 (2021). https://doi.org/10.1109/TEVC.2020.3004027
9. Li, H., Xiong, P., Fan, H., Sun, J.: DFANet: deep feature aggregation for real-time semantic segmentation. In: 2019 IEEE/CVF Conference on Computer Vision and Pattern Recognition (CVPR), Long Beach, CA, USA, pp. 9514–9523. IEEE (2019). https://doi.org/10.1109/CVPR.2019.00975
10. Luc, P., Couprie, C., Chintala, S., Verbeek, J.: Semantic Segmentation using Adversarial Networks (2016). arXiv:1611.08408
11. Romera, E., Álvarez, J.M., Bergasa, L.M., Arroyo, R.: ERFNet: efficient residual factorized ConvNet for real-time semantic segmentation. IEEE Trans. Intell. Transp. Syst. 19(1), 263–272 (2018). https://doi.org/10.1109/TITS.2017.2750080

12. Wang, Y., et al.: LEDNet: a lightweight encoder-decoder network for real-time semantic segmentation. In: 2019 IEEE International Conference on Image Processing (ICIP), pp. 1860–1864 (2019). https://doi.org/10.1109/ICIP.2019.8803154. ISSN 2381-8549

13. Wang, Z., Cao, L., Lin, W., Jiang, M., Tan, K.C.: Robust graph meta-learning via manifold calibration with proxy subgraphs. In: Proceedings of the AAAI Conference on Artificial Intelligence, vol. 37, pp. 15224–15232 (2023). https://doi.org/10.1609/aaai.v37i12.26776

14. Wu, T., Tang, S., Zhang, R., Cao, J., Zhang, Y.: CGNet: a light-weight context guided network for semantic segmentation. IEEE Trans. Image Process. **30**, 1169–1179 (2021). https://doi.org/10.1109/TIP.2020.3042065

15. Xie, E., Wang, W., Yu, Z., Anandkumar, A., Alvarez, J.M., Luo, P.: SegFormer: simple and efficient design for semantic segmentation with transformers. In: Ranzato, M., Beygelzimer, A., Dauphin, Y., Liang, P.S., Vaughan, J.W. (eds.) Advances in Neural Information Processing Systems, vol. 34, pp. 12077–12090. Curran Associates, Inc. (2021)

16. Ye, C., Jiang, M., Luo, Z.: Smoke segmentation based on weakly supervised semantic segmentation. In: 2022 12th International Conference on Information Technology in Medicine and Education (ITME), pp. 348–352 (2022). https://doi.org/10.1109/ITME56794.2022.00082

17. Yuan, F., Li, K., Wang, C., Fang, Z.: A lightweight network for smoke semantic segmentation. Pattern Recogn. **137**, 109289 (2023). https://doi.org/10.1016/j.patcog.2022.109289

18. Yuan, F., Zhang, L., Xia, X., Huang, Q., Li, X.: A wave-shaped deep neural network for smoke density estimation. IEEE Trans. Image Process. **29**, 2301–2313 (2020). https://doi.org/10.1109/TIP.2019.2946126

19. Yuan, F., Zhang, L., Xia, X., Wan, B., Huang, Q., Li, X.: Deep smoke segmentation. Neurocomputing **357**, 248–260 (2019). https://doi.org/10.1016/j.neucom.2019.05.011

20. Zhang, F., Pan, T., Wang, B.: Semi-supervised object detection with adaptive class-rebalancing self-training. In: Proceedings of the AAAI Conference on Artificial Intelligence, vol. 36, no. 3, pp. 3252–3261 (2022). https://doi.org/10.1609/aaai.v36i3.20234

21. Zhao, H., Shi, J., Qi, X., Wang, X., Jia, J.: Pyramid scene parsing network. In: 2017 IEEE Conference on Computer Vision and Pattern Recognition (CVPR), Honolulu, HI, pp. 6230–6239. IEEE (2017). https://doi.org/10.1109/CVPR.2017.660

22. Zhou, Z., Rahman Siddiquee, M.M., Tajbakhsh, N., Liang, J.: UNet++: a nested U-Net architecture for medical image segmentation. In: Stoyanov, D., et al. (eds.) DLMIA/ML-CDS -2018. LNCS, vol. 11045, pp. 3–11. Springer, Cham (2018). https://doi.org/10.1007/978-3-030-00889-5_1

EKGRL: Entity-Based Knowledge Graph Representation Learning for Fact-Based Visual Question Answering

Yongjian Ren[1,2], Xiaotang Chen[1,2,3], and Kaiqi Huang[1,2,3(✉)]

[1] School of Artificial Intelligence, University of Chinese Academy of Sciences, Beijing, China
[2] Institute of Automation, Chinese Academy of Sciences, Beijing, China
[3] Center for Excellence in Brain Science and Intelligence Technology, Chinese Academy of Sciences, Beijing, China
renyongjian2022@ia.ac.cn, {xtchen,kaiqi.huang}@nlpr.ia.ac.cn

Abstract. Fact-based Visual Question Answering (FVQA) is a task aiming at answering question based on given image and external knowledge associated with it. The reasoning abilities of current FVQA models including query-based and joint learning methods are insufficient. To achieve stronger reasoning ability, we propose an entity-based knowledge graph representation learning (EKGRL) method. Our model achieves state-of-the-art performance on FVQA dataset. Furthermore, we build a psychological fact-based VQA dataset (PFVQA) containing 6129 questions from six different types, which is, as far as we know, the first VQA dataset built on psychological knowledge. We demonstrate that EKGRL continues to achieve state-of-the-art performance on PFVQA, showing the ability to maintain a good performance on reasoning and knowledge representation based on external knowledge from both commonsense and psychological domains.

Keywords: FVQA · Representation Learning · Psychological Knowledge

1 Introduction

Visual Question Answering (VQA) is proposed to examine the perceptual and reasoning ability of AI systems, which uses an image and corresponding natural language questions as input and outputs the answers automatically. At present, it is widely used in medical treatment, advertising and visual assistance for blind people. In traditional VQA work [1,2,9,10,12,17,24], researchers expect systems to answer a broad range of vision questions like humans do. Therefore, most of the visual questions generated by humans or machines are beyond the visual information of the pictures themselves and need external knowledge as support. The research of Fact-based VQA tasks [4,14,15,18,20,21,25] aims to

© The Author(s), under exclusive license to Springer Nature Singapore Pte Ltd. 2024
Q. Liu et al. (Eds.): PRCV 2023, LNCS 14430, pp. 485–496, 2024.
https://doi.org/10.1007/978-981-99-8537-1_39

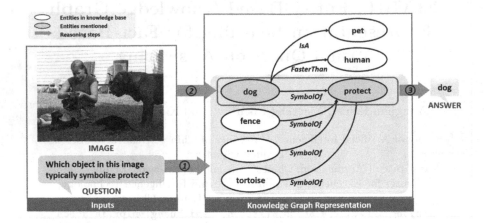

Fig. 1. The main idea of reasoning process. The orange arrows represent the reasoning steps, where 1 and 2 indicate that the scope of the answer is limited to the entity "dog" and the entity "protect" by reasoning, and 3 indicates that the answer entity is further inferred to be "dog".

combine structured external facts to answer visual questions, which is committed to solving the problem of information loss inherent in traditional VQA work.

FVQA work can be roughly divided into two classes: query-based and joint learning methods. For query-based methods [14,15,18,20,21], supporting facts are rarely used for reasoning. The step of filtering numerours facts is inefficient and the pipeline working mode leads to the accumulation of errors. For joint learning methods [4,25], although the knowledge space is embedded in reasoning process, it is still independent to the answer space, resulting in the lack of reasoning ability.

In conclusion, the existing FVQA methods do not fully represent knowledge space for reasoning, resulting in insufficient reasoning ability. In this paper, we propose an **entity-based knowledge graph representation learning (EKGRL)** method to learn the mapping of concepts in the knowledge space. In the EKGRL model, we perform representation learning in the entity space to express the features of knowledge graph. And also a gating mechanism is used to further infer the specific answer from the concepts. The main idea of reasoning process is shown in Fig. 1.

Furthermore, we build a **psychological fact-based VQA (PFVQA)** dataset based on professional psychological facts containing 2190 images and 6129 psychological question and answer (Q&A) pairs. PFVQA requires models to answer questions about psychological meanings of objects in an image, which is a both practical and challenging task while current FVQA datasets are mainly built on commonsense knowledge. On one hand, some phychological assessment and psychotherapy methods, such as sandplay therapy, require diagnosing the mental state of individuals based on external psychological knowledge and visual scenes created by subjects. On the other hand, the task of "answering the psychological meaning of objects in an image" demands models to perform multi-hop reasoning on visual questions, relying on both general and psychological external knowledge. This task tests the perceptual and reasoning abilities of AI models in understanding visual scenes.

In general, our contributions are as follows:

- We introduce a novel model EKGRL that can answer visual questions more efficiently. Performing representation learning on knowledge graph along with a gating mechanism makes EKGRL have strong reasoning and representation ability on knowledge graph.
- We build a new dataset PFVQA for verifying the representation ability of VQA methods on knowledge space. As an extension of FVQA dataset, PVQA adds 2,150 psychological Q&A pairs and 1,247 psychological facts, introducing the knowledge of psychological field into FVQA.
- Quantitatively, we demonstrate the effectiveness of our method on FVQA and PFVQA dataset, outperforming state-of-the-art methods by (6%) on Top-1 accuracy.

2 Related Works

Visual Question Answering (VQA). The VQA task [12] is first proposed for a deeper understanding of visual scenes. Since then, new fusion methods [9,17,24] of image features and question features have been proposed. Existing fusion models [5,19,23] have been able to fully exploit the contextual information between objects in images. In particular, the attention mechanism methods based on region proposals have significantly improved the performance of visual question answering, such as the Bottom-Up and Top-Down Attention VQA [1] and the Bilinear Attention Networks [10]. At the same time, researchers propose a small-scale dataset DAQUAR [12] for indoor complex scenes, a VQA dataset [2] aiming at open and free question answering, a VQA v2.0 dataset [6] and a COCO-QA dataset [17] with more evenly distributed answers. However, in these traditional work and dataset, most questions are too open-ended, especially in the VQA dataset [2] and the DAQUAR dataset [12].

Fact-Based Visual Question Answering (FVQA). The early method [22] retrieves the relevant unstructured description information from the external knowledge base DBpedia according to the visual attributes. Another work [13] uses ArticleNet model to find the most possible answer from external text, which is obtained by retrieving Wikipedia. However, there is a lot of noise in the unstructured external knowledge text, which introduces errors.

VQA work based on structured external knowledge mainly refer to the FVQA work, where the external knowledge consists of facts. On one hand, early FVQA methods were query-based methods [14,15,18,20], in which answers were filtered out step by step. Therefore, the reasoning ability of such methods is relatively

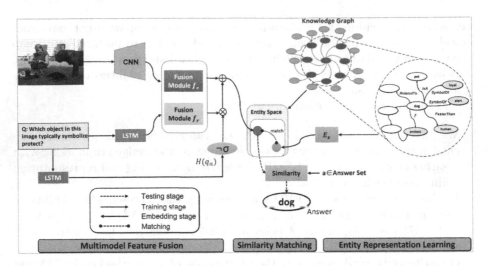

Fig. 2. Overview of our framework. EKGRL model consists of three parts: multimodal feature fusion, entity representation learning and similarity matching. Firstly, we use a CNN module (ResNet [7]) to extract features from images and a LSTM [8] module to extract features from questions. Secondly, we use a fusion module f_e to fuse image features and question features, learning the embedding representation to entity space. Meanwhile, we use another fusion module f_r to learn the embedding representation to relation space. Finally, we use a gating mechanism $\neg\sigma$ and another LSTM module to learn the position of answer entity in the supporting fact triplet.

weak because the external knowledge is not fully utilized. Furthermore, the inherent pipeline working mode in such methods leads to the accumulation of errors [4]. On the other hand, joint learning approaches [4,25] aim to enhance the representation of knowledge spaces. Although some methods use graph neural networks to represent the knowledge space and improve the accuracy of the answer [25], the reasoning ability is limited due to the independence of the knowledge space and the answer space. Another method maps image-question features to answer space, relation space, and knowledge space through three fusion models [4], but the answer space and external knowledge space are still independent. And the external knowledge is only used as a mask to filter candidate answers by affecting the similarity. In general, due to the insufficient representation of the external knowledge space and the independence of the external knowledge and answer space, the joint learning methods also lacks sufficient reasoning ability.

3   Methods

3.1   Main Idea

Our model is inspired by the process of human answering visual questions. When given a question related to an image such as the example in Fig. 1, People can quickly think of entities related to the question. Next, if the image in Fig. 1 is provided, people will point out the position of "dog" in the image and answer. Inspired by such process, we regard all concepts in VQA dataset as entities in knowledge space, including concepts in images (goals, behaviors, scenes), implicit or displayed concepts in questions and answers. To avoid error accumulation, we give up the pipeline working mode and use a series of subspaces to characterize the features of different modals. Our EKGRL model is established based on three parts as shown in Fig. 2: multimodal feature fusion, entity representation learning and similarity matching.

3.2 Preliminary

Different from the open-answer VQA setting defined in OK-VQA [13], we define the answer set $\mathcal{A} = \{a|(a, r, e)\,or\,(e, r, a),\ r \in \mathcal{R}_\mathcal{S}\ and\ e \in \mathcal{E}_\mathcal{S}\}$, where $\mathcal{R}_\mathcal{S}$ and $\mathcal{E}_\mathcal{S}$ respectively represent the relation set and entity set in supporting fact set KB_S. The supporting fact set is defined as $KB_S = \{f_1, f_2, \cdots, f_n\}$.

3.3 Representation Learning in Entity Space

The introduction of representation learning on knowledge graph enhances the representational and reasoning abilities of EKGRL. Our **entity representation learning** module is inspired by TransE [3], where triplets in knowledge graph are expressed as Formula 1.

$$\boldsymbol{h} + \boldsymbol{r} = \boldsymbol{t} \tag{1}$$

where $\boldsymbol{h}$ is the head entity in the triplet, $\boldsymbol{r}$ is the relation in the triplet and $\boldsymbol{t}$ is the tail entity in the triplet.

In our EKGRL model, answer entity embedding is obtained by Formula 2:

$$A(i_n, q_n) = \begin{cases} F_{\theta e}(i_n, q_n), & \text{if } H(q_n) = 1 \\ F_{\theta e}(i_n, q_n) + F_{\theta r}(i_n, q_n), & \text{if } H(q_n) = 0 \end{cases} \tag{2}$$

where $H(q_n)$ judges whether the answer entity is the head entity or the tail entity, $F_{\theta e}$ and $F_{\theta r}$ learn the mapping from the fusion features of the image i_n and the question q_n to the entity space and the relation space.

During actual training process, Formula 2 is expressed as Formula 3:

$$A(i_n, q_n) = F_{\theta e}(i_n, q_n) + F_{\theta r}(i_n, q_n) \cdot (1 - H(q_n)) \tag{3}$$

where $H(q_n) = 1$ indicates the answer is head entity and $H(q_n) = 0$ indicates the answer is tail entity, according to Formula 1. $H(q_n) = 1$ is calculated by Formula 4.

$$H(q_n) = \sigma(\text{LSTM}(q_n)) \tag{4}$$

Based on probabilistic model of compatibility (PMC) method defined in [4], we obtain the prediction probability of answer by Formula 5:

$$P(a|i_n, q_n) = \frac{\exp(A(i_n, q_n)\top \cdot E(a)/\tau)}{\sum_{a' \in \mathcal{A}} \exp(A(i_n, q_n)\top \cdot E(a')/\tau)} \tag{5}$$

where loss temperature τ is benefit for optimization process, and $E(a)$ is the entity embedding representation pretrained on external knowledge base.

We maximize the probability of PMC model, our loss function is defined as:

$$l_\alpha = -\sum_n^N \sum_{b \in \mathcal{A}} \alpha(a, b) log P(b|i_n, q_n) \tag{6}$$

where $\alpha(a, b)$ is a binary classification weight function. When answer a is predicted correctly, that is, when $a = b$, the value of $\alpha(a, b)$ is 1, otherwise it is 0.

Our EKGRL model finally gets the answer through the following Formula:

$$\hat{a} = \arg\max_{a \in \mathcal{A}} A(i_n, q_n)\top \cdot E(a). \tag{7}$$

For the input questions, we use the pretrained word vector GloVe embedding [16] to express each word, while for the answer vocabulary, we use the pretrained knowledge graph to express the mapping $E_e(a)$.

4   PFVQA Dataset

As mentioned in previous sections, almost all the existing FVQA datasets are built on world knowledge or common knowledge, enabling corresponding FVQA models to answer commonsense-based questions. Psychological visual question answering (PFVQA) requires models to answer questions related to the psychological meaning of objects in an image, which is a both practical and challenging task requiring the model's perceptual and reasoning abilities in understanding visual scenes. According to unstructured psychological knowledge, we build PFVQA dataset as an expansion of FVQA, containing 2190 images and 6129 question-answer pairs in total.

4.1   Image and Psychological Facts

Our PFVQA dataset use the images in FVQA, with a total of 2190 images and 326 types of objects. Psychological experts provide us with unstructured psychological symbol description texts, including 419 kinds of common Artificial and natural objects and corresponding psychological mapping descriptions. We summarize 1247 psychological facts from these texts and store them in the form of RDF triples as the psychology knowledge base. The form of every fact is (*goal*, SymbolOf, *psychological symbolic concept*). The psychology knowledge base contains 333 kinds of objects as head entities and 606 kinds of different psychological concept as tail entities.

4.2   Construction of Q&A Set and Dataset Statistics

Psychological question and answer pairs of PFVQA are generated by three annotators. After being provided with images and psychological support facts, the

Table 1. Examples and number of questions for each type.

Type	Example	Number
Property	what does the vehicle in this picture usually symbolize?	895
Location	In this picture, where is the location of the object symbolizing peace?	853
Simularity	Which animal is closer to the one in image in symbolic meaning, elephant or wolf?	902
Object	what animal pictured in this image is the symbol of loyalty?	1455
Category	which category does the object in the image belong to?	976
Valence	Does the red symbol typically symbolize a positive, negative, or neutral sentiment?	1048

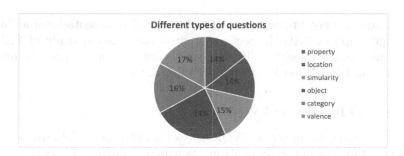

Fig. 3. A statistic result on different types of questions in our PFVQA dataset. The proportions of six separate question types are presented in the chart.

annotators are asked to generate Q&A pairs that belong to one of the following six types: property, location, simularity, object, category and valence.

To ensure the balance and no-bias across different types in PFVQA dataset, we instruct the annotators to generate question-answer pairs that fall into one of the six specified types. We also filtered out questions that exhibited clear scene bias and those that could be answered without the external psychological knowledge. The PFVQA dataset contains 2,190 images and 6129 psychological Q&A pairs, with an average question length of 14.4. The example, number and proportion of each Q&A type is shown in Table 1 and Fig. 3. We also compare PFVQA dataset with various VQA dataset in Table 2, based on the number of images and questions, and the source of external knowledge.

5   Results

To evaluate the reasoning ability and representational ability on knowledge graph of EKGRL model, we validate our method on FVQA and PFVQA datasets. Section 5.1 and 5.2 show the overall results of our model on FVQA and PFVQA

Table 2. Comparison of several major VQA datasets. The last three columns represent the characteristics of the external knowledge base.

Dataset	#Images	#Questions	Image source	Support Fact	Common Sense	Psychology knowledge
DAQUAR	1,449	12,468	NYU-Depth	-	-	-
COCO-QA	69,172	117,684	COCO	-	-	-
VQA-abstract	50,000	150,000	Clipart	-	-	-
VQA	204,721	614,163	COCO	-	-	-
Visual 7w	47,300	327,939	COCO	-	-	-
Visual Genome	108,000	1,445,322	COCO	-	-	-
VQA-2	204,721	1,105,904	COCO	-	-	-
TallyQA	165,443	287,907	Visual genome + COCO	-	-	-
KB-VQA	700	2,402	COCO+ImgNet	-	-	-
KVQA	24,602	183,007	Wikipedia	-	✓	-
FVQA	2,190	5,826	COCO+ImgNet	✓	✓	-
Ours (PFVQA)	2,190	6129	COCO+ImgNet	✓	✓	✓

datasets respectively, proving reasoning ability and representational ability on knowledge graph of EKGRL. Section 5.3 gives the ablation study of EKGRL, which evaluates the impact of different components. Figure 4 shows the qualitative results of EKGRL on FVQA and PFVQA dataset.

5.1   Overall Results on FVQA

To evaluate the performance on FVQA dataset of different models, we use Top-1, Top-3 and Top-10 accuracy in experiments, where Top-K indicates the ground-truth answer ranks in range of top K answers provided by models. We use the average accuracy of the 5 test splits as the overall accuracy.

Table 3 shows the performance of EKGRL and other VQA models on FVQA dataset. The methods we used as comparision including Hie-Question+Image+Pre-VQA [11], FVQA (top-3-QQmaping) [20], FVQA(Ensemble) [20], Straight to the Facts [15], Out of the Box [14], Multi-Layer Cross-Modal Knowledge Reasoning [25] and Zero-shot FVQA [4]. The comparision methods above can represent query-based FVQA models and joint learning models respectively.

Among all these FVQA methods, our EKGRL model gains the best performance with 2.23% boost on Top-1 accuracy and 1.27% boost on Top-3 accuracy. Mucko model gains the smallest accuracy gap because the model manages to perform joint reasoning by using modality-aware heterogeneous graph convolutional network. Compared with joint learning models, our method explicitly reflects the reasoning process from image-problem pair to external knowledge, making the reasoning and associative process clearer and more interpretable.

5.2   Overall Results on PFVQA

PVQA dataset includes 2190 images and 6129 psychological Q&A pairs. We use Top-1, Top-3 and Top-10 accuracy in experiments as metrics. Table 3 shows the performance of EKGRL and other VQA models on PFVQA dataset.

Table 3. Performance of different methods on FVQA (left) and PFVQA (right) dataset. Our method performance the best on Top-1, Top-3 and Top-10 accuracy

Method	Overall Accuracy		
	Top-1	Top-3	Top-10
Hie-Q+I+Pre-VQA	43.14	59.44	66.08
FVQA (top-3-QQmaping)	56.91	64.65	66.35
FVQA (Ensemble)	58.76	68.14	77.20
Straight to the Facts (STTF)	62.20	75.60	81.32
Out of the Box (OB)	69.35	80.25	84.56
Mucko	73.06	85.94	91.24
ZS-FVQA	58.27	76.51	88.49
Our method (EKGRL)	**75.29**	**87.21**	**92.08**

Method	Overall Accuracy		
	Top-1	Top-3	Top-10
Hie-Q+I+Pre-VQA	40.85	57.41	69.18
FVQA (top-3-QQmaping)	54.89	60.44	63.37
FVQA (Ensemble)	56.64	66.56	74.60
Straight to the Facts (STTF)	58.84	70.66	79.49
Out of the Box (OB)	64.82	75.10	81.42
Mucko	68.01	79.41	88.47
ZS-FVQA	57.42	76.51	83.84
Our method (EKGRL)	**74.78**	**85.51**	**93.19**

Table 4. Ablation study of key components of EKGRL.

Method	Overall Accuracy		
	Top-1	Top-3	Δ Top-1
EKGRL	**74.78**	**85.51**	-
w/o Entity embedding	62.76	73.03	−12.02
w/o Entity space mapping	64.02	76.74	−10.76
w/o Relation space mapping	63.88	76.50	−10.90
w/o Representation learning	55.79	68.47	**−18.99**
fusion model (MLP)	67.02	78.33	−7.76
fusion model (BAN)	73.71	81.72	−1.07

On PFVQA dataset, our EKGRL model obtains the best performance among all these VQA methods, achieving 6.77% boost on Top-1 accuracy, 6.10% boost on Top-3 accuracy and 4.72% boost on Top-10 accuracy. After adding psychological knowledge and Q&A pairs, the accuracy of other VQA methods reduces while EKGRL remains the same performance against change of dataset and expansion of knowledge graph. The psychological relationships introduced in PFVQA dataset contain multiple relationship chains between different entities, bringing implicit connections between different entities that are not related in common semantic space. These hidden relationship chains bring challenges to the reasoning process of the model because the expansion of knowledge graph tests the representation ability on knowledge graph. The results on PFVQA dataset prove the representation and generalization ability of our model. Our EKGRL model can provide a new research route to solve VQA tasks and perform knowledge graph representation.

5.3 Ablation Study

The results of ablation study is shown in Table 4. We firstly evaluate the influence of **multimodel feature fusion** in different spaces by removing entity embedding, entity space mapping and relation space mapping. Compared with

Fig. 4. Qualitative results of EKGRL. The first two columns ((a), (b), (e), (f)) show the examples of psychological VQA based on psychological supporting fact. The last two columns ((c), (d), (g), (h)) show the examples on common sense VQA. The first row shows the samples whose question-image entities are head entities, while the second row shows the ones with tail entities as question-image entities. (d) and (h) are two failure examples of prediction. The reason for (d) to fail is finding the wrong supporting fact and the reason for (h) to fail is confusion of similar entities. Especially, **(f)** shows an example of PFVQA on a sandplay scene created on 3D electronic sandplay platform, showing the potential of EKGRL in phychological assesssment and psychotherapy.

full model, the Top-1 accuracy reduce 12.02%, 10.76% and 10.90% respectively, which indicates the positive effect of entity space and relation space, providing valuable information for answer inference. We also change our **fusion model** from Stacked attention (SAN) to MLP or BAN, resulting in 7.76% and 1.07% reduction of Top-1 accuracy. We assess the value of proposed **Representation learning** by removing answer-entity embedding, which results in a 18.99% decrease in Top-1 accuracy, higher than any other items in ablation study. The significant decrease cause by removing answer-entity embedding can prove the value and effectiveness of **representation learning**.

6   Conclusions

In this work, we have proposed the EKGRL model to learn the mapping of concepts in the knowledge space and solve FVQA task. We have built a PFVQA dataset containing psychological questions to verify the representational and reasoning ability of EKGRL in psychological domain. As an expansion to FVQA dataset, the PFVQA dataset contains psychological Q&A pairs and supporting facts, introducing psychological attributes to entities which can test the abilities of reasoning and knowledge representation of models. Our EKGRL model has achieved state-of-the-art performance on both FVQA and PFVQA datasets. The experiment results have proved that the entity representation learning and the gating mechanism in EKGRL enable the model to have strong multi-hop reasoning and representation abilities on knowledge graph.

References

1. Anderson, P., He, X., Buehler, C., Teney, D., Johnson, M., Gould, S., Zhang, L.: Bottom-up and top-down attention for image captioning and visual question answering. In: Proceedings of the IEEE Conference on Computer Vision and Pattern Recognition (CVPR) (2018)
2. Antol, S., et al.: VQA: visual question answering. In: Proceedings of the IEEE International Conference on Computer Vision (ICCV) (2015)
3. Bordes, A., Usunier, N., Garcia-Duran, A., Weston, J., Yakhnenko, O.: Translating embeddings for modeling multi-relational data. In: Burges, C., Bottou, L., Welling, M., Ghahramani, Z., Weinberger, K. (eds.) Advances in Neural Information Processing Systems, vol. 26. Curran Associates, Inc. (2013). https://proceedings.neurips.cc/paper/2013/file/1cecc7a77928ca8133fa24680a88d2f9-Paper.pdf
4. Chen, Z., Chen, J., Geng, Y., Pan, J.Z., Chen, H.: Zero-shot visual question answering using knowledge graph (2021)
5. Fukui, A., Park, D.H., Yang, D., Rohrbach, A., Darrell, T., Rohrbach, M.: Multimodal compact bilinear pooling for visual question answering and visual grounding. In: EMNLP (2016)
6. Goyal, Y., Khot, T., Summers Stay, D., Batra, D., Parikh, D.: Making the V in VQA matter: elevating the role of image understanding in visual question answering, pp. 6325–6334 (2017). https://doi.org/10.1109/CVPR.2017.670
7. He, K., Zhang, X., Ren, S., Sun, J.: Deep residual learning for image recognition. In: 2016 IEEE Conference on Computer Vision and Pattern Recognition (CVPR), pp. 770–778 (2015). https://api.semanticscholar.org/CorpusID:206594692
8. Hochreiter, S.: Long short term memory (1995). https://api.semanticscholar.org/CorpusID:261210188
9. Kafle, K., Kanan, C.: Answer-type prediction for visual question answering. In: 2016 IEEE Conference on Computer Vision and Pattern Recognition (CVPR), pp. 4976–4984 (2016). https://doi.org/10.1109/CVPR.2016.538
10. Kim, J.H., Jun, J., Zhang, B.T.: Bilinear attention networks. In: Bengio, S., Wallach, H., Larochelle, H., Grauman, K., Cesa-Bianchi, N., Garnett, R. (eds.) Advances in Neural Information Processing Systems, vol. 31. Curran Associates, Inc. (2018). https://proceedings.neurips.cc/paper/2018/file/96ea64f3a1aa2fd00c72faacf0cb8ac9-Paper.pdf
11. Lu, J., Yang, J., Batra, D., Parikh, D.: Hierarchical question-image co-attention for visual question answering (2016)
12. Malinowski, M., Fritz, M.: A multi-world approach to question answering about real-world scenes based on uncertain input. In: Ghahramani, Z., Welling, M., Cortes, C., Lawrence, N., Weinberger, K. (eds.) Advances in Neural Information Processing Systems, vol. 27. Curran Associates, Inc. (2014). https://proceedings.neurips.cc/paper/2014/file/d516b13671a4179d9b7b458a6ebdeb92-Paper.pdf

13. Marino, K., Rastegari, M., Farhadi, A., Mottaghi, R.: OK-VQA: a visual question answering benchmark requiring external knowledge. In: Proceedings of the IEEE/CVF Conference on Computer Vision and Pattern Recognition (CVPR) (2019)
14. Narasimhan, M., Lazebnik, S., Schwing, A.: Out of the box: reasoning with graph convolution nets for factual visual question answering. In: Bengio, S., Wallach, H., Larochelle, H., Grauman, K., Cesa-Bianchi, N., Garnett, R. (eds.) Advances in Neural Information Processing Systems, vol. 31. Curran Associates, Inc. (2018). https://proceedings.neurips.cc/paper/2018/file/c26820b8a4c1b3c2aa868d6d57e14a79-Paper.pdf
15. Narasimhan, M., Schwing, A.G.: Straight to the facts: learning knowledge base retrieval for factual visual question answering. In: Proceedings of the European Conference on Computer Vision (ECCV), pp. 451–468 (2018)
16. Pennington, J., Socher, R., Manning, C.: GloVe: global vectors for word representation. In: Proceedings of the 2014 Conference on Empirical Methods in Natural Language Processing (EMNLP), Doha, Qatar, pp. 1532–1543. Association for Computational Linguistics (2014). https://doi.org/10.3115/v1/D14-1162. https://aclanthology.org/D14-1162
17. Ren, M., Kiros, R., Zemel, R.S.: Exploring models and data for image question answering. In: Proceedings of the 28th International Conference on Neural Information Processing Systems - Volume 2, NIPS 2015. pp. 2953–2961. MIT Press, Cambridge (2015)
18. Shah, S., Mishra, A., Yadati, N., Talukdar, P.P.: KVQA: knowledge-aware visual question answering. In: Proceedings of the AAAI Conference on Artificial Intelligence, vol. 33, no. 01, pp. 8876–8884 (2019). https://doi.org/10.1609/aaai.v33i01.33018876. https://ojs.aaai.org/index.php/AAAI/article/view/4915
19. Shih, K.J., Singh, S., Hoiem, D.: Where to look: focus regions for visual question answering. In: 2016 IEEE Conference on Computer Vision and Pattern Recognition (CVPR), pp. 4613–4621 (2016). https://doi.org/10.1109/CVPR.2016.499
20. Wang, P., Wu, Q., Shen, C., Dick, A., van den Hengel, A.: FVQA: fact-based visual question answering. IEEE Trans. Pattern Anal. Mach. Intell. 40(10), 2413–2427 (2018). https://doi.org/10.1109/TPAMI.2017.2754246
21. Wang, P., Wu, Q., Shen, C., Dick, A., Van Den Henge, A.: Explicit knowledge-based reasoning for visual question answering. In: Proceedings of the 26th International Joint Conference on Artificial Intelligence, IJCAI 2017, pp. 1290–1296. AAAI Press (2017)
22. Wu, Q., Wang, P., Shen, C., Dick, A., van den Hengel, A.: Ask me anything: free-form visual question answering based on knowledge from external sources. In: Proceedings of the IEEE Conference on Computer Vision and Pattern Recognition (CVPR) (2016)
23. Yang, Z., He, X., Gao, J., Deng, L., Smola, A.: Stacked attention networks for image question answering. In: Proceedings of the IEEE Conference on Computer Vision and Pattern Recognition (CVPR) (2016)
24. Zhou, B., Tian, Y., Sukhbaatar, S., Szlam, A., Fergus, R.: Simple baseline for visual question answering. CoRR abs/1512.02167 (2015). http://arxiv.org/abs/1512.02167
25. Zhu, Z., Yu, J., Wang, Y., Sun, Y., Hu, Y., Wu, Q.: Mucko: multi-layer cross-modal knowledge reasoning for fact-based visual question answering. In: Bessiere, C. (ed.) Proceedings of the Twenty-Ninth International Joint Conference on Artificial Intelligence, IJCAI 2020, pp. 1097–1103. International Joint Conferences on Artificial Intelligence Organization (2020). https://doi.org/10.24963/ijcai.2020/153

Disentangled Attribute Features Vision Transformer for Pedestrian Attribute Recognition

Caihua Liu[1,3], Jiaxian Guo[2(✉)], Sichu Chen[1], and Xia Feng[1,3]

[1] School of Computer Science and Technology,
Civil Aviation University of China, Tianjin, China
{chliu,2022051004,xfeng}@cauc.edu.cn
[2] Geophysical Exploration Center, China Earthquake Administration,
Zhengzhou, China
guojx@gec.ac.cn
[3] Key Laboratory of Intelligent Airport Theory and System, CAAC, Tianjin, China

Abstract. Pedestrian attribute recognition, as a sub-task of multi-label classification, is influenced by the joint and coupled relationships among the numerous labels of the samples. This impact affects the model's ability to judge labels and results in a loss of recognition accuracy for both overall and individual labels in unfamiliar scenes. In this paper, we propose a Disentangled Attribute Features Vision Transformer (DAF-ViT) model for pedestrian attribute recognition, aiming to enhance the discriminative ability of correlated features in pedestrian samples and suppress the discriminative ability of irrelevant features. The decoupling model consists of three parallel self-attention modules: a correlation self-attention module (CSA) for extracting correlated attribute features, a exclusive self-attention module (ESA) for extracting mutually exclusive attribute features, and a random sequence self-attention module (RSA) for extracting unrelated attribute features. The final predicted attributes are obtained by combining the results of these modules using a multi-level result fusion module. Compared to the baseline methods, our approach achieves a 2.8% and 2.5% increase in mA accuracy on the PETA and PA-100K datasets, respectively. This effectively improves the average accuracy of the model on various attributes.

Keywords: Pedestrian attribute recognition · Self-attention mechanism · Disentangled features

1 Introduction

Pedestrian attribute recognition, characterized by the unique properties of pedestrian attributes, presents several novel challenges in algorithmic model design

This work was supported by the Scientific Research Project of Tianjin Educational Committee under Grant 2021KJ037.

compared to traditional multi-label image classification tasks. One such challenge pertains to the influence of coupling relationships among pedestrian attributes on multi-label classification. Coupling relationships refer to the inter-dependencies and interactions between multiple attributes of the same pedestrian. These relationships often manifest through two phenomena: attribute co-occurrence and attribute exclusivity. Attribute co-occurrence describes the higher probability of two or more attributes appearing simultaneously, such as skirt and female, shorts and short sleeves, and so on. Attribute exclusivity, on the other hand, denotes the higher probability of two or more attributes not co-occurring, for instance, long hair and male, shorts and overcoat, and so forth. Furthermore, the co-occurrence patterns of other attribute pairs exhibit no discernible regularity, such as glasses and long sleeves, hat and backpack, among others.

The coupling relationships among pedestrian attributes specifically refer to semantic connections that induce attribute co-occurrence or attribute exclusivity phenomena in pedestrian datasets. These relationships can serve as prior knowledge to aid models in improving prediction accuracy and speed for regular samples across most training and testing scenarios. However, when faced with a small proportion of outlier data, they may lead to prediction biases. For example, when a model tends to associate and jointly predict the attributes of long hair and female, it may exhibit high accuracy in training and testing datasets, as well as in most application scenarios. However, when encountering rare test samples, such as males with long hair or females with short hair, the model's judgments may be influenced by prior knowledge, leading to prediction errors. Therefore, excessive reliance on such coupling relationships as prior knowledge for pedestrian attribute recognition can enhance the training and inference speed of the model, perform well on specific test sets, but may compromise the model's robustness and generalization, thereby hindering its deployment in more complex environmental applications.

In order to address the influence of coupling relationships among pedestrian attributes on attribute classification, this paper presents a novel model called the Disentangled Attribute Features Vision Transformer (DAF-ViT). DAF-ViT extends the Vision Transformer [3] with three self-attention modules: the Correlation self-attention (CSA) module, responsible for extracting relevant attribute features; the Exclusive self-attention (ESA) module, responsible for extracting mutually exclusive attribute features; and the Random sequence attention (RSA) module, responsible for extracting unrelated attribute features. These parallel self-attention modules perform their respective tasks, and after individually classifying the decoupled features, the multi-level result fusion module [4] is utilized to obtain the final result.

The contributions of this work can be summarized as follows:

- An effective Disentangled Attribute Features Vision Transformer (DAF-ViT) is proposed for pedestrian attribute recognition, which uses sequential latent feature attention to generate multiple latent tokens dynamically. Aiming to enhance the discriminative ability of correlated features in pedestrian samples and suppress the discriminative ability of irrelevant features.

- Three parallel self-attention modules are proposed: a correlation self-attention module (CSA) for extracting correlated attribute features, a exclusive self-attention module (ESA) for extracting mutually exclusive attribute features, and a random sequence self-attention module (RSA) for extracting unrelated attribute features.
- Comprehensive experiments show that DAF-ViT effectively improves the average accuracy of the model on various attributes based on PETA and PA-100K pedestrian datasets.

2   Related Works

2.1   Pedestrian Attribute Recognition

Early pedestrian attribute recognition mainly employed handcrafted methods targeting low-level features in classification algorithms. However, they exhibited limitations when dealing with samples with different feature distributions. Currently, the majority of solutions for pedestrian attribute recognition employ deep learning-based multi-label classification models. To this end, Bourdev [1] propose the Poselets model to split the image into a set of parts, each capturing a salient pattern corresponding to a given viewpoint and local pose. Moreover, Tang [16] proposes Attribute Localization Module (ALM) to adaptively discover the most discriminative attention regions and learn the regional features for each attribute at multiple scales. But these existing methods still run into insufficient dynamic mapping and insufficient hierarchical aggregation problems.

2.2   Disentangled Features Learning

Coupling relationships refer to the inter-dependencies and interactions between multiple attributes of the same pedestrian. These relationships often manifest through two phenomena: attribute co-occurrence and attribute exclusivity. Attribute co-occurrence describes the higher probability of two or more attributes appearing simultaneously. Attribute exclusivity, on the other hand, denotes the higher probability of two or more attributes not co-occurring. Many studies have approached the multi-class problem by considering the correlation relationships among attributes. Singh [13] use the class activation maps (CAM) as weak location annotation, to de-correlate feature representations of a category from its co-occurring context. They achieve this by learning a feature subspace that explicitly represents categories occurring in the absence of context along side a joint feature sub-space that represents both categories and context.

3   Methods

3.1   Overview

In order to address the impact of the coupling relationship between pedestrian attributes on attribute classification, we introduce the Disentangled Attribute

Features Vision Transformer (DAF-ViT). As shown in Fig. 1, the DAF-ViT model utilizes the T2T-ViT [18] framework as backbone network and incorporates three self-attention modules: Correlation self-attention (CSA), Exclusive self-attention (ESA), and Random sequence attention (RSA). These three attention modules are connected in parallel to the operations of the original self-attention.

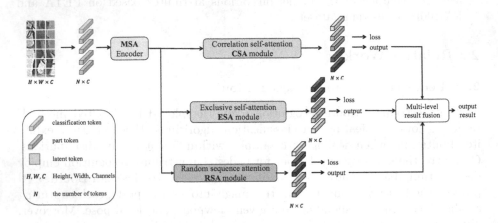

Fig. 1. The pipeline of the DAF module consists of Correlation self-attention (CSA), Exclusive self-attention (ESA), and Random sequence attention (RSA).

Upon receiving pedestrian RGB images as input, the DAF-ViT model employs a Vision Transformer encoder for initial feature extraction. The self-attention mechanism is used to learn the similarity between image patches. Subsequently, the feature sequence obtained from the self-attention is simultaneously fed into three branch modules: the correlation self-attention module, the exclusive self-attention module, and the random sequence attention module. The weight parameters of these three modules are not shared and are trained independently. Each module performs multi-label classification individually. Finally, the results from the three modules are fused using the multi-level result fusion, resulting in the final classification outcome.

3.2   Correlation Self-attention Module

The Correlation self-attention module and the original Transformer module share the same self-attention mechanism, which employs similarity calculations through self-attention to capture the feature relationships between blocks of pedestrian images. Throughout this process, the image blocks are sequentially connected, ensuring the preservation of the positional relationships between them.

As shown in Fig. 2, in the subsequent self-attention computation, the tokens in the image sequence are first multiplied by three different matrices to obtain

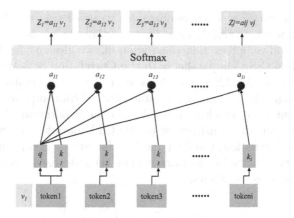

Fig. 2. The detailed workflow of the DAF module consists of Correlation self-attention (CSA), Exclusive self-attention (ESA), and Random sequence attention (RSA).

the query vectors q, key vectors k, and value vectors v. Subsequently, the query vector q_i of each token is used to calculate the similarity with the key vectors k_j of other tokens, resulting in attention weights a_{ij}. The calculation formula for the attention weights is as follows:

$$a_{ij} = softmax\left(\frac{q_i \cdot k_j^T}{\sqrt{D}}\right) \tag{1}$$

where D represents the dimension of q and k. To prevent numerical explosion caused by the increase in dimension, the values are normalized by dividing them by D.

The Softmax function is then applied to map the computed results to the range $[0, 1]$, representing the probability distribution of attention similarity between different tokens. After obtaining the attention weights a_{ij}, they are multiplied with all the original tokens to obtain updated token values. The calculation is expressed as follows:

$$z_j = a_{ij} \cdot x_j \tag{2}$$

By multiplying the attention weights with the original tokens, the attention feature distribution in the sequence can represent the extracted feature vector of the sample image. The Correlation self-attention module learns the similarity relationships between sequential image blocks and can be utilized to explore the co-occurrence information among pedestrian attributes. In most datasets and application scenarios, these coupling relationships can serve as prior knowledge to facilitate the subsequent classification process.

3.3 Exclusive Self-attention Module

Although the coupling relationships of attribute co-occurrence can enhance the efficiency of multi-label classification for pedestrians in most datasets and application scenarios, they can become obstacles to predicting attributes in certain testing environments. To simultaneously strengthen the phenomenon of attribute co-occurrence and suppress this phenomenon, we design the Exclusive Self-Attention module (ESA) in parallel with the Correlation Self-Attention module.

The key idea of the ESA module is to learn the dissimilarity within the sequence of image blocks, aiming to reduce the relationships among the blocks in the feature space as much as possible. So the generation of q, k, and v in the ESA module remains consistent with the CSA module. However, in terms of attention value calculation, the ESA module focuses on computing the dissimilarity between two image blocks to represent their degree of exclusiveness. The calculation formula is as follows:

$$a_{ij} = 1 - softmax \left(\frac{q_i \cdot k_j^T}{\sqrt{D}} \right) \tag{3}$$

where the softmax function normalizes the results to obtain values between 0 and 1, indicating the similarity and relevance of an image block to the remaining blocks. Subtracting the normalized result from 1 represents the exclusiveness and irrelevance of an image block relative to the other blocks. Through multiple iterations, the image blocks in the sequence are pushed further apart in the feature space, thereby capturing the differences between individual blocks.

This calculation method of exclusive attention cannot guarantee the accuracy of the classification results for each image block, and in some cases, it may even have lower overall accuracy compared to the correlation attention. However, in certain outlier data (e.g., males with long hair), the exclusive attention can provide more accurate results than other methods. In the subsequent result fusion module, it is sufficient to extract the higher-predicted results from the exclusive attention, while the lower-accuracy results can be replaced by the results from other branch modules.

3.4   Random Sequence Attention Module

The relationships between various attributes of pedestrian samples are not only manifested in the semantic relationships between labels but also in the positional relationships corresponding to each attribute. When the sample image is encoded into a 1D sequence of image blocks, the relative information of different body parts of the pedestrian is maintained, meaning that the feature of the head always comes before the feature of the body, and the feature of the legs always comes after the feature of the body.

Therefore, one of the reasons for the co-occurring and mutually exclusive phenomena of pedestrian attributes is the fixed relative positions of these attributes in the sample. Therefore, the problem of attribute co-occurrence and mutual exclusivity among pedestrian sample attributes addressed by transforming it into the influence of fixed relative positions of pedestrian attributes on attribute learning.

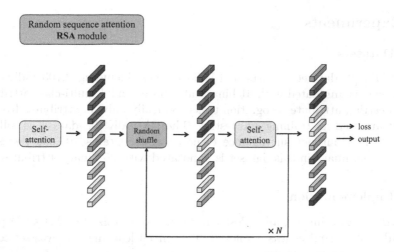

Fig. 3. The workflow of the Random sequence attention (RSA) module.

To eliminate this influence as much as possible, a Random sequence attention (RSA) module is designed to operate in parallel with the related self-attention module. The RSA module has a similar overall computation process to the CSA module. The difference lies in the step before each round of self-attention computation, where the sequence of image blocks is randomly shuffled before being fed into the subsequent self-attention computation process. And the positions of the elements in the image block sequence are replaced accordingly based on the new index. This shuffling process achieves the goal of disordering the image block sequence, which is then passed to the self-attention weight calculation process and the assignment process. The specific algorithmic flow of the Random sequence attention (RSA) is shown in Fig. 3.

Similar to the application of the mutual exclusivity attention results, the computation results of the RSA cannot guarantee the accuracy of each attribute's classification result. The computation of the Random Self-Attention is only highly accurate for a few samples such as outliers or rare samples. In most datasets and application scenarios (e.g., when attribute co-occurrence promotes the accuracy of attribute classification), the related self-attention algorithm may be more effective. Therefore, when obtaining the final classification results, a result fusion module is used to select the attribute with the highest accuracy from the results of the mutual exclusivity attention as its predicted result. The lower-accuracy results can be replaced by the results from other modules (such as the CSA and ESA modules). By leveraging the strengths of different modules, the final attribute prediction results can be obtained.

4   Experiments

4.1   Datasets

The PETA [2] dataset consists of 19,000 images, including 8,705 individuals. Each image is annotated with 61 binary attributes and 4 multi-class attributes. For pedestrian attribute recognition tasks, typically only 35 attributes from the dataset are used for training and testing. The PA-100K [11] dataset is collected from 598 real outdoor surveillance cameras and contains 100,000 pedestrian images. Each image in this dataset is annotated with 26 binary attributes.

4.2   Implementation

Prior to the experiments, all pedestrian images were resized to 224×224 pixels. The AdamW optimizer was employed. The initial learning rate was set to 1e-5 with cosine learning rate decay.

4.3   Loss Function

We adopt the weighted binary cross-entropy loss function (Weighted BCE loss) [8] in DAF-ViT. Weighted BCE loss can effectively alleviate the problem of imbalanced data and labels and formulated as follow:

$$Loss = -\frac{1}{N} \sum_{i=1}^{N} \sum_{j=1}^{N} w_j(\hat{y_{ij}}log(y_{ij}) + (1 - \hat{y_{ij}})log(1 - y_{ij})) \qquad (4)$$

$$w_j = exp(-r/\alpha^2), \qquad (5)$$

where i and j are the number of images and attributes, y_{ij} is the prediction result and the $\hat{y_{ij}}$ is the ground truth label, w_j is the loss weight for j-th attribute, r_j is the positive ratio of j-th attribute in the training set, α is a tuning parameter which is set as 1 in experiments.

4.4   Comparison with State-of-the-Art Methods

Due to the limited availability of pedestrian attribute recognition models specifically designed for attribute decoupling prior to this study, there is a lack of comparative conditions. Therefore, the models included in the comparison were categorized based on their backbone networks, including AlexNet, InceptionNet, ResNet, LSTM, and ViT for pedestrian attribute recognition. Table 1 presents the experimental results on the PETA dataset, while Table 2 presents the experimental results on the PA-100K dataset.

From Table 1, it can be observed that the DAF-ViT model achieves the highest level of accuracy (mA) compared to the T2T-ViT model with a 2.8% improvement. Additionally, it shows a 1% improvement in recall rate compared to the baseline network. However, the DAF-ViT model exhibits significant disadvantages in terms of accuracy (Acc), precision (Prec), and F1 score. The same phenomenon can be observed in the experimental results on the PA-100K dataset

Table 1. Comparison of models on the PETA dataset

Model	Backbone	mA	Acc	Prec	Recall	F1
ACN [14]	AlexNet	81.2	-	84.1	81.2	82.6
DeepMAR [8]	AlexNet	82.9	75.1	83.7	83.1	83.4
PGDM [9]	AlexNet	82.9	78.1	86.9	84.7	85.8
HP-Net [11]	InceptionNet	81.8	76.1	84.9	83.2	84.0
JLPLS-PAA [15]	InceptionNet	84.9	79.5	87.4	86.3	86.9
ALM [16]	InceptionNet	86.3	79.5	85.7	88.1	86.9
SSC [6]	ResNet-50	86.5	78.9	86.0	87.1	86.9
MsVAA [12]	ResNet-50	84.6	78.9	86.8	86.1	86.4
JRL [17]	LSTM	82.1	-	82.6	82.1	82.0
JRL-V2 [17]	LSTM	85.6	-	86.0	85.3	85.4
ViT [3]	ViT	69.7	50.2	63.2	70.8	66.8
C-Trans [7]	ViT	84.6	78.7	85.9	86.6	86.3
SimViT [10]	ViT	84.9	79.1	86.9	86.5	86.5
T2T-ViT [18]	ViT	83.7	78.5	84.6	88.2	86.3
DAF-ViT (ours)	ViT	86.5	78.2	83.0	89.2	86.0

Table 2. Comparison of models on the PA-100K dataset

Model	Backbone	mA	Acc	Prec	Recall	F1
DeepMAR [8]	AlexNet	72.7	70.4	82.2	80.4	81.3
PGDM [9]	AlexNet	74.9	73.1	84.4	82.3	83.3
HP-Net [11]	InceptionNet	74.2	72.2	82.9	82.1	82.6
JLPLS-PAA [15]	InceptionNet	81.6	78.9	86.8	87.7	87.3
VAC [5]	ResNet-50	79.2	79.4	88.9	86.2	87.6
SSC [6]	ResNet-50	81.8	78.8	85.9	89.1	86.8
ViT [3]	ViT	67.7	49.3	60.2	68.3	64.2
C-Trans [7]	ViT	81.6	75.9	83.3	82.6	83.6
SimViT [10]	ViT	82.3	76.8	85.1	84.6	85.8
T2T-ViT [18]	ViT	81.7	76.3	82.5	85.7	84.3
DAF-ViT (ours)	ViT	84.2	69.6	73.8	90.9	81.0

presented in Table 2. The reasons behind these observations will be analyzed and discussed in the next section, considering the results of ablation experiments.

4.5   Ablation Study

To verify the effectiveness of the Contextual Self-Attention module (CSA), Exclusive Self-Attention module (ESA), and Random Self-Attention module

(RSA), a series of ablation experiments were conducted. Table 3 presents the results of these ablation experiments on the PETA dataset using the DAF-ViT model. The Baseline refers to the original network (referred to as B). $B + CSA$ represents the addition of the CSA module to the Baseline. To ensure fair comparisons and eliminate the influence of attention layers and parameter counts on the experimental results, the total number of attention layers was kept consistent.

Table 3. Ablation study results on PETA dataset

Model	Attention Layers	mA	Acc	Prec	Recall	F1	Para. (M)
Baseline	14	83.71	78.53	84.62	88.24	86.32	25.92
B+CSA	14 + 12	86.26	77.13	81.25	90.31	85.54	41.83
B+ESA	14 + 12	86.17	76.29	79.66	91.64	85.23	41.83
B+RSA	14 + 12	86.22	75.53	78.68	91.66	84.67	41.83
B+CSA+ESA	14 + 6 + 6	86.27	77.06	80.67	91.21	85.62	40.35
B+CSA+RSA	14 + 6 + 6	86.20	75.77	79.28	91.36	84.89	40.35
B+ESA+RSA	14 + 6 + 6	86.50	76.96	80.83	90.63	85.45	40.35
B+CSA+ESA+RSA	14 + 4 + 4 + 4	86.38	78.17	82.60	90.08	86.17	38.87
B+CSA+ESA+RSA	14 + 12 + 12 + 12	86.54	78.17	82.97	89.17	85.96	74.3

From Table 3, it can be observed that compared to the Baseline model, CSA, ESA, and RSA have improved the mA accuracy by 2.55%, 2.46%, and 2.51% respectively. However, there is a slight decrease in Acc, Prec, and F1 accuracy, while Recall accuracy has improved. Both ESA and RSA modules have lower mA, Acc, Prec, and F1 metrics compared to the CSA module. Although the CSA, ESA, and RSA modules individually improve the average accuracy, they also lead to more false negatives and false positives. This phenomenon may arise from the fact that only a subset of features exhibits coupling relationships, while some features have no apparent correlations. When performing exclusive or random attention calculations, the model treats all features as a single feature sequence, causing unnecessary decoupling computations involving features that do not require decoupling, thereby affecting the accuracy of attribute classification.

When combining the Baseline model with two decoupling modules, the above conclusions generally hold. The $B + ESA + RSA$ combination achieves the highest mA accuracy, improving by 2.79% compared to the Baseline, slightly outperforming the $B + ESA$ combination by 0.33% and the $B + RSA$ combination by 0.28%. However, Acc, Prec, and F1 accuracy didn't improve or even decreased.

To investigate the impact of different levels of attention on the performance of the model, two experiments were designed by combining two Baseline models with all three decoupling modules: one with a total attention layer count of $14 + 4 + 4 + 4$ and the other with $14 + 12 + 12 + 12$ attention layers.

In the first experiment, the total attention layer count remained consistent with the previous experiments, but the attention layer count per module was reduced to 4 layers. The mA accuracy in this experiment improved by 0.11% and 0.18% compared to the $B + CSA + ESA$ and $B + CSA + RSA$ combinations respectively, but decreased by 0.12% compared to the $B + ESA + RSA$ experiment. However, in terms of Acc, Prec, and F1 accuracy, it showed the highest and significant improvements, indicating that the $B + CSA + ESA + RSA$ combination effectively addresses the issues of false negatives and false positives, with the ESA and RSA modules making larger contributions.

In the second experiment, the attention layers of the three sub-modules in $B + CSA + ESA + RSA$ were all set to 12 layers. This configuration allows all modules to fully leverage self-attention learning and explore the maximum learning capacity of the DAF-ViT model. The experiment showed significant improvements in all evaluation metrics, with an increase of 2.83% in mA accuracy compared to the baseline model and a 0.16% improvement compared to the $12 + 4 + 4 + 4$ layer combination.

Based on the comprehensive comparisons, it can be concluded that when the total attention layers of the model remain constant, the $Baseline + ESA + RSA$ combination achieves the highest accuracy. This is because among the numerous attribute features, some have coupled information while others do not. When the exclusive self-attention mechanism and the random self-attention mechanism decouple the features from different perspectives, the baseline model can retain the learning of uncoupled information features as much as possible and obtain the optimal results from each module during the final result fusion.

5   Conclusion

While the co-occurrence and mutual exclusion phenomena among pedestrian attributes can serve as prior knowledge to help the model improve the prediction accuracy and speed for regular samples in most training and testing scenarios, they may lead to prediction biases when dealing with a small number of outlier data. We propose a Disentangled Attribute Features Vision Transformer (DAF-ViT) model which extends the Vision Transformer with three self-attention modules: the Correlation self-attention (CSA), responsible for extracting relevant attribute features; the Exclusive self-attention (ESA), responsible for extracting mutually exclusive attribute features; and the Random sequence attention (RSA), responsible for extracting unrelated attribute features. These parallel self-attention modules perform their respective tasks, and after individually classifying the decoupled features, the multi-level result fusion module is utilized to obtain the final result. The experimental results demonstrate a significant improvement in the average accuracy of DAF-ViT on pedestrian dataset.

References

1. Bourdev, L., Maji, S., Malik, J.: Describing people: a poselet-based approach to attribute classification. In: 2011 International Conference on Computer Vision, pp. 1543–1550. IEEE (2011)
2. Deng, Y., Luo, P., Loy, C.C., Tang, X.: Pedestrian attribute recognition at far distance. In: Proceedings of the 22nd ACM International Conference on Multimedia, pp. 789–792 (2014)
3. Dosovitskiy, A., et al.: An image is worth 16×16 words: transformers for image recognition at scale. arXiv preprint arXiv:2010.11929 (2020)
4. Feng, X., Guo, J., Liu, C.: Latent dynamic token vision transformer for pedestrian attribute recognition. In: 2023 IEEE SmartWorld, Ubiquitous Intelligence & Computing, Advanced & Trusted Computing, Scalable Computing & Communications, Internet of People and Smart City Innovation (SmartWorld/SCALCOM/UIC/ATC/IOP/SCI). IEEE (2023)
5. Guo, H., Zheng, K., Fan, X., Yu, H., Wang, S.: Visual attention consistency under image transforms for multi-label image classification. In: Proceedings of the IEEE/CVF Conference on Computer Vision and Pattern Recognition, pp. 729–739 (2019)
6. Jia, J., Chen, X., Huang, K.: Spatial and semantic consistency regularizations for pedestrian attribute recognition. In: Proceedings of the IEEE/CVF International Conference on Computer Vision, pp. 962–971 (2021)
7. Lanchantin, J., Wang, T., Ordonez, V., Qi, Y.: General multi-label image classification with transformers. In: Proceedings of the IEEE/CVF Conference on Computer Vision and Pattern Recognition, pp. 16478–16488 (2021)
8. Li, D., Chen, X., Huang, K.: Multi-attribute learning for pedestrian attribute recognition in surveillance scenarios. In: 2015 3rd IAPR Asian Conference on Pattern Recognition (ACPR), pp. 111–115. IEEE (2015)
9. Li, D., Chen, X., Zhang, Z., Huang, K.: Pose guided deep model for pedestrian attribute recognition in surveillance scenarios. In: 2018 IEEE International Conference on Multimedia and Expo (ICME), pp. 1–6. IEEE (2018)
10. Li, G., Xu, D., Cheng, X., Si, L., Zheng, C.: SimViT: exploring a simple vision transformer with sliding windows. arXiv preprint arXiv:2112.13085 (2021)
11. Liu, X., et al.: HydraPlus-Net: attentive deep features for pedestrian analysis. In: Proceedings of the IEEE International Conference on Computer Vision, pp. 350–359 (2017)
12. Sarafianos, N., Xu, X., Kakadiaris, I.A.: Deep imbalanced attribute classification using visual attention aggregation. In: Ferrari, V., Hebert, M., Sminchisescu, C., Weiss, Y. (eds.) ECCV 2018. LNCS, vol. 11215, pp. 708–725. Springer, Cham (2018). https://doi.org/10.1007/978-3-030-01252-6_42
13. Singh, K.K., Mahajan, D., Grauman, K., Lee, Y.J., Feiszli, M., Ghadiyaram, D.: Don't judge an object by its context: Learning to overcome contextual bias. In: Proceedings of the IEEE/CVF Conference on Computer Vision and Pattern Recognition, pp. 11070–11078 (2020)
14. Sudowe, P., Spitzer, H., Leibe, B.: Person attribute recognition with a jointly-trained holistic CNN model. In: Proceedings of the IEEE International Conference on Computer Vision Workshops, pp. 87–95 (2015)
15. Tan, Z., Yang, Y., Wan, J., Hang, H., Guo, G., Li, S.Z.: Attention-based pedestrian attribute analysis. IEEE Trans. Image Process. **28**(12), 6126–6140 (2019)

16. Tang, C., Sheng, L., Zhang, Z., Hu, X.: Improving pedestrian attribute recognition with weakly-supervised multi-scale attribute-specific localization. In: Proceedings of the IEEE/CVF International Conference on Computer Vision, pp. 4997–5006 (2019)
17. Wang, J., Zhu, X., Gong, S., Li, W.: Attribute recognition by joint recurrent learning of context and correlation. In: Proceedings of the IEEE International Conference on Computer Vision, pp. 531–540 (2017)
18. Yuan, L., et al.: Tokens-to-token ViT: training vision transformers from scratch on ImageNet. In: Proceedings of the IEEE/CVF International Conference on Computer Vision, pp. 558–567 (2021)

A High-Resolution Network Based on Feature Redundancy Reduction and Attention Mechanism

Yuqing Pan[✉], Weiming Lan, Feng Xu, and Qinghua Ren

School of Computer Science and Communication Engineering, Jiangsu University,
Zhenjiang 212013, China
panyq@ujs.edu.cn

Abstract. Model lightweighting is an essential aspect of computer vision tasks. We found that HRNet achieves superior performance by maintaining high resolution through multiple parallel subnetworks, but it also introduces many unnecessary redundant features, resulting in high complexity. Many methods replace modules in the backbone network with lightweight ones, often based on MobileNet, Sandglass modules, or ShuffleNet. However, these methods often suffer from significant performance degradation. This paper proposes GA-HRNet, a lightweight high-resolution human pose estimation network with an integrated attention mechanism, built upon the HRNet framework. The network structure of HRNet is restructured with reference to the G-Ghost Stage, introducing GPU-efficient cross-layer cheap operations to reduce inter-block feature redundancy, significantly lowering complexity while preserving high accuracy. Moreover, to compensate for the accuracy loss due to reconstruction, an attention module is designed and introduced to emphasize more important information in the channel and spatial dimensions, thereby improving network precision. Experimental results on the COCO and COCO-WholeBody datasets demonstrate that the proposed GA-HRNet is lighter and yet more accurate than HRNet. Moreover, it outperforms several state-of-the-art methods in terms of performance.

Keywords: human pose estimation · lightweight · attention mechanism · high-resolution network

1 Introduction

High-resolution representation is essential for achieving high performance in human pose estimation [20]. Due to limited computational resources, developing efficient high-resolution models is of paramount importance. HRNet has demonstrated superior capabilities in position-sensitive tasks within large-scale networks, such as semantic segmentation, human pose estimation, and object detection. HRNet has a large computational complexity. Although some progress has been made in model lightweight, such as Lite-HRNet [28], they often suffer from significant accuracy loss. In addition, existing lightweight methods

Q. Liu et al. (Eds.): PRCV 2023, LNCS 14430, pp. 510–521, 2024.
https://doi.org/10.1007/978-981-99-8537-1_41

designed specifically for CPUs introduce GPU-inefficient operations and do not make a good trade-off between GPU latency and computational complexity. The proposed approach reduces feature redundancy by reconstructing the network structure, significantly reducing complexity during the inference process while barely compromising accuracy. We avoid introducing GPU-inefficient operations, enabling the network to leverage the parallel computing capabilities of GPUs effectively.

Building upon HRNet, we design a lightweight network called G-HRNet by incorporating the structure of G-Ghost Stage from G-GhostNet [9]. Experimental results indicate that G-HRNet significantly reduces complexity while preserving high accuracy. The attention mechanism can bring performance improvement to various neural networks, where channel & spatial attention further combines the advantages of channel attention and spatial attention. To further improve performance, we introduce a lightweight Channel-Spatial Attention Module (CSAM) that combines information across the channel and spatial dimensions, which significantly enhances information representation. The feature maps are encoded with the Channel Attention Module and the Spatial Attention Module in order. The resulting network is referred to as GA-HRNet.

The main contributions of this paper are as follows:

1. We propose a lightweight network G-HRNet, which is reconstructed by reducing feature redundancy. Compared to the baseline HRNet, our method significantly reduces complexity while preserving high accuracy.
2. Building upon G-HRNet, we propose a lightweight high-resolution network, GA-HRNet, which incorporates an efficient Channel-Spatial Attention Module (CSAM).
3. We evaluate GA-HRNet on the COCO and the COCO-WholeBody datasets, and experimental results demonstrate that our approach achieves a better balance between accuracy and complexity.

2   Related Work

Lightweight. In recent years, a series of lightweight model architectures have been proposed. MobileNets [1,10,18] first reduced the complexity by deep separable convolutions, then introduced inverted residual blocks and utilized AutoML techniques to obtain better performance. ShuffleNet [15,30] introduces channel shuffle operation and proposes design guidelines for efficient models in terms of both memory consumption and GPU parallelism. C-GhostNet [8] aims to reduce input redundancy by processing input features with fewer convolution filters, resulting in reduced FLOPs, parameters, and memory movement within the network.

Due to significant hardware architecture differences between CPUs and GPUs, some operations with lower FLOPs (particularly depthwise convolutions and channel shuffle) may not perform optimally on GPU devices and other throughput-oriented processing units, resulting in slower inference speeds compared to traditional convolutional neural networks. G-GhostNet [9] investigates inter-layer feature redundancy in convolutional structures and introduces GPU-efficient cross-layer inexpensive operations to reduce computational complexity while minimizing memory data movement.

Attention Mechanism. Humans can naturally and effectively identify salient regions in complex scenes. Inspired by this observation, attention mechanisms have been introduced in computer vision to mimic this aspect of the human visual system. Attention mechanisms have yielded remarkable success in various computer visual tasks. Existing attention methods can be categorized into four basic categories: channel attention, spatial attention, temporal attention, and branch attention, as well as two mixed categories: channel & spatial attention and spatial & temporal attention [7]. In neural networks, different channels represent different objects [2], and Channel attention assigns weights to all channels, which can be seen as a process of object selection: deciding what to focus on. Hu et al. [11] first proposed channel attention in SENet, where the core is an SE module. The SE module consists of a squeeze module and an excitation module. GSoP-Net [6] uses global second-order pooling in the squeeze module to achieve better performance, by improving the excitation module, ECANet [21] reduces the complexity, and SRM [12] improves both the squeeze module and excitation module. Spatial attention is an adaptive mechanism for selecting spatial regions.

Channel & spatial attention have both channel attention and spatial attention advantages. It focuses on both important objects and regions [2]. CBAM [23] and BAM [17] improve efficiency by introducing global average pooling and separating attention in the channel and spatial dimensions. DANet [5] and RGA [31] combine self-attention mechanisms with channel & spatial attention to study pairwise interactions.

3   Approach

3.1   G-HRNet

HRNet. HRNet enhances the high-resolution representation by generating feature maps in parallel to obtain high performance, while at the same time, the numerous multi-resolution fusions bring more redundant features [24]. As shown in Table 1. To alleviate this problem, we rethink the internal structure of HRNet, it can be found that HRNet consists of stacked structures with many identical blocks, referred to as "block stacks" for convenience. The intermediate feature sizes within each block stack are the same. To get the output features, the information is processed through a series of blocks. However, this is not necessary and will generate a lot of redundant features.

G-Ghost Stage. G-Ghost Stage [9] is a general block stack structure used for constructing GPU-efficient network architectures. In any block stack, the intermediate feature sizes are the same, and feature similarity and redundancy exist

Table 1. Architecture of HRNet_W48

Layer	Operate Module	Module Repeat	Module Branch	Block Repeat	Output Channels
Input image					3
Stage 1	conv2d	2			64
	Layer (bottleneck)	1	1	4	
Stage 2	HRModule (basic block)	1	2	4,4	48,96
Stage 3	HRModule (basic block)	4	3	4,4,4	48,96,192
Stage 4	HRModule (basic block)	3	4	4,4,4,4	48,96,192,384

across multiple layers within that block stack. G-Ghost Stage leverages the original block stack with fewer channels to generate intrinsic features. It uses inexpensive operations and information-additive aggregation of intrinsic and ghost features to generate ghost features. Finally, the intrinsic and ghost features are combined. By adapting the CNN structure using the G-Ghost stage, a significant reduction in computational complexity can be achieved while maintaining high accuracy.

G-HRNet. G-HRNet is constructed using HRNet_W48 as a reference. As shown in Table 1, the initial stage of HRNet comprises 4 residual units. The second, third, and fourth stages consist of 1, 4, and 3 HRModules. Each HRModule contains 4 residual units. We have applied the G-Ghost Stage modification to all residual units in all stages, resulting in G-HRNet.

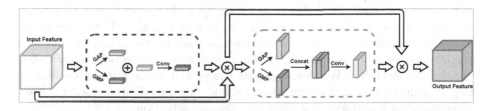

Fig. 1. The overall structure of CSAM.

3.2 GA-HRNet

The SE module has achieved significant success in computer vision, and it can easily bring large performance gains to CNN-based networks. However, it only considers the channel dimension and not the spatial dimension, which loses a lot of spatial information. In addition, its excitation module is not efficient enough. specifically, the dimensionality reduction operation in the SE module can reduce the model complexity, but this also destroys the direct correspondence between channels and their weights, which can negatively affect the prediction of channel attention [21]. Although the later CBAM and ECA modules solve the above two problems respectively and achieve better results, they do not provide a good solution for the other problem. To reduce the accuracy loss caused by structural reconstruction, we propose a new attention module called Channel-Spatial Attention Module (CSAM), which introduces spatial attention and optimizes the channel attention module compared with the SE module. The two modules are combined in parallel and sequentially, Sanghyun et al. [23] have shown that the sequential arrangement of channel preference has better results, and we follow this basic structure. Given the intermediate feature map $F \in R^{C \times H \times W}$ of the input, CSAM sequentially infers a one-dimensional channel attention map $M_c \in R^{C \times 1 \times 1}$ and a two-dimensional spatial attention map $M_s \in R^{1 \times H \times W}$, as shown in Fig. 1. The entire attention process can be summarized as:

$$F' = M_c(F) \bigotimes F \tag{1}$$

$$F'' = M_s(F') \bigotimes F' \tag{2}$$

where $\bigotimes$ represents element-wise multiplication. F' is the intermediate output and F'' is the final output. The details of each attention module are described below.

Channel Attention Module. We generate the channel attention map by utilizing the inter-channel relationships of the features. To effectively compute channel attention, we compress the spatial dimension of the input feature map. Aggregating spatial information is currently commonly done using average pooling. However, max pooling gathers another important information in the feature map to achieve finer channel-wise attention [23]. Thus, we use both average pooling and max-pooling features. The degradation of the SE module in the excitation module destroys the direct correspondence between channels and their weights, and we avoid degradation in the channel attention section to learn effective channel attention. We first aggregate the spatial information of the feature map by using an average pooling operation and a max pooling operation. Then a 1D convolution is performed to generate a channel attention map: $M_c \in R^{C \times 1 \times 1}$. In short, channel attention is calculated as:

$$M_c(F) = \sigma(f_1(AvgPool(F) + MaxPool(F))) \tag{3}$$

where σ represents the sigmoid function and f_1 represents a 1D convolution operation.

Spatial Attention Module. In addition to the channel dimension, the spatial dimension information is also important for neural networks. Spatial attention focuses on "where", and generating spatial attention maps through the spatial interrelationship of features can complement the information neglected by channel attention. Consistent with CBAM, In short, spatial attention is calculated as:

$$M_s(F) = \sigma(f^{7 \times 7}([AvgPool(F); MaxPool(F)])) \tag{4}$$

where $f^{7 \times 7}$ represents a convolution operation with a kernel size of 7×7.

GA-HRNet. We build GA-HRNet based on G-HRNet. The first stage of G-HRNet contains 4 residual units, and the second, third, and fourth stages contain 1, 4, and 3 exchange blocks, respectively. An exchange block contains 4 residual units. We add CSAM modules at the end of the 4 residual units in the first stage and all residual blocks in the second, third, and fourth stages to enhance the network's information aggregation capabilities and obtain GA-HRNet.

Table 2. Results on COCO val. Comparison on the COCO val. We only report the input size, parameters, FLOPs, and mAP for the pose estimation model, and ignore the object detection model. "-" indicates data not provided in the original paper.

method	Input size	Params(MB)	GFLOPs	AP	AP^{50}	AP^{75}	AP^M	AP^L	AR
SimpleBaseline [25]	256 × 192	68.6	15.7	72	89.3	79.8	68.7	78.9	77.8
SimpleBaseline [25]	384 × 288	68.6	35.6	74.3	89.6	81.1	70.5	79.7	79.7
TransPose-H-A6 [27]	256 × 192	17.5	21.8	75.8	-	-	-	-	80.8
HRNet_W32 [19]	256 × 192	28.5	7.1	73.4	89.5	80.7	70.2	80.1	78.9
HRNet_W32 [19]	384 × 288	28.5	16	75.8	90.6	82.7	71.9	82.8	81
EvoPose2D-S [16]	256 × 192	2.53	1.07	70.2	88.9	77.8	66.5	76.8	76.9
EvoPose2D-M [16]	384 × 288	7.34	5.59	75.1	90.2	81.9	71.5	81.7	81
EvoPose2D-L [16]	512 × 384	14.7	17.7	76.6	90.5	83	72.7	83.4	82.3
UniFormer-S [13]	488 × 320	25.2	14.8	76.2	90.6	83.2	68.6	79.4	81.4
UniFormer-B [13]	256 × 192	53.5	9.2	75	90.6	83	67.8	77.7	80.4
UniFormer-B [13]	384 × 288	53.5	22.1	76.7	90.8	84	69.3	79.7	81.9
HRNet_W48 [19]	256 × 192	63.6	15.77	74.8	90.3	82	71.3	81.6	80.2
G-HRNet_W48	256 × 192	43.38	10.57	75	90.3	82.1	71.6	81.8	80.5
A-HRNet_W48	256 × 192	63.61	15.8	76.2	90.9	83.2	72.1	83	80.9
GA-HRNet_W48	256 × 192	43.39	10.6	76.1	90.5	82.5	72.4	82.9	81.1
HRNet_W48 [19]	384 × 288	63.6	35.48	76.3	90.8	82.9	72.3	83.4	81.2
GA-HRNet_W48	384 × 288	43.39	23.84	**76.9**	**91**	83.8	**73.1**	**83.7**	**82.7**

4 Experiments

Based on MMPose [3], we conducted comparative experiments and ablation studies on two datasets: COCO [14] and COCO-WholeBody [26].

4.1 COCO

Setting. The COCO (Common Objects in Context) [14] dataset is mainly used in computer vision. Our model is trained on train2017 (including 118K images and over 200K person instances) and evaluated on val2017 (including 5K images) and test-dev2017 (including 40K images). The COCO dataset annotations include 17 full-body keypoints.

Training. The experimental environment is configured as follows: Ubuntu 18.04 LST 64-bit system, two GeForce RTX 3090 graphics cards, using PyTorch1.10.2 deep learning framework. The number of workers per GPU is

2. Adam is used as the optimizer during network training with an initial learning rate of 5e−4. Linear warm-up is used with a warm-up iteration count of 500. The network is trained for a total of 210 rounds.

Testing. The testing environment is the same as the training environment and tests are conducted on both the COCO val and the COCO test-dev dataset.

Evaluation metrics. This paper uses the standard evaluation metrics for human pose estimation tasks on the COCO dataset: Mean Average Precision (mAP) based on Object Keypoint Similarity (OKS) verification standard. We evaluate performance using standard average precision and average recall scores including AP, AP^{50}, AP^{75}, AP^M, AP^L, and AR.

Table 3. Results on COCO test-dev. Comparison on COCO test-dev. We only report the input size, parameters, FLOPs, and mAP of the pose estimation model and ignore the object detection model. "-" indicates that data was not provided in the original paper.

Method	Input Size	Params(MB)	GFLOPs	AP	AP^{50}	AP^{75}	AP^M	AP^L	AR
Lite-HRNet-30 [28]	384 × 288	1.8	0.7	69.7	90.7	77.5	66.9	75	75.4
SimpleBaseline [25]	384 × 288	68.6	35.6	73.7	91.9	81.1	70.3	80	79
TransPose-H-A6 [27]	256 × 192	17.5	21.8	75	92.2	82.3	71.3	81.1	-
HRNet_W32 [19]	384 × 288	28.5	16	74.9	92.5	82.8	71.3	80.9	80.1
EvoPose2D-L [16]	512 × 384	14.7	17.7	75.7	91.9	83.1	72.2	81.5	**81.7**
HRNet_W48 [19]	256 × 192	63.6	15.77	73.8	92.1	82	70.5	79.6	79.2
GA-HRNet_W48	256 × 192	43.39	10.6	75.1	92.3	82.9	71.8	80.8	80.3
HRNet_W48 [19]	384 × 288	63.6	35.48	75.5	92.5	**83.3**	71.9	81.5	80.5
GA-HRNet_W48	384 × 288	43.39	23.84	**76**	**92.6**	83.1	**72.5**	**81.8**	**81.7**

Results on COCO val. Table 2 and Fig. 2 show the performance of our proposed method and other methods on the COCO val. Our GA-HRNet reduces the complexity of the baseline model HRNet by 1/3 at both 256 × 192 and 384 × 288 resolutions while achieving significant improvements in accuracy (1.3% and 0.6% AP respectively).

When trained with an input size of 256 × 192, GA-HRNet always outperforms other methods at the same resolution. Specifically, our GA-HRNet is 0.3% higher than TransPose-H-A6 on AP, 5.9% higher than EvoPose2D-S, and 1.1% higher than UniFormer-B. When the input size is 384 × 288, our GA-HRNet also achieves a good balance between complexity and accuracy. We are 2.6% higher than SimpleBaseline on AP, 1.8% higher than EvoPose2D-M, and 0.2% higher than UniFormer-B (76.9% vs 76.7%). In addition, GA-HRNet performs better than EvoPose2D-L at a resolution of 512 × 384 (76.9% vs 76.6%).

Results on COCO test-dev. Table 3 shows the performance of our network and other methods on the COCO test-dev. Our GA-HRNet achieves the highest AP of 69.0%. Compared to TransPose-H-A6 with the same input size, our small network achieves a 0.1% AP improvement with a clear advantage in complexity;

it is significantly better than EvoPose2D-L with a larger input size. Compared to the baseline model, GA-HRNet achieves significant accuracy improvements while significantly reducing complexity.

4.2   COCO-WholeBody

Setting. The COCO-WholeBody [26] dataset is the first large-scale benchmark dataset for whole-body pose estimation. It is an extension of the COCO2017 dataset with the same training/validation set split as COCO. It has 133 keypoints. Our model is trained on the train2017 dataset and validated on val2017.

Training. All configurations are consistent with COCO.

Testing. The testing environment is the same as the training environment and tests are conducted on the COCO-WholeBody validation set.

Evaluation metrics. Consistent with COCO, use the standard evaluation metrics for human pose estimation tasks on the COCO dataset including five types of AP and AR: body, foot, face, hand, and whole-body.

Table 4. Results on COCO-WholeBody val. Comparison on COCO-wholeBody val, we only report the input size, parameters, FLOPs, and 5 mAPs of the pose estimation model and ignore the object detection model. Among the above mAPs, whole-body AP is the main metric we evaluate. "-" indicates that data was not provided in the original paper.

Method	Input Size	Params (MB)	GFLOPs	Body AP	Body AR	Foot AP	Foot AR	Face AP	Face AR	Hand AP	Hand AR	Whole-body AP	Whole-body AR
SimpleBaseline [25]	384 × 288	-	20.42	66.6	74.7	63.5	76.3	73.2	81.2	53.7	64.7	57.3	67.1
TCFormer [29]	256 × 192	27.78	6.62	69.1	77	69.8	**81.3**	64.9	74.6	53.5	65	57.2	67.8
PVT [22]	384 × 288	-	19.65	67.3	76.1	66	79.4	74.5	82.2	54.5	65.4	58.9	68.9
FastPose50-dcn-si [4]	256 × 192	-	6.1	70.6	75.6	**70.2**	77.5	77.5	82.5	45.7	53.9	59.2	66.5
ZoomNet [26]	384 × 288	-	28.49	**74.5**	81	60.9	70.8	**88**	92.4	57.9	73.4	63	74.2
HRNet_W48 [19]	256 × 192	63.6	15.79	69.6	77.4	65.3	77.2	65.3	74.3	52.8	63	57.4	67.7
G-HRNet_W48	256 × 192	43.38	10.59	70.8	78.9	66.9	78.6	64.4	73.9	47.1	57.9	56.7	67.7
A-HRNet_W48	256 × 192	63.61	15.81	72.7	77.3	62.4	75.7	74.4	84.4	53	63.3	60.6	70.5
GA-HRNet_W48	256 × 192	43.39	10.6	72.4	79.5	60.6	69.6	75.8	88.8	52.1	62.0	59.9	70.1
HRNet_W48 [19]	384 × 288	63.6	35.52	72.2	79	69.4	79.9	77.7	83.4	58.7	67.9	63.1	71.6
GA-HRNet_W48	384 × 288	43.39	23.88	73.5	**81**	63.5	75.2	87.5	**92.9**	**59**	**74**	**64**	**75**

Results on COCO-WholeBody val. As shown in Table 4 and Fig. 2(b), GA-HRNet achieves superior performance and well balances accuracy and complexity. Specifically, GA-HRNet achieves higher accuracy (AP increased by 2.5% and 0.9% respectively) and lower complexity (GFLOPs and parameters both decreased by about 1/3) compared to the baseline model HRNet. Compared to mainstream methods, it has significant advantages in terms of accuracy and complexity (at the same resolution, we are 2.7% higher than TCFormer on AP; 0.7% higher than FastPose50-dcn-si; 5.1% higher than PVT; 1.0% higher than ZoomNet and 4.61 GFLOPs lower).

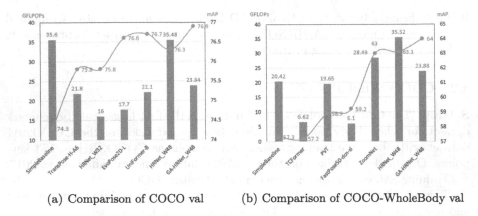

(a) Comparison of COCO val   (b) Comparison of COCO-WholeBody val

Fig. 2. Illustration of the complexity and accuracy comparison on the COCO val and COCO-WholeBody val.

4.3   Ablation Study

Consistent with Lite-HRNet [28], We perform ablation on the validation set of two datasets: COCO [14] and COCO-WholeBody [26], and report the results. The input size of both COCO and COCO-Wholebody is 256×192.

G-Ghost Stage. Tables 2 and 4 show the comparison between G-HRNet and HRNet_W48. We can see that the complexity of G-HRNet is about 2/3 of that of HRNet_W48 while the accuracy is almost unchanged. Specifically, on COCO val, the AP of G-HRNet_W48 is 0.2% higher than that of HRNet_W48; on COCO-Wholebody val, it is reduced by 0.7%. The results show that compressing the network by reducing feature redundancy is feasible and highly efficient. This can bring a considerable reduction in complexity to the network at a very small cost in accuracy.

Attention Module. Tables 2 and 4 show the comparison between A-HRNet and HRNet_W48. Compared to HRNet_W48, A-HRNet brings a significant improvement in accuracy while only adding negligible complexity. Specifically, on COCO val, it improves by 1.4%, adding 0.03 GFLOPs and 0.01 M parameters; on COCO-WholeBody val, it improves by 3.2%, adding 0.02 GFLOPs and 0.01 M parameters.

The results show that the proposed attention module is lightweight and efficient, able to fully combine channel attention and spatial attention to bring significant performance improvements to the network.

Fig. 3. Example images. The top side is the result in COCO format and the bottom side is the result in COCO-WholeBody format.

5 Conclusion

In this paper, based on HRNet, we propose a high-resolution network GA-HRNet for human pose estimation that produces accurate and spatially precise keypoint heatmaps. As shown in Fig. 3, the proposed network produces accurate results. Compared to HRNet, we have made two improvements: (1) optimized the network structure to significantly reduce computational complexity and parameter count; (2) proposed an attention module to improve network accuracy. Experiments on two benchmark datasets show that our proposed method significantly improves the baseline model and achieves competitive performance on several human-centric visual tasks (pose estimation and whole-body pose estimation).

We envision that our proposed method is general and can be applied to a wide range of visual tasks such as object detection and semantic segmentation. Future work will focus on exploring the effectiveness of our method on these visual tasks.

Acknowledgement. This work was supported in part by the National Natural Science Foundation of China under Grant 62202207; in part by the Natural Science Foundation of Jiangsu Higher Education Institutions under Grant 22KJB520015; and in part by the Scientific Research Foundation of Jiangsu University under Grant 21JDG051.

References

1. Andrew, H., et al.: Searching for MobileNetV3. In: Proceedings of the IEEE International Conference on Computer Vision, pp. 1314–1324 (2019)
2. Chen, L., et al.: SCA-CNN: spatial and channel-wise attention in convolutional networks for image captioning. In: Proceedings of the IEEE Conference on Computer Vision and Pattern Recognition, pp. 5659–5667 (2017)

3. MMPose Contributors: OpenMMLab pose estimation toolbox and benchmark (2020)
4. Fang, H.S., et al.: AlphaPose: whole-body regional multi-person pose estimation and tracking in real-time. IEEE Trans. Pattern Anal. Mach. Intell. **46**(6), 7157–7173 (2022)
5. Fu, J., et al.: Dual attention network for scene segmentation. In: Proceedings of the IEEE/CVF Conference on Computer Vision and Pattern Recognition, pp. 3146–3154 (2019)
6. Gao, Z., Xie, J., Wang, Q., Li, P.: Global second-order pooling convolutional networks. In: Proceedings of the IEEE/CVF Conference on Computer Vision and Pattern Recognition, pp. 3024–3033 (2019)
7. Guo, M.H., et al.: Attention mechanisms in computer vision: a survey. Comput. Visual Media **8**(3), 331–368 (2022)
8. Han, K., Wang, Y., Tian, Q., Guo, J., Xu, C., Xu, C.: GhostNet: more features from cheap operations. In: Proceedings of the IEEE/CVF Conference on Computer Vision and Pattern Recognition, pp. 1580–1589 (2020)
9. Han, K., et al.: GhostNets on heterogeneous devices via cheap operations. Int. J. Comput. Vis. **130**(4), 1050–1069 (2022)
10. Howard, A.G., et al.: MobileNets: efficient convolutional neural networks for mobile vision applications. arXiv preprint arXiv:1704.04861 (2017)
11. Hu, J., Shen, L., Sun, G.: Squeeze-and-excitation networks. In: Proceedings of the IEEE Conference on Computer Vision and Pattern Recognition, pp. 7132–7141 (2018)
12. Lee, H., Kim, H.E., Nam, H.: SRM: a style-based recalibration module for convolutional neural networks. In: Proceedings of the IEEE/CVF International Conference on Computer Vision, pp. 1854–1862 (2019)
13. Li, K., et al.: Uniformer: unifying convolution and self-attention for visual recognition. IEEE Trans. Pattern Anal. Mach. Intell. **45**, 12581–12600 (2023)
14. Lin, T.-Y., et al.: Microsoft COCO: common objects in context. In: Fleet, D., Pajdla, T., Schiele, B., Tuytelaars, T. (eds.) ECCV 2014. LNCS, vol. 8693, pp. 740–755. Springer, Cham (2014). https://doi.org/10.1007/978-3-319-10602-1_48
15. Ma, N., Zhang, X., Zheng, H.-T., Sun, J.: ShuffleNet V2: practical guidelines for efficient CNN architecture design. In: Ferrari, V., Hebert, M., Sminchisescu, C., Weiss, Y. (eds.) Computer Vision – ECCV 2018. LNCS, vol. 11218, pp. 122–138. Springer, Cham (2018). https://doi.org/10.1007/978-3-030-01264-9_8
16. McNally, W., Vats, K., Wong, A., McPhee, J.: EvoPose2D: pushing the boundaries of 2D human pose estimation using accelerated neuroevolution with weight transfer. IEEE Access **9**, 139403–139414 (2021)
17. Park, J., Woo, S., Lee, J.Y., Kweon, I.S.: BAM: bottleneck attention module. arXiv preprint arXiv:1807.06514 (2018)
18. Sandler, M., Howard, A., Zhu, M., Zhmoginov, A., Chen, L.C.: MobileNetV2: inverted residuals and linear bottlenecks. In: Proceedings of the IEEE Conference on Computer Vision and Pattern Recognition, pp. 4510–4520 (2018)
19. Sun, K., Xiao, B., Liu, D., Wang, J.: Deep high-resolution representation learning for human pose estimation. In: Proceedings of the IEEE/CVF Conference on Computer Vision and Pattern Recognition, pp. 5693–5703 (2019)
20. Wang, J., et al.: Deep high-resolution representation learning for visual recognition. IEEE Trans. Pattern Anal. Mach. Intell. **43**(10), 3349–3364 (2020)
21. Wang, Q., Wu, B., Zhu, P., Li, P., Zuo, W., Hu, Q.: ECA-Net: efficient channel attention for deep convolutional neural networks. In: Proceedings of the IEEE/CVF Conference on Computer Vision and Pattern Recognition, pp. 11534–11542 (2020)

22. Wang, W., et al.: Pyramid vision transformer: a versatile backbone for dense prediction without convolutions. In: Proceedings of the IEEE/CVF International Conference on Computer Vision, pp. 568–578 (2021)

23. Woo, S., Park, J., Lee, J.-Y., Kweon, I.S.: CBAM: convolutional block attention module. In: Ferrari, V., Hebert, M., Sminchisescu, C., Weiss, Y. (eds.) ECCV 2018. LNCS, vol. 11211, pp. 3–19. Springer, Cham (2018). https://doi.org/10.1007/978-3-030-01234-2_1

24. Wu, H., Liang, C., Liu, M., Wen, Z.: Optimized HRNet for image semantic segmentation. Expert Syst. Appl. **174**, 114532 (2021)

25. Xiao, B., Wu, H., Wei, Y.: Simple baselines for human pose estimation and tracking. In: Ferrari, V., Hebert, M., Sminchisescu, C., Weiss, Y. (eds.) ECCV 2018. LNCS, vol. 11210, pp. 472–487. Springer, Cham (2018). https://doi.org/10.1007/978-3-030-01231-1_29

26. Xu, L., et al.: ZoomNAS: searching for whole-body human pose estimation in the wild. IEEE Trans. Pattern Anal. Mach. Intell. **45**(4), 5296–5313 (2022)

27. Yang, S., Quan, Z., Nie, M., Yang, W.: Transpose: keypoint localization via transformer. In: Proceedings of the IEEE/CVF International Conference on Computer Vision, pp. 11802–11812 (2021)

28. Yu, C., et al.: Lite-HRNet: a lightweight high-resolution network. In: Proceedings of the IEEE/CVF Conference on Computer Vision and Pattern Recognition, pp. 10440–10450 (2021)

29. Zeng, W., et al.: Not all tokens are equal: human-centric visual analysis via token clustering transformer. In: Proceedings of the IEEE/CVF Conference on Computer Vision and Pattern Recognition, pp. 11101–11111 (2022)

30. Zhang, X., Zhou, X., Lin, M., Sun, J.: ShuffleNet: an extremely efficient convolutional neural network for mobile devices. In: Proceedings of the IEEE Conference on Computer Vision and Pattern Recognition, pp. 6848–6856 (2018)

31. Zhang, Z., Lan, C., Zeng, W., Jin, X., Chen, Z.: Relation-aware global attention for person re-identification. In: Proceedings of the IEEE/CVF Conference on Computer Vision and Pattern Recognition, pp. 3186–3195 (2020)

Author Index

Printed in the United States
by Baker & Taylor Publisher Services